ENCYCLOPEDIA OF MATHEMATICS AND ITS APPLICATIONS

FOUNDED BY G.-C. ROTA

Volume 112

The Classical Fields:
Structural Features of the
Real and Rational Numbers

ENCYCLOPEDIA OF MATHEMATICS AND ITS APPLICATIONS

ENCYCLOPEDIA OF MATHEMATICS AND ITS APPLICATIONS

The Classical Fields

Structural Features of the Real and Rational Numbers

H. SALZMANN

University of Tübingen

T. GRUNDHÖFER

University of Würzburg

H. HÄHL

University of Stuttgart

R. LÖWEN

Technical University of Braunschweig

CAMBRIDGE
UNIVERSITY PRESS

CAMBRIDGE UNIVERSITY PRESS
Cambridge, New York, Melbourne, Madrid, Cape Town, Singapore, São Paulo

Cambridge University Press
The Edinburgh Building, Cambridge CB2 8RU, UK

Published in the United States of America by Cambridge University Press,
New York

www.cambridge.org
Information on this title: www.cambridge.org/9780521865166

First published 2007

Printed in the United Kingdom at the University Press, Cambridge

A catalog record for this publication is available from the British Library

ISBN   978-0-521-86516-6 hardback

Contents

Preface

The rational numbers, the real numbers, the complex numbers and the p-adic numbers are classical fields. These number systems are the topic of this book.

The real numbers, which are basic and indispensable for most parts of mathematics, comprise several rich and intimately interwoven structures, namely the algebraic structure as a field, the topological structure and the ordering structure. Each of these structures, as well as their particular blend, is beautifully adapted to the intended use of numbers (for counting, computing, taking measurements, comparing sizes and modelling physical space and time). It is the purpose of this book to consider these structures separately, and to analyse the interaction and the interdependencies between these structures. The real numbers are characterized in various categories by simple abstract properties. Each of these characterization results is a possible answer to the question: why exactly are the real numbers so fundamentally important?

The ordering and the topology of the real numbers are rooted deeply in our geometric intuition about points on a line. The algebraic operations of addition and multiplication describe the isometries and the similarities of the one-dimensional geometry of a line. (In fact, one-dimensional geometry becomes interesting only by imposing some additional structure on the set of points of the only line.)

Apart from the real numbers, we also treat the rational numbers (in Chapter 3) and the p-adic numbers (in Chapter 5). The complex numbers are considered in Section 14 (to some extent also in Section 13), and Chapter 2 deals with non-standard numbers. We study the structural components of each of these fields and their interactions; we also describe the pertaining automorphism groups and typical substructures and quotients.

ix

Of course, also finite fields are classical number systems. However, finite fields are purely algebraic objects, they have no interesting ordering or topology, and their algebraic features are closely related to number theory. Therefore finite fields appear only incidentally in this book.

* * *

The first and longest chapter considers the field $\mathbb{R}$ of real numbers. We study the additive and the multiplicative group of $\mathbb{R}$, and then $\mathbb{R}$ as an ordered set, as a topological space, as a measure space, as an abstract field and as a topological field. The additive group $(\mathbb{R}, +)$ is considered as an ordered group and as a topological group. Algebraic properties of the field $\mathbb{R}$ lead to the Artin–Schreier theory of formally real fields.

The last section of Chapter 1 treats the complex numbers $\mathbb{C}$; many structural features of $\mathbb{C}$ can be inferred from the description of $\mathbb{C} = \mathbb{R}(\sqrt{-1})$ as a quadratic field extension of $\mathbb{R}$. However, the existence of discontinuous field automorphisms and of unexpected subfields is a peculiar property of $\mathbb{C}$.

According to Pontryagin, $\mathbb{R}$ and $\mathbb{C}$ are the only topological fields that are locally compact and connected; see Theorem 13.8.

It is not our main task to construct the real numbers $\mathbb{R}$, we rather take them for granted (constructions can be found in the books mentioned at the end of this preface). Still, we describe constructions of $\mathbb{R}$ in Section 23 (by means of an ultrapower of the field $\mathbb{Q}$ of rational numbers) and in 42.11 and 44.11.

Non-standard numbers are the theme of Chapter 2. These number systems can be constructed easily via ultrapowers. Contrasting $\mathbb{R}$ with its non-standard counterpart $^*\mathbb{R}$ sheds additional light on the particular role of $\mathbb{R}$. The additive and the multiplicative groups of $\mathbb{R}$ and $^*\mathbb{R}$ are isomorphic, and $^*\mathbb{R}(\sqrt{-1}) \cong \mathbb{C}$ (see 24.2, 24.4 and 24.6), but $\mathbb{R}$ and $^*\mathbb{R}$ are not isomorphic as fields, and $\mathbb{R}$ and $^*\mathbb{R}$ are quite different as topological spaces. The natural embedding of $\mathbb{R}$ into $^*\mathbb{R}$ leads to some basic notions of non-standard analysis.

In Chapter 3 we treat the system $\mathbb{Q}$ of rational numbers in a similar way as the real numbers. The different structural components of $\mathbb{Q}$ are less tightly related; in particular, the additive group and the multiplicative group of $\mathbb{Q}$ are quite different. As $\mathbb{Q}$ is the field of fractions of the ring $\mathbb{Z}$ of integers, number theory plays an important role in this chapter. In many respects $\mathbb{R}$ is simpler than $\mathbb{Q}$; for example, $\mathbb{R}$ has only two square classes and an easy theory of quadratic forms, whereas $\mathbb{Q}$ has infinitely many square classes and a rich theory of quadratic forms (see CASSELS 1978). Moreover, the natural topology of $\mathbb{R}$ is locally compact and connected, in contrast to the topology of $\mathbb{Q}$. This explains why in this book we treat first $\mathbb{R}$ and then $\mathbb{Q}$, despite the fact that $\mathbb{R}$ can be obtained by completing $\mathbb{Q}$.

A field is said to be complete, if its additive group is complete with respect to a given ordering or topology. In Chapter 4 we discuss relevant completion procedures. First we complete chains, and then ordered groups and fields. Next we construct the (essentially unique) completion of a topological abelian group. A complete ordered group is also complete in the topology determined by the ordering (see 43.10). The results are finally applied to topological rings and fields.

In Chapter 5 we deal with the p-adic numbers $\mathbb{Q}_p$ as relatives of the real numbers. Indeed, $\mathbb{Q}_p$ can be obtained by completing $\mathbb{Q}$ with respect to the p-adic metric; this metric reflects the divisibility by powers of the prime p. As a consequence, the p-adic topology has a rather algebraic flavour; note that the non-zero ideals of the ring $\mathbb{Z}_p \subset \mathbb{Q}_p$ of p-adic integers form a neighbourhood basis at 0 for the topology of $\mathbb{Q}_p$. This topology is locally compact and totally disconnected. We consider the additive and the multiplicative group of $\mathbb{Q}_p$, and we study the squares of $\mathbb{Q}_p$. The field $\mathbb{Q}_p$ cannot be made into an ordered field (compare 54.2). Like $\mathbb{R}$, the field $\mathbb{Q}_p$ admits no automorphism except the identity (see 53.5).

The properties of $\mathbb{Q}_p$ are placed in a more general context by considering absolute values, valuations and topologies of valuation type in Sections 55, 56, 57. The last section of Chapter 5 deals with field extensions of $\mathbb{Q}_p$ and with the classification of all locally compact (skew) fields. We prove that $\mathbb{R}$ and $\mathbb{Q}_p$ are the only non-discrete locally compact fields that contain $\mathbb{Q}$ as a dense subfield (58.7).

Note that $\mathbb{R}$ and $\mathbb{Q}_p$ are encoded in the additive group $(\mathbb{Q}, +)$, and hence in the semigroup $(\mathbb{N}, +)$ of positive integers: the field $\mathbb{Q}$ is the endomorphism ring of $(\mathbb{Q}, +)$ (see 8.28), and $\mathbb{R}$ and the fields $\mathbb{Q}_p$ are the completions of $\mathbb{Q}$ with respect to the absolute values of $\mathbb{Q}$ (compare 44.11).

In an Appendix we collect some facts on ordinal and cardinal numbers and on topological groups, we summarize the duality theory of locally compact abelian groups, and we present basic facts and constructions of field theory.

Most sections end with a few exercises, of different character and degree of difficulty. The chapter 'Hints and solutions' provides a solution or at least a clue for each exercise.

* * *

There is a vast literature on number theory, and there exist many books which deal with the real numbers and the rational numbers. Several of these texts explain the successive construction of the number systems $\mathbb{N}$, $\mathbb{Z}$, $\mathbb{Q}$, $\mathbb{R}$ and $\mathbb{C}$; typical examples are DEDEKIND 1872, LANDAU 1930, COHEN–EHRLICH 1963, FEFERMAN 1964. The historical development is presented in FLEGG 1983, EHRLICH 1994 and LÓPEZ PELLICER 1994. The three volumes by FELSCHER 1978/79 on 'Naive Mengen und abstrakte Zahlen' emphasize logical and set-theoretic aspects. The classical division algebras $\mathbb{H}$ (Hamilton's quaternions) and $\mathbb{O}$ (octonions) are treated in SALZMANN *et al.* 1995 Chapter 1, EBBINGHAUS *et al.* 1991 Part B; see also CONWAY–SMITH 2003, BAEZ 2002, WARD 1997. None of these books has much overlap with the present text.

Our book is based on lectures given by H. Salzmann in Tübingen in 1971/72 and on a two-volume set of lecture notes (prepared by R. Löwen and H. Hähl) with the title 'Zahlbereiche'. These lecture notes had been available for a short while in mimeographed form: SALZMANN 1971 and SALZMANN 1973.

We would like to thank Nils Rosehr for technical support, and Joachim Gräter for helpful discussions. We are grateful to the friendly staff of Cambridge University Press for their professional help and advice in publishing this book.

The authors

Notation

As usual, $\mathbb{N}$, $\mathbb{Z}$, $\mathbb{Q}$, $\mathbb{R}$, $\mathbb{C}$ denote the natural, integer, rational, real and complex numbers, respectively. By convention, $0 \notin \mathbb{N}$ and $\mathbb{N}_0 := \mathbb{N} \cup \{0\}$.

We use $:=$ for equality by definition and $:\Leftrightarrow$ for equivalence by definition. The symbols $\wedge$ and $\vee$ are the logical connectives 'and' and 'or'.

We write $A \cong B$ if two structures A and B are isomorphic, and $A < B$ if A is a proper substructure of B.

The notation $X \approx Y$ means that the topological spaces X and Y are homeomorphic (compare 5.51).

$\mathbb{F}_q$	finite field of order q
$\mathbb{P}$	prime numbers
$\mathbb{R}_{\mathrm{pos}}$	positive real numbers (as a multiplicative group)
$\mathbb{R}_{\mathrm{alg}}$	real algebraic numbers
$\mathbb{S}_n$	sphere of dimension n
$\mathbb{T} := \mathbb{R}/\mathbb{Z}$	torus (1.20)
$\mathbb{L}$, $\mathbb{L}^+$	long line, long ray (5.25)
$\mathcal{C}$	Cantor set (5.35, as a topological space)
$\mathbb{H}$	quaternions (Section 13, Exercise 6, and 34.17)
$\mathbb{Q}_p$	p-adic numbers (44.11 and Chapter 5)
$\mathbb{Z}_p$	p-adic integers (51.6)

$a \mid b$	a divides b
$\gcd(a, b)$	greatest common divisor of a, b
$[\, c_0; c_1, c_2, \ldots\,]$	continued fraction (4.1)

2^S	power set of S
$\bigcup \mathfrak{S}$	union $\{\, x \mid \exists_{S \in \mathfrak{S}} : x \in S \,\}$ of a system $\mathfrak{S}$ of sets
$\bigcap \mathfrak{S}$	intersection $\{\, x \mid \forall_{S \in \mathfrak{S}} : x \in S \,\}$ of a system $\mathfrak{S}$ of sets

id	identity mapping
im	image
Y^X	set $\{f \mid f : X \to Y\}$ of mappings
Fix	set of all fixed elements
supp	support (5.61, 64.22)
$\times_i S_i$	Cartesian product of sets (with structure) S_i
$\mathcal{O}$	ordinal numbers (Section 61)
card S	cardinality of a set S (61.8)
$\aleph_0$	cardinality of $\mathbb{N}$
$\aleph = 2^{\aleph_0}$	cardinality of $\mathbb{R}$ (1.10)
C_n	cyclic group of order n
C_{p^∞}	Prüfer group (1.26)
Sym S	symmetric (permutation) group of the set S
$\mathrm{GL}_n F$	general linear group of F^n
$\mathrm{PGL}_2 F$	projective quotient of $\mathrm{GL}_2 F$ (11.16, 64.19)
$\mathcal{H} = \mathcal{H}(\mathbb{R})$	homeomorphism group of $\mathbb{R}$ (5.51)
$\mathcal{H}(\mathbb{Q})$	homeomorphism group of $\mathbb{Q}$ (33.12)
Aut X	automorphism group of X
$\mathrm{Aut}_c X$	group of continuous automorphisms of X
End X	endomorphisms of X
$\mathrm{End}_c X$	continuous endomorphisms of X
$\mathrm{Hom}(X, Y)$	homomorphisms $X \to Y$
Cs	centralizer
A^*	character group of A (63.1)
$\bigoplus_i G_i$	direct sum of groups G_i (1.16)
$E\|F$	field extension $F \subseteq E$ (64.1)
$[E : F]$	degree of $E\|F$ (64.1)
F^+	additive group of F
$F^\times$	multiplicative group (of units) of F
$F^\square$	set $\{x^2 \mid 0 \neq x \in F\}$ of squares of $F^\times$
$F^\natural$	algebraic closure of F (64.13)
char(F)	characteristic of F (64.4)
trdeg	transcendency degree (64.20)
tr	trace
$\mathrm{Gal}_F E$, $\mathrm{Aut}_F E$	Galois group, relative automorphism group (64.17)
$F[t]$	polynomial ring
$F(t)$	field of fractions of $F[t]$

$F[[t]]$ ring of power series (64.22)
$F((t))$ field of Laurent series (64.23)
$F((t^{1/\infty}))$ field of Puiseux series (64.24)
$F((\Gamma))$, $F((\Gamma))_1$ fields of Hahn power series (64.25)

$\lim \mathfrak{B}$ limit of a filterbase $\mathfrak{B}$ (43.2)
$\widehat{\mathfrak{C}}$ minimal concentrated filter (43.15)

Non-standard objects (Chapter 2):

S^Ψ ultrapower (21.4)
$^*\mathbb{Q}$ non-standard rationals (Section 22)
$^*\mathbb{R}$ non-standard reals (Section 24)
$\mathfrak{T}$ $^*\mathbb{R}$ as a topological space (Section 24)
h^Ψ, $^*h$ extension of a map h (21.10, 25.0)
$^\circ a$ standard part of a (23.9)
$x \approx y$ $y - x$ is infinitely small (23.9, Sections 25–27)

1

Real numbers

This chapter is devoted to various aspects of the structure of $\mathbb{R}$, the field of real numbers. Since we do not intend to give a detailed account of a construction of the real numbers from the very beginning, we need to clarify the basis of our subsequent arguments. What we shall assume about the real numbers is that they form an ordered field whose ordering is complete, in the sense that every non-empty bounded set of real numbers has a least upper bound. All these notions will be explained in due course, but presumably they are familiar to most readers. It is well known and will be proved in Section 11 that the properties just mentioned characterize the field of real numbers.

Historically, a satisfactory theory of the real numbers was obtained only at the end of the nineteenth century by work of Weierstraß, Cantor and Dedekind (see FLEGG 1983, EHRLICH 1994 and LÓPEZ PELLICER 1994). Starting from the rational numbers, they used different approaches, namely, Cauchy sequences on the one hand and Dedekind cuts on the other.

In Sections 42 and 44, we shall actually show how to obtain the real numbers from the rational numbers via completion. Another construction in the context of non-standard real numbers will be given in Section 23. We mention also the approach of CONWAY 1976, whose 'surreal numbers' go beyond non-standard numbers. These ideas were carried further by GONSHOR 1986, ALLING 1987; see also EHRLICH 1994, 2001 and DALES–WOODIN 1996.

Several methods have been proposed for constructing $\mathbb{R}$ directly from the ring $\mathbb{Z}$ of integers, without using the rational numbers as an intermediate step; compare Section 6, Exercise 2 (which is related to FALTIN *et al.* 1975; see also Section 51, Exercise 3, for the p-adic analogue). DE BRUIJN 1976 defines the ordered additive group of real numbers via certain mappings $f : \mathbb{Z} \to \{0, 1, \ldots, b-1\}$, with the idea that a non-negative real number is represented by $\sum_{n \in \mathbb{Z}} f(n) b^{-n}$. A'CAMPO 2003 considers all maps $f : \mathbb{Z} \to \mathbb{Z}$ which are 'slopes' (or 'near-endomorphisms') in the sense that $\{ f(x+y) - f(x) - f(y) \mid x, y \in \mathbb{Z} \}$ is a finite set, and he constructs $\mathbb{R}$ by identifying two slopes f, g if $f - g$ has finite image; see also ARTHAN 2004, GRUNDHÖFER 2005 for details pertaining to this construction.

1 The additive group of real numbers

The first feature of the field $\mathbb{R}$ to be examined is its additive group $(\mathbb{R}, +)$. The following is the essential fact about this group (which actually characterizes it): $(\mathbb{R}, +)$ is a vector space over the rational numbers, and the dimension of this vector space is given by the cardinality of the real numbers.

We shall derive various consequences from this rational vector space structure; in particular, we consider subgroups and quotient groups and characterize them as groups where possible. The axiom of choice will be used in many places in this section because we rely on the existence of bases for infinite dimensional vector spaces.

We do not at this point go into the obvious question as to what can be said about the additive structure of the rational numbers themselves. This will be deferred to Section 31.

1.1 The additive group of real numbers From the construction of the real numbers, we take the following facts for granted.

(a) The real numbers under addition form an *abelian group* denoted $(\mathbb{R}, +)$, or briefly $\mathbb{R}^+$, with neutral element 0.

(b) By repeated addition of the multiplicative unit 1 we exhaust the set $\mathbb{N} = \{1, 2, 3 \ldots\}$ of natural numbers; together with their additive inverses and 0 they form the infinite cyclic group of integers, which is a subgroup $\mathbb{Z}^+ \leq \mathbb{R}^+$.

Here, the word 'cyclic' means 'generated by a single element', namely, by 1 (or by -1). In other words, $\mathbb{Z}^+$ is the smallest subgroup of $\mathbb{R}^+$ containing 1. That this subgroup is infinite is a consequence of the ordering: from $0 < 1$ one obtains, by induction, that $0 < n < n+1$ for all natural numbers n.

We start our investigation of the group $\mathbb{R}^+$ by examining its subgroups, starting with the smallest possible ones. First, we turn to a general consideration.

1.2 Cyclic subgroups of arbitrary groups In an additively written group $(G, +)$, we consider the multiples $ng := g + \cdots + g$ of an element $g \in G$, where $n \in \mathbb{N}$ is the number of summands. In addition, we set $0g := 0$ and $zg := (-z)(-g)$ for negative integers z. The *order* $\#g$ of g is, by definition, the smallest natural number k such that $kg = 0$; we set $\#g = \infty$ if no such number k exists.

Fix $g \in G$. The mapping defined by $\varphi(z) := zg$ is a homomorphism $\varphi : (\mathbb{Z}, +) \to (G, +)$ which maps $\mathbb{Z}$ onto the smallest subgroup $\langle g \rangle$ containing g. The kernel $\ker \varphi$ equals $k\mathbb{Z}$ if $k = \#g$ is finite, and $\ker \varphi = \{0\}$ otherwise. It follows that $\langle g \rangle$ is isomorphic to the factor group $\mathbb{Z}/\ker \varphi$, and thus to the cyclic group C_k if $k \in \mathbb{N}$ and to $\mathbb{Z}^+$ otherwise. Clearly, all homomorphisms $\mathbb{Z} \to G$ arise in this manner (just set $g := \varphi(1)$).

The group G is said to be *torsion free* if all its elements except 0 have infinite order; equivalently, if all its non-trivial cyclic subgroups are infinite or if all non-trivial homomorphisms $\mathbb{Z} \to G$ are injective.

1.3 Theorem: Cyclic subgroups of $\mathbb{R}^+$ *The group homomorphisms $\mathbb{Z}^+ \to \mathbb{R}^+$ are precisely the maps $\varphi_s : z \mapsto zs$ for arbitrary $s \in \mathbb{R}$. Every non-zero element $r \in \mathbb{R}$ generates an infinite cyclic subgroup $\langle r \rangle = \mathbb{Z}r \cong \mathbb{Z}$, and the group $\mathbb{R}^+$ is torsion free.*

Proof It suffices to remark that, by virtue of the distributive law, the multiple zs can also be obtained as $z \cdot s$, using the multiplication of real numbers. Now the distributive law shows that φ_s is a homomorphism; if $r \neq 0$, then φ_r is injective by the absence of zero divisors. Instead of these arguments, one could use the same reasoning (based on the ordering) as in 1.1.   □

1.4 Theorem *Every non-trivial subgroup $H \leq \mathbb{R}^+$ is either cyclic or dense in $\mathbb{R}$.*

Proof Let $r \in \mathbb{R}$ be the infimum (the greatest lower bound) of the set $\{h \in H \mid h > 0\}$. If $r \neq 0$ and $r \in H$, then the cyclic group $\mathbb{Z}r$ coincides with H. Indeed, every $g \in H$ belongs to the interval $[zr, (z+1)r[$ for some $z \in \mathbb{Z}$, which implies that $g - zr \in [0, r[\cap H$ and $g = zr$. (We have used, somewhat informally, the fact that $\mathbb{R}^+$ is an Archimedean ordered group; compare 7.4 and 7.5.)

On the other hand, if $r = 0$ or $r \notin H$, then r is a cluster point of H, and H contains pairs with arbitrarily small differences. Since H is a group, it contains those differences as elements. Now if $g \in \,]0, \varepsilon[\cap H$, then every closed interval of length ε contains some integer multiple zg,

$z \in \mathbb{Z}$, and H is dense. (This argument used the Archimedean property again.) □

Having treated completely the subgroups generated by a single element, we turn now to subgroups generated by pairs of elements. As a preparation, we need the following simple result of number theory.

1.5 Bézout's Theorem *The greatest common divisor* $\gcd(m,n)$ *of two integers is an integral linear combination of* m *and* n, *i.e., there are integers* x, y *such that* $\gcd(m,n) = xm + yn$. *In particular, if* m, n *are relatively prime, then* 1 *is a linear combination:* $1 = xm + yn$.

Proof The subgroup $I = \{\, xm + yn \mid x, y \in \mathbb{Z} \,\} \leq \mathbb{Z}^+$ is generated by any element $d \in I \smallsetminus \{0\}$ of smallest absolute value. This is shown by applying the Euclidean algorithm: an element $z \in I$ can be written as $z = bd + r$ with $b \in \mathbb{Z}$ and $|r| < |d|$; it follows from the equation that $r \in I$, and we must have $r = 0$ by the definition of d.

Now if $I = \langle d \rangle$, then d divides m and n and is a linear combination of the two; hence every common divisor of m, n divides d. — We remark that d, x and y can be computed explicitly by making full use of the Euclidean algorithm. This is needed, for example, in order to compute multiplicative inverses in the field $\mathbb{Z}/p\mathbb{Z}$, where p is a prime. □

Now we apply this fact to subgroups of $\mathbb{R}$.

1.6 Theorem (a) *A subgroup* $\langle a, b \rangle \leq \mathbb{R}^+$ *generated by two non-zero elements* a, b *is cyclic if, and only if, the quotient* a/b *is a rational number. More precisely, if* $a/b = m/n$ *with* m, n *relatively prime, then* $\langle a, b \rangle = \langle b/n \rangle$.

  (b) *If* a/b *is irrational, then* $\langle a, b \rangle$ *is dense in* $\mathbb{R}$.

  (c) *The additive group* $\mathbb{Q}^+$ *of rational numbers is locally cyclic, i.e., any finite subset generates a cyclic subgroup.*

Proof (a) If $a/b = m/n$, then $xa + yb = b(xm + yn)/n$ for $x, y \in \mathbb{Z}$. From Bézout's Theorem 1.5 it follows that $b/n \in \langle a, b \rangle$. On the other hand, both b and $a = ba/b = bm/n$ are integer multiples of b/n. Conversely, if a and b belong to a cyclic group $\langle c \rangle$, then $a = xc$ and $b = yc$ for some integers x, y, and $a/b = x/y$.

  (b) follows from (a) together with 1.4, and (c) is obtained by repeated application of (a). □

There is a more general (and less easy) result behind the density assertion 1.6b. See 5.69 for a statement and proof of this theorem due to Kronecker.

Our next aim is the characterization of $\mathbb{R}^+$ given in 1.14. It uses the notion of divisible group, which we introduce first.

1.7 Definition: Divisible groups A group $(G, +)$ is said to be *divisible* if for every $g \in G$ and every $n \in \mathbb{N}$ there is $h \in G$ such that $nh = g$. Here, the integer multiple nh is taken in the sense of 1.2. An abelian group that is both divisible and torsion free is *uniquely divisible*, i.e., the element h is uniquely determined by g and n. Indeed, $g = nh = nh'$ implies $n(h - h') = 0$, and then $h - h' = 0$ because G is torsion free. The unique h satisfying $nh = g$ will then be denoted g/n.

In particular, we have:

1.8 Theorem *The group $\mathbb{R}^+$ is uniquely divisible.*   □

1.9 Theorem *A uniquely divisible abelian group G carries a unique structure as a rational vector space.*

Proof It follows from the vector space axioms that multiplication by a scalar $z \in \mathbb{Z}$ is the one defined in 1.2, and then multiplication by the scalar $1/n$ is the operation introduced in 1.7. This proves uniqueness. On the other hand, we can always introduce on G the structure of a vector space over $\mathbb{Q}$ in this way.   □

The structure of uniquely divisible non-abelian groups is more complicated: GUBA 1986 shows that there exists a group of this kind that is generated by two elements.

Any bijection between given bases of two vector spaces over the same field extends to an isomorphism between the spaces, thus a vector space over a given field is determined up to isomorphism by the cardinality of a basis, i.e., by its (possibly transfinite) dimension; compare Exercise 1. Hence we can characterize the uniquely divisible group $\mathbb{R}^+$ if we determine its dimension as a vector space over the rationals. Before we can do this, we need to determine the cardinality of $\mathbb{R}$ itself.

1.10 Theorem *The set $\mathbb{R}$ has cardinality $\aleph := \operatorname{card} \mathbb{R} = 2^{\aleph_0} > \aleph_0$.*

Proof We use the fact that every real number between 0 and 1 has a unique binary expansion $\sum_{n \in \mathbb{N}} c_n 2^{-n}$ with $c_n \in \{0, 1\}$ and $c_n = 0$ for infinitely many n. By 61.14, the last condition excludes only countably many coefficient sequences $(c_n)_{n \in \mathbb{N}}$, and there remain $2^{\aleph_0}$ admissible sequences. Thus, $\operatorname{card} [0, 1[= 2^{\aleph_0}$ and, since $\mathbb{R}$ decomposes into countably many intervals $[m, m + 1[$, it follows that $\operatorname{card} \mathbb{R} = \aleph_0 \cdot 2^{\aleph_0} = 2^{\aleph_0}$; see 61.12. According to 61.11, we have $2^{\aleph_0} > \aleph_0$.

There is a slightly faster way to establish the equation $\mathrm{card}[0,1] = 2^{\aleph_0}$ using the Cantor set $\mathcal{C} = \{\sum_{\nu=1}^{\infty} c_\nu 3^{-\nu} \mid c_\nu \in \{0,2\}\}$; compare 5.35ff. Indeed, $\mathcal{C}$ is a subset of the unit interval, and on the other hand, $\mathcal{C}$ maps onto $[0,1]$ via $\sum_{\nu=1}^{\infty} c_\nu 3^{-\nu} \mapsto \sum_{\nu=1}^{\infty} c_\nu 2^{-\nu-1}$.   □

One naturally wonders what is the precise relationship between the two cardinalities $\mathrm{card}\,\mathbb{Q} = \aleph_0$ and $\mathrm{card}\,\mathbb{R} = \aleph$; are there any other cardinalities in between or not? The question thus raised is known as the *continuum problem*; see 61.17 for a brief introduction.

1.11 The real numbers as a rational vector space The vector space structure on $\mathbb{R}^+$ that we determined in 1.9 can be described more easily. In fact, the product qr of a scalar $q \in \mathbb{Q}$ and a vector $r \in \mathbb{R}$ is just their product as real numbers. This follows from uniqueness of the vector space structure by observing that multiplication in $\mathbb{R}$ does define such a structure.

What is the dimension of this vector space? The cardinality of a basis $B \subseteq \mathbb{R}$ cannot exceed that of the space $\mathbb{R}$ itself. To obtain a lower estimate for $\mathrm{card}\,B$, we count the finite rational linear combinations of B. Every linear combination is determined by a finite subset of $\mathbb{Q} \times B$. By 61.14, the set of all these subsets has the same cardinality as $\mathbb{Q} \times B$ itself. Thus we have $\mathrm{card}\,B \le \mathrm{card}\,\mathbb{R} = \aleph \le \mathrm{card}(\mathbb{Q} \times B) = \max\{\aleph_0, \mathrm{card}\,B\}$; for the last equality use 61.12. This proves the following.

1.12 Theorem *Any basis of the vector space $\mathbb{R}$ over $\mathbb{Q}$ has the same cardinality as $\mathbb{R}$ itself, that is, $\dim_{\mathbb{Q}} \mathbb{R} = \aleph = \mathrm{card}\,\mathbb{R}$.*   □

Incidentally, we have proved that the concept of transfinite dimension is meaningful in the given situation, independently of Exercise 1. Examining the proof, we see that countability of $\mathbb{Q}$ is essential. In fact we have shown that a basis of any infinite-dimensional rational space V has the same cardinality as V itself. This is not true for real vector spaces.

1.13 Hamel bases A basis of the rational space $\mathbb{R}$ is usually referred to as a *Hamel basis*. No one has ever written down such a basis, nor probably ever will. Yet such bases have several applications. They permit, for example, the solution of the functional equation $f(x+y) = f(x) + f(y)$; this is precisely what Hamel invented them for (HAMEL 1905, compare also ACZÉL 1966). Moreover, Hamel bases allow one to construct subsets of $\mathbb{R}$ that behave strangely with respect to Lebesgue measure; see 10.8ff. One should therefore keep in mind that the existence of Hamel bases depends on the Axiom of Choice (AC); compare the introduction to Section 61. A large proportion of the subsequent results in this section therefore need AC, namely 1.14, 1.15, 1.17, 1.19, 1.24. In addition, 1.23 uses AC (directly).

In spite of the elusiveness of Hamel bases, there are many ways of constructing large $\mathbb{Q}$-linearly independent sets of real numbers. A simple example is the set of all logarithms of prime numbers (Exercise 6). More refined techniques yield sets that are uncountable (BRENNER 1992). This is surpassed by VON NEUMANN 1928 and KNESER 1960, who show that certain sets of cardinality $\aleph$ are even algebraically independent over $\mathbb{Q}$ (as defined in 64.20). Kneser's set consists of the numbers

$$\sum_{n\geq 1} 2^{-\lfloor n^{n+s}\rfloor}, \quad 0 \leq s < 1,$$

where $\lfloor t\rfloor$ denotes the largest integer not exceeding $t \in \mathbb{R}$. Other examples of this kind are given by DURAND 1975 and ELSNER 2000.

There is a nice survey on this topic by WALDSCHMIDT 1992, and there are several other contributions by the same author. Finally, we mention LACZKOVICH 1998, who shows that there are $\mathbb{Q}$-linearly independent subsets of $\mathbb{R}$ that are G_δ-sets (that is, intersections of countably many open sets).

Observe that the only information about $\mathbb{R}$ that we needed in order to compute the dimension was the cardinality of $\mathbb{R}$ itself. Thus we have proved the following.

1.14 Characterization Theorem *An abelian group is isomorphic to the additive group $\mathbb{R}^+$ of real numbers if, and only if, it is torsion free and divisible and has the same cardinality as $\mathbb{R}$.* □

1.15 Consequences   A few surprising (at first sight) consequences of the characterization are worth pointing out. The additive group of any finite dimensional real vector space, e.g., of $\mathbb{R}^n$ or of $\mathbb{C}$, satisfies the conditions that characterize $\mathbb{R}^+$. Thus, all these *vector groups* are isomorphic. By contrast, two real vector spaces of different finite dimensions are not isomorphic, and their additive groups are not isomorphic as topological groups; indeed, every additive map between such vector spaces is $\mathbb{Q}$-linear (by 1.9), hence $\mathbb{R}$-linear if it is continuous; alternatively, we could use Theorem 8.6, which implies that $\mathbb{R}^n$ contains closed discrete subgroups isomorphic to $\mathbb{Z}^n$, but no closed discrete subgroup isomorphic to $\mathbb{Z}^{n+1}$.

A Hamel basis has many subsets of the same cardinality, giving rise to many vector subspaces that are isomorphic to $\mathbb{R}$ both as rational vector spaces and as additive groups. These subgroups may also be obtained as factor groups; just factor out a complementary vector subspace.

Now we determine the cardinality of the set $\mathcal{G}$ of all subgroups of $\mathbb{R}^+$: the set 2^B of all subsets of a Hamel basis B injects into $\mathcal{G}$, because a subset $C \subseteq B$ generates a vector subspace V_C, and $C = V_C \cap B$. Hence the cardinality $2^\aleph$ of 2^B is a lower bound for card $\mathcal{G}$. But $2^\aleph$ is also the cardinality of the set of all subsets of $\mathbb{R}$, which is an upper bound for card $\mathcal{G}$. *Thus we see that there are $2^\aleph = 2^{2^{\aleph_0}}$ distinct subgroups of $\mathbb{R}^+$.* (This does not say anything about the number of isomorphism types of subgroups, which depends on how many cardinalities there are between $\aleph_0$ and $\aleph$; compare 61.17.)

We proceed to examine decompositions of $\mathbb{R}^+$ as a direct sum. First we present the basic notions of direct sum and direct product.

1.16 Definition: Direct sums and products Given a family of additively written groups G_i (abelian or otherwise), indexed by a set I, we form a group

$$\bigtimes_{i \in I} G_i ,$$

called the *direct product* of the G_i, as follows: the elements of this group are the indexed families $(g_i)_{i \in I}$ such that $g_i \in G_i$, and the group operation is defined componentwise, i.e., $(g_i)_{i \in I} + (h_i)_{i \in I} = (g_i + h_i)_{i \in I}$.

At the moment, we are more interested in the subgroup formed by those families $(g_i)_{i \in I}$ that satisfy $g_i = 0$ with only finitely many exceptions. This group is called the *direct sum* of the G_i and denoted

$$\bigoplus_{i \in I} G_i .$$

Of course, a difference between sums and products exists only if I is infinite; compare, for example, 1.30. Note that every summand (or factor) G_i is isomorphic to a subgroup of the direct sum or product, respectively. Likewise, the sum or product of the same G_i taken over any subset of the index set is contained in the total sum or product, respectively.

In the special case where $G_i = G$ for all i, we refer to the direct product as a *power* of G; its elements can be thought of as functions $I \to G$. We use the simplified notation

$$\bigtimes_{i \in I} G = G^I \quad \text{and} \quad \bigoplus_{i \in I} G = G^{(I)} .$$

If V is any vector space over a field F and $B \subseteq V$ is a basis, then every element of V has a unique representation as a *finite* linear combination

$v = \sum_{b \in B} f_b b$, thus $f_b = 0$ with finitely many exceptions. We can view v as an element $(f_b)_{b \in B}$ of a direct sum of copies of the additive group F^+, indexed by the set B. Therefore, we have an isomorphism

$$V^+ \cong \bigoplus_{b \in B} F_b^+ \, ,$$

where $F_b^+ = F^+$ for all b. In particular,

1.17 Theorem *The group $\mathbb{R}^+$ is a direct sum of $\aleph$ copies of $\mathbb{Q}^+$.* $\square$

The above decomposition cannot be refined any further, as the following theorem shows.

1.18 Theorem *The group $\mathbb{Q}^+$ cannot be decomposed as a direct product of two non-trivial subgroups G, H.*

Proof Any two non-zero elements $a/b \in G$, $c/d \in H$ have a non-zero common multiple $cb(a/b) = ca = ad(c/d) \in G \cap H$, which is a contradiction to $G \cap H = \{0\}$. $\square$

1.19 Theorem *The group $\mathbb{R}^+$ admits a decomposition as a direct sum of indecomposable subgroups. This decomposition is unique up to isomorphism.*

Proof Existence is obtained from 1.17 and 1.18; it remains to prove uniqueness. Suppose we have two decompositions of the specified kind. Unique divisibility of $\mathbb{R}^+$ implies that every summand is uniquely divisible; remember that addition is done componentwise. Hence, every summand is a vector space over $\mathbb{Q}$. Indecomposability implies that the summands are in fact one-dimensional vector spaces. Choosing a non-zero element from each summand of one decomposition, we construct a Hamel basis of $\mathbb{R}^+$. The two bases so obtained can be mapped onto each other by a vector space isomorphism. This shows that the two direct sums are isomorphic by an isomorphism that maps the summands of one onto those of the other. $\square$

The remainder of this section is devoted to the study of a prominent factor group of $\mathbb{R}^+$, the *torus group* $\mathbb{T} = \mathbb{R}^+/\mathbb{Z}$. We shall return to this in 5.16 and in Section 8. First we show why this group plays an important role both in analysis and in geometry.

1.20 Theorem *The following three groups are isomorphic:*
(a) *The factor group $\mathbb{T} = \mathbb{R}^+/\mathbb{Z}$*
(b) *The multiplicative group $\mathbb{S}_1$ of complex numbers of absolute value 1*
(c) *The group $\mathrm{SO}_2\mathbb{R}$ of rotations of the plane $\mathbb{R}^2$.*

Proof The exponential law $e^{z+w} = e^z e^w$ for the complex exponential function implies that the map $\varphi : t \mapsto e^{2\pi i t}$ is a group homomorphism of $\mathbb{R}^+$ into the multiplicative group of complex numbers. From $|\varphi(t)|^2 = \varphi(t)\overline{\varphi(t)} = \varphi(t)\varphi(-t) = 1$ we see that $\varphi(t) \in \mathbb{S}_1$. The Euler relation $e^{ix} = \cos x + i \sin x$ implies that φ maps $\mathbb{R}$ onto $\mathbb{S}_1$ and that $\ker \varphi = \mathbb{Z}$. It follows that $\mathbb{R}^+/\mathbb{Z} \cong \mathbb{S}_1$.

By definition, $SO_2\mathbb{R}$ consists of the real orthogonal 2×2 matrices of determinant 1 (or of the linear maps of $\mathbb{R}^2$ defined by these matrices). For $z \in \mathbb{S}_1$, let $\gamma(z)$ be the $\mathbb{R}$-linear map $\mathbb{C} \to \mathbb{C}$ defined by $w \mapsto zw$. This map preserves the norm $|w|$ and, hence, the scalar product. The matrix of $\gamma(e^{ix})$ with respect to the basis $1, i$ of $\mathbb{C}$ is

$$\begin{pmatrix} \cos x & -\sin x \\ \sin x & \cos x \end{pmatrix};$$

its determinant is 1. From the fact that the first column vectors exhaust $\mathbb{S}_1$ it follows that we have constructed a surjective map $\gamma : \mathbb{S}_1 \to SO_2\mathbb{R}$. It is immediate from the definition that γ is an injective group homomorphism. This completes the proof. □

The reader might wonder why this group is called a torus. The name originally refers to the direct product $\mathbb{S}_1 \times \mathbb{S}_1$, which is a topological torus (doughnut). More generally, the product of $n \geq 1$ copies of $\mathbb{S}_1$ is called an *n*-torus, so $\mathbb{S}_1$ itself is the 1-torus.

We shall now examine the structure of $\mathbb{T}$ in a similar way as we did for $\mathbb{R}^+$. As a tool, we need the concept of an injective abelian group.

1.21 Definition An abelian group S is said to be *injective* if every homomorphism of abelian groups $G \to S$ extends to any abelian group H containing G. What we need here is the following consequence of the definition.

1.22 Theorem *Let S be an injective subgroup of an abelian group G. Then S is a direct summand of G, i.e., there is a subgroup $T \leq G$ such that the direct sum $S \oplus T$ is isomorphic to G via the map $(s, t) \mapsto s + t$.*

Proof By injectivity, the identity map of S extends to a homomorphism $\rho : G \to S$ (a *retraction*). Define $T := \ker \rho$ and observe that $\rho \circ \rho = \rho$. Therefore, $\tau(g) := g - \rho(g) \in T$ for every $g \in G$, and the identity $g = \rho(g) + \tau(g)$ shows that $g \mapsto (\rho(g), \tau(g))$ is an isomorphism of G onto $S \oplus T$ with inverse $(s, t) \mapsto s + t$. □

1.23 Theorem *An abelian group is injective if, and only if, it is divisible.*

Proof (1) Suppose that S is injective and let $s \in S$, $n \in \mathbb{N}$. Define a homomorphism $n\mathbb{Z} \to S$ by $n \mapsto s$. By injectivity, this extends to a homomorphism $\alpha : \mathbb{Z} \to S$, and we have $n\alpha(1) = \alpha(n) = s$, hence S is divisible.

(2) Conversely, let S be divisible, and let $\varphi : G \to S$ be a homomorphism. We have to extend φ to a given abelian group H containing G. We consider the set of all pairs (N, ρ) consisting of a group N between G and H together with an extension $\rho : N \to S$ of φ. We say that (N, ρ) precedes (N', ρ') in this set if $N \subseteq N'$ and ρ' extends ρ. It is easily seen that the ordering defined in this way satisfies the hypothesis of Zorn's Lemma (compare Section 61), so there is an extension $\psi : M \to S$ of φ that cannot be extended any further. We shall show that in case $M \neq H$, further extension *is* possible; the conclusion then is that $M = H$.

Consider any element $h \in H \smallsetminus M$. If $\langle h \rangle$ intersects M trivially, then we can extend ψ to the (direct) sum $M + \langle h \rangle$ by setting $\psi(zh) = 0$ for all integers z. If the intersection is non-trivial, then $M \cap \langle h \rangle = \langle kh \rangle$ for some $k \in \mathbb{N}$. The image $\psi(kh) = s$ is already defined, and we choose $\psi(h) \in S$ such that $k\psi(h) = s$; this is possible by divisibility. It is easy to check that this yields a well-defined homomorphism $M + \langle h \rangle \to S$ extending ψ. $\square$

We are now ready to examine the structure of the torus group.

1.24 Theorem *There is an isomorphism* $\mathbb{T} = \mathbb{R}^+/\mathbb{Z} \cong \mathbb{R}^+ \oplus (\mathbb{Q}^+/\mathbb{Z})$.

Proof There exists a Hamel basis B such that $1 \in B$. The remainder $B' := B \smallsetminus \{1\}$ has the same cardinality as B, hence the rational vector space R generated by B' is isomorphic to $\mathbb{R}$. We have $\mathbb{R}^+ \cong R^+ \oplus \mathbb{Q}^+$, and factorization modulo $\mathbb{Z} \leq \mathbb{Q}$ yields the result. $\square$

1.25 Consequences By 1.15, we may substitute $(\mathbb{R}^{n+1})^+ \cong \mathbb{R}^+$ for the left summand of $\mathbb{T}$, and we obtain after regrouping that

$$\mathbb{T} \cong (\mathbb{R}^n)^+ \oplus \mathbb{T} .$$

We continue by examining $\mathbb{Q}/\mathbb{Z}$; compare also Section 31. First we remark that $\mathbb{Q}/\mathbb{Z}$ is a *torsion group*, i.e., every element has finite order, because every rational number has some multiple belonging to $\mathbb{Z}$.

Given any abelian group G and a prime p, the *p-primary component* $G_p \leq G$ is defined as the set of all elements of G whose order is a power of p. This is in fact a subgroup; indeed, if $\#g = p^k$ and $\#h = p^l$, then $p^{\max(k,l)}(g + h) = 0$, hence $\#(g + h)$ is a power of p. Let us examine the primary components of $G = \mathbb{Q}/\mathbb{Z}$.

1.26 Prüfer groups The primary components of the group $G = \mathbb{Q}/\mathbb{Z}$ are called Prüfer groups. We denote them by:

$$C_{p^\infty} := (\mathbb{Q}/\mathbb{Z})_p .$$

Consider the coset $ab^{-1} + \mathbb{Z}$ of a rational number ab^{-1}, where a, b are relatively prime. The coset belongs to C_{p^∞} if, and only if, there is a power p^k such that $p^k ab^{-1}$ is an integer, that is, if, and only if, the denominator b is a power of p. It follows that C_{p^∞} is the union (or the direct limit) of the groups $C_{p^k} = (p^{-k}\mathbb{Z})/\mathbb{Z} \cong \mathbb{Z}/p^k\mathbb{Z}$, the cyclic groups of order p^k:

$$C_{p^\infty} = \bigcup_{k \in \mathbb{N}} C_{p^k} .$$

We note the following consequence of this representation. If a subgroup $G \leq C_{p^\infty}$ contains elements of arbitrarily large order, then G is the entire group. If the orders of the elements of G are bounded, then G is one of the subgroups C_{p^k}. In particular, every proper subgroup of C_{p^∞} is cyclic and is uniquely determined by its order.

1.27 Theorem C_{p^∞} *is a divisible group. Moreover, multiplication by a natural number q not divisible by p defines an automorphism of C_{p^∞}.*

Proof In order to divide an element of C_{p^k} by p, we have to use $C_{p^{k+1}}$; we have $ap^{-k} = p(ap^{-k-1})$ for $a \in \mathbb{Z}$. Division by q takes place within the group C_{p^k}; indeed, multiplication by q is an injective endomorphism of that finite group, hence an automorphism. Passing to the union over all k, we obtain an automorphism of C_{p^∞}.   □

1.28 Theorem *There is a decomposition $\mathbb{Q}^+/\mathbb{Z} \cong \bigoplus_p C_{p^\infty}$, where the sum is taken over all primes.*

Proof We have to show that every element $x = ab^{-1} + \mathbb{Z}$ has a unique representation $x = g_{p_1} + \cdots + g_{p_k}$, where $g_{p_i} \in C_{p_i \infty}$. To prove existence, it suffices to write ab^{-1} as a sum of fractions with prime power denominators. This can be done by induction. If $b = p^l q$, where p is a prime not dividing q, write $up^{-l} + vq^{-1} = (uq + vp^l)(p^l q)^{-1}$ and use Bézout's Theorem 1.5 to find u, v such that $uq + vp^l$ equals a.

The proof of uniqueness reduces quickly to proving that $x = 0$ has only the trivial representation. Writing $-g_{p_1} = g_{p_2} + \cdots + g_{p_k}$, we see that the order of the left-hand side is a power of p_1 while the order of the right-hand side is a product of powers of the remaining primes. This shows that both sides are zero if the given primes are all distinct, and uniqueness follows by induction.   □

Virtually the same proof shows that every abelian torsion group splits as the direct sum of its primary components (Exercise 2). The following result can be obtained from 1.27 together with 1.28, or directly from the fact that $\mathbb{Q}^+/\mathbb{Z}$ is an epimorphic image of the divisible group $\mathbb{Q}^+$.

1.29 Corollary *The group $\mathbb{Q}^+/\mathbb{Z}$ is divisible.*   □

This group will be studied more closely in Section 31.

1.30 Theorem *There is an isomorphism $\bigtimes_p C_{p^\infty} \cong \mathbb{R}^+ \oplus \bigoplus_p C_{p^\infty}$, where p ranges over all primes.*

Proof By definition, the direct sum of all Prüfer groups is contained in their direct product. Both groups are divisible, hence injective (1.23), and 1.22 yields a decomposition $\bigtimes_p C_{p^\infty} = R \oplus \bigoplus_p C_{p^\infty}$. We shall use the characterization 1.14 of $\mathbb{R}^+$ in order to show that $R \cong \mathbb{R}^+$. Being a summand of a divisible group, R is divisible. Moreover, R is torsion free. Indeed, consider any element $x = (g_p)_p \in R \smallsetminus \{0\}$, where $g_p \in C_{p^\infty}$. Then x does not belong to the direct sum of the C_{p^∞}, hence g_{p_i} is non-zero for an infinite sequence of primes p_i. If the order of x were a finite number n, then all p_i would divide n, which is impossible.

It remains to check that R has the right cardinality. The product of all Prüfer groups is a Cartesian product of countably many countable sets, hence its cardinality is $\aleph_0^{\aleph_0} = 2^{\aleph_0}$; see 61.15. On the other hand, the right summand is the countable group $\mathbb{Q}/\mathbb{Z}$; see 1.28. From $\bigtimes_p C_{p^\infty} = R \oplus \mathbb{Q}/\mathbb{Z}$ we get $2^{\aleph_0} = \aleph_0 \cdot \operatorname{card} R = \operatorname{card} R$ by 61.12.   □

1.31 Corollary *The torus group $\mathbb{R}^+/\mathbb{Z}$ is isomorphic to the direct product $\bigtimes_p C_{p^\infty}$ of all Prüfer groups.*

Proof Combine the results 1.24, 1.28, and 1.30.   □

We conclude this section by looking at automorphisms of $\mathbb{R}^+$.

1.32 Theorem *The automorphism group $\Lambda = \operatorname{Aut} \mathbb{R}^+$ consists of the $\mathbb{Q}$-linear bijections of $\mathbb{R}$, and $\operatorname{card} \Lambda = 2^{\aleph} > \aleph$.*

Proof If λ is an automorphism of $\mathbb{R}^+$ and if $r \in \mathbb{Q}$, then $\lambda(r \cdot x) = r \cdot \lambda(x)$ because $\mathbb{R}^+$ is uniquely divisible; compare 1.9. Therefore, λ is $\mathbb{Q}$-linear.

A Hamel basis B of $\mathbb{R}$ has the same cardinality $\aleph$ as $\mathbb{R}$ itself; see 1.12. Therefore, B and $\mathbb{R}$ have the same transfinite number $\operatorname{card} \operatorname{Sym} B = 2^{\aleph}$ of permutations; compare 61.16. Every permutation of B gives rise to an automorphism of $\mathbb{R}^+$, which is, of course, a permutation of $\mathbb{R}$. Thus, we have $2^{\aleph} = \operatorname{card} \operatorname{Sym} B \leq \operatorname{card} \Lambda \leq \operatorname{card} \operatorname{Sym} \mathbb{R} = 2^{\aleph}$. The last part of the assertion follows from 61.11.   □

1.33 Corollary *The automorphism group* Aut $\mathbb{T}$ *of the torus group has the same cardinality* $2^{\aleph}$ *as* Aut $\mathbb{R}^+$.

Proof By 1.25, we have $\mathbb{T} \cong \mathbb{R}^+ \oplus \mathbb{T}$. Thus, every automorphism of $\mathbb{R}$ extends to an automorphism of $\mathbb{T}$. We infer that $2^{\aleph} \leq \operatorname{card} \operatorname{Aut}(\mathbb{R}^+) \leq \operatorname{card} \operatorname{Aut} \mathbb{T} \leq \operatorname{card} \operatorname{Sym} \mathbb{T} = 2^{\aleph}$.   □

This corollary should be contrasted with the fact that the torus group has only two continuous automorphisms; see Theorem 8.27. We infer that the torus group has uncountably many discontinuous group automorphisms.

Exercises

(1) Suppose that V is a vector space having two *infinite* bases B and B'. Show that card B = card B'.

(2) Prove that every abelian torsion group splits as the direct sum of all its primary components.

(3) Show that $\mathbb{R}^+$ contains subgroups isomorphic to $\mathbb{R}^+ \times \mathbb{Z}^n$ for arbitrary $n \in \mathbb{N}$. Show moreover that these groups are not isomorphic to $\mathbb{R}^+$.

(4) The set $\mathcal{H}$ of hyperplanes in the rational vector space $\mathbb{R}$ has cardinality card $\mathcal{H} = 2^{\aleph}$: there are more hyperplanes than one-dimensional subspaces.

(5) The Cantor set $\{ \sum_{\nu=1}^{\infty} c_{\nu} 3^{-\nu} \mid c_{\nu} \in \{0,2\} \}$ (compare 5.35) is not contained in any proper subgroup of $\mathbb{R}$.

(6) The numbers $\log p$, p a prime in $\mathbb{N}$, are linearly independent over $\mathbb{Q}$.

(7) Let F be a field of characteristic 0. Show that the additive group F^+ has no maximal subgroup.

(8) Let $r \in \mathbb{R} \smallsetminus \mathbb{Q}$. Does the subgroup $\mathbb{Z} + r\mathbb{Z}$ of $\mathbb{R}^+$ admit an automorphism of order 5 ?

(9) Determine the isomorphism type of the factor group $\mathbb{R}/\mathbb{Q}$ of additive groups.

2   The multiplication of real numbers, with a digression on fields

The multiplicative group of real numbers has 'almost' the same structure as the additive group. This fact will be established quickly (2.2), and after that we conduct a systematic search for similar phenomena in other fields. This will culminate in the construction of a field whose multiplicative group is actually isomorphic to the additive group of real numbers (see 2.11). The construction makes use of formal power series, which will be treated more systematically later in this book. Some readers may therefore prefer to skip this topic on first reading.

By $\mathbb{R}^\times$ we denote the multiplicative group $(\mathbb{R} \smallsetminus \{0\}, \cdot)$ of real numbers. It contains two notable subgroups, the cyclic group $\{1, -1\} \cong C_2$ of order 2 and the group $\mathbb{R}^\times_{\mathrm{pos}}$ of positive real numbers. A real number $r \neq 0$ is uniquely expressed as a product of $\operatorname{sign} r \in \{1, -1\}$ and its absolute value $|r| \in \mathbb{R}^\times_{\mathrm{pos}}$. This proves the following.

2.1 Theorem   *There is a direct sum decomposition* $\mathbb{R}^\times = \mathbb{R}^\times_{\mathrm{pos}} \oplus C_2$. $\square$

The following theorem is the structural interpretation of the functional equation $\exp(x + y) = \exp(x) \exp(y)$.

2.2 Theorem   *The exponential function is an isomorphism* $\mathbb{R}^+ \cong \mathbb{R}^\times_{\mathrm{pos}}$. *Therefore,* $\mathbb{R}^\times \cong \mathbb{R}^+ \oplus C_2$. $\square$

The close relationship between the additive and the multiplicative group of real numbers exhibited by 2.2 is so important that it seems worthwhile to look systematically for other fields sharing this property. (As always in this book, fields are commutative by definition.)

Let us consider a few examples, starting with the complex numbers. There, the exponential function is not injective. In fact, the multiplicative group $\mathbb{C}^\times$ contains a group isomophic to $\mathbb{R}^+/\mathbb{Z}$ (see 1.20), which has a large torsion subgroup $\mathbb{Q}^+/\mathbb{Z}$ (see 1.24ff). Hence, $\mathbb{C}^\times$ is not isomorphic to the torsion free group $\mathbb{C}^+$ or to $\mathbb{C}^+ \oplus C_2$, nor to any subgroup of these groups.

The rational numbers form another negative example. Indeed, the additive group $\mathbb{Q}^+$ is locally cyclic (1.6c), and $\mathbb{Q}^\times$ is not. (In fact, $\mathbb{Q}^\times$ is a direct sum of C_2 and a countably infinite number of infinite cyclic groups; compare 32.1.)

Both examples seem to indicate that the isomorphism 2.2 is a rare phenomenon for fields in general. However, the exact answer depends on how we make our question precise. The following result answers a rather coarse form of the question.

2.3 Proposition   *There is no (skew) field F such that the additive group F^+ is isomorphic to the entire multiplicative group $F^\times$.*

Proof We prove this by counting involutions. An *involution* in a group is an element g of order 2, i.e., $g \neq 1 = g^2$ in multiplicative notation or $g \neq 0 = 2g$ written additively.

The non-zero elements of the additive group F^+ of a field F all have the same order, depending on the characteristic $\operatorname{char} F$; compare 64.4. This common order is equal to $\operatorname{char} F$ if $\operatorname{char} F$ is a prime, and infinite if $\operatorname{char} F = 0$. Thus, F^+ contains involutions only if $\operatorname{char} F = 2$, and then all elements except 0 are involutions.

On the other hand, if $s \in F^\times$ is an involution, then $(s-1)(s+1) = s^2 - 1 = 0$. This equation has either no solution $s \neq 1$ (if char $F = 2$) or one such solution (if char $F \neq 2$). In both cases, the numbers of involutions do not match. $\qquad\square$

This was disappointing, so we modify our question and ask for pairs of fields F, G such that F^+ is isomorphic to $G^\times$. To formulate the answer in the finite case, we need a number theoretic notion.

2.4 Definition: Mersenne primes   A prime number of the form $p = 2^k - 1$, $k \in \mathbb{N}$, is called a *Mersenne prime*. Of course, $2^k - 1$ is not always prime. A necessary (but not sufficient) condition is that the exponent k be prime. Indeed, for $k = mn$, we have

$$2^k - 1 = (2^{k-n} + 2^{k-2n} + \cdots + 2^{k-mn})(2^n - 1) \, .$$

More information on Mersenne primes, including references, will be given in 32.10.

2.5 Example   Let q be a prime power. We denote by $\mathbb{F}_q$ the unique finite field of order q (for a proof of uniqueness see COHN 2003a 7.8.2, JACOBSON 1985 p. 287 or LANG 1993 V 5.1). We have

$$\mathbb{F}_p^+ \cong \mathbb{F}_{p+1}^\times$$

if $p = 2$ or if p is a prime such that $p+1$ is a power of 2; in the latter case, p is a Mersenne prime. Indeed, the conditions ensure that fields of the given orders exist; moreover, the two groups have the same order, and they are both cyclic (the multiplicative group of a finite field is always cyclic; see Exercise 1 of Section 64).

2.6 Theorem   *The pairs $\mathbb{F}_p$, $\mathbb{F}_{p+1}$ where $p = 2$ or p is a Mersenne prime are the only pairs of fields F, G such that char $F \neq 0$ and $F^+ \cong G^\times$.*

Proof   Let $p = \operatorname{char} F$, and note that F^+ contains at least $p-1$ non-zero elements, all of order p, while $G^\times$ cannot have more than $p-1$ elements of order p (the solutions $x \neq 1$ of $x^p - 1 = 0$). Now $F^+ \cong G^\times$ implies that $F = \mathbb{F}_p$ and $G = \mathbb{F}_{p+1}$. $\qquad\square$

In order to formulate a first result for the case char $F = 0$, we need the following.

2.7 Definition   We say that a multiplicative group $(A, \cdot)$ *has unique roots* if the mapping $a \mapsto a^n$ is a bijection of A for all $n \in \mathbb{N}$. This is just the equivalent in multiplicative language of the notion of unique divisibility (1.7). We say that a field has unique roots if its multiplicative group has this property. Note that only fields of characteristic two can have unique roots, because $(-1)^2 = 1^2$ in any field.

2.8 Proposition   *Given a field G, there exists a field F of characteristic zero such that $F^+ \cong G^\times$ if, and only if, (char $G = 2$ and) G has unique roots.*

Proof If char $F = 0$, then F is a vector space over its prime field $\mathbb{Q}$, hence the additive group F^+ is uniquely divisible. Thus, a necessary condition for the existence of F is that G has unique roots. Conversely, assume that this is the case. Then $G^\times$ is isomorphic to the additive group of a rational vector space V; see 1.9. Both this vector space and its additive group are determined, up to isomorphism, by the cardinality card B of a basis.

If this cardinality is finite, say card $B = n$, then we take F to be an algebraic extension of degree n over $\mathbb{Q}$ (e.g., $F = \mathbb{Q}(\sqrt[n]{2})$) to ensure that $F^+ \cong G^\times$. If card B is infinite, then we use a purely transcendental extension $F = \mathbb{Q}(T)$ (compare 64.19), where card $T = $ card B. In order to prove that $F^+ \cong V^+ \cong G^\times$, we have to show that $\dim_{\mathbb{Q}} \mathbb{Q}(T) = $ card T if T is infinite. This is proved in 64.20.   □

Proposition 2.8 raises the question as to which uniquely divisible groups occur as the multiplicative groups of fields. This question is answered completely by CONTESSA *et al.* 1999 5.3 and 5.5; they show that a uniquely divisible abelian group A is the multiplicative group of some field if, and only if, the dimension of A as a vector space over $\mathbb{Q}$ is infinite. These groups also occur as the additive groups of fields, as we have shown in 2.8.

Here we shall be content to give examples of fields G having unique roots. The following lemma allows us to obtain roots in the power series ring $F[[t]]$, which is defined in 64.22.

2.9 Lemma   *Let F be a field. If $m \in \mathbb{N}$ is not a multiple of the characteristic of F, then every element $a \in 1 + tF[[t]]$ admits an mth root $c \in 1 + tF[[t]]$, that is, $c^m = a$.*

Proof Put $c_0 = 1$. Then $c_0^m = 1 \equiv a \bmod t$; in general, a congruence $x \equiv y \bmod t^n$ means that t^n divides $x - y$ in the ring $F[[t]]$.

Assume that we have found elements $c_0, c_1, \ldots, c_{k-1} \in F$ such that $p_{k-1}^m \equiv a \bmod t^k$, where $p_{k-1} := \sum_{i=0}^{k-1} c_i t^i$. Then

$$p_{k-1}^m \equiv a + bt^k \bmod t^{k+1}$$

for some $b \in F$. By our assumption on m, we can define $c_k \in F$ by

$mc_k = -b$. We show that the polynomial $p_k := p_{k-1} + c_k t^k$ satisfies $p_k^m \equiv a \bmod t^{k+1}$.

We have $p_{k-1} \equiv c_0 \equiv 1 \bmod t$, which implies $p_{k-1}^{m-1} \equiv 1 \bmod t$ and $p_{k-1}^{m-1} t^k \equiv t^k \bmod t^{k+1}$. Therefore the following congruences $\bmod t^{k+1}$ hold:

$$p_k^m = (p_{k-1} + c_k t^k)^m \equiv p_{k-1}^m + mc_k p_{k-1}^{m-1} t^k \equiv a + bt^k - bp_{k-1}^{m-1} t^k \equiv a \ .$$

Now the formal power series $c := \sum_{i \geq 0} c_i t^i$ satisfies the congruences $c^m - a \equiv p_{k-1}^m - a \equiv 0 \bmod t^k$ for every $k \in \mathbb{N}$, hence $c^m - a = 0$.   □

We shall now apply this lemma to *fields of Puiseux series* $F((t^{1/\infty}))$ as introduced in 64.24, in order to produce examples of fields having unique roots. A Puiseux series is a formal sum

$$a = \sum_{i \geq k} a_i t^{i/n} \ ,$$

where $n \in \mathbb{N}$, $k \in \mathbb{Z}$, and the coefficients a_i belong to a given field F.

2.10 Theorem *If F is a field (necessarily of characteristic 2) which has unique roots, e.g., $F = \mathbb{F}_2$, then the field $F((t^{1/\infty}))$ of Puiseux series has unique roots, as well.*

In particular, the multiplicative group of $F((t^{1/\infty}))$ is isomorphic to the additive group of some other field, by 2.8.

Proof (1) Each non-zero Puiseux series can be written uniquely as a product $at^r(1+b)$, where $a \in F^\times$, $r \in \mathbb{Q}$ and b is a Puiseux series which involves only powers of t with positive exponents, i.e., $b \in t^{1/n} F[[t^{1/n}]]$ for some $n \in \mathbb{N}$.

The Puiseux series of the form $1 + b = 1 + \sum_{i \geq 1} b_i t^{i/n}$ with $b_i \in F$ form a subgroup of the multiplicative group $F((t^{1/\infty}))^\times$ (note that the geometric series $\sum_{i \geq 0} b^i = (1+b)^{-1}$ makes sense and shows that $(1+b)^{-1}$ is again of that form; compare also 64.24), and $F((t^{1/\infty}))^\times$ is the direct product of $F^\times$, $t^{\mathbb{Q}}$ and this subgroup.

Therefore it suffices to examine these three factors separately. The element $a \in F^\times$ has unique roots by our hypothesis, and $t^{r/m}$ is the unique mth root of t^r. It remains to show that $1 + b = 1 + \sum_{i \geq 1} b_i t^{i/n}$ has a unique mth root for every $m \in \mathbb{N}$, and we may assume that m is a prime number.

(2) If $m \neq 2 = \mathrm{char}(F)$, then the existence of such a root is a consequence of Lemma 2.9, since the subring $F[[t^{1/n}]]$ of $F((t^{1/\infty}))$ is isomorphic to the power series ring $F[[t]]$. For $m = 2$ we compute directly that $(1 + \sum_{i \geq 1} \sqrt{b_i} t^{i/(2n)})^2 = 1 + b$.

(3) For the proof of uniqueness, it suffices to show that $1 = (1 + b)^m$ implies $b = 0$. Assume that $b \neq 0$ and let $k := \min\{i \in \mathbb{N} \mid b_i \neq 0\} \geq 1$. By binomial expansion we have

$$1 = (1+b)^m = 1 + \binom{m}{1}b + \cdots + b^m = 1 + mb_k t^{k/n} + \textstyle\sum_{i>k} c_i t^{i/n}$$

with $c_i \in F$. We obtain $mb_k = 0 \neq b_k$, hence $m = 2$ and $1 = (1 + b)^2 = 1 + b^2$, a contradiction to $b \neq 0$.   □

2.11 Corollary   *The multiplicative group of the field $F = \mathbb{F}_2((t^{1/\infty}))$ is isomorphic to the additive group $\mathbb{R}^+$.*

Proof By 2.10, the group $F^\times$ is divisible and torsion free. Moreover $\mathbb{F}_2^{\mathbb{N}} \subseteq F \subseteq \mathbb{F}_2^{\mathbb{Q}}$, hence $\operatorname{card} F = 2^{\aleph_0} = \operatorname{card} \mathbb{R}$. Now 1.14 gives the assertion.   □

Using different methods (compare 2.12), this was shown by CONTESSA *et al.* 1999 5.6 (Moreover they prove that, in contrast, the additive group of rational numbers is not isomorphic to the multiplicative group of any field, loc. cit., 5.3).

Furthermore the fields $F = \mathbb{F}_2((\mathbb{Q}^n))$ and $F = \mathbb{F}_2((\mathbb{R}))$ of Hahn power series (see 64.25) have the property that $F^\times \cong \mathbb{R}^+$.

The field $\mathbb{F}_2((t^{1/\infty}))$ of Puiseux series is not an algebraic extension of $\mathbb{F}_2$, as t is transcendental. In fact, a proper algebraic extension of $\mathbb{F}_2$ contains finite subfields distinct from $\mathbb{F}_2$, hence it contains non-trivial roots of unity. However, we show next that it is possible to obtain fields having unique roots by algebraic extension from fields containing no roots of unity. The result is taken from CONTESSA *et al.* 1999 4.3.

2.12 Theorem   *Let L be a field of characteristic 2 that does not contain any non-trivial roots of unity. If L is not the prime field $\mathbb{F}_2$, then there is an algebraic extension field M of L that has unique roots.*

For example, the field L may be any purely transcendental extension of $\mathbb{F}_2$.

Proof The problem is to adjoin roots of all degrees for every $a \in L \setminus \{0, 1\}$ without adjoining any root of unity. Consider an algebraic closure $L^{\natural}$ and the collection $\mathcal{M}$ of all fields F with $L \leq F \leq L^{\natural}$ such that F does not contain any non-trivial root of unity. The union of every chain in $\mathcal{M}$ belongs to $\mathcal{M}$. By Zorn's Lemma, there is a maximal element $M \in \mathcal{M}$, and we proceed to show that M contains a pth root of a for every prime p and every $a \in M$. We may assume that $a \notin \{0, 1\}$. Our claim is a consequence of maximality together with the following lemma.   □

2.13 Lemma *Let M be a field of characteristic 2 that does not contain any non-trivial roots of unity. Suppose that the polynomial $q(x) = x^p - a$ has no root in M for some prime p and some $a \in M \smallsetminus \{1\}$, and consider the algebraic extension $M(\vartheta)$ by a root ϑ of $q(x)$. Then $M(\vartheta)$ does not contain any non-trivial roots of unity.*

Proof (1) Suppose that $p = 2$ and that for some $b, c \in M$, the element $\eta = b + c\vartheta \in M(\vartheta)$ satisfies $\eta^k = 1$, where $k > 1$. Then $\eta^2 = b^2 + c^2\vartheta^2 = b^2 + c^2 a \in M$ as we are in characteristic 2. Since $\eta^{2k} = 1^2$ and since M does not contain any roots of unity other than 1, we conclude that $\eta^2 = 1$ and, hence, that $\eta = 1$. For the remainder of the proof, we may assume that p is odd.

(2) We claim that the polynomial $q(x)$ is irreducible over M (compare COHN 2003a 7.10.8 or LANG 1993 VI §9). The roots of $q(x)$ in $M^\natural$ are of the form $\zeta_\nu \vartheta$, where $\zeta_0, \zeta_1, \ldots, \zeta_{p-1}$ are the pth roots of unity in $M^\natural$. If $q(x)$ splits over M as $q(x) = r(x)s(x)$, then $r(x)$ is a product of some linear factors $x + \zeta_\nu \vartheta$, hence the constant term $b \in M$ of $r(x)$ has the form $\delta\vartheta^\mu$, where $\delta^p = 1$ and $0 < \mu < p$. We have $b^p = a^\mu$, and Bézout's Theorem 1.5 provides $m, n \in \mathbb{Z}$ such that $m\mu + np = 1$; this yields $a = a^{m\mu + np} = b^{mp}a^{np}$, hence $b^m a^n \in M$ is a root of $q(x)$, a contradiction.

We have shown that the degree $[M(\vartheta) : M]$ equals the prime p, hence the degree formula 64.2 implies that there are no fields properly between M and $M(\vartheta)$.

(3) Suppose that $M(\vartheta)$ contains some root of unity $\eta \neq 1$. Let $k > 1$ be the multiplicative order of η. Then $x^k - 1 = \prod_{i=0}^{k-1}(x - \eta^i)$. Thus, $M(\eta) \subseteq M(\vartheta)$ is the splitting field of the polynomial $x^k - 1$ and is a normal extension field of M; see 64.10. Now step (2) implies that $M(\eta) = M(\vartheta)$; therefore, $q(x)$ splits into linear factors in $M(\vartheta)$, and the linear factors are as shown in step (2). It follows that $M(\vartheta)$ contains all pth roots of unity $\zeta_0, \zeta_1, \ldots, \zeta_{p-1}$. We have $\zeta_i \neq 1$ for some index i, otherwise $x^p - 1 = (x - 1)^p$ in $M[x]$, which leads to the excluded case $p = 2$. As before, we conclude that $M(\zeta_i) = M(\vartheta)$, but the minimal polynomial of ζ_i over M divides $x^{p-1} + x^{p-2} + \ldots + 1$ and hence has degree less than p, a contradiction. $\square$

The reader may feel that we still have not treated the 'right' question. We are looking for fields whose behaviour is similar to that of the real numbers – but $\mathbb{R}^+$ is not isomorphic to $\mathbb{R}^\times$. To come closer to the real case, we should consider ordered fields F (necessarily of characteristic zero) whose additive group is isomorphic to the multiplicative group of *positive* elements. For the notion of an ordered field, compare 11.1. We shall not determine all fields having the above property, but the *groups* appearing as F^+ and $F_{\mathrm{pos}}^{\times}$ for the same ordered field F will be determined up to isomorphism. The following result was presented by G. Kaerlein at a conference at Bad Windsheim, Germany, in 1980.

2.14 Theorem *For an abelian group $(H, +)$, the following conditions are equivalent.*

(a) *H is uniquely divisible and the dimension of H as a rational vector space (compare 1.9) is infinite.*

(b) *H is a direct sum of infinitely many copies of $\mathbb{Q}^+$.*

(c) *There is an ordered field F such that $H \cong F^+ \cong F_{\mathrm{pos}}^{\times}$.*

Proof (1) We know from Section 1 that conditions (a) and (b) are equivalent, so we have to show that they are necessary and sufficient for (c).

(2) Necessity. An ordered field F has characteristic 0, hence F is a vector space over its prime field $\mathbb{Q}$, and F^+ is uniquely divisible. We have to show that $\dim_{\mathbb{Q}} F$ is infinite. Now $F_{\mathrm{pos}}^{\times} \cong F^+$ has unique roots. This implies that F contains elements that are algebraic over $\mathbb{Q}$ of arbitrarily large degrees, and $\dim_{\mathbb{Q}} F$ is an upper bound for these degrees (see 64.5), hence $\dim_{\mathbb{Q}} F$ is infinite.

(3) Sufficiency. As in the proof of 2.8, we find a purely transcendental extension $\mathbb{Q}(T)$ whose additive group is isomorphic to H, and we proceed to turn $\mathbb{Q}(T)$ into an ordered field. We use a total ordering of the transcendency basis T and define a lexicographic ordering on monomials: For $t_1 > \cdots > t_k$ and integers $n_i \geq 0$, $m_i \geq 0$, we set $t_1^{n_1} \ldots t_k^{n_k} > t_1^{m_1} \ldots t_k^{m_k}$ if, and only if, there is an index i_0 such that $n_i = m_i$ for $i < i_0$ and $n_{i_0} > m_{i_0}$. A polynomial in the indeterminates $t \in T$ is positive by definition if the coefficient of the largest monomial is positive. A quotient of two polynomials is said to be positive if either both polynomials are positive or both are negative. The proof that this makes the field of fractions $\mathbb{Q}(T)$ an ordered field is left to the reader.

From the fact that the field $\mathbb{Q}(T)$ is ordered, we deduce that it is formally real; see 12.1. By 12.16, it has an algebraic extension F that is real closed. Now 12.10(ii) together with 7.3 yields that $F_{\mathrm{pos}}^{\times}$ has unique roots. The cardinality of F is the same as that of $\mathbb{Q}(T)$; see 64.5. Again the arguments of the proof of 2.8 show that $F^+ \cong H \cong F_{\mathrm{pos}}^{\times}$.　□

Remembering that the real exponential function is monotone, we may sharpen the previous question once again. We can ask for ordered fields such that F^+ and $F_{\mathrm{pos}}^{\times}$ are isomorphic as *ordered* groups. This leads to the notion of *exponential fields*. The question, which belongs to the realm of ordered fields, will be discussed in Section 11; see 11.10.

Exercises

(1) Supply the details for the proof of 2.14 by proving that the definition given there turns a purely transcendental extension $\mathbb{Q}(T)$ into an ordered field.

(2) The group $\mathbb{R}^+$ has infinitely many extensions by the cyclic group C_2 of order 2. Only one of them is commutative.

3   The real numbers as an ordered set

There is an *ordering* relation (or *order*) $<$ on the set of real numbers, which makes $\mathbb{R}$ a *chain* (or totally ordered set). This means that for $r, s \in \mathbb{R}$, precisely one of the relations $r < s$, $s < r$, $s = r$ holds. In addition, we have the usual 'transitivity' property of an ordering: $r < s$ and $s < t$ together imply $r < t$. The following examples explain our notation for *intervals* in a chain C:

$$[a, b[\,= \{\, c \in C \mid a \le c < b \,\}$$
$$]a,\ [\,= \{\, c \in C \mid a < c \,\}\ .$$

We note that the ordering induced on an interval $[n, n+1[$ between consecutive integers coincides with the lexicographic ordering obtained by comparing binary expansions.

The following notions will be used to characterize the chain $\mathbb{R}$ of real numbers and its subchains $\mathbb{Z}$ (the integers) and $\mathbb{Q}$ (rational numbers). In a later chapter, this will be applied in order to characterize the topological space of real numbers; see 5.10.

3.1 Completeness, density, separability   A chain C (or its ordering) is said to be *complete* if every non-empty subset $B \subseteq C$ which is bounded above has a *least upper bound* (or *supremum*) $\sup B$. The chain of real numbers is complete. In fact, the usual way to obtain the real numbers is by completion of the rational numbers. Compare 42.11 and 44.11. The idea of completion will be studied thoroughly in Chapter 4.

A subset A of a chain C is said to be *coterminal* if A contains lower and upper bounds for every $c \in C$. A coterminal subset $A \subseteq C$ is said to be *weakly dense* in C if every $c \in C$ satisfies $c = \inf\{\, a \in A \mid c \le a \,\}$ and $c = \sup\{\, a \in A \mid a \le c \,\}$. Thus A is weakly dense in the chain C if, and only if, for all $c_1 < c_2$ in C there exist elements $a_1, a_2, a_3, a_4 \in A$ such that $a_1 \le c_1 \le a_2 < a_3 \le c_2 \le a_4$. This implies that every closed

interval $[c, d]$ of C with $c < d$ contains an element of A. For the purposes
of the present section, the latter condition would suffice, but in Sections
41 and 42 we shall need the notion of weak density as defined here.

A subset $A \subseteq C$ is called *strongly dense* if for all pairs $c < d$ in
C, the set A contains elements x, y, z such that $x < c < y < d < z$.
For example, the rational numbers form a strongly dense subset of $\mathbb{R}$.
Observe that a strongly dense subset of C is infinite unless C has only
one element. Note also that strong density implies weak density.

If A is strongly dense in C, then C has no extremal (smallest or
largest) element and does not contain any gaps, i.e., there is no empty
open interval $]c, d[$. If A is only weakly dense, then extremal elements
may exist, but they have to belong to A, and gaps $]c, d[$ may exist, but
then c and d must belong to A. On the other hand, if $]c, d[$ is non-empty,
say $x \in]c, d[$, then $[x, d[$ must contain an element of A.

A chain C is called weakly or strongly *separable* if it has a weakly or
strongly dense subset A, respectively, which is at most countable. The
density of $\mathbb{Q}$ implies that $\mathbb{R}$ is strongly separable.

3.2 Topology induced by an ordering   Every chain C is a topological space in a natural way. The *topology induced by the ordering* or
order topology of C is the smallest topology such that all open intervals
(including $]a, [$ and $] , a[$) are open sets. In this topology, a set is open
if, and only if, it is the union of (perhaps infinitely many) open intervals.

Let us compare the notions of density for chains to their topological
analogues. In a topological space X, density means that every non-
empty open set of X contains an element of A. In a chain, an open
interval can be empty. Therefore, a subset can be dense in the topology
induced by the ordering without being strongly dense in the chain. For
example, the pair of chains $\mathbb{Z} \subseteq \mathbb{Z}$ is topologically dense and weakly
but not strongly dense, and the pair $]0, 1] \subseteq [0, 1]$ is topologically dense
but not even weakly dense. Conversely, weak density implies topological
density, because only non-empty open intervals are involved. Thus we
have the implications

   strongly dense $\Rightarrow$ weakly dense $\Rightarrow$ topologically dense,

none of which is reversible.

A topological space X is said to be *separable* if it has a countable dense
subset A, and similar remarks as above hold for this notion. Thus, $\mathbb{Z}$ is
a separable topological space, but not a strongly separable chain; it is,
however, weakly separable.

The following is a standard fact.

3.3 Proposition *For a chain C, the following two conditions are equivalent.*

(1) *The topology induced by the ordering of C is connected.*
(2) *The chain C is complete, and each pair $c < d$ in C defines a nonempty interval $]c, d[$.*

The condition about intervals is satisfied if C is strongly dense in itself, but it does not exclude the existence of a largest (or a smallest) element.

Proof Suppose that condition (2) is violated; we show that C is not connected. If the interval $]c, d[$ is empty, then C is the disjoint union of the two open sets $] , d[$ and $]c, [$, and is not connected. On the other hand, if C is not complete, let $A \subseteq C$ be a bounded subset without a supremum. Then the set B of all upper bounds of A is open, as well as the union U of all intervals $] , a[, a \in A$. The sets U and B are disjoint and cover C.

Next suppose that (2) is satisfied; we show that C is connected. Thus we assume that C is the disjoint union of two non-empty open sets U, V, and we derive a contradiction. There are elements $a \in U$ and $b \in V$, and we may assume that $a < b$. We consider the bounded set B obtained as the union of all intervals $] , u[$, where $u \in U$ and $u < b$. We show that every open interval I containing $s = \sup B$ meets both U and V, contradicting the assumption that one of these disjoint open sets contains s.

We prove that $I \cap V$ is non-empty. We may assume that $s < b$, and then the non-empty set $]s, b[$ does not contain any $u \in U$; or else, $]s, u[$ would be contained in B, contrary to the definition of s.

Finally, I contains some interval $]c, s]$, which in turn contains some element $x \in B$. By definition of B, there is an element $u \in U$ such that $x < u < b$; in fact, $u \leq s$ because we have seen that $]s, b[\subseteq V$. Thus, $u \in I \cap U$. $\square$

We say that map $\varphi : C \to C'$ between two chains *preserves the ordering*, or is *order-preserving*, if for all $x, y \in C$ such that $x \leq y$ one has $\varphi(x) \leq \varphi(y)$. Note that such a map need not be injective. An order-preserving map φ is injective precisely if $x < y$ implies $\varphi(x) < \varphi(y)$.

An order-preserving bijection $\varphi : C \to C'$ is called an *order isomorphism*; if such a bijection exists, then the chains C and C' are called *order isomorphic*.

The following characterization of the chain $\mathbb{Q}$ is due to CANTOR 1895 §9; compare HAUSDORFF 1914 p. 99, BIRKHOFF 1948 p. 31. It is a crucial step in the proof of the subsequent characterization of $\mathbb{R}$.

3.4 Theorem   *A chain A is order isomorphic to the chain $\mathbb{Q}$ of rational numbers if, and only if, A is countable and strongly dense in itself.*

Proof Clearly, $\mathbb{Q}$ has the properties mentioned. Conversely, suppose that A has the same properties. We use enumerations $a_1, a_2, \ldots$ and $q_1, q_2, \ldots$ of A and $\mathbb{Q}$, respectively, in order to construct an order isomorphism $f : A \to \mathbb{Q}$ via induction.

We set out by defining $f_1(a_1) = q_1$ and $A_1 = \{a_1\}$, $\mathbb{Q}_1 = \{q_1\}$. We define two sequences of subsets $A_m \subseteq A$ and $\mathbb{Q}_m \subseteq \mathbb{Q}$, both of cardinality m, and order isomorphisms $f_m : A_m \to \mathbb{Q}_m$. Alternatingly, we perform two different inductive steps, each time adding single elements a_μ and q_ν to the sets A_m and $\mathbb{Q}_m$, respectively, and at the same time extending f_m to an isomorphism $A_{m+1} \to \mathbb{Q}_{m+1}$. If m is odd, we choose ν to be the smallest number such that $q_\nu \notin \mathbb{Q}_m$. By strong density of A, there is an element $a_\mu \in A \smallsetminus A_m$ such that the desired extension is possible, and we insist that μ be chosen minimal in order to make a definite choice. If m is even, we interchange the roles of A and $\mathbb{Q}$; in other words, we begin by choosing $a_\nu \notin A_m$ with ν minimal, and then select q_μ such that the extension of f_m is possible and μ is minimal. This alternating strategy together with the minimality condition for ν ensures that every element of A and of $\mathbb{Q}$ is used at some point, hence by taking unions we obtain a bijection $A \to \mathbb{Q}$ which preserves the ordering.   □

The basic idea of this proof can be used to show that the chain $\mathbb{Q}$ embeds into every chain A that is dense in itself. The direction of the map f is of course reversed, and it is not necessary to alternate the roles of A and $\mathbb{Q}$. Instead of an enumeration of A, one uses a well-ordering.

3.5 Theorem   *The following conditions are mutually equivalent for a chain R having more than one element.*
   (a) *R is order isomorphic to the chain $\mathbb{R}$ of real numbers.*
   (b) *R is strongly separable and complete.*
   (c) *R is strongly separable and the topology induced by the ordering of R is connected.*

Proof We know that (b) and (c) are properties of the real numbers, and we proved in 3.3 that (b) and (c) are equivalent. We shall finish the proof by showing that a complete strongly separable chain R can be mapped order isomorphically onto the chain $\mathbb{R}$. By strong separability,

R contains a countably infinite chain A that is strongly dense in R and, hence, in itself. By 3.4, there is an order isomorphism $f : A \to \mathbb{Q}$. We show how to extend f to an order isomorphism $R \to \mathbb{R}$.

Given $r \in R$, let $A_r := \{ a \in A \mid a \leq r \}$ and choose $b \in A$ such that $b > r$. Then $f(A_r) < f(b)$ is a bounded subset of the complete chain $\mathbb{R}$, and we may define $f(r) := \sup f(A_r)$. This map extends the given one, and it has an inverse, defined in the same way; note that $r = \sup A_r$. Moreover, f preserves the ordering. Indeed, if $r < s$, then $f(A_r) \subseteq f(A_s)$, hence $f(r) = \sup f(A_r) \leq \sup f(A_s) = f(s)$.   □

Note that the same arguments prove that a chain is isomorphic to a subchain of $\mathbb{R}$ if, and only if, it is weakly separable; compare BIRKHOFF 1948 p. 32.

3.6 Definition   A chain is said to be *homogeneous*, if its group of automorphisms is transitive, that is, if every element can be mapped onto every other one by an automorphism.

Examples of homogeneous chains are $\mathbb{Z}$ (the integers), $\mathbb{Q}$ (rational numbers), and $\mathbb{R}$ (real numbers). For two of these examples, we have the following characterization, taken from BIRKHOFF 1948 III.8.

3.7 Theorem   *Let C be a chain that is complete, homogeneous, and weakly separable as defined in 3.1. Then C is order isomorphic to the chain $\mathbb{Z}$ of integers or to the chain $\mathbb{R}$ of real numbers, or C has only one element.*

Proof (1) If C has a smallest or largest element, then C is a singleton by homogeneity. Henceforth, we assume that this is not the case.

(2) Assume that every interval $]a, b[$ with $a < b$ is non-empty. This case leads to the real numbers. Indeed, $]a, b[$ contains a closed interval, and weak separability of C implies strong separability, whence we may apply Theorem 3.5.

(3) If $a < b$ and $]a, b[= \emptyset$, then every element of C has an upper neighbour and a lower neighbour, by homogeneity. Inductively, we may construct an isomorphism of $\mathbb{Z}$ onto a subchain $Z \subseteq C$ by insisting that the 'upper neighbour' relation be preserved. This subchain cannot be bounded, because its supremum or infimum could not have any lower or upper neighbours, respectively. Therefore, every $c \in C$ must coincide with its least upper bound in Z. It follows that $Z = C$, and the proof is complete.   □

We remark that $\mathbb{R}$ contains other subchains with rather strange properties: VAN MILL 1992 constructs an example of a homogeneous subchain

$X \subseteq \mathbb{R}$ which is dense in itself and *rigid* in the sense that the identity is the only automorphism fixing a point; the automorphism group of X is then sharply transitive on X.

A striking characterization of the real line is obtained in GUREVICH–HOLLAND 1981; compare also GLASS 1981. They prove the existence of a formula in the elementary language of groups such that $\mathbb{R}$ is the only homogeneous chain whose automorphism group satisfies this formula.

There is a similar result for the chain of rational numbers, but with a difference. The chain of real numbers is obtained from the chain of rational numbers by completion (see 42.11), hence the automorphism group of the chain $\mathbb{Q}$ can be identified with the group of all automorphisms of $\mathbb{R}$ that leave $\mathbb{Q}$ invariant. The same argument holds for the chain $\mathbb{R} \smallsetminus \mathbb{Q}$ of irrational numbers, hence the chains $\mathbb{Q}$ and $\mathbb{R} \smallsetminus \mathbb{Q}$ cannot be distinguished by properties of their automorphism groups.

3.8 Souslin's problem A weakly separable chain C satisfies *Souslin's condition* (also called the *countable chain condition*, SOUSLIN 1920): there is no uncountable set of pairwise disjoint, non-empty open intervals in C. Indeed, each of those intervals would contain an element of a fixed countable dense subset of C.

It has been conjectured that this condition is equivalent to topological separability (*Souslin's hypothesis*; see BIRKHOFF 1948 III.8, ALVAREZ 1999). Then we would obtain a variation of the preceding characterization of $\mathbb{R}$ (Theorem 3.7) with the Souslin condition in place of weak separability. It turned out soon that Souslin's hypothesis cannot be decided on the basis of a 'weak' set theory. Later, it was even shown that there are models of set theory satisfying the axiom of choice and the generalized continuum hypothesis (see 61.16), where Souslin's hypothesis is false (TENNENBAUM 1968 and RUDIN 1969).

The Souslin hypothesis is equivalent to the non-existence of *Souslin trees*, which are defined as follows. A Souslin tree is a partially ordered set $(T, \leq)$ of cardinality $\aleph_1$ such that (1) for each $x \in T$, the set $\{\, y \in T \mid y < x \,\}$ is well-ordered with respect to $\leq$, and (2) if $X \subseteq T$ is a chain (totally ordered by $\leq$) or an antichain (containing no pair of $\leq$-comparable elements), then Y is at most countable. For more information on the Souslin problem, see DEVLIN–JOHNSBRÅTEN 1974. In his article, FELGNER 1976 treats the Souslin hypothesis in the context of ordered algebraic structures. Among other things, he proves that an abelian ordered group that is dense in itself and satisfies Souslin's condition is separable. He also obtains some negative results.

Exercises

(1) Leaving the given ordering unchanged, split each real number x into a pair $\{\check{x}, \hat{x}\}$ with $\check{x} < \hat{x}$. Compare the properties of the chain $\check{\mathbb{R}}$ of the split reals with those of $\mathbb{R}$.

(2) Consider the set $\mathbb{R}^2$ with lexicographic ordering, that is, $(x, y) < (u, v)$ means that $x < u$ or ($x = u$ and $y < v$). Is there an injection $\mathbb{R}^2 \to \mathbb{R}$ which preserves the ordering?

4   Continued fractions

Continued fractions are a useful tool for a better understanding of the embedding of $\mathbb{Q}$ into $\mathbb{R}$, the approximation of irrational numbers by rational ones, the topology of $\mathbb{R} \smallsetminus \mathbb{Q}$, and related phenomena. The booklet by KHINCHIN 1964 gives an easy introduction to the subject; see also Chapter 5 of NIVEN 1956, Chapter X of HARDY–WRIGHT 1971, or the more recent text by ROCKETT–SZÜSZ 1992.

4.1 Definition   For each real number ζ, the process $(\star)$ defined by the conditions

$$\zeta = \zeta_0 = c_0 + 1/\zeta_1, \ c_0 \in \mathbb{Z}, \ 1 < \zeta_\nu = c_\nu + 1/\zeta_{\nu+1}, \ c_\nu \in \mathbb{N} \text{ for } \nu > 0$$

uniquely determines the (finite or infinite) sequences of integers c_ν and real numbers ζ_ν. If $\zeta \notin \mathbb{Q}$, then none of the ζ_ν is rational, and the process $(\star)$ does not terminate. For $\zeta \in \mathbb{Q} \smallsetminus \mathbb{Z}$, however, $\zeta_1 = a_1/a_2$ is a quotient of two coprime natural numbers and the c_ν are given by the Euclidean algorithm $a_\nu = c_\nu a_{\nu+1} + a_{\nu+2}$ with $a_\nu > a_{\nu+1} \in \mathbb{N}$. Hence $(\star)$ stops at some finite index n with $\zeta_n = c_n$, and one obtains a representation of ζ as a *finite continued fraction*

$$[c_0; c_1, c_2, \ldots, c_n] := c_0 + \cfrac{1}{c_1 + \cfrac{1}{c_2 + \cfrac{1}{\cdots + \cfrac{1}{c_n}}}} = p_n/q_n \ ,$$

where $p_n \in \mathbb{Z}$ and $q_n \in \mathbb{N}$ are integers without common factor. For $\zeta \notin \mathbb{Q}$ it will be seen below that the rational numbers p_n/q_n converge to ζ in a particularly nice manner.

The definition of $[c_0; c_1, c_2, \ldots, c_n]$ makes sense also if $c_n \geq 1$ is an arbitrary real number.

4.2   Observe that $[c_0; c_1, \ldots, c_{n+1}]$ is obtained from $[c_0; c_1, \ldots, c_n]$ by writing $c_n + 1/c_{n+1}$ in place of c_n. Let pro forma $p_{-1} = 1$, $q_{-1} = 0$

and put $p_0 = c_0$, $q_0 = 1$. Then the following recursion formulae can be proved by induction:

$$p_{\nu+1} = c_{\nu+1}p_\nu + p_{\nu-1}, \qquad q_{\nu+1} = c_{\nu+1}q_\nu + q_{\nu-1}$$
$$\text{and} \qquad p_{\nu+1}q_\nu - p_\nu q_{\nu+1} = (-1)^\nu \ . \tag{$\dagger$}$$

In fact, if ($\dagger$) is true for smaller indices, then

$$\frac{p_{\nu+1}}{q_{\nu+1}} = \frac{(c_\nu + c_{\nu+1}^{-1})p_{\nu-1} + p_{\nu-2}}{(c_\nu + c_{\nu+1}^{-1})q_{\nu-1} + q_{\nu-2}} = \frac{p_\nu + p_{\nu-1}/c_{\nu+1}}{q_\nu + q_{\nu-1}/c_{\nu+1}} = \frac{c_{\nu+1}p_\nu + p_{\nu-1}}{c_{\nu+1}q_\nu + q_{\nu-1}} \ .$$

The last part of ($\dagger$) follows inductively from the first. It shows that p_ν and q_ν are indeed coprime.

Equivalently,

$$\begin{pmatrix} p_\nu & p_{\nu-1} \\ q_\nu & q_{\nu-1} \end{pmatrix} = \begin{pmatrix} c_0 & 1 \\ 1 & \end{pmatrix} \begin{pmatrix} c_1 & 1 \\ 1 & \end{pmatrix} \cdots \begin{pmatrix} c_\nu & 1 \\ 1 & \end{pmatrix} \ . \tag{$\ddagger$}$$

The determinant of ($\ddagger$) yields the last part of ($\dagger$) once more. For later use, we rewrite this as

$$p_\nu/q_\nu - p_{\nu-1}/q_{\nu-1} = (-1)^{\nu-1}/q_\nu q_{\nu-1} \ .$$

Note that for $\nu \geq 1$ the q_ν form a strictly increasing sequence. Moreover, again by induction, $q_\nu/q_{\nu-1} = [c_\nu; c_{\nu-1}, \ldots, c_1]$.

The recursion formulae can also be expressed as follows: if

$$|c_1, c_2, \ldots, c_\nu| = \begin{vmatrix} c_1 & 1 & & & & & \\ -1 & c_2 & 1 & & & & \\ & -1 & c_3 & 1 & & & \\ \cdots & \cdots & \cdots & \cdots & \cdots & \cdots & \cdots \\ & & & & -1 & c_{\nu-1} & 1 \\ & & & & & -1 & c_\nu \end{vmatrix},$$

then $p_\nu = |c_0, c_1, \ldots, c_\nu|$ and $q_\nu = |c_1, c_2, \ldots, c_\nu|$.

4.3 Convergence   If $\zeta \in \mathbb{R} \setminus \mathbb{Q}$, if the c_ν are defined by ($\star$), and if $[c_0; c_1, \ldots, c_n] = s_n$, then $s_{2\nu} < \zeta < s_{2\nu+1}$: in fact, we have $\zeta = [c_0; c_1, \ldots, c_{n-1}, \zeta_n]$, and continued fractions are monotone in each variable, increasing for even indices, decreasing for odd indices. In 4.2 it has been shown that $s_{2\nu+1} - s_{2\nu} = 1/q_{2\nu+1}q_{2\nu} < q_{2\nu}^{-2} \leq (2\nu)^{-2}$ converges to zero. Hence the s_ν converge alternatingly to ζ, and one may write ζ as an *infinite continued fraction* $\zeta = [c_0; c_1, c_2, \ldots]$.

For an arbitrary sequence $(c_\nu)_\nu \in \mathbb{Z} \times \mathbb{N}^{\mathbb{N}}$, let $s_\nu = [c_0; c_1, \ldots, c_\nu] = p_\nu/q_\nu$ as before. Then again $|s_{\nu+1} - s_\nu| < q_\nu^{-2}$, and the *approximating fractions* s_ν converge to some real number γ. If p/q is any approximating fraction, then $0 < |q\gamma - p| < q^{-1}$, as we have seen. Thus 1 and γ generate a dense subgroup of $(\mathbb{R}, +)$ and γ is irrational; see 1.6a. Moreover, the process $(\star)$ applied to γ produces the sequence $(c_\nu)_\nu$. This follows by induction, using the obvious fact that $[c_0; c_1, c_2, \dots] = c_0 + [c_1; c_2, \dots]^{-1}$.

4.4 Example If $\zeta = \sqrt{3}$, then $\zeta_1 = (\sqrt{3} - 1)^{-1} = (\sqrt{3} + 1)/2$, $\zeta_2 = 2(\sqrt{3} - 1)^{-1} = \sqrt{3} + 1$, $\zeta_3 = \zeta_1$, $\zeta_4 = \zeta_2$, and hence $\zeta = [1; 1, 2, 1, 2, 1, \dots]$, $s_7 = 97/56$, $s_7 - \sqrt{3} < 10^{-4}$. Conversely, let $z = [0; 1, 2, 1, 2, \dots]$. Then $z = 1/(1 + 1/(2 + z)) = (z + 2)/(z + 3)$, hence $(z + 1)^2 = 3$.

4.5 Remark The example is a typical case of the following theorem: *A continued fraction $c = [c_0; c_1, c_2, \dots]$ is finally periodic if, and only if, $\mathbb{Q}(c)$ is quadratic over $\mathbb{Q}$*; see KHINCHIN 1964 §10 or ROCKETT–SZÜSZ 1992 Chapter III for proofs.

4.6 Theorem (Hurwitz 1891) *One of any three consecutive approximating fractions satisfies even* $|\zeta - p_\nu/q_\nu| < \left(\sqrt{5}\, q_\nu^2\right)^{-1}$.

For a *proof* see KHINCHIN 1964 §7 Theorem 20 or ROCKETT–SZÜSZ 1992 p. 80; compare also BENITO–ESCRIBANO 2002 and Exercise 3.

The approximation of real numbers by rational numbers has been studied by many mathematicians, including Thue, Siegel, Dyson, Gelfond and T. Schneider. The following famous result says that only transcendental numbers admit many rational approximations that are better than 4.6. This result is due to ROTH 1955 (see also ROTH 1960 and, for a detailed proof, CASSELS 1957 Chapter VI).

4.7 Theorem (Roth 1955) *For any real algebraic number α and any real number $\mu > 2$, there are only finitely many rational numbers p/q such that $|\alpha - p/q| < q^{-\mu}$.*

In contrast, for $\mu \le 2$ and every irrational number α there are infinitely many rational numbers p/q such that $|\alpha - p/q| < q^{-\mu}$. This can be obtained for instance by expanding α into a continued fraction; see 4.3.

4.8 Ordering For infinite continued fractions a and b, the alternating lexicographic ordering

$$a < b \rightleftharpoons \begin{cases} a_\mu < b_\mu \wedge \mu \in 2\mathbb{Z} \\ a_\mu > b_\mu \wedge \mu \notin 2\mathbb{Z} \end{cases}, \quad \mu = \mu(a, b) = \min\{\nu \ge 0 \mid a_\nu \ne b_\nu\},$$

gives the usual ordering of the irrational numbers; it can be extended to the rationals by writing $[c_0; c_1, \dots, c_n] = [c_0; c_1, \dots, c_n, \infty]$, where

$\mathbb{R} < \infty$. Thus, mapping $[c_0; c_1, c_2, \ldots]$ to its limit is a homeomorphism of the set of all infinite continued fractions onto the set $\mathfrak{I}$ of the irrational reals with respect to the order topologies.

4.9 Proposition *A lower bound for the distance of an irrational number ζ and an approximating fraction is given by* $|\zeta - p_\nu/q_\nu| > q_\nu^{-1}(q_\nu + q_{\nu+1})^{-1}$.

Proof Let $\zeta = [c_0; c_1, c_2, \ldots]$. If ν is an even integer, then $p_\nu/q_\nu = [c_0; c_1, \ldots, c_\nu] < \zeta = [c_0; c_1, \ldots, c_\nu, \zeta_{\nu+1}]$. The interval $[p_\nu/q_\nu, \zeta]$ contains the number $[c_0; c_1, \ldots, c_\nu, c_{\nu+1} + 1]$, which by 4.2($\dagger$) is equal to $(p_{\nu+1} + p_\nu)/(q_{\nu+1} + q_\nu)$, and the assertion follows. The odd case is treated analogously. $\square$

Now we explain how continued fractions can be used to describe the topology of $\mathfrak{I} = \mathbb{R} \setminus \mathbb{Q}$.

4.10 Topology Consider the set of all infinite continued fractions as a subspace of the (Tychonoff) power $\mathbb{Z}^{\mathbb{N}}$, where $\mathbb{Z}$ carries the discrete topology. This means that a typical neighbourhood of $[c_0; c_1, c_2, \ldots]$ consists of all $x = [x_0; x_1, x_2, \ldots]$ such that $x_\kappa = c_\kappa$ for $\kappa \leq m$ and some natural number m. If m is even, this is equivalent to

$$[c_0; c_1, \ldots, c_m] < x < [c_0; c_1, \ldots, c_{m-1}, c_m + 1];$$

analogous relations hold for odd m. Therefore, $\mathbb{Z}^{\mathbb{N}}$ induces the same topology as the ordering, and $\mathfrak{I}$ is homeomorphic to $\mathbb{Z} \times \mathbb{N}^{\mathbb{N}}$. As both $\mathbb{Z}$ and $\mathbb{N}$ are countable, discrete spaces and hence $\mathbb{Z} \approx \mathbb{N}$, the space $\mathfrak{I}$ is also homeomorphic to $\mathbb{N}^{\mathbb{N}}$.

4.11 Metric The discrete topology on the νth factor of $\mathbb{Z}^{\mathbb{N}}$ is given by the metric d_ν that has only the values 0 and 1. The space $\mathbb{Z}^{\mathbb{N}}$ is then easily seen to be complete with respect to the metric $d = \sum_\nu 2^{-\nu} d_\nu$; note that d induces the product topology on $\mathbb{Z}^{\mathbb{N}}$. Together with 4.10, this yields a complete metric for the subspace $\mathfrak{I}$ of $\mathbb{R}$.

Remark A more general sort of continued fractions (with numerators b_ν instead of 1) is discussed in BEARDON 2001.

Exercises

(1) Let $[0; c_1, \ldots, c_\nu] = p_\nu/q_\nu$ with integers $p_\nu, q_\nu \in \mathbb{N}$ and $\nu \geq 4$. Then $q_\nu \geq 5 \cdot (3/2)^{\nu-4}$ and $q_{\nu-1}q_\nu \geq 6 \cdot (5/2)^{\nu-3}$.

(2) Show that $p_\nu q_{\nu-2} - p_{\nu-2}q_\nu = (-1)^\nu c_\nu$.

(3) Prove that one of any two consecutive approximating fractions for an irrational number ζ satisfies $|\zeta - p/q| < (2q^2)^{-1}$.

5   The real numbers as a topological space

The chain $\mathbb{R}$ can be endowed with the topology induced by the ordering
(see 3.1), and thus becomes a topological space. The open intervals form
a basis of this topology; in other words, every open set is a union of open
intervals. The same topology can be generated by the metric $d(x,y) =$
$|x - y|$. This means that the open balls $U_\varepsilon(y) = \{\, x \in \mathbb{R} \mid d(x,y) < \varepsilon \,\}$
for all $y \in \mathbb{R}$ and all $\varepsilon > 0$ form a basis, as well.

Like every metric space, $\mathbb{R}$ is a *Hausdorff space*, i.e., every two points
possess disjoint neighbourhoods. In particular, each point $r \in \mathbb{R}$ forms
a closed set, and $\mathbb{R}$ is a T_1-*space*. In passing, we mention some stronger
consequences of being metric, namely, normality and paracompactness;
see, for example, DUGUNDJI 1966 Section IX.5.

Density of the rational numbers in the chain $\mathbb{R}$ (see 3.1) implies that
each point in $U_\varepsilon(y)$ belongs to some ball $U_r(q)$ contained in $U_\varepsilon(y)$ and
having rational centre q and rational radius r. In other words, we have
found another basis for the topology of $\mathbb{R}$ consisting of countably many
open sets. This property is stronger than separability (existence of a
countable dense subset in the topological sense, see 3.2; the rational
numbers form such a set). Whenever we speak of the real numbers as
a topological space without further specification, we mean this 'natural'
topology. We record the properties just recognized.

5.1 Proposition   *The topological space $\mathbb{R}$ of real numbers has a countable basis; in particular, $\mathbb{R}$ is separable.*   □

We call a topological space *compact*, if every covering of the space by
open sets contains a finite subcovering; the Hausdorff separation property is not required.

In order to determine the compact subsets of $\mathbb{R}$, we characterize the
chains that are compact with respect to their order topologies. We call
a chain *bounded* if it has a smallest and a greatest element. A chain that
is both bounded and complete is *unconditionally complete*, i.e., every
non-empty subset has a least upper bound and a greatest lower bound.

5.2 Theorem   *A chain C is compact with respect to the topology
induced by the ordering if, and only if, it is unconditionally complete.*

Proof Suppose that C is compact. In search of a least upper bound of a
non-empty set $A \subseteq C$, we consider a set of open subsets of C, consisting
of the intervals $]\,,a[$ for all $a \in A$ together with the intervals $]b,\,[$ for all
$b \geq A$. Note that an element of C not contained in any of these intervals
is a least upper bound of A. Therefore, we may assume that we have

constructed an open cover of C. There is a finite subcover, and then in fact $C =]\,,a_0[\,\cup\,]b_0,\,[$. However, neither of the two intervals contains a_0, which is a contradiction. Similarly, A has a greatest lower bound.

Now suppose that C is unconditionally complete and let $\mathcal{U}$ be an open cover of C. We have to show the existence of a finite subcover. The smallest element $s \in C$ belongs to some $U \in \mathcal{U}$, and U contains an interval $[s, x[$. *A fortiori*, this interval can be covered by finitely many sets from $\mathcal{U}$, and the non-empty set X of all x' sharing this property has a least upper bound y. This element, too, belongs to an open interval I contained in some set $V \in \mathcal{U}$. That interval I is contained in X, because we can add V to some finite cover of $[s, t[$ for $t \in I \cap X$. It follows that $y \in X$ and that y must be the greatest element of C, or else we would have a contradiction to the definition of y. We have proved that $C = [s, y]$ is covered by finitely many sets from $\mathcal{U}$.   □

A topological space is said to be *locally compact* if every point has arbitrarily small compact neighbourhoods.

5.3 Theorem   *A set $A \subseteq \mathbb{R}$ of real numbers is compact if, and only if, it is closed and bounded. In particular, $\mathbb{R}$ is locally compact, but not compact.*

Proof For an unbounded set A, the sets $A \cap \,]-n, n[$, $n \in \mathbb{N}$, form an open cover without a finite subcover. If A is not closed and x is a boundary point not belonging to A, then the open cover formed by the sets $A \smallsetminus [x - 1/n, x + 1/n]$, $n \in \mathbb{N}$, lacks a finite subcover. This completes the first half of the proof.

Conversely, let A be bounded and closed. Then A is a closed subset of some interval $I = [x, y]$, which is compact according to 5.2. Let $\mathcal{U}$ be an open cover of A. Each $U \in \mathcal{U}$ is the intersection of some open set $U' \subseteq I$ with A, and the sets U', $U \in \mathcal{U}$, together with $I \smallsetminus A$, form an open cover $\mathcal{U}'$ of I. A finite subcover of $\mathcal{U}'$ yields a finite subcover of A by taking intersections with A.   □

Next we turn to the connectedness properties of $\mathbb{R}$. A topological space is said to be *locally connected* if every point has arbitrarily small connected neighbourhoods. In the following theorem, the term *interval* refers to all sorts of intervals, closed, open, and half open, bounded or unbounded. In particular, $\mathbb{R}$ itself is considered as an interval.

5.4 Theorem   *The connected subsets of $\mathbb{R}$ are precisely the intervals. In particular, $\mathbb{R}$ is connected and locally connected.*

Proof Completeness of the chain $\mathbb{R}$, together with strong density, implies that every interval is connected; see 3.3. Conversely, a set A that is not an interval contains elements a, b separated by some $x \notin A$, i.e., $a < x < b$. Then $A = (A \cap \,] \,, x[) \cup (A \cap]x, \,[)$ is not connected.   □

5.5 Connected components By definition, a *connected component* of a topological space X is a connected subset that is maximal, i.e., not contained in a bigger one. Since the closure of every connected set is connected, all connected components are closed. Distinct components do not intersect, or else their union would be connected. Hence every space is the disjoint union of all its connected components. If X happens to be locally connected, then the connected component containing a point x contains every connected neighbourhood of x, hence the components of such a space are open sets. As a consequence, we obtain a description of the open subsets of $\mathbb{R}$.

5.6 Theorem *Every open subset of $\mathbb{R}$ is a disjoint union of at most countably many pairwise disjoint open intervals.*

Proof An open subset $U \subseteq \mathbb{R}$ is locally connected, just like $\mathbb{R}$ itself. Therefore, all connected components of U are open sets. By 5.4, these sets are (open) intervals, and they are pairwise disjoint (being connected components). Each of these intervals contains a rational number. In view of their disjointness, this implies that there can be only countably many of them.   □

Characterizing the real line, the arc, and the circle

We remark first that the real line $\mathbb{R}$ is homeomorphic to any open interval. For instance, a homeomorphism

$$\mathbb{R} \approx \,]{-}\pi, \pi[$$

is given by the function arctan. Apart from the real line, we shall be concerned with some other spaces. The *arc* is the closed unit interval $I = [0, 1] \subseteq \mathbb{R}$ or any space homeomorphic to it; also the half open interval $[0, 1[$ will be of interest to us. The *circle* is the subset of $\mathbb{R}^2$ given by

$$\mathbb{S}_1 = \{\, (x, y) \in \mathbb{R}^2 \mid x^2 + y^2 = 1 \,\}.$$

Our aim is to characterize these spaces among all topological spaces, using the properties we have detected so far. Besides their apparent similarity, we have more reasons to treat the spaces together. The intervals are subspaces of the real line, and, as we shall show later (5.16), the circle can be considered as a quotient space of $\mathbb{R}$.

Some of the results we prove are 'best possible' in the sense that every relaxation of conditions leads to counter-examples; we shall demonstrate this in the next subsection. Other results are not best possible. The need to optimize the setup as a whole made it impossible to optimize every single result, so that in some cases there will be a superfluous hypothesis. There, we shall give references to the literature to make up for the defects. Before we can prove theorems, we need some auxiliary notions and results. For better understanding, the reader is advised to visualize each of these in the cases $X = \mathbb{R}$ and $X = [0, 1]$.

5.7 Definition A subset A in a connected topological space X is said to be *separating*, if $X \smallsetminus A$ is disconnected. In particular, we speak about a separating point x when $A = \{x\}$ is separating. Other points are said to be *non-separating*, or we call them *end points*. We say that a point x *separates* two other points if these lie in different connected components of $X \smallsetminus \{x\}$.

Every disconnected space admits a *separation*, that is, the space can be represented as a disjoint union of two non-empty open subsets U, V. Somewhat loosely, we speak of 'a separation $X = U \cup V$'. We note that in a locally connected space, every connected component is open. The union of all other components is open, too, hence every component U of a locally connected space X gives rise to a separation $X = U \cup V$.

5.8 Lemma *Suppose that X is a connected space, $A \subseteq X$ is any subset, and $X \smallsetminus A = U \cup V$ is a separation.*
  (a) *If A is connected, then $U \cup A$ is connected.*
  (b) *If A is closed, then A contains the boundary of U relative to X.*
  (c) *If $A = \{a\}$ is a singleton and X is a T_1-space, then $\overline{U} = U \cup \{a\}$, and this set is connected.*

Proof Suppose that A is connected and $U \cup A = S \cup T$ is a separation. The connected set A must be contained in S or in T, let us say, in S; then $T \subseteq U$. Therefore, T is closed and open in each of the following spaces: in $U \cup A$, in U (which is closed and open in $X \smallsetminus A$) and hence in $X \smallsetminus A$, and finally in the union of these spaces, X; compare Exercise 1. This contradiction proves (a).

If A is closed, then U and V are open in X, and assertion (b) follows. Finally, (a) and (b) together imply (c).   □

The following shows a first glimpse of how the ordering of the real line reflects in its topology.

5.9 Lemma *Consider two points x, y in a connected T_1-space X, and let $X \smallsetminus \{x\} = U_x \cup V_x$ and $X \smallsetminus \{y\} = U_y \cup V_y$ be separations such that $y \in U_x$ and $x \in U_y$. Then $V_x \subseteq U_y$ and $V_y \subseteq U_x$, and no other inclusions occur among these sets.*

Proof By Lemma 5.8, the closure $\overline{V}_x = V_x \cup \{x\}$ is a connected set not containing y, hence it is contained in U_y or in V_y. Since $x \notin V_y$, only the first possibility can occur. Likewise, U_x contains V_y. The inclusion is proper since $y \in U_x \smallsetminus V_y$, hence U_x is not contained in either of the two sets U_y, V_y. □

We are ready for the first characterization result. It is easy to distinguish between the spaces $]0, 1[\approx \mathbb{R}$, $[0, 1]$, and $]0, 1]$ (by the number of non-separating points). Therefore, it is quite enough to characterize the three spaces together, as is done in the following theorem due to KOWALSKY 1958. Other, similar results will follow.

5.10 Theorem *A topological space X is homeomorphic to one of the spaces $]0, 1[\approx \mathbb{R}$, $[0, 1]$, or $]0, 1]$ if, and only if, it has at least two points and has the following four properties.*

 (i) *X is separable.*
 (ii) *X is a T_1-space, that is, all points of X are closed.*
 (iii) *X is connected and locally connected.*
 (iv) *Among any three connected proper subsets of X, there are two which fail to cover X.*

Proof (1) The necessity of the first three conditions has already been shown (5.1, 5.4). A connected proper subset of any of the three types of intervals is a subinterval which has a non-trivial bound on at least one side. Among any three such subintervals, some two are bounded on the same side and, hence, fail to cover the entire interval. In the remainder of this proof, we shall be concerned with the sufficiency of the four conditions.

(2) *For each $x \in X$, the closure of any connected component C of $X \smallsetminus \{x\}$ equals $C \cup \{x\}$ and is connected.*

By (ii), the complement $X \smallsetminus \{x\}$ is open, and we may assume that $C \neq X \smallsetminus \{x\}$. By local connectedness, there is a separation $X \smallsetminus \{x\} = C \cup D$. The assertion now follows from 5.8c.

(3) *The complement of a point in X has at most two connected components.*

Here we use condition (iv). If $X \smallsetminus \{x\}$ has more than two components, we choose C_1, C_2 among them and denote the union of all the others by Y. Then we obtain three proper subsets, any two of which cover X, by taking $X \smallsetminus C_1$, $X \smallsetminus C_2$, and $X \smallsetminus Y$. By 5.8c, each of these sets is connected, contrary to (iv).

Applying (iv) to three connected point complements, we obtain the next observation.

(4) *At most two points have connected complements.*

In particular, there are separating points, because the connected T_1-space X cannot have only two points.

(5) The connected components of $X \smallsetminus \{x\}$ will be called *x-parts* for brevity. For a separating point x, the two x-parts form a separation of $X \smallsetminus \{x\}$. If x is an end point, then also $\{x\}$ will be considered as an x-part, so that each point x defines precisely two x-parts. If $x, y \in X$ are two distinct points, then we denote the x-part containing y by $T(x, y)$, and the other x-part by $A(x, y)$. The letters T and A here stand for 'pointing *towards* y' and 'pointing *away* from y', respectively. The inclusion relations among these sets will eventually lead to an ordering relation on X. The following holds trivially for end points and follows from 5.9 in the case of separating points.

(6) If $x, y \in X$ are distinct points, then $A(x, y) \subseteq T(y, x)$ and $A(y, x) \subseteq T(x, y)$, and no other inclusion relations hold among the x-parts and the y-parts.

(7) Two parts defined by distinct points will be called *compatible* if one of them contains the other. (6) may be rephrased by saying that, given any x-part P and a point $y \neq x$, there is always precisely one y-part compatible with P. We choose an arbitrary point $e \in X$ and an e-part P_e. For every other point x, we let P_x denote the unique x-part compatible with P_e. We claim:

(8) *The parts P_x are all mutually compatible.*

For x, y, e distinct from each other, we have to show that P_x and P_y are compatible. We know that both of them are compatible with P_e. If $P_x \subseteq P_e \subseteq P_y$, then no argument is needed. The difficult cases are that P_x and P_y are both contained in, or both contain, P_e.

We assume that none of P_x, P_y contains the other, and we look at the two sets $M_1 = P_x \cup P_y^{\complement}$ and $M_2 = P_y \cup P_x^{\complement}$, where $Y^{\complement}$ denotes the complement of $Y \subseteq X$. Both sets are non-empty and connected (for instance, P_x must meet $P_y^{\complement}$). Moreover, M_1 is a proper subset of X, or else we have $P_y \subseteq P_x$. Likewise, we also obtain that M_2 is a proper subset.

If $P_x, P_y \subseteq P_e$, then the three sets M_1, M_2 and P_e contradict condition (iv). If $P_e \subseteq P_x, P_y$, then the same is true for M_1, M_2 and $P_e^{\complement}$, and this proves (8).

(9) We have shown that the set $\mathcal{P}$ of all parts P_x, $x \in X$, is totally ordered by inclusion. We have a bijection $x \mapsto P_x$ from $\mathcal{P}$ onto X (injectivity follows from (2)), and hence we can carry the total ordering to X:

$$x \leq y \;\rightleftharpoons\; P_x \subseteq P_y \;.$$

From (6) it follows that $x < y \Leftrightarrow y \notin P_x \cup \{x\} \Leftrightarrow x \in P_y \smallsetminus \{y\}$.

(10) *The topology of X coincides with the topology obtained from the ordering $\leq$.*

The order topology is generated by the intervals of type $]x, [$ and $] , x[$, where $x \in X$ is a separating point. By (9), these intervals coincide with the connected components of the complement $X \smallsetminus \{x\}$, and these are open sets of X because X is a locally connected T_1 space. This proves that the identity is a continuous map from X with the given topology to X with the order topology.

Conversely, we show that every open neighbourhood U of $x \in X$ with respect to the given topology contains a neighbourhood of x with respect to the order topology. We may assume that U is connected. By continuity of the identity map, U is also connected with respect to the order topology. In every order topology, a connected set must be an interval. Suppose that the interval U contains one of its end points, y. If $y \neq x$, then we may replace U by $U \smallsetminus \{y\}$. If $U =]a, x]$ and x is not an end point of X, then $X =] , x] \cup]x, [$ is disconnected. This shows that U is open in the order topology and proves step (10).

(11) It remains to determine the isomorphism type of X as a chain. Here we use the separability of X for the first time. Any end points of X will be deleted at first. After the remainder of X has been identified as $]0, 1[$, the end points may be restored as smallest or largest elements of the chain.

We know that the order topology is connected, and we shall show that the chain X is strongly separable. Then condition (c) of the characterization theorem 3.5 is satisfied and our proof is complete. Separability of the order topology implies that every non-empty open interval contains an element of some countable set $A \subseteq X$. By connectedness, and because we deleted end points, the open intervals $] , x[,]x, [$ and $]x, y[$ for $x < y$ are non-empty and contain elements of A. This means strong separability of the chain X. $\square$

The conditions characterizing $\mathbb{R}$ according to the last theorem require information about every connected subset of X, and this may be asking a bit much. We therefore give now another characterization that uses information on the connected components of point complements only. This result was proved directly in the lecture notes SALZMANN 1971, without using Kowalsky's theorem 5.10.

5.11 Theorem *A topological space X is homeomorphic to one of the spaces $]0,1[\approx \mathbb{R}$, $[0,1]$, or $]0,1]$ if, and only if, it has at least four points and has the following four properties.*

(i) *X is separable.*

(ii) *X is a T_1-space, that is, all points of X are closed.*

(iii) *X is connected and locally connected.*

(iv') *Among any three points of X, there is one which separates the other two (see 5.7).*

Proof We show that a space X with these four properties satisfies the conditions of 5.10.

(1) The closure of any connected component C of $X \smallsetminus \{x\}$ is $\overline{C} = C \cup \{x\}$, as in step (2) of the last proof. From (iv') it follows immediately that there are at most two non-separating points. Next, we prove that $X \smallsetminus \{x\}$ has at most two components. Indeed, if C_1, C_2, C_3 are distinct components, then no point $x_1 \in C_1$ can separate $x_2 \in C_2$ from $x_3 \in C_3$, because these points belong to the connected set $\overline{C}_2 \cup \overline{C}_3$, which does not contain x_1.

(2) The last fact allows us to introduce the same notation $A(x,y)$ and $T(x,y)$ as in the last proof (step (5)). We also have the property $A(x,y) \subseteq T(y,x)$ obtained from 5.9. We conclude that the only possibility to cover X by an x-part together with a y-part is to use $T(x,y)$ together with $T(y,x)$.

(3) We show that X satisfies condition (iv) of 5.10. A proper connected subset A of X is contained in an open x-part, where $x \notin A$. Hence we may assume that the three sets appearing in condition (iv) are three open parts with respect to points x, y, z. Two open parts with respect to the same point do not cover X, hence we may assume that we have three distinct points. If condition (iv) is violated, then step (2) shows that each of the three given parts contains both of the points defining the other parts. But by (iv'), one of the points, say y, separates the other two, i.e., x and z lie in different parts with respect to y. This contradiction completes the proof. □

If compactness is assumed, then the separation conditions (iv) or (iv′) can be weakened considerably. This is based on the following auxiliary result, taken from CHRISTENSON–VOXMAN 1977.

5.12 Lemma *Suppose that p is a separating point of a compact, connected T_1-space X and let $X \smallsetminus \{p\} = U \cup V$ be a separation. Then both U and V contain at least one non-separating point of X.*

Proof We assume that all points of U are separating and aim for a contradiction. For each $x \in U$, choose a separation (U_x, V_x) such that $p \in V_x$. By 5.9 we have $U_x \subseteq U$. The set $\mathcal{U} = \{U_x \mid x \in U\}$ is partially ordered by inclusion. By Hausdorff's maximal chain principle (or equivalently, by Zorn's lemma, see the beginning of Section 61), there exists a maximal subchain $\mathcal{W} \subseteq \mathcal{U}$. We shall show (by an indirect argument) that $\bigcap \mathcal{W} = \emptyset$ and then derive a final contradiction from that result.

We set $W = \{x \in U \mid U_x \in \mathcal{W}\} \subseteq U$. For $z \in \bigcap \mathcal{W}$ and $w \in W$, we have $z \in U_w \subseteq U$. Hence, we may form U_z, and we shall show that $U_z \subseteq U_w \smallsetminus \{z\}$. This tells us that $\mathcal{W} \cup \{U_z\}$ is a chain larger than $\mathcal{W}$, contrary to maximality. We know that $z \in U_w$, whence the connected set $V_w \cup \{w\}$ is contained in $X \smallsetminus \{z\} = U_z \cup V_z$. The former set meets V_z (in p), hence it is contained in V_z and, in particular, $w \in V_z$. Therefore $U_z \cup \{z\}$ is contained in $X \smallsetminus \{w\}$ and meets U_w (in z), so that finally $U_z \subseteq U_w \smallsetminus \{z\}$.

We have just seen that $\bigcap \mathcal{W} = \emptyset$, and we deduce next that the chain of closed sets $\{U_w \cup \{w\} \mid U_w \in \mathcal{W}\}$ has empty intersection, too — which contradicts compactness. An element belonging to that intersection can only be a point $w \in W$. The set U_w is not a smallest element of $\mathcal{W}$, and there is a proper subset U_v of U_w. Then $w \notin U_v \cup \{v\}$, and our chain of closed sets has empty intersection. □

5.13 Theorem *A topological space X of cardinality ≥ 2 is homeomorphic to the arc $[0, 1]$ if, and only if, X is compact, T_1, separable, connected and locally connected, and X contains at most two non-separating points.*

Proof From Lemma 5.12, we deduce that (i) there are exactly two non-separating points $a, b \in X$, and that (ii) every other point $x \in X \smallsetminus \{a, b\}$ separates a from b. It follows that no proper connected subset can contain both a and b, hence the space has property (iv) and the assertion follows from 5.10. □

5.14 Theorem   *A topological space X of cardinality ≥ 2 is homeomorphic to the circle $\mathbb{S}_1$ if, and only if, X is compact, T_1, separable, connected and locally connected, and every pair of distinct points separates X.*

Proof (1) According to Lemma 5.12, there exist two non-separating points $p, q \in X$. Let A be a connected component of $X \smallsetminus \{p, q\}$. Since X is locally connected, there is a separation (A, B) of $X \smallsetminus \{p, q\} = (X \smallsetminus \{p\}) \smallsetminus \{q\}$. Using Lemma 5.8a, we conclude that $A \cup \{q\}$ is connected. Likewise, $A \cup \{p\}$ and, hence, $\widehat{A} = A \cup \{p, q\}$ are connected.

(2) *If $\widehat{A}$ is separated by each point $a \in A$, then there is a homeomorphism $[0,1] \to \widehat{A}$ sending $\{0,1\}$ to $\{p,q\}$.* We cannot use 5.13 directly in order to show this, as $\widehat{A}$ might not be locally connected at p or at q. But A itself is connected and locally connected, and each $a \in A$ separates p from q; this follows from 5.12, applied to the space $\widehat{A}$, which is compact by 5.8. Therefore, no proper connected subset of A accumulates at both p and q, hence A has property (iv), and 5.10 shows that $A \approx \,]0, 1[$.

We identify A with $]0, 1[$, and we show next that p and q are attached to A in the right way. We have seen that for each $a \in \,]0, 1[$, there is an (open) separation $\widehat{A} \smallsetminus \{a\} = U \cup V$ such that $p \in U$ and $q \in V$. Each connected component of $]0, 1[\, \smallsetminus \{a\}$ is contained in U or in V. Hence, we may adjust the notation such that $U = \{p\} \cup \,]0, a[$ and $V = \,]a, 1[\, \cup \{q\}$. Therefore, the standard neighbourhoods of 0 and 1 correspond to open sets of $\widehat{A}$. Conversely, we note that $K = \widehat{A} \smallsetminus U = [a, 1[\, \cup \{q\}$ is compact. If W is an open neighbourhood of q, then $K \smallsetminus W$ is a compact subset of $[a, 1[$, and hence W contains a standard neighbourhood.

(3) If B is a connected component of $X \smallsetminus \{p, q\}$ that is not separated by some $b \in B$, then all other connected components C satisfy the assumptions of step (2), or else there is a point $c \in C$ such that $\widehat{C} \smallsetminus \{c\}$ is connected; denoting by E the union of all sets $\widehat{D}$ obtained from the remaining components, we infer that $X \smallsetminus \{b, c\} = (\widehat{B} \smallsetminus \{b\}) \cup (\widehat{C} \smallsetminus \{c\}) \cup E$ is connected, contrary to our hypothesis. It follows that $\widehat{C} \approx [0, 1]$ in this case. However, this implies that the connected components of $\widehat{C} \smallsetminus \{c\}$ all intersect $\widehat{B} \smallsetminus \{b\}$ (in p or in q), and that $X \smallsetminus \{b, c\}$ is connected. This is impossible.

(4) We have shown in (3) that $\widehat{B} \approx [0, 1]$ for each connected component of $X \smallsetminus \{p, q\}$, where the points $0, 1$ correspond to p, q. Clearly, the hypothesis implies that there are precisely two such components, and mapping each of the two intervals onto one half of the circle in a compatible way we obtain a homeomorphism $X \to \mathbb{S}^1$. $\square$

5.15 Remarks and variations The last two results hold for metric spaces without the assumption of local connectedness. Note that a compact space is metrizable if, and only if, it is a Hausdorff space with a countable basis for its topology. In this form, 5.13 was given by MOORE 1920; see CHRISTENSON–VOXMAN 1977 9.A.8 for a proof in modern language. The methods are very similar to the ones used here in the proof of 5.10, but it seems to be impossible to obtain all these results simultaneously. The variation of 5.14 without local connectedness is an exercise in CHRISTENSON–VOXMAN 1977. Our proof is more complicated since we need local connectedness when we apply 5.13.

KOWALSKY 1958 proves a weak form of 5.13. A characterization of the real line in the spirit of 5.13 is given by WARD 1936. He requires that the complement of any point has exactly two connected components, and that the whole space is metric, separable, and locally connected. This result is certainly not true without the condition of metrizability; compare Example 5.23. The real line has also been characterized in terms of the cardinalities of the boundaries of connected open subsets; see JONES 1939. His result implies Kowalsky's theorem 5.10 in the case of Hausdorff spaces.

In the context of manifolds, we shall now present another characterization of $\mathbb{R}$ and of the circle.

5.16 The circle as a quotient of $\mathbb{R}$ There is a surjective map $p :$ $\mathbb{R} \to \mathbb{S}_1$ defined by $p(t) = (\cos 2\pi t, \sin 2\pi t)$. If we identify $\mathbb{R}^2$ with $\mathbb{C}$, the set of complex numbers, via $(x, y) \mapsto x + iy$, then we may write $p(t) = e^{2\pi it}$. The map p relates the topologies of $\mathbb{R}$ and $\mathbb{S}_1$ in a very efficient way. We say that p is an *identification map* or that $\mathbb{S}_1$ carries the *quotient topology* with respect to p, and this means that a subset of $\mathbb{S}_1$ is open if, and only if, its inverse image under p is open in $\mathbb{R}$. The 'only if' part of this statement expresses the well known continuity of p. The 'if' part follows from the fact that the restriction of p to any open interval of length 1 is a homeomorphism onto its (open) image, as can be seen from the existence of complex logarithms.

This last fact also shows that $\mathbb{S}_1$ is a *one-dimensional manifold* (or 1-*manifold* for short). A Hausdorff space is said to be an n-manifold if each point has an open neighbourhood homeomorphic to $]0, 1[^n$ (or, equivalently, to $\mathbb{R}^n$). We shall prove the following.

5.17 Theorem *Every separable, connected one-dimensional manifold is homeomorphic to $\mathbb{R}$ or to $\mathbb{S}_1$.*

We remark that separability is essential. In 5.25 below, we shall introduce the 'long line', a space which is a non-separable 1-manifold (see 5.26). The *proof* of 5.17 requires some auxiliary notions and results; it will be given after 5.20. First recall that an arc in a topological space X is a subset $A \subseteq X$ homeomorphic to the closed interval $[0, 1]$. The non-separating points of A (corresponding to 0 and 1) are called the end points of the arc.

5.18 Lemma *Suppose that a topological space X contains an arc A with end point a and an arc B with end point $b \neq a$. If the intersection of the two arcs is non-empty, then their union contains an arc with end points a, b.*

Proof Let $f : [0, 1] \to A$ be a homeomorphism which sends 0 to a. By compactness, there is a smallest $t \in [0, 1]$ such that $f(t) \in B$. Then the subarc $f([0, t])$ together with the unique subarc C of B containing $f(t)$ and b form the desired arc from a to b. To show that we really get an arc, we replace f on the interval $[t, 1]$ with a parametrization of C running from $f(t)$ to b. — This argument will be used tacitly in the future.   □

5.19 Definition A topological space X is said to be *arcwise connected* if every two distinct points $x, y \in X$ are 'joined' by an arc $A \subseteq X$ with end points x, y. In general, the maximal arcwise connected subsets of X are called the *arc components* of X. It follows from 5.18 that two arc components are either disjoint or equal. If a space is *locally arcwise connected* (every point has arbitrarily small arcwise connected neighbourhoods), then every arc component of this space is open. The union of the other arc components is open, as well, hence arc components are open and closed in this case. In particular, a connected, locally arcwise connected space is arcwise connected.

5.20 Lemma: Domain invariance property of 1-manifolds
A 1-manifold Y contained in another 1-manifold X is always open.

Proof Every point $y \in Y$ has a neighbourhood $U \approx \mathbb{R}$ in X, and U contains a neighbourhood $V \approx \mathbb{R}$ with respect to Y. The connected subset $V \subseteq U$ is an interval by 5.4; more precisely, an open interval as it is homeomorphic to $\mathbb{R}$. This shows that y is an interior point of Y.   □

We are now ready to prove 5.17. We shall invoke the characterization theorem 5.11 in this proof. This will save us some of the technical labours involved in a direct proof such as CHRISTENSON–VOXMAN 1977

Chapter 5A. In particular, separability is exploited in a very simple way. We believe that our proof nonetheless gives appropriate insight.

Proof of 5.17. Every manifold is locally arcwise connected, hence the given space X is arcwise connected by 5.19. We distinguish between two cases.

Case I: There are points $u, v \in X$ that can be joined by two distinct arcs A, B. Choose a point lying on only one of the arcs, say $x \in A \smallsetminus B$. Then A contains exactly two minimal arcs joining x to points $b, c \in B$. These two arcs have only the point x in common. Together with the subarc of B joining b to c, they form a subset $Y \subseteq X$ homeomorphic to $\mathbb{S}_1$. The compact 1-manifold Y is closed in X and also open by 5.20, hence $X = Y \approx \mathbb{S}_1$.

Case II: Every pair of distinct points $x, y \in X$ is joined by a unique arc having end points x, y. We know that X has all the characteristic properties of $\mathbb{R}$ required in 5.11, with the possible exception of the separation property (iv$'$). We end our proof by showing that X does have the separation property. Observe that the other spaces characterized by these conditions (closed and half open intervals) are not manifolds.

Suppose that an arc A in X joins the points a, b, and that x is a third point on the arc. Then we claim that x separates a from b. If not, then a, b belong to the same connected component $U \subseteq X \smallsetminus \{x\}$. We know that U is connected, open in X, and locally arcwise connected. According to 5.19, there is an arc in U joining a to b, contrary to the assumption of Case II.

Now let three points $x, y, z \in X$ be given. If one of them lies on the arc joining the other two, then we can finish our proof by applying the previous paragraph. Assume therefore that this does not happen and consider the two arcs A and B joining x to y and to z, respectively. According to 5.18, the union $A \cup B$ contains an arc C joining y to z, and $C \smallsetminus \{y, z\}$ is open in X by 5.20. But, following the arc A from x, there must be a first point where we hit C, and this point is distinct from y and z. This is a contradiction, and the proof is complete. $\square$

Independence of characteristic properties

Our next aim is to prove independence of the conditions characterizing $\mathbb{R}$ according to 5.11. To do this, we shall exhibit examples of spaces that have three of the properties but not the fourth. Moreover, in all our examples the complement of any point will have exactly two connected components. For instance, we are not satisfied with $X = \mathbb{R}^2$ as an example of a space violating only the separation property (iv′).

It is relatively easy to give an example X_{iii} that is (connected but) not locally connected, and has all the other properties. One takes the set of real numbers and defines the space X_{iii} by refining the topology of $\mathbb{R}$ as follows. A set A is open by definition if, and only if, each point $a \in A$ has a neighbourhood U in the usual topology of $\mathbb{R}$ such that $U \cap \mathbb{Q} \subseteq A$. The verifications are left to the reader. Compare also counter-example 68 in STEEN–SEEBACH 1978.

A space X_{ii} without the T_1 property but satisfying the other conditions is obtained from the subspace $]\,,-1] \cup \{0\} \cup [1,\,[\,\subseteq \mathbb{R}$ by a slight modification of the topology. The change consists in the omission of certain open sets, namely, those that contain one of the points 1 or -1 without containing 0. In particular, $\{0\}$ is not a closed subset. The other verifications are again left to the reader.

The remaining examples are more interesting and will be considered in detail. In order to construct a space that violates only the separation property (iv′), we begin by constructing an auxiliary space that is countable but resembles the circle in many respects. This space is taken from the beautiful book of topological counter-examples STEEN–SEEBACH 1978 (counter-example 61). Originally, it was described by GOLOMB 1959 and KIRCH 1969.

5.21 Example We construct a topological space with point set $\mathbb{N}$, the natural numbers (0 excluded). In order to do so, we specify a basis $\mathcal{B}$ for the topology, i.e., the open sets will be the unions of arbitrarily many sets from $\mathcal{B}$. There is a condition that we have to observe if this is supposed to work. The intersection of two basis sets has to be a union of other basis sets. Our basis consists of all sets of the form $\mathbb{N} \cap (a + b\mathbb{Z})$, where $a, b \in \mathbb{N}$ are relatively prime, that is, $\gcd(a, b) = 1$, and b is *square free*, i.e., no square > 1 divides b. It is not necessary to make use of the possibility $b = 1$. Then we can always assume that $a < b$, and write the corresponding basis set as $a + b\mathbb{N}_0$, where $\mathbb{N}_0 = \mathbb{N} \cup \{0\}$. Let us note that *every* element of this set is relatively prime to b.

We verify the condition for a basis in the case of two basis sets $a + b\mathbb{N}_0$ and $c + d\mathbb{N}_0$. We consider a point x in the intersection, $x = a + bu = c + dv$ and look for a basis set containing x and lying within the intersection. Such a set is given by $\mathbb{N} \cap (x + e\mathbb{Z})$, where e is the least common multiple of b and d, which is square free. The choice of x implies that x is relatively prime to both b and d and, hence, to e.

5.22 Proposition *The space* $\mathbb{N}$ *with the topology constructed above is separable (being countable), connected and locally connected, and is a Hausdorff space, but not metrizable. Moreover, every point in this space has a connected complement.*

Proof We show first that the basis sets $A = a + b\mathbb{N}_0 \in \mathcal{B}$ are connected. Thus we assume that $A = S \cup T$ where S, T are open and non-empty, and we show that $S \cap T \neq \emptyset$. There are basis sets $s + bc\mathbb{N}_0 \subseteq S$ and $t + bd\mathbb{N}_0 \subseteq T$; the special form is enforced since these sets are contained in A. We set $u = cd$; then b and u are relatively prime, because both bc and bd are square free. By 1.5, there are integers f, z such that

$$a + bz = uf.$$

This yields $a + b(z + nu) = u(f + nb)$ for every $n \in \mathbb{N}$, and we see that we may choose f and z to be positive. The displayed equation shows that uf belongs to A and, hence, to S or to T. The roles played by S and T in the construction of this element are interchangeable, hence it does not matter which of the two sets it belongs to – say $uf \in T$. There exists a neighbourhood $uf + bg\mathbb{N}_0 \subseteq T$ of this element, where $1 = \gcd(uf, bg) = \gcd(cdf, bg) = \gcd(c, g)$. Since s and uf belong to A, we have

$$uf - s = be \in b\mathbb{Z}.$$

We use 1.5 again to write $cx = e + gy$, where $x, y \in \mathbb{Z}$; as before, we may take both these integers to be positive. We finally obtain the element $s + bcx = s + be + bgy = uf + bgy \in S \cap T$.

We may conclude that $\mathbb{N}$ is locally connected, and we show next that every point $z \in \mathbb{N}$ has a connected complement. As a by-product, this will imply that $\mathbb{N}$ itself is connected and, hence, not metrizable (note that the distance function on a connected metric space has to take a continuum of values). Given $x, y \in \mathbb{N} \smallsetminus \{z\}$, we propose to construct two basis sets $A = \mathbb{N} \cap (x + p\mathbb{Z})$ and $B = \mathbb{N} \cap (y + q\mathbb{Z})$ not containing z such that $A \cap B \neq \emptyset$. Then $A \cup B$ is connected and contains x and y, proving that $\mathbb{N} \smallsetminus \{z\}$ is connected. At the same time, we will obtain the Hausdorff property (just knowing that $x \in A \subseteq \mathbb{N} \smallsetminus \{z\}$).

We choose p and q to be distinct primes such that $p > |x - z|$ and $q > |y - z|$. This is possible because there are infinitely many primes (see 32.7), and it ensures that $z \notin A \cup B$. Once more by 1.5, there are positive integers m, n such that $x - y = qm - pn$, and then $x + pn = y + qm \in A \cap B$. $\square$

5.23 Example   We construct a new space X_{iv} (looking like a brush) from the 'countable circle' ($\mathbb{N}$ with the topology defined in 5.21) by attaching a bristle to each of its points. Formally, $X_{iv} = \mathbb{N} \times [0,1[$ with the topology defined by the following basis. For $n \in \mathbb{N}$ and an open set $U \subseteq {]0,1[}$ (not containing 0 !), the set $\{n\} \times U$ is a basis set, and for an open set $A \subseteq \mathbb{N}$ and $\varepsilon > 0$, the set $A \times [0, \varepsilon[$ is another one. The proof that this is the basis of a topology having the properties stated in the next proposition is left to the reader.

5.24 Proposition   *The space X_{iv} constructed above is a connected and locally connected, separable Hausdorff space. The complement of any point has exactly two connected components. However, a point in $\mathbb{N} \times \{0\}$ can never separate two other points belonging to that subset.*

□

Our next aim is to construct a space X_i having all characteristic properties of $\mathbb{R}$ except separability. In the proof of Theorem 5.10, only the last step made use of separability. Therefore, the space we are looking for must be a non-separable chain endowed with the order topology. The starting point of the construction is the observation that $\mathbb{Z} \times [0,1[$ with the lexicographic ordering is a chain isomorphic to $\mathbb{R}$, and one tries to replace $\mathbb{Z}$ with an uncountable chain that is not too big.

For this purpose, we use the first uncountable ordinal ω_1. For information on ordinals, see the Appendix 61 and 61.7 in particular. We need the properties that ω_1 is an uncountable well-ordered chain and that a subset of this chain is countable if, and only if, it is bounded. The verification of the latter property is the only occasion where we need to know that every ordinal can be identified, somewhat paradoxically, with the set of all smaller ordinals (61.6), and that every set $\mathcal{M}$ of ordinals has a least upper bound in the ordered class of all ordinals, obtained simply as the union $\bigcup \mathcal{M}$ (61.5). The elements of ω_1 are precisely all countable ordinals (61.7), and it follows that every bounded set in ω_1 is countable. Conversely, if we have a countable set of countable ordinals, then their union is again a countable set (61.13) and an ordinal, hence it is an element of ω_1. Henceforth, we shall ignore the original meaning of the ordering relation for ordinals and write $<$ for this relation.

5.25 Example: Alexandroff's long line   The long line $\mathbb{L}$ will be composed of the long ray $\mathbb{L}^+$ and its mirror image $-\mathbb{L}^+$, the details being as follows. The *long ray* is defined as the chain

$$\mathbb{L}^+ = \omega_1 \times [0,1[,$$

with the lexicographic ordering relation

$$(\mu, s) < (\nu, t) \;\; \leftrightharpoons \;\; (\mu < \nu) \vee \big((\mu = \nu) \wedge (s < t) \big) \, .$$

For brevity, we shall write 0 instead of $(0,0) \in \mathbb{L}^+$. For each $x \in \mathbb{L}^+$, we introduce a mirror image $-x$, such that $x \mapsto -x$ becomes a bijective correspondence $\mathbb{L}^+ \to -\mathbb{L}^+ = \{ -x \mid x \in \mathbb{L}^+ \}$. The sets $\mathbb{L}^+$ and $-\mathbb{L}^+$ are supposed to be disjoint except for $0 = -0$. For $x < y$ in $\mathbb{L}^+$ we set $-y < -x < y$ and thus turn $\mathbb{L} = \mathbb{L}^+ \cup -\mathbb{L}^+$ into a chain. This chain is the *long line*. Both $\mathbb{L}$ and $\mathbb{L}^+$ are endowed with the topology induced by their ordering, and both $\mathbb{L}$ and $\mathbb{L}^+ \smallsetminus \{0\}$ can serve as our example X_{i} of a space having all the characteristic properties of $\mathbb{R}$ except separability, as it will turn out later.

The involutory map $\iota : \mathbb{L} \to \mathbb{L}$ that interchanges x and $-x$ is an anti-automorphism of the chain $\mathbb{L}$ and hence a homeomorphism of the long line. However, appearances can be deceptive, and if $\nu + 1$ denotes the successor of an ordinal, then the map $(\nu, s) \mapsto (\nu + 1, s)$ from $\mathbb{L}^+$ into itself is not continuous. Indeed, if ν runs over all natural numbers, then both $(\nu, 0)$ and $(\nu + 1, 0)$ converge to $(\omega_0, 0)$, but $(\omega_0, 0)$ is not fixed by the map. For similar reasons, we refrained from describing all of $\mathbb{L}$ by a single lexicographic product; this would not allow for a simple description of the map ι.

The long line is a more prominent example than those discussed so far. It was first presented in ALEXANDROFF 1924. The existence of several non-isomorphic analytic structures both on $\mathbb{L}$ and on $\mathbb{L}^+$ was shown in KNESER–KNESER 1960. More recently, NYIKOS 1992 constructed $2^{\aleph_1}$ smooth structures on each of them, whose tangent bundles are not homeomorphic.

5.26 Proposition *Every interval $[a, b]$ in the long line $\mathbb{L}$ is order isomorphic and, hence, homeomorphic to the unit interval $[0, 1] \subseteq \mathbb{R}$.*

Proof We treat the case $a = 0$ in detail. After that one sees, using the order isomorphism ι introduced in 5.25, that $[-b, b] = [-b, 0] \cup [0, b] \cong [-1, 0] \cup [0, 1] \subseteq \mathbb{R}$, and the proposition follows. First we obtain an isomorphism of the open intervals, $]0, b[\cong]0, 1[$, using the characterization of the chain $\mathbb{R}$ given in 3.5b. Such an isomorphism uniquely extends to the closed intervals.

If $b = (\nu, t)$, then the interval $[0, \nu]$ of ordinals (that is, the ordinal $\nu + 1$) is countable, and $([0, \nu] \times \mathbb{Q}) \cap]0, b[$ is a countable, strongly dense subset of $]0, b[$. It remains to be shown that the given interval is complete.

So let a bounded subset $A \subseteq]0, b[$ be given, $A \leq (\eta, s) < b$, and consider the projection $p_1 : A \to \omega_1 : (\nu, t) \mapsto \nu$. The well-ordering of ω_1 implies that $p_1(A)$ has a least upper bound $\xi \leq \eta$ in ω_1. If $\xi \notin p_1(A)$, then the least upper bound of A is $\sup A = (\xi, 0) \leq (\eta, s)$. If $\xi \in p_1(A)$, then we look for $\sup A$ in the complete interval $[(\xi, 0), (\xi + 1, 0)] \cong [0, 1]$ and find $\sup A \leq (\eta, s)$. Greatest lower bounds are treated in the same manner. $\square$

5.27 Definition   A topological space X is said to be *n-homogeneous*, $n \in \mathbb{N}$, if any sequence $x_1, \ldots, x_n$ of n distinct points of X can be mapped onto any other such sequence, $y_1, \ldots, y_n$, by a homeomorphism of $f : X \to X$. To be precise, it is required that $f(x_i) = y_i$ for each i.

The real line $\mathbb{R}$ is 2-homogeneous (but not 3-homogeneous): using affine maps $x \mapsto ax + b$, every pair of real numbers can be mapped onto every other pair (but a homeomorphism cannot move the middle point of a triple to a position where it does not separate the images of the other two). We show that the long line shares this property of $\mathbb{R}$.

5.28 Corollary   *The long line $\mathbb{L}$ is 2-homogeneous.*

Proof Let two pairs of points x_1, x_2 and y_1, y_2 be given as in 5.27. By applying the map ι from 5.25, if necessary, to one or to both pairs, we may arrange that $x_1 < x_2$ and $y_1 < y_2$. We find an open interval $I \subseteq \mathbb{L}$ containing the four points. By 5.26, this interval I is homeomorphic to the real interval $]0, 1[\approx \mathbb{R}$. The remarks preceding the corollary show that I admits a homeomorphism f sending x_i to y_i, and f is monotone due to our special arrangement of the two pairs of points. It follows that f extends to a homeomorphism of $\mathbb{L}$ that induces the identity on the complement of I. To solve the original task, this map f may have to be multiplied on one side or on both by the map ι. $\square$

5.29 Corollary   *Both the long line and the long ray satisfy all conditions of 5.11 except separability.*

Proof The two cases are very similar, and we concentrate on the long line itself. Every chain is a Hausdorff space in the induced topology. The long line is the union of all its intervals $[-a, a]$, which are connected by 5.26. This shows connectedness of $\mathbb{L}$, and local connectedness is even more obvious. The separation condition (iv′) again holds in every chain. We note that, as with our other counter-examples, the complement of any point has exactly two connected components. Finally, $\mathbb{L}$ is not separable; this follows from the following stronger assertion. $\square$

5.30 Proposition *Every countable subset of the long line is bounded.*

Proof This is an immediate consequence of the fact that ω_1 has the same property; see the remarks preceding 5.25.   □

The following is perhaps the most striking feature of the long line.

5.31 Proposition *If $f : \mathbb{L} \to \mathbb{R}$ is a continuous map, then there are $a, b \in \mathbb{L}$ such that f is constant on $] \,, a]$ and on $[b, \, [$.*

Proof We concentrate on the restriction of f to $\mathbb{L}^+$. For a set X of real numbers, we define the diameter $\operatorname{diam} X = \sup\{ |x - y| \mid x, y \in X \} \in \mathbb{R} \cup \{\infty\}$. Our first claim is that for each $n \in \mathbb{N}$ we can find $c_n \in \mathbb{L}$ such that $\operatorname{diam} f([c_n, \, [) \leq 1/n$. Suppose that this is impossible. Then we can find a positive real δ and an increasing sequence $\{ a_k \mid k \in \mathbb{N} \}$ in $\mathbb{L}^+$ such that $|f(a_k) - f(a_{k+1})| \geq \delta$ for each k. The sequence is bounded by 5.30, hence it is contained in a subchain isomorphic to $[0, 1]$ by 5.26. In this subchain, every increasing sequence converges, but $f(a_k)$ does not converge — a contradiction to continuity.

The sequence $\{ c_n \mid n \in \mathbb{N} \}$ has an upper bound $c \in \mathbb{L}$, too, and we conclude that $\operatorname{diam} f([c, \, [) \leq \operatorname{diam} f([c_n, \, [) \leq 1/n$ for each n. This means that f is constant on $[c, \, [$.   □

Properties 5.30 and 5.31 of the long line are actually compactness properties. We make that explicit after the following.

5.32 Definition A topological space X is said to be *pseudo-compact* if every continuous map $f : X \to \mathbb{R}$ produces a compact image $f(X)$. The space X is said to be *sequentially compact* if every sequence in X has a convergent subsequence.

For further reading on notions of compactness, we recommend the book CHRISTENSON–VOXMAN 1977. We mention that a compact space is both sequentially compact and pseudo-compact, and that a sequentially compact space is *countably compact*, that is, every countable open cover $\{ U_i \mid i \in \mathbb{N} \}$ has a finite subcover. Indeed, if the first n sets U_i fail to cover, we select a point x_n that is not covered. If this happens for every n, there results a sequence, which has a convergent subsequence — but the limit cannot belong to any U_i, a contradiction.

5.33 Proposition *The long line $\mathbb{L}$ is not compact, but it is locally compact, sequentially compact, countably compact, and pseudo-compact.*

Proof The long line is covered by all its open intervals $]a, b[$, and this cover has no finite subcover. Thus, $\mathbb{L}$ is not compact. On the other hand, every interval $[a, b]$ is compact according to 5.26. This implies local compactness of $\mathbb{L}$. By 5.30, every sequence is contained in a compact interval and, hence, has an accumulation point. Countable compactness follows from sequential compactness (5.32). For a continuous real-valued map f, we learn from 5.31 that some compact interval has the same image as all of $\mathbb{L}$, which implies peudocompactness. $\qquad\square$

5.34 Corollary *There is no metric generating the topology of the long line.*

Proof In metric spaces, sequential compactness is the same as compactness; see, for example, CHRISTENSON–VOXMAN 1977 3.A.11 p. 83. More easily we can show that a sequentially compact metric space is separable and then apply 5.29. So suppose that X is a sequentially compact space. If, for every $n \in \mathbb{N}$, there is a finite subset $F_n \subseteq X$ such that every point of X lies at distance $< 1/n$ from some member of F_n, then the union of all F_n is a countable (61.13) dense set. If no finite set has the properties required for some particular F_n, then by induction we find a sequence $\{\, x_k \mid k \in \mathbb{N} \,\}$ such that each x_k is at distance $\geq 1/n$ from all its predecessors. Such a sequence has no convergent subsequence. $\qquad\square$

As a last remark on the long line, we note that it does not satisfy the Souslin condition (compare 3.8): there are uncountably many disjoint open intervals $\{\nu\} \times\,]0, 1[$ in $\mathbb{L}$.

Subspaces and continuous images of the real line

We recall that one important continuous image of the real line is the circle; see 5.16. Several characterizations of the circle have been given (5.14, 5.17). Among subsets of the real line, a class that has received special attention is formed by the totally imperfect subsets, i.e., sets that do not contain any uncountable closed set. We shall not treat this topic but refer to the literature (MILLER 1984, 1993).

We continue by discussing the Cantor set (which is closed and uncountable). At first sight, this looks like a rather far-fetched example having weird properties. However, it turns out that the Cantor set is of crucial importance in our context. We show that it can be used in the construction of space-filling curves, and this is closely related to the role it plays in the proof of the Hahn–Mazurkiewicz theorem characterizing the continuous images of the arc. Furthermore, the continuous images of the Cantor set itself form an important class of topological spaces.

In order to define the Cantor set, we use the triadic expansion of real numbers $x \in [0, 1]$:

$$x = \sum_{n=1}^{\infty} a_n 3^{-n} \ ,$$

where $a_n \in \{0, 1, 2\}$. This expansion is not unique; typically, $0.1\bar{0} = 0.0\bar{2}$, where we write, as in the usual decimal notation, $x = 0.a_1 a_2 a_3 \ldots$ for the above expansion, and periods are indicated by overlining.

5.35 Definition The *Cantor set* $\mathcal{C} \subseteq \mathbb{R}$ is defined as the set of all $x \in [0, 1]$ admitting a (unique) triadic expansion using only 0 and 2 as digits. The following alternative description may be more familiar; we leave it to the reader to prove the equivalence with our first description.

One starts with the unit interval $I = I_0$ and forms $I_1 = I \setminus \left]\frac{1}{3}, \frac{2}{3}\right[= [0, \frac{1}{3}] \cup [\frac{2}{3}, 1]$. In other words, I_1 is obtained by deleting the open middle third from the unit interval. We continue in the same way, forming I_{n+1} by deleting the open middle thirds from each of the 2^n intervals constituting I_n. This gives a descending chain of compact subsets of I, and the (compact, non-empty) intersection of this chain is the Cantor set. Since the intervals forming I_n are of length 3^{-n}, we see immediately that the total length $(\frac{2}{3})^n$ of I_n tends to zero, and we have obtained two important features of $\mathcal{C}$ (compare 10.6):

5.36 Proposition *The Cantor set $\mathcal{C} \subseteq [0, 1]$ is a compact set of Lebesgue measure zero.* □

Note, however, that the measure is not an invariant property of the topological space $\mathcal{C}$; compare 10.6(c). Compactness follows once more from our next result, where the space $\{0, 2\}$ is taken with the discrete topology.

5.37 Proposition *The Cantor set is homeomorphic to the compact product space $\{0, 2\}^{\mathbb{N}}$.*

Proof We show that a homeomorphism is obtained by reading the co-efficient sequence of the triadic expansion with digits $a_n \in \{0, 2\}$ as an element of the product space. The product topology is defined using a basis, where a typical basis element consists of all sequences $(a_k)_k$ with a fixed initial part $a_1, \ldots, a_n$. The corresponding points of I form the open and closed set $\mathcal{C} \cap \left(\sum_{k=1}^{n} a_k 3^{-k} + [0, 3^{-n}]\right)$. This proves that we have a homeomorphism. Since $\{0, 2\}$ is a compact space, it follows that the product space is compact, as well. □

5.38 Corollary *The Cantor set $\mathcal{C}$ is homeomorphic to its own square $\mathcal{C} \times \mathcal{C}$.*

Proof There is a homeomorphism $\{0,2\}^{\mathbb{N}} \to \{0,2\}^{\mathbb{N}} \times \{0,2\}^{\mathbb{N}}$ defined by 'opening the zipper', that is, the sequence $a_1, b_1, a_2, b_2, \ldots$ is mapped to the pair of sequences $(a_1, a_2, \ldots ; b_1, b_2, \ldots)$. □

5.39 Proposition   *There is a continuous surjection $\mathcal{C} \to I = [0,1]$.*

Proof Such a map may be defined by reading the coefficient sequence of a triadic expansion as the coefficient sequence of a dyadic one, replacing the digit a_n by $a_n/2$. □

5.40 Proposition   *There is a continuous surjection $I \to I^2$ of the arc onto a square.*

Proof We use the homeomorphism $\varphi : \mathcal{C} \to \mathcal{C}^2$ of 5.38 and the surjection $\rho : \mathcal{C} \to I$ of 5.39. We form the surjection $\rho \times \rho : \mathcal{C}^2 \to I^2$ sending (x, y) to $(\rho(x), \rho(y))$. The composition $(\rho \times \rho) \circ \varphi : \mathcal{C} \to I^2$ is continuous and surjective. Its domain is a closed subset of I, and it can be extended to all of I by Tietze's extension theorem; see CHRISTENSON–VOXMAN 1977 4.B.8. In our concrete situation, it is not necessary to invoke that theorem, since it suffices to define the map on each of the middle third intervals deleted from I in the course of the construction of $\mathcal{C}$. This can be done directly using the convexity of the unit square I^2; one simply maps each of these intervals to the segment joining the images of its end points, using a suitable affine map $x \mapsto ax + b$.

A different construction of a surface-filling curve is given in Exercise 5. □

5.41 Remarks   Clearly, the last result can be generalized; every power I^n, $n \in \mathbb{N}$, is a continuous image of the arc. But this is only a shadow of what can be proved: there is a comprehensive theorem, the Hahn–Mazurkiewicz theorem, which characterizes exactly the spaces that are continuous images of the arc; see 5.42 below. The basic idea is the same as for the proof of 5.40; in order to obtain a continuous surjection of I onto a space X, one first constructs a surjection $\mathcal{C} \to X$ and then extends it over I. Therefore, the characterization of the continuous images of $\mathcal{C}$ is a key ingredient; see 5.43 below. The extension is not as easy as in the last proof. In order to define it on one missing interval, one uses the topological characterization of the arc 5.13, but in the form that does not involve local connectedness; see 5.15. We refer the interested reader to CHRISTENSON–VOXMAN 1977 for details and proofs, but we state the precise form of the results.

5.42 Theorem   *A Hausdorff space is a continuous image of the arc if, and only if, it is a compact, connected and locally connected metric space.*   □

Such spaces are called Peano continua in honour of G. Peano and his space-filling curves. In the course of the proof, it is shown that every compact, connected and locally connected metric space is arcwise connected. We might mention here that, of course, the circle is a special case; compare 5.16. We return to our systematic study of the Cantor set by stating the result mentioned in 5.41.

5.43 Theorem   *Every compact metric space is a continuous image of the Cantor set.*   □

The Cantor set plays a prominent role also in lattice theory. Indeed, the collection of all subsets of $\mathcal{C}$ that are both open and closed forms a lattice, which is a countable, atomless Boolean algebra (atomless, because $\mathcal{C}$ has no isolated points). This Boolean algebra is determined up to isomorphism by the properties mentioned; usually, this fact is reduced, via the Stone representation theorem, to the topological characterization of $\mathcal{C}$ given in 5.48; see SIKORSKI 1964 §9C p. 28; compare also §14B p. 44. An elementary, direct proof is suggested in Exercise 9.

The Cantor set is markedly more homogeneous than the real line; compare the remark following 5.27. In DROSTE–GÖBEL 2002, it is shown that the homeomorphism group (compare 5.51 below) of $\mathcal{C}$ is quite large in the sense that it cannot be obtained as the union of a countable ascending chain of proper subgroups. In fact, the size of the homeomorphism group is such that it contains an isomorphic copy of every countably infinite group; see HJORTH–MOLBERG 2006. Moreover, the homeomorphism group of the Cantor set is a simple group. Together with similar results about the space of rational numbers and the space of irrational numbers, this is proved by ANDERSON 1958. Here we are content to prove the following.

5.44 Proposition   *The Cantor set is n-homogeneous for each $n \in \mathbb{N}$.*

Proof (1) First we prove that $\{0,2\}^{\mathbb{N}} \approx \mathcal{C}$ is 1-homogeneous. Let $(a_n)_n$ and $(b_n)_n$ be elements of $\{0,2\}^{\mathbb{N}}$. For each $n \in \mathbb{N}$, there is a bijection f_n of $\{0,2\}$ such that $f_n(a_n) = b_n$. Applying f_n to the nth term of each sequence in $\{0,2\}^{\mathbb{N}}$, we obtain a bijection f which maps the sequence $(a_n)_n$ to $(b_n)_n$. In order to verify continuity of f we have to check, according to the universal property of products, that the nth member

d_n of the sequence $f(x)$ depends continuously on the sequence x. This is true, since $d_n = f_n(c_n)$, where c_n is the nth member of x.

(2) From the second description of the Cantor set (as the intersection of a descending chain of compact sets I_n) it follows that every permutation of the intervals forming some particular I_n gives rise to a homeomorphism of the Cantor set. One has to observe that for each of these intervals J, the intersection $J \cap C$ is open in C and is a scaled version of C itself, and that any two of these sets are homeomorphic by a translation.

Now let two finite sequences $x_1, \ldots, x_k$ and $y_1, \ldots, y_k$ of distinct points in C be given. Choose n so large that no two of the x_i or of the y_i occupy the same interval in I_n. Permute the intervals so that the ith interval contains the image of x_i for $i = 1, \ldots, n$ and use a homeomorphism of $J \cap C$ as in step (1) for each of these intervals, so that x_i is sent to the smallest element in $J \cap C$. Together, this yields a homeomorphism g of C onto itself. There is a similar homeomorphism h for the y_i, and the quotient $h^{-1} \circ g$ sends x_i to y_i for each i.   □

5.45 Definition A topological space X is said to be *totally disconnected* if every connected component of X is reduced to a point. Equivalently, X contains no connected subsets except points and the empty set.

A space X is said to be *perfect* if it has no isolated points, that is, no singleton $\{x\} \subseteq X$ is an open set.

5.46 Proposition *A 2-homogeneous topological space is either connected or totally disconnected.*

Proof If $A \subseteq X$ is connected and contains two distinct points a, b, then any two points $x, y \in X$ belong to a connected set $f(A) \subseteq X$, where f is a homeomorphism of X sending $\{a, b\}$ to $\{x, y\}$. It follows that X is connected.   □

5.47 Proposition *The Cantor set is totally disconnected and perfect.*

Proof The Cantor set is 2-homogeneous and not connected, hence it is totally disconnected. The definition of the product topology on $\{0, 2\}^{\mathbb{N}}$ makes it clear that no open set can consist of a single point.   □

There is a nice characterization of the Cantor set on the basis of the properties that we have proved. Again we record this result without proof; see CHRISTENSON–VOXMAN 1977 6.C.11.

5.48 Theorem   *Every compact, totally disconnected, perfect metric space is homeomorphic to the Cantor set.*   □

We turn now to the subspace $\mathfrak{I} = \mathbb{R} \smallsetminus \mathbb{Q} \subseteq \mathbb{R}$, the space of irrational numbers. One feels that there is an immense gap separating the spaces $\mathbb{R}$ and $\mathfrak{I}$. Clearly, $\mathfrak{I}$ contains no non-trivial interval of the real line, hence it is totally disconnected by 5.4. On the other hand, it was shown in 4.11 that $\mathfrak{I}$ admits a complete metric generating the topology. The proof relies on the result 4.10, which we repeat here, and which will allow us to prove an unexpected, close relationship between $\mathfrak{I}$ and $\mathbb{R}$ (5.50).

5.49 Theorem   *The space $\mathfrak{I} = \mathbb{R} \smallsetminus \mathbb{Q}$ is homeomorphic to the product space $\mathbb{N}^{\mathbb{N}}$.*   □

5.50 Theorem   *There is a continuous bijection $\mathfrak{I} = \mathbb{R} \smallsetminus \mathbb{Q} \to \mathbb{R}$.*

Proof (1) We shall construct (in step (2) below) a continuous bijection $\alpha : \mathbb{N}^{\mathbb{N}} \to [0, 1[$. Together with the homeomorphism $\beta : \mathfrak{I} \to \mathbb{N}^{\mathbb{N}}$ of 5.49 this yields a continuous bijection $\gamma = \beta \circ \alpha : \mathfrak{I} \to [0, 1[$. This is not quite what we want; we have to get rid of the element 0 in the image of this map. This can be done by showing that $\mathfrak{I} \smallsetminus \{\gamma^{-1}(0)\}$ is homeomorphic to $\mathfrak{I}$. Indeed, $\mathbb{Q} \cup \{\gamma^{-1}(0)\}$ is chain isomorphic to $\mathbb{Q}$ by 3.4, and the proof of 3.5 shows that any isomorphism of these chains extends to a chain automorphism φ of $\mathbb{R}$, and φ induces the homeomorphism we need.

A more direct construction is this: The space $\mathfrak{I}$ is the disjoint union of the open subsets $\mathfrak{I} \cap [n, n + 1]$, $n \in \mathbb{Z}$. Under the homeomorphism $\mathfrak{I} \to \mathbb{Z} \times \mathbb{N}^{\mathbb{N}}$ constructed in 4.10 using continued fractions, this interval corresponds to $\{n\} \times \mathbb{N}^{\mathbb{N}}$, hence it is homeomorphic to $\mathfrak{I}$ itself. Thus, each of these intervals can be mapped continuously and bijectively onto $[n, n + 1[$, and together these maps yield the desired continuous bijection $\mathfrak{I} \to \mathbb{R}$.

(2) In order to construct α, we read an element $(a_n)_n \in \mathbb{N}^{\mathbb{N}}$ as an instruction for the definition of a sequence of digits $b_n \in \{0, 2\}$: two consecutive digits equal to 0 in this sequence shall be separated by a_n digits equal to 2. Thus, the zero digits $b_n = 0$ are those whose index has the form $n = a_1 + a_2 + \cdots + a_k - 1$, for some $k \in \mathbb{N}$. This defines a map ψ of $\mathbb{N}^{\mathbb{N}}$ into the Cantor set $\mathcal{C} \approx \{0, 2\}^{\mathbb{N}}$ (see 5.37), in fact, a homeomorphism of $\mathbb{N}^{\mathbb{N}}$ onto $\psi(\mathbb{N}^{\mathbb{N}})$ because both spaces carry the product topology. The image $\psi(\mathbb{N}^{\mathbb{N}})$ consists precisely of the sequences having infinitely many digits equal to 0, and on this set the surjection $\mathcal{C} \to [0, 1]$ constructed in 5.39 induces a continuous bijection onto $[0, 1[$.   □

We remark that this theorem together with 5.42 allows us to map the irrationals continuously onto any Peano continuum.

The space $\mathfrak{J}$ is clearly not homeomorphic to its complement $\mathbb{Q}$ (the two spaces even have different cardinalities; more about the topology of $\mathbb{Q}$ will be said in Section 33). It is possible, however, to exhibit subsets of $\mathbb{R}$ that are homeomorphic to their complements. One can even construct a situation where both complementary sets are homogeneous, and this can be done in essentially different ways; see VAN MILL 1982a, 1982b.

Homeomorphisms of the real line

5.51 The homeomorphism group   We have frequently used the notion of homeomorphism in this section. In the particular case of a homeomorphism of the topological space $\mathbb{R}$ onto itself, we are dealing with a bijective map $f : \mathbb{R} \to \mathbb{R}$ such that both f and the inverse map f^{-1} are continuous.

Various characterizations of homeomorphisms are known as standard facts of real analysis (but note that they do not hold for general topological spaces). For instance, it follows from the intermediate value theorem that an injective continuous map of $\mathbb{R}$ into itself preserves or reverses the ordering (i.e., it is strictly monotone or antitone). Conversely, it is immediate that every order-preserving or order-reversing bijection maps each open interval onto an open interval, hence its inverse map is continuous. Therefore, the homeomorphisms of $\mathbb{R}$ are precisely the bijections which preserve or reverse the ordering.

The same arguments, put together differently, show that a continuous bijection of $\mathbb{R}$ preserves or reverses the ordering and therefore has a continuous inverse. In other words, the continuity of f^{-1} need not be required in the definition of homeomorphisms of the real line. This is true for maps between any two connected manifolds of the same dimension. The proof of this general result is much harder and uses Brouwer's theorem on the invariance of domain; see, for example, DUGUNDJI 1966 XVII.3.1.

Clearly, the homeomorphisms of $\mathbb{R}$ form a group under composition, which will be denoted $\mathcal{H}(\mathbb{R})$ or, briefly, $\mathcal{H}$. It has a normal subgroup of index two, formed by the monotone (or order-preserving) bijections and denoted $\mathcal{H}^+(\mathbb{R})$ or $\mathcal{H}^+$. In other words, the subgroup $\mathcal{H}^+ \leq \mathcal{H}$ has precisely two cosets, consisting of the monotone and antitone bijections, respectively.

Our aims are to show that both the fixed point free elements and the involutions form a conjugacy class, and that every element of $\mathcal{H}$ is a product of at most four involutions. Moreover, we determine the commutator subgroups and we describe all normal subgroups of $\mathcal{H}$ and of $\mathcal{H}^+$. There are very few of them, thus the two groups are very close to being simple. The material is taken from FINE–SCHWEIGERT 1955. The more difficult proofs will be only sketched, the reader being referred to the original article for full detail. A more special result about subgroups consisting of fixed point free homeomorphisms will also be included (5.57).

We remark that we treat $\mathcal{H}$ as an abstract group, without a topology, although there is a topology making $\mathcal{H}$ a topological transformation group of $\mathbb{R}$, namely, the compact-open topology; see DUGUNDJI 1966 Chapter XII. We also point out that we often omit the symbol for composition in $\mathcal{H}$ and write fg instead of $f \circ g$.

5.52 Definition: Special homeomorphisms Any element $i \in \mathcal{H}$ of order two will be called an *involution*. We shall show that each involution is conjugate to the *standard involution* i_0 defined by $i_0(x) = -x$. (Recall that two elements g, h of a group G are said to be conjugate if there is an $f \in G$ such that $h = f^{-1}gf$. This is an equivalence relation.) By a *translation* of $\mathbb{R}$ we mean an element $t \in \mathcal{H}$ without fixed points, that is, $t(x) \neq x$ for all x. The name will be justified when we show that each translation is conjugate to the *standard translation* t_1 defined by $t_1(x) = x + 1$. Applying the intermediate value theorem to $t - \mathrm{id}$, we see that a translation satisfies either $t(x) < x$ for all x or $x < t(x)$ for all x. Accordingly, we call t a *left translation* or a *right translation*.

The long line $\mathbb{L}$ does not admit any translations; compare Exercise 8.

5.53 Theorem *Every involution $i \in \mathcal{H}$ is conjugate to the standard involution i_0, where $i_0(x) = -x$. In particular, i reverses the ordering.*

Proof Choose $a \in \mathbb{R}$ arbitrarily. The involution i interchanges a and $i(a)$ and therefore reverses the ordering. Moreover, i maps the interval with end points a, $i(a)$ onto itself. Applying the intermediate value theorem to the map $x \mapsto i(x) - x$ on this interval, we see that there is a fixed point x_0 of i. Replacing i by its conjugate i', where $i'(x) = i(x + x_0) - x_0$, we reduce the problem to the case $x_0 = 0$.

Define a homeomorphism $f \in \mathcal{H}^+$ by $f(x) = x$ for $x \leq 0$ and $f(x) = -i'(x)$ for $x \geq 0$. Then $f^{-1}i_0f = i'$ may be verified by showing that $-f(x) = f(i'(x))$ both for $x \leq 0$ and for $x \geq 0$. $\square$

5.54 Theorem *Every translation (fixed point free homeomorphism)*
$t \in \mathcal{H}$ *is conjugate to the standard translation* t_1, *where* $t_1(x) = x + 1$.
In particular, t preserves the ordering.

Proof The idea is similar to that of the last proof. If t is a left translation, we pass to $t'(x) = -t(-x)$; in other words, we conjugate t by i_0.
Thus, we may assume that t is a right translation. Then the two-ended
sequence $\{t^n(x) \mid n \in \mathbb{Z}\}$ is strictly monotone. A least upper bound
of this sequence would be a fixed point of t, hence the sequence is unbounded above and, likewise, unbounded below. Thus, $\mathbb{R}$ can be written
as the union of the intervals $I_n = [t^n(0), t^{n+1}(0)]$ for all $n \in \mathbb{Z}$. Clearly,
t induces a homeomorphism of I_n onto I_{n+1}. We set $f(x) = t^n(x - n)$
for $n \leq x \leq n + 1$, $n \in \mathbb{Z}$. This defines a homeomorphism $f \in \mathcal{H}$, and
the conjugation identity $f(t_1(x)) = t(f(x))$ is easily verified.   □

We remark that the conjugating element in the last theorem may be
antitone. If we consider conjugacy in the group $\mathcal{H}^+$, then the first step
(conjugation with i_0) in the proof of 5.54 is not needed. Hence, the
same proof shows that there are two classes of translations, the right
translations and the left translations.

5.55 Definition A *one-parameter group* of homeomorphisms of $\mathbb{R}$ or
a *flow* on $\mathbb{R}$ is a group homomorphism $r \mapsto g_r$ from the additive group
$\mathbb{R}$ into $\mathcal{H}$. Thus, we have the identity $g_{r+s} = g_r g_s = g_s g_r$. Note that
$g_r = g_{r/2} g_{r/2}$ is always monotone. Conversely, we have the following.

5.56 Theorem *Every monotone homeomorphism* $g \in \mathcal{H}^+$ *can be embedded as* g_1 *in a flow* $g_r, r \in \mathbb{R}$.

Proof The proof is easy in the case of a translation $g = t$. Write $g =
f^{-1} t_1 f$ as in 5.54 and set $g_r = f^{-1} t_r f$, where $t_r(x) = x + r$ is the
standard flow. In general, a monotone $g \in \mathcal{H}$ has a fixed point set F,
and the complement of F splits up into at most countably many disjoint
open intervals (the connected components of $\mathbb{R} \smallsetminus F$). On each of these
intervals, g induces a translation, to which the previous argument may be
applied. The resulting flows combine into one global flow which induces
the identity on F.   □

The following result is taken from SALZMANN 1958. Its proof introduces an ordering on a certain group of homeomorphisms. The entire
group $\mathcal{H}$ cannot be made into a chain in a reasonable way, but it does
have partial orders. The system of all partial orders has some striking
properties; compare HOLLAND 1992.

5.57 Theorem *Let $G \leq \mathcal{H}$ be a group of homeomorphisms acting freely on $\mathbb{R}$, that is, every element $g \in G \smallsetminus \{\mathrm{id}\}$ is fixed point free. Then G is isomorphic, as an abstract group, to a subgroup of the additive group $\mathbb{R}$. In particular, G is commutative.*

Proof Every element of $G \smallsetminus \{\mathrm{id}\}$ is either a left or right translation. For distinct elements $g, h \in G$, we set $g < h$ if $h^{-1}g$ is a left translation. This means that the evaluation map $g \mapsto g(x_0)$ of G onto an arbitrary orbit $G(x_0) = \{\, g(x_0) \mid g \in G \,\}$ is an order isomorphism. This definition makes G an ordered group, that is, $g < h$ implies $fg < fh$ and $gf < hf$ for all $f, g, h \in G$.

Moreover, this ordered group is Archimedean, which means that for any two elements $g, h \in G$ such that $\mathrm{id} < g < h$ there exists a natural number n such that $h < g^n$. To verify this, just observe that the sequence $g^n(x_0)$ tends to infinity for every x_0 (use 5.54 or look into the proof of that result).

Now every Archimedean ordered group is isomorphic to a subgroup of the ordered group of real numbers with addition. We shall give a proof of this embedding theorem in 7.8. □

5.58 Warning The last result should not be misunderstood. It does *not* say that a freely acting group of homeomorphisms of $\mathbb{R}$ *acts like* a subgroup of $\mathbb{R}$ considered as the group of standard translations $x \mapsto x+r$. If G does not happen to be cyclic, then it is true that G is a dense subgroup of $\mathbb{R}$ by 1.4, and every orbit $G(x)$ is order isomorphic to this dense subgroup of $\mathbb{R}$. This does not allow us to conclude that $G(x)$ is dense in $\mathbb{R}$. (If this is the case, then the action of G on $\mathbb{R}$ is indeed the standard action.) For counter-examples and for a complete classification of the groups G as transformation groups, see LÖWEN 1985.

5.59 Theorem *In the group $\mathcal{H}$ of homeomorphisms of $\mathbb{R}$, every element is a product of no more than four involutions. The number four is best possible. Every monotone homeomorphism is a product of two translations (in the sense of 5.52).*

Proof (1) The standard translation $x \mapsto x + 1$ is the product of the two involutions $x \mapsto -x$ and $x \mapsto 1-x$. Since every translation t is conjugate to this one (5.54), it follows that t is a product of two involutions as well.

(2) We show that every monotone homeomorphism g is a product of two translations (and, hence, of four involutions). On each connected component of the complement of its fixed point set, g induces a translation. Since all translations are conjugate, we see that g is conjugate to

a homeomorphism h such that $|h(x) - x|$ is bounded by 1. Then $t = ht_2$ is fixed point free, where $t_2(x) = x + 2$. Thus $h = tt_2^{-1}$ is a product of two translations, and this carries over to the conjugate g of h.

(3) If $h \in \mathcal{H}$ is antitone, consider the antitone homeomorphism $g = t_{-1}h$ and note that $g(x) < h(x)$ for all x. The antitone continuous map defined by $i(x) = \min\{g(x), g^{-1}(x)\}$ is an involution $i \in \mathcal{H}$. Indeed, if $g(x) \le g^{-1}(x)$, for instance, then $gg(x) \ge x$, hence $ii(x) = i(g(x)) = x$. For all x we have $i(x) \le g(x) < h(x)$; therefore, $t = ih$ is fixed point free, and $h = it$ is a product of three involutions.

(4) Let $h(x) = x$ for $x \le 0$ and $h(x) = 2x$ for $x \ge 0$. If h is a product of fewer than four involutions, then their number must be two, because h is monotone. Suppose that $h = ij$ is a product of two involutions. Then the conjugate $ihi = ji = h^{-1}$ has fixed point set $F_{ihi} = i(F_h) = i(]\,, 0])$. This set is an interval of the form $[a, [$. But clearly h and h^{-1} should have the same fixed point set, a contradiction.   □

The elements of $\mathcal{H}$ that can be written as a product of two involutions have been characterized by YOUNG 1994. Moreover, he shows that there are three involutions that generate a dense subgroup of $\mathcal{H}$.

5.60 Theorem   *The centre of $\mathcal{H}$ is trivial, and the commutator subgroup of both $\mathcal{H}$ and $\mathcal{H}^+$ is $\mathcal{H}' = \mathcal{H}^+$.*

Proof (1) The map h_a defined by $h_a(x) = 2x - a$ has $a \in \mathbb{R}$ as its unique fixed point. Any element $g \in \mathcal{H}$ commuting with h_a must fix a.

(2) All commutators $hgh^{-1}g^{-1}$ are monotone. Conversely, we show that every (right) translation is a commutator of two monotone homeomorphisms. Since the translations generate the group $\mathcal{H}^+$ by 5.59, all assertions about commutator groups follow. So let t be a right translation. By 5.54, the two right translations t and t^2 are conjugate; in fact, the remark following 5.54 shows that $t^2 = f^{-1}tf$ for some monotone homeomorphism f. The desired relation $t = t^{-1}f^{-1}tf$ follows.   □

Our next aim is to show that the only normal subgroups of $\mathcal{H}^+$ are the trivial ones $\{\mathrm{id}\}$ and $\mathcal{H}^+$ itself, and the groups $\mathcal{H}_{-\infty}$, $\mathcal{H}_\infty$ and $\mathcal{H}_c$ which we define next.

5.61 Definition   We continue to denote the fixed point set of $h \in \mathcal{H}$ by F_h. The closure $\overline{\mathbb{R} \smallsetminus \{F_h\}}$ of the set of non-fixed points is called the *support* of h.

The groups $\mathcal{H}_{-\infty}$, $\mathcal{H}_\infty$ and $\mathcal{H}_c$ consist of those homeomorphisms of $\mathbb{R}$ whose support is bounded below or bounded above or compact, respectively. Thus, h belongs to $\mathcal{H}_{-\infty}$ or to $\mathcal{H}_\infty$ if, and only if, F_h contains some interval $]\,,a]$ or $[a,\,[$, respectively, and $\mathcal{H}_c = \mathcal{H}_{-\infty} \cap \mathcal{H}_\infty$. Each of these groups is a normal subgroup of $\mathcal{H}^+$, and $\mathcal{H}_c$ is also normal in $\mathcal{H}$ itself.

5.62 Lemma *Every proper normal subgroup* $\mathcal{N} \leq \mathcal{H}^+$ *is contained in* $\mathcal{H}_{-\infty}$ *or in* $\mathcal{H}_\infty$.

Proof (1) We assume that $\mathcal{N}$ contains elements whose support is unbounded below and others whose support is unbounded above. Our aim is to show that then $\mathcal{N}$ contains a right translation. This implies that $\mathcal{N}$ contains all right translations (5.54) and their inverses, the left translations. Finally, 5.59 shows that that $\mathcal{N} = \mathcal{H}^+$.

(2) We examine the tools available for our construction. Every monotone homeomorphism $h \in \mathcal{H}^+$ has a fixed point set F_h, whose complement is a disjoint union of open intervals. On each interval I of this kind, h induces a (left or right) translation. By passing to a conjugate fhf^{-1}, $f \in \mathcal{H}^+$, we may manipulate the fixed point set, which becomes $f(F_h)$. Instead of doing this, we can conjugate the translation induced on an interval I and replace it with any other translation having the same direction (left or right), see Theorem 5.54. This can be done simultaneously for all intervals I; the conjugating homeomorphisms together define a monotone homeomorphism of $\mathbb{R}$ that induces the identity on the set F_h.

Our strategy in applying these tools is to construct, in this order, elements of $\mathcal{N}$ whose fixed point set is bounded above, elements with compact fixed point set, elements with at most one fixed point, and finally translations.

(3) Assume that the support of $h \in \mathcal{N}$ is unbounded above. We shall construct an element $g \in \mathcal{N}$ whose fixed point set is bounded above. If h itself does not have this property, then there is an unbounded, increasing sequence of open intervals I_n, $n \in \mathbb{N}$, on which h induces translations in the same direction (say, right translations). The gap between I_n and I_{n+1} is a closed interval J_n, which contains fixed points and perhaps some other points that may be moved to the left or to the right. Note that J_n is mapped onto itself. Using the tools exhibited in step 2, we may construct a conjugate f of h having 'gap intervals' $J_n = [2n - 1/5, 2n + 1/5]$ and satisfying $\varphi_f(x) = |f(x) - x| > 2/5$ for all $x \in [2n + 1 - 1/5, 2n + 1 + 1/5]$, all $n \in \mathbb{N}$. Note that $\varphi_f(x) < 2/5$ for $x \in J_n$. Now we form the conjugate $f' = t_{-1}ft_1$, for which the roles of the odd and even numbers are interchanged. It is easily verified that the product $g = f'f$ moves every $x \geq 2$ to the right.

(4) If the support of the element g constructed in step 3 is unbounded below, then we can repeat the procedure of step 3 in a symmetrical fashion and obtain an element with compact fixed point set. The same can be achieved if the support of g is bounded below: by assumption, there is an element whose support is unbounded below, and we can do everything over again to obtain an element g' whose fixed point set is bounded below and whose support is bounded above. Then the fixed point set of gg' is compact.

(5) We have obtained an element $k \in \mathcal{N}$ that induces translations on two unbounded intervals $]\,,a[$ and $]b,\,[$. If these translations have the same direction, then the ideas of step 3 can be applied to obtain a translation belonging to $\mathcal{N}$. If the directions are not the same, then it is possible to construct an element with precisely one fixed point by combining the ideas of step 3 with those of the next step 6. The details of this argument are left to the reader (and can be found in FINE–SCHWEIGERT 1955).

(6) Suppose that $l \in \mathcal{N}$ fixes precisely one point, say 0, and moves all $x < 0$ to the right and all $x > 0$ to the left. After taking the inverse, if necessary, this is what we get from step 5. Using the tools of step 2, we obtain a conjugate l' of l that fixes the point 1 and satisfies $l(x) < l'(x)$ for all x. Then $l^{-1}l'$ is a right translation, and this ends the proof. $\square$

5.63 Theorem (a) *The group $\mathcal{H}_c$ of all homeomorphisms of $\mathbb{R}$ with compact supports is simple, that is, it has only the trivial normal subgroups: $\{\mathrm{id}\}$ and $\mathcal{H}_c$ itself.*

   (b) *The only non-trivial normal subgroup of $\mathcal{H}_{-\infty}$ and of $\mathcal{H}_\infty$ is $\mathcal{H}_c$.*

   (c) *The non-trivial normal subgroups of $\mathcal{H}^+$ are precisely $\mathcal{H}_{-\infty}$, $\mathcal{H}_\infty$, and $\mathcal{H}_c$.*

   (d) *The non-trivial normal subgroups of $\mathcal{H}$ are precisely $\mathcal{H}_c$ and $\mathcal{H}^+$.*

Proof (1) For $u < v$ in $\mathbb{R} \cup \{-\infty, \infty\}$, we introduce the group $\mathcal{H}_{u,v}$ consisting of all $h \in \mathcal{H}$ that induce the identity on $]\,,u] \cup [v,\,[$. These groups are all isomorphic to $\mathcal{H}^+$. An isomorphism $\psi : \mathcal{H}_{u,v} \rightarrow \mathcal{H}^+$ is obtained by 'conjugation' with a homeomorphism $f : \,]u,v[\, \rightarrow \mathbb{R}$. Clearly, the groups $\mathcal{H}_{u,v}$ for $u, v \in \mathbb{R}$ are all conjugate.

(2) If $\mathcal{N} < \mathcal{H}_c$ is a normal subgroup and $h \in \mathcal{N}$ is not the identity, then there is a minimal closed interval $[u, v] \subseteq \mathbb{R}$ such that $h \in \mathcal{H}_{u,v}$. The appropriate isomorphism ψ takes $\mathcal{N} \cap \mathcal{H}_{u,v}$ to a normal subgroup $\mathcal{M}$ of $\mathcal{H}^+$. The support of $\psi(h)$ is unbounded above and below, and we conclude from 5.62 that $\mathcal{M} = \mathcal{H}^+$ and that $\mathcal{N}$ contains $\mathcal{H}_{u,v}$. Hence, it

contains $\mathcal{H}_c$, which is the union of all conjugates of $\mathcal{H}_{u,v}$. This proves assertion (a); the proof of (b) is entirely similar and is omitted.

(3) A proper normal subgroup of $\mathcal{H}^+$ is a normal subgroup of $\mathcal{H}_{-\infty}$ or of $\mathcal{H}_\infty$ by 5.62. Hence, assertion (c) follows from (b). Finally, consider a normal subgroup $\mathcal{N} < \mathcal{H}$ which is not contained in $\mathcal{H}^+$. Then there is an antitone homeomorphism $h \in \mathcal{N}$. Given any right translation t, the commutator $h^{-1}(t^{-1}ht)$ belongs to $\mathcal{N}$. Now $h^{-1}(t^{-1}ht) = (h^{-1}t^{-1}h)t$ is a product of two right translations, and such a product is again a right translation. Hence, $\mathcal{N}$ contains all translations and their products, which exhaust all of $\mathcal{H}^+$; see 5.54 and 5.59.   □

A characterization of the group $\mathcal{H}$ has been mentioned after 3.7. Compare also Section 6 of FINE–SCHWEIGERT 1955. For more general results about simplicity of homeomorphism groups, we refer to EPSTEIN 1970 and references given there.

In relation to the structure of the homeomorphism group $\mathcal{H}$, there are some more results in the literature that we would like to mention.

The first such result, due to ABEL 1980, is about the subgroup $\mathcal{P} \leq \mathcal{H}$ consisting of all monotone, piecewise linear bijections and contrasts the simplicity results about $\mathcal{H}$. In fact, Abel shows that the group $\mathcal{P}$ contains an uncountable collection of normal subgroups.

The following example given by COHEN–GLASS 1997 is of interest: if p is an odd prime, then the elements $f, g \in \mathcal{H}$ defined by $f(x) = x+1$ and $g(x) = x^p$ generate a free subgroup of $\mathcal{H}$. BENNETT 1997 constructs a free subgroup of $\mathcal{H}$ with n free generators, for every $n \in \mathbb{N}$. Actually, $\mathcal{H}$ contains a free subgroup with a free generator set having the cardinality of the continuum. For a proof, see BLASS–KISTER 1986.

Banach proved the following strange property of the group of ordinary translations of the real line. There is a decomposition of $\mathbb{R}$ into disjoint sets M_1, M_2, each having the same cardinality as $\mathbb{R}$, such that for each $x \in \mathbb{R}$ and each $i \in \{1, 2\}$ the cardinality of the symmetric difference $(M_i \cup (M_i + x)) \setminus (M_i \cap (M_i + x))$ is strictly smaller than that of $\mathbb{R}$. For a recent proof, see ABEL–MISFELD 1990, where it is also shown that the corresponding statement for $\mathbb{Q}$ is false.

Weird topologies on the real line

Despite their strange properties, the topologies described in this subsection occur quite naturally. The first of them represents a pathology that causes unavoidable trouble in the theory of Lie groups (see 5.66, 5.68).

5.64 Definition: The torus   Recall the surjective local homeomorphism $p : \mathbb{R} \to \mathbb{S}_1$ introduced in 5.16. (A *local homeomorphism* is a map of topological spaces $f : X \to Y$ such that each $x \in X$ has an open neighbourhood that is mapped homeomorphically onto some open neighbourhood of $f(x)$.) We consider the product space $\mathbb{T}^2 = \mathbb{S}_1 \times \mathbb{S}_1$, which can be embedded in $\mathbb{R}^3$ as the surface of a doughnut, and the surjective map

$$q = p \times p : \mathbb{R}^2 \to \mathbb{T}^2 : (x, y) \mapsto \left(e^{2\pi i x}, e^{2\pi i y}\right) .$$

If $I, J \subseteq \mathbb{R}$ are any two open intervals of length 1, then the restriction of q to the open square $S = I \times J$ is a homeomorphism onto its image $q(S)$, and $q(S)$ is open in $\mathbb{T}^2$. This follows from the corresponding properties of p; compare 5.16. We conclude that $\mathbb{T}^2$ is a 2-manifold and that the surjection q is continuous and open. In particular, $\mathbb{T}^2$ carries the quotient topology with respect to q. From the periodicity of p it follows that the inverse images $q^{-1}(u)$ of the points $u \in \mathbb{T}^2$ are precisely the cosets of the subgroup $\mathbb{Z}^2 \leq \mathbb{R}^2$. Since q is a group homomorphism, it follows that $\mathbb{T}^2$ may be considered as the factor group:

$$\mathbb{T}^2 = \mathbb{S}_1 \times \mathbb{S}_1 = \mathbb{R}^2/\mathbb{Z}^2 .$$

5.65 Example: The torus topology on $\mathbb{R}$ The restriction of the map q (see 5.64) to the subgroup $\mathbb{R}(1, \sqrt{2})$ is injective because $\sqrt{2}$ is irrational. Therefore, we can identify $\mathbb{R}$ as a set with its image under the map $t \mapsto q(t, t\sqrt{2})$. The torus topology is defined as the subspace topology induced on this copy of $\mathbb{R}$ by the torus $\mathbb{T}^2$. We shall write T for the set of real numbers with this topology.

We note that $T \subseteq \mathbb{T}^2$ is a dense subset; this follows from a well known theorem of Kronecker, which asserts that even the cyclic subsemigroup of $\mathbb{T}^2$ generated by $q(\sqrt{3}, \sqrt{6})$ is dense. We shall prove Kronecker's theorem in 5.69. Each coset of $T + v$ arises as the image of T under the homeomorphism $u \mapsto u + v$ of $\mathbb{T}^2$, hence $T + v$ is dense, as well. In particular, the complement of T is dense.

5.66 Remark   The torus is a topological group (addition and subtraction are continuous, compare Section 8) and, in fact, a Lie group (the group operations are even differentiable). The continuity of group operations carries over from $\mathbb{T}^2$ to the subgroup T, i.e., T is a topological subgroup. The study of this subgroup is forced on us by Lie theory. Indeed, the map q can be viewed as the exponential map from the Lie algebra $\mathbb{R}^2$ onto $\mathbb{T}^2$, and T is the exponential image of the Lie subalgebra

$\mathbb{R}(1, \sqrt{2}) \le \mathbb{R}^2$. It follows from the fundamental theorems of Lie theory that T is a Lie group itself, but unfortunately, since T is not closed in $\mathbb{T}^2$, the topology making T into a Lie group is not the topology inherited from $\mathbb{T}^2$. This is a widespread phenomenon among Lie groups; compare also 5.68 below.

5.67 Proposition *The torus topology on $\mathbb{R}$ is homogeneous and connected, but not locally connected and not locally compact. All points of T are non-separating, and T is locally homeomorphic to $\mathbb{R} \times \mathbb{Q}$.*

Proof Since T is a topological group (5.66), the maps $u \to u + v$ are homeomorphisms of T for each $v \in T$, and homogeneity follows. The continuity of the surjection $q : \mathbb{R} \to T$ shows that T is connected.

The next two in our list of properties follow from the local homeomorphism to $\mathbb{R} \times \mathbb{Q}$, which is obtained as follows. The inverse image $A = q^{-1}(T)$ equals $V + \mathbb{Z}^2$, where $V = \mathbb{R}(1, \sqrt{2})$. The map q is a local homeomorphism from this set onto T because $q : \mathbb{R}^2 \to \mathbb{T}^2$ is a local homeomorphism (5.64). Therefore, it suffices to consider A. Using a real vector space decomposition $\mathbb{R}^2 = V \times W$, we obtain that $A = V \times (W \cap A)$, and $W \cap A \cong A/V$ is a non-cyclic subgroup of $W \cong \mathbb{R}$ generated by two elements (namely, the elements corresponding to two generators of $\mathbb{Z}^2$, compare 1.6). By 1.4 or 1.6b, this subgroup is dense, and by 3.5 it is chain isomorphic and, hence, homeomorphic, to the set of rationals, $\mathbb{Q}$.

It remains to be shown that no point separates T. By homogeneity, it suffices to do this for the neutral element $0 \in T$. The proof of the density of $T \subseteq \mathbb{T}^2$ shows in fact that $q(P(1, \sqrt{2}))$, where $P =]0, [$, is dense in $\mathbb{T}^2$. In particular, this connected set is dense in $T \smallsetminus \{0\}$, which shows the connectedness of the latter set. $\qquad\square$

5.68 Remark: Retrieving the original topology of $\mathbb{R}$ The fact that T is locally homeomorphic to $\mathbb{R} \times \mathbb{Q}$ allows us to reconstruct the standard topology of $\mathbb{R}$. In a neighbourhood homeomorphic to $\mathbb{R} \times \mathbb{Q}$ (every open set in $\mathbb{R} \times \mathbb{Q}$ contains such a neighbourhood), the sets of the form $\mathbb{R} \times \{r\}$, $r \in \mathbb{Q}$, are precisely the arcwise connected components. Under the map q, these sets correspond to open intervals in the real line. Therefore, a method to reconstruct the original topology is to define the arcwise connected components of all open sets of T to be open. This method works for non-closed Lie subgroups of Lie groups in general; see GLEASON–PALAIS 1957, HOFMANN–MORRIS 1998 Appendix 4, STROPPEL 2006 §39.

This ends our treatment of the torus topology, but before we start another topic we want to supply a statement and proof of Kronecker's Theorem, which we relied on in 5.65. In the course of the proof it will become clear in which sense the theorem generalizes the result 1.6b on subgroups of $\mathbb{R}^+$ having two generators. The proof given here is taken from BOURBAKI 1966 VII §1.3. A very direct proof is given by ADAMS 1969 4.3. We formulate Kronecker's theorem (in its first version) for the n-dimensional torus $\mathbb{T}^n = \mathbb{R}^n/\mathbb{Z}^n$. Let $q : \mathbb{R}^n \to \mathbb{T}^n$ denote the quotient map.

5.69 Kronecker's Theorem *Let $w = (w_1, \ldots, w_n) \in \mathbb{R}^n$ be given and suppose that the numbers $1, w_1, \ldots, w_n$ are linearly independent over $\mathbb{Q}$. Then the semigroup generated by $q(w)$ is dense in $\mathbb{T}^n$.*

Proof (1) First we look at the cyclic group generated by $q(w)$ rather than the semigroup; in other words, we consider also multiples $zq(w)$ with negative $z \in \mathbb{Z}$. By continuity of q, it suffices to show that the group $H \leq \mathbb{R}^n$ generated by w together with $\ker q = \mathbb{Z}^n$ is dense. (In the case $n = 1$, this is the result 1.6b.)

If H is not dense, then we apply the classification of all closed subgroups of $\mathbb{R}^n$; see 8.6. It implies that there is a basis $v_1, \ldots, v_n$ of $\mathbb{R}^n$ such that the closure $\overline{H}$ consists of all linear combinations $v = \sum_{i=1}^{n} r_i v_i$ satisfying $r_i \in \mathbb{Z}$ for $i > i_1$ and $r_i = 0$ for $i > i_2$, where $i_1 \leq i_2 \leq n$. We have $i_1 < n$ since $\overline{H}$ is a proper subgroup of $\mathbb{R}^n$, and $i_2 = n$ because $\mathbb{Z}^n \leq H$ generates the real vector space $\mathbb{R}^n$. In particular, there is a non-trivial linear form $f(v) = r_n$ on $\mathbb{R}^n$ which takes only integer values on H. The standard basis vectors $e_1, \ldots, e_n$ belong to $\mathbb{Z}^n$, hence the values $f(e_i)$ are integers, not all equal to zero. We find that $f(w) = \sum_{i=1}^{n} w_i f(e_i) \in \mathbb{Z}$, and this contradicts our assumption of linear independence.

(2) We have shown in particular that the integral multiples of $q(w)$ accumulate at $0 \in \mathbb{T}^n$. This implies (by continuity of subtraction in $\mathbb{T}^n$) that there is a strictly increasing sequence of natural numbers n_k such that $n_k q(w) \to 0$ in $\mathbb{T}^n$. Now we use a metric d on $\mathbb{T}^n$. Suppose we have $d(zq(w), t) < \varepsilon$ for some negative integer z and $t \in \mathbb{T}^n$. By continuity of addition, $(z + n_k)q(w) \to zq(w)$. Thus, we can find $n_k > -z$ such that $d((z + n_k)q(w), zq(w)) < \varepsilon$, and then $d((z + n_k)q(w), t) < 2\varepsilon$ as desired. □

We use this opportunity to rephrase Kronecker's Theorem in terms of Diophantine approximations.

5.70 Kronecker's Theorem, second version *Let $w_1, \ldots, w_n$ be real numbers such that $1, w_1, \ldots, w_n$ are linearly independent over $\mathbb{Q}$. For any numbers $v_1, \ldots, v_n \in \mathbb{R}$ and $\varepsilon > 0$ there exist integers $k > 0$ and $z_1, \ldots, z_n$ such that for each $i \in \{1, \ldots, n\}$ we have*

$$|v_i - kw_i - z_i| < \varepsilon \ .$$

Proof Let $w = (w_1, \ldots, w_n)$ and $v = (v_1, \ldots, v_n)$. Let $U \subseteq \mathbb{R}^n$ be the open set $U = \{ x - z \mid x \in \mathbb{R}^n, z \in \mathbb{Z}^n, \|v - x\| < \varepsilon \}$. Here we use the norm $\|y\| = \max\{|y_1|, \ldots, |y_n|\}$ on $\mathbb{R}^n$. With respect to the quotient map $q : \mathbb{R}^n \to \mathbb{T}^n$, the set U is saturated, that is, $U = q^{-1}q(U)$, and therefore $q(U)$ is open in $\mathbb{T}^n$. According to 5.69, the set $q(U)$ contains some multiple $kq(w)$, $k \in \mathbb{N}$. This means exactly what is stated in the assertion. We remark that these arguments can be read backwards to show that the present version of the theorem implies the earlier, group theoretic one.   □

5.71 Uniform distribution The following sharper result is due to Bohl, Sierpiński and Weyl: under the assumptions of 5.70, the sequence $kw \in \mathbb{R}^n$, $k \in \mathbb{N}$, is *uniformly distributed modulo* 1. This means that for every box $B(\varepsilon) = \{y \in \mathbb{R}^n \mid \|y - v\| < \varepsilon\}$, the proportion of elements kw contained in $B(\varepsilon) + \mathbb{Z}^n$ among the first N members of the sequence converges to the box volume $(2\varepsilon)^n$ as $N \to \infty$. For a proof and more information, see HLAWKA *et al.* 1991 Proposition 4 on p. 26 or KUIPERS–NIEDERREITER 1974.

5.72 Example: The topologies of semicontinuity Let X be a topological space. A lower semicontinuous map $f : X \to \mathbb{R}$ is a map such that for each $b \in \mathbb{R}$ the inverse image of $]b, \, [$ is an open set. This property can also be described as continuity with respect to the topology on $\mathbb{R}$ generated by a basis consisting of all the intervals $]b, \, [$ with $b \geq -\infty$. In fact, the basis coincides with the topology in this case, and the open sets form a chain with respect to inclusion.

From this it follows that the topology of lower semicontinuity is connected (even every subset is connected), and the subspaces $[x, \, [$ are compact. Points are not closed sets, hence there is no metric generating this topology. There is a countable basis: just restrict b to be a rational number. The ordinary topology of $\mathbb{R}$ is the coarsest one that contains this topology together with its mirror image, the topology of upper semicontinuity. This expresses the familiar fact that a map is continuous if, and only if, it is both upper and lower semicontinuous.

5.73 Example: The Sorgenfrey topology   The *Sorgenfrey line S*
is the set $\mathbb{R}$ endowed with the topology generated by all intervals of the
form $]a, b]$. For a metric space X, continuity of a map $f : S \to X$ means
that the left-sided limit $f(s-)$ exists for each $s \in S$ and coincides with
$f(s)$. The Sorgenfrey line obviously is separable, but the topology does
not have a countable basis. Indeed, a basis must contain at least one
interval $]a, b]$ for each $b \in \mathbb{R}$. A separable metric space has a countable
basis, consisting of all balls having rational radius, and centre belonging
to a given countable dense set; compare 61.13. Consequently, S is not
metrizable.

The sets $]\ , x]$ and $]x, [$ are open in S, and S is not connected. However, S can be embedded as a dense subspace of some connected and
locally pathwise connected space; see FEDELI–LE DONNE 2001. The sequence $(1/n)_n$ has no convergent subsequence. Every open set contains
a translate of this sequence (up to finitely many elements), hence no
neighbourhood of any point is sequentially compact and S is not locally
compact.

In SORGENFREY 1947, this topology on $\mathbb{R}$ was introduced as an example of a paracompact (and hence normal) space whose Cartesian square
$S \times S$ is not normal; compare also VAN DOUWEN–PFEFFER 1979.

5.74 Note   We mention some other results pertaining to our present
topic. In BAKER *et al.* 1978, it is shown that there are many topologies
on the real line that produce the same set of convergent sequences as
the standard topology. The infimum of these topologies is the cofinite
topology (precisely the complements of all finite sets are open); this
topology itself does not have the property mentioned. In GUTHRIE *et al.*
1978 a maximal connected topology on $\mathbb{R}$ is constructed that is finer than
the standard topology. VAN DOUWEN–WICKE 1977 give an example of a
topology on $\mathbb{R}$ having an especially strange cocktail of properties. Finally
we mention the article CH'ÜAN–LIU 1981 on topologies making $\mathbb{R}^+$ into
a topological group; we shall return to this in Section 8.

5.75 Note: Ordering and topology on $\mathbb{R}$ Following the general
pattern of this book, the reader might expect to see a section on the
relationships of the ordering and the topology on the real line, such
as the one on ordering and group structure which follows (Section 7).
However, there is no such section, and this is due to the results proved in
the present section, which say that not only does the ordering determine
the topology (3.1), but also the topology determines the ordering up
to reversal (that is, up to interchanging $<$ with $>$); this was the key

step in the proof of the characterization theorem 5.10. One cannot expect something new to arise from combining two structures that are essentially the same. Compare, however, 5.51.

Exercises

(1) Consider a topological space $Z = X \cup Y$. A subset $A \subseteq X \cap Y$ which is open (closed) in both X and Y is open (closed) in Z.

(2) Consider the split reals (Section 3, Exercise 1) with the topology induced by the ordering and compare the topological properties of this space to those of $\mathbb{R}$.

(3) Prove the properties of the counter-examples described before 5.21 and in 5.23.

(4) Show that the space $\mathfrak{I} = \mathbb{R} \smallsetminus \mathbb{Q}$ of irrational numbers is n-homogeneous for each $n \in \mathbb{N}$.

(5) Divide the triangle T with sides of length 3, 4 and 5 by an altitude into two similar triangles T_0 and T_1, where T_0 is the smaller one. Proceed by induction: divide each T_k by an altitude into similar triangles T_{k0}, T_{k1}, the smaller one being T_{k0}. Write each real number in the interval $[0, 1]$ as a binary expansion $c = 0.c_1 c_2 \ldots$, put $c|\nu = c_1 c_2 \ldots c_\nu$, and map c onto $T_c = \bigcap_\nu T_{c|\nu}$. Show that this map is well-defined and is a continuous surjection $\varphi : [0, 1] \to T$.

(6) Give an example of a subchain S of $\mathbb{R}$ such that the order topology of S is different from the subspace topology.

(7) Give an example of a subspace of $\mathbb{R}$ which is complete with respect to a compatible metric, but not complete with respect to the ordering.

(8) Show that every continuous, increasing self-map of the long line $\mathbb{L}$ has a fixed point.

(9) Use back and forth induction as in 3.4 to show that any two countable, atomless Boolean algebras are isomorphic. (A Boolean algebra is the same thing as a complemented distributive lattice.)

(10) Let $\mathfrak{I} = \mathbb{R} \smallsetminus \mathbb{Q}$. Show that the subset $S := \mathbb{Q}^2 \cup \mathfrak{I}^2$ of $\mathbb{R}^2$ is arcwise connected (REN 1992).

6   The real numbers as a field

The real numbers equipped with their addition and multiplication form a field. The field structure has already appeared explicitly in this book (Section 2), and was used implicitly on several occasions, for example, in Section 3. The present section will focus on properties concerning the field structure alone. There will be other sections on $\mathbb{R}$ as an ordered field (Section 11) or as a topological field (Section 13). In order to emphasize the roles played by ordering and topology, we treat these aspects in separate sections although the ordering and, hence, the order topology of the real number field are completely determined by the algebraic structure (the positive real numbers are precisely the non-zero squares).

Following the general pattern of this book, we begin by examining the precise relationship between addition and multiplication in a field; in other words, we scrutinize the field axioms; compare also Section 64.

6.1 The field axioms   A *skew field* consists of two group structures $(F, +)$, with neutral element 0, and $F^\times = (F \smallsetminus \{0\}, \cdot)$, with neutral element 1, having almost the same underlying set. Multiplication has to be defined for all pairs in $F \times F$, and the two structures are connected by the two *distributive laws*

$$a(x + y) = ax + ay$$
$$(x + y)a = xa + ya \ .$$

Usually, *commutativity* of addition $x + y = y + x$ is stipulated as an additional axiom, but this can be proved by computing $(1 + x)(1 + y)$ in two ways, using the left and right distributive laws one after the other, in different orders.

Commutativity of multiplication $xy = yx$ is another matter and holds only for a subclass of all skew fields, the *fields*, which of course include the real numbers. The standard example of a proper (non-commutative) skew field is provided by the quaternions $\mathbb{H}$; see Section 13, Exercise 6, or 34.17.

It is often intriguing to beginners that the neutral element 0 has to be excluded from the multiplicative group, which means essentially that *division by zero* is banned. This is unavoidable because we do not want to accept fields consisting of 0 alone: the distributive law enforces that $0x = 0$ for all x (indeed, $0x = (0 + 0)x = 0x + 0x$); if we allow division by 0, we end up with $1 = 00^{-1} = 0$ and $x = 1x = 0x = 0$ for all x. (As a ring, $\{0\}$ is usually accepted, but not as a field.)

To highlight the importance of the distributive laws, we interpret them as follows. We consider the *left multiplication maps* $\lambda_a : x \mapsto ax$, $a \in F$. The first distributive law asserts that λ_a is an endomorphism of the additive group $(F, +)$, and the second one says that the map sending a to λ_a is a homomorphism from $(F, +)$ into the semigroup $\operatorname{End} F$ of all endomorphisms of $(F, +)$. A similar statement, with the roles of the distributive laws interchanged, holds for the right multiplication maps $\rho_a : x \mapsto xa$.

While students are sometimes tempted to divide by zero, ignoring the fundamental lack of symmetry between 0 and 1, the much stronger asymmetry of the distributive laws usually goes unnoticed, and no one misses the *'dual distributive law'* $(ab) + c = (a + c)(b + c)$ among the field axioms. This law does hold in Boolean algebras (where one usually writes $\vee$ and $\wedge$ for the operations of addition and multiplication), but there we do not have additive or multiplicative inverses. In the presence of the other field axioms, the dual distributive law becomes contradictory: evaluating $(1 + 1)(0 + 1)$ in two ways one would obtain that $1 = 0$.

Interesting *generalizations of the skew field axioms* are obtained if one dismisses the associative laws of addition and multiplication and the distributive laws wholly or in part. From the group axioms, one retains that the equations $ax = b$, $xa = b$ for $a \neq 0$ and $a + x = b$, $x + a = b$ for arbitrary a have unique solutions. This leads to notions such as nearfield, quasifield, semifield. These structures are important in geometry because the weak axioms, together with a condition on the unique solvability of certain types of 'linear' equations, still ensure that $F \times F$ with lines defined by the equations of type $y = ax + b$ or $x = \text{const.}$ forms an affine plane, obeying Euclid's parallel postulate. A detailed discussion and numerous examples can be found in SALZMANN *et al.* 1995. Nearfields (lacking only one distributive law) are also important because of their relationship with sharply 2-transitive permutation groups; see DIXON–MORTIMER 1996 7.6.

6.2 Independence of the field operations (a) The additive group of real numbers does not determine the multiplication of the field $\mathbb{R}$. Indeed, the field of complex numbers has an additive group isomorphic to that of the real numbers (see 1.15), and $\mathbb{C}^\times \not\cong \mathbb{R}^\times$, because $\mathbb{C}^\times$ has a large torsion subgroup $\mathbb{Q}^+/\mathbb{Z}$ (compare 1.20 and 1.24).

(b) Conversely, suppose that F is a field such that $F^\times \cong \mathbb{R}^\times \cong C_2 \times \mathbb{R}^+$; compare 2.2. If char $F = 0$, then $F^+ \cong \mathbb{R}^+$ by 1.14. (We shall return to this case in part (d) below.) If char $F = p$ is a prime, then $p \neq 2$ and $F^\times \cong \mathbb{R}^\times$ contains the multiplicative group $\mathbb{F}_p^\times$ of the prime field, hence $p = 3$.

(c) Now we can produce an example of a field F such that $F^\times \cong \mathbb{R}^\times$ and $\mathbb{F}^+ \not\cong \mathbb{R}^+$. Using model theory, an example with similar properties is constructed by CONTESSA *et al.* 1999 6.2. As in 2.10, we consider the field of Puiseux series

$$F = \mathbb{F}_3((t^{1/\infty}));$$

compare 64.24. We have char $F = 3$, hence all non-zero elements of F^+ have order 3, and $F^+ \not\cong \mathbb{R}^+$. Every series $a \in F^\times$ can be written uniquely as $a = a_0 t^r (1 + b)$, where $r = v(a) \in \mathbb{Q}$ and $a_0 \in \mathbb{F}_3^\times$. Here, v

is the natural valuation of F, as in 64.24. It follows that $1 + b$ belongs to the subgroup $F_1 = \{1 + c \in F \mid v(c) > 0\} \le F^\times$, and we obtain a direct product decomposition of the multiplicative group

$$F^\times \cong \mathbb{F}_3^\times \times t^\mathbb{Q} \times F_1.$$

As in the proof of 2.10, it can be shown using Lemma 2.9 that F_1 has unique roots; in steps (2) and (3) of the proof, the prime 2 has to be replaced with 3. This is possible since F has characteristic 3, whence $\left(\sum_{i \ge k} c_i t^{i/n}\right)^3 = \sum_{i \ge k} c_i^3 t^{3i/n}$. It follows that $F^\times \cong C_2 \times H$, where the group H has unique roots.

We want to apply 1.14 in order to show that $H \cong \mathbb{R}^+$, hence we need to determine $\operatorname{card} H = \operatorname{card} F$. The inclusions $\mathbb{F}_3^\mathbb{N} \subseteq F \subseteq \mathbb{F}_3^\mathbb{Q}$ show that $\operatorname{card} F = 2^{\aleph_0} = \operatorname{card} \mathbb{R}$ (compare 1.10), and 2.2 implies that $F^\times \cong C_2 \times \mathbb{R}^+ \cong \mathbb{R}^\times$.

See 14.7 and 34.2 for analogous questions about the fields $\mathbb{C}$ and $\mathbb{Q}$. (d) Finally, we give an example of a field F such that $F^+ \cong \mathbb{R}^+$ and $F^\times \cong \mathbb{R}^\times$ are separately isomorphic to their counterparts in $\mathbb{R}$ but $F \not\cong \mathbb{R}$ as a field. See 24.4f for other examples with these properties, and 14.7 for an analogous example concerning the field $\mathbb{C}$.

The field $\mathbb{R}$ may be described as an algebraic extension of a purely transcendental extension $\mathbb{Q}(T)$; in other words, T is a transcendency basis of $\mathbb{R}$ over $\mathbb{Q}$; compare 64.20. Now let $S \subseteq T$ be a proper subset having the same cardinality as T. Then $\mathbb{Q}(S) \cong \mathbb{Q}(T)$, and this field has the same cardinality as its algebraic closure, $\mathbb{C}$; see 64.16. From the inclusion $S \subseteq T$ we obtain a proper inclusion of the algebraic closures $\mathbb{Q}(S)^\natural \subseteq \mathbb{Q}(T)^\natural = \mathbb{C}$ (proper, because elements of $T \smallsetminus S$ do not belong to $\mathbb{Q}(S)^\natural$; compare 64.20), and we define F to be the field

$$F = \mathbb{Q}(S)^\natural \cap \mathbb{R}$$

of all real numbers that are algebraic over $\mathbb{Q}(S)$. The additive group F^+ is uniquely divisible (since $\operatorname{char} F = 0$) and has the same cardinality as $\mathbb{R}$, hence $F^+ \cong \mathbb{R}^+$ by 1.14. The multiplicative group is a direct product $F^\times = \langle -1 \rangle \times F_{\mathrm{pos}}^\times$, where $F_{\mathrm{pos}}^\times$ consists of the positive elements. By our construction, the latter group has unique roots, and 1.14 shows that it is isomorphic to $\mathbb{R}^+$. By 2.2, this implies that $F^\times \cong \mathbb{R}^\times$. As a field, however, F is not isomorphic to $\mathbb{R}$. Indeed, the positive elements of F are precisely the non-zero squares, hence a field isomorphism $F \to \mathbb{R}$ would be an order isomorphism (compare 11.7), but F is a dense proper subchain of $\mathbb{R}$, hence F is not complete.

6.3 Subfields of $\mathbb{R}$ Every subfield $F \leq \mathbb{R}$ contains the element 1 and, hence, the subfield generated by 1, the *prime field* $\mathbb{Q}$ of $\mathbb{R}$. Thus F may be described as an extension field of $\mathbb{Q}$; compare Section 64. Easy examples are the fields $\mathbb{Q}(\sqrt{2}) = \{q + r\sqrt{2} \mid q, r \in \mathbb{Q}\}$, and the simple transcendental extensions $\mathbb{Q}(t)$ (e.g., $t = \pi$), consisting of all rational expressions $p_1(t)/p_2(t) \in \mathbb{R}$, where p_1, p_2 are polynomials with rational coefficients and $p_2 \neq 0$. Other nice subfields include the field $\mathbb{R} \cap \mathbb{Q}^\natural$ of all real algebraic numbers, the Euclidean closure and the Pythagorean closure of $\mathbb{Q}$; see 12.8. A remarkable fact is that the vector space dimension of $\mathbb{R}$ over any proper subfield is infinite, as a consequence of Theorem 12.15.

Following our general scheme, we look for homomorphic images of the field $\mathbb{R}$, but we do not find much. This is because, for every homomorphism $\varphi : F \to H$ between fields, the kernel $\varphi^{-1}(0)$ is an ideal of F, and since F is a field, its only ideals are $\{0\}$ and F; compare 64.3. This means that φ is a monomorphism (so the image field is nothing new) or $\varphi = 0$. Of course, in the latter case $\varphi(1) \neq 1$, so φ is usually not considered to be admissible as a homomorphism between fields.

For *endomorphisms* $\varphi : \mathbb{R} \to \mathbb{R}$, we can say even more, but we need the inextricable tie between the field structure and the ordering of $\mathbb{R}$, which we shall study systematically later on. At first, it seems conceivable that $\mathbb{R}$ is isomorphic to some proper subfield of itself or that $\mathbb{R}$ has non-trivial automorphisms, but none of this is true; the following result is due to DARBOUX 1880.

6.4 Rigidity Theorem (Darboux) *The only non-zero endomorphism of the field* $\mathbb{R}$ *is the identity. In particular, the automorphism group of* $\mathbb{R}$ *is trivial:* $\operatorname{Aut}\mathbb{R} = \{\operatorname{id}\}$.

Proof A non-zero endomorphism φ is injective by the preceding remarks. From $\varphi(1) \neq 0$ and $\varphi(1)^2 = \varphi(1^2)$ we conclude that $\varphi(1) = 1$. It follows that φ induces the identity on $\mathbb{Q}$.

Now we use the fact that the positive real numbers are precisely the non-zero squares, and its consequence, that every endomorphism sends positive numbers to positive numbers and, hence, preserves the ordering; compare 11.7. If there is a number r such that $\varphi(r) \neq r$, then there is a rational number q between r and $\varphi(r)$, and φ cannot preserve the ordering of the pair q, r. $\qquad\square$

A skew field without any non-trivial automorphisms is said to be *rigid*, thus we have proved that $\mathbb{R}$ is rigid. A well known non-rigid field is the field of complex numbers with its conjugation automorphism. There are subfields of $\mathbb{R}$ admitting the same type of automorphism, for example, the field $\mathbb{Q}(\sqrt{2})$ has the automorphism $q + r\sqrt{2} \mapsto q - r\sqrt{2}$. The simple transcendental extension $\mathbb{Q}(t)$ has automorphisms sending the element t to $(at+b)/(ct+d)$, where $ad-bc \neq 0$; see 64.19 and references given there. Apart from non-trivial automorphisms, the field of complex numbers has proper (i.e., non-surjective) endomorphisms; see 14.9. We remark here that $\mathbb{C}$ has rigid extension fields. In fact, DUGAS–GÖBEL 1987 show that every field can be embedded in an extension field with prescribed automorphism group.

Exercises

(1) The field $\mathbb{R}$ has an uncountable, strictly increasing family of subfields.

(2) We denote by $\mathbb{Z}[[x]]^{\flat} := \{\sum_{n \geq 0} a_n x^n \mid a_n \in \mathbb{Z} \wedge \exists_{C,k \in \mathbb{N}} \forall_{n \in \mathbb{N}} |a_n| \leq Cn^k\}$ the set of all integral power series with polynomially bounded coefficients. Show that $\mathbb{Z}[[x]]^{\flat}$ is a subring of the integral power series ring $\mathbb{Z}[[x]]$ and that the quotient ring $\mathbb{Z}[[x]]^{\flat}/(1 - 2x)$ is isomorphic to the field $\mathbb{R}$. (See also Section 51, Exercise 3.)

7   The real numbers as an ordered group

We take it as a known fact that the real numbers with their usual addition and their usual ordering form an ordered group in the sense of the following definition.

7.1 Definition   An *ordered group* is a set G equipped with two structures, a group structure $(G, +)$ and a chain structure $(G, <)$, related by the *law of monotonicity*

$$x < y \quad \Longrightarrow \quad x + c < y + c \text{ and } c + x < c + y,$$

which holds for all $x, y, c \in G$. Even though we write $+$ for the operation in G (because we think of the example $\mathbb{R}^+$), *we do not assume that G is commutative.*

The law of monotonicity admits an interpretation reminiscent of the one we gave for the distributive laws in a skew field, see 6.1. We consider the *left and right addition maps* $\lambda_c, \rho_c : G \to G$, defined by $\lambda_c(x) = c+x$ and $\rho_c(x) = x + c$. The law of monotonicity expresses that these two maps are endomorphisms (order-preserving maps) of the chain $(G, <)$. Observing that, for instance, $\lambda_c \lambda_{-c} = \lambda_{-c} \lambda_c = \mathrm{id}_G$, we see that these maps are in fact automorphisms (order-preserving bijections). Observing moreover that $\lambda_{y-x}(x) = y$, we obtain the following proposition. In passing, we also note that the inversion map $g \mapsto -g$ is antitone in every ordered group.

7.2 Proposition *The chain underlying an ordered group is homogeneous, i.e., it has a transitive automorphism group.* □

Thus the group influences the chain underlying an ordered group. Conversely, if $0 < x \in G$, then $x < x + x$ by monotonicity, and an easy induction shows that the multiples nx with $n \in \mathbb{N}$ form a strictly increasing sequence. In particular, they are all distinct, and the same holds for $x < 0$ by a symmetric argument; these remarks prove the following.

7.3 Proposition *Every ordered group is torsion free.* □

Conversely, a torsion free group which has a (transfinite) descending central series with intersection $\{0\}$ can be ordered; see NEUMANN 1949b, PRIESS-CRAMPE 1983 I §4 Satz 14. In particular, this holds for torsion free abelian groups and for non-abelian free groups.

We mention that every divisible ordered abelian group is isomorphic to an initial subgroup of the additive group of Conway's surreal numbers; see EHRLICH 2001.

The next definition introduces a particularly important feature of the ordered group of real numbers.

7.4 Definition An ordered group G is said to be *Archimedean* if for any two elements $a, b \in G$ such that $0 < a$, there is a natural number $n \in \mathbb{N}$ such that $b \leq na$.

Examples of Archimedean ordered groups are common enough; compare 7.5 below. For examples of non-Archimedean ordered groups, see 7.14 and the exercises.

For brevity, we call an ordered group with a complete underlying chain a *completely ordered group*.

7.5 Theorem *Every completely ordered group is Archimedean. In particular, the ordered group of real numbers is Archimedean.*

Proof If G is not Archimedean, then there is a positive $a \in G$ such that the increasing sequence na, $n \in \mathbb{N}$, is bounded. There exists $c = \sup \mathbb{N}a = \sup(\mathbb{N}a + a)$, but monotonicity implies that $\sup(\mathbb{N}a + a) = (\sup \mathbb{N}a) + a$, which is a contradiction. □

7.6 Proposition *Let G be an ordered group.*

(a) *If G has no smallest positive element, then for every positive $g \in G$ and every $n \in \mathbb{N}$, there exists a positive $h \in G$ such that $nh < g$.*

(b) *If G is Archimedean and has a smallest positive element e, then G is the infinite cyclic group with generator e, endowed with the natural ordering.*

Proof (a) If no smallest positive element exists and $0 < g \in G$ is given, we find an element x with $0 < x < g$. We write $g = x + y$ and find a positive element $z < \min(x, y)$. Then $2z < g$, and assertion (a) follows by induction.

(b) Suppose that e is a smallest positive element and that G is Archimedean. The cyclic group $\mathbb{Z}e$ generated by e exhausts G because for each $g \in G$ there is an $n \in \mathbb{Z}$ such that $ne \leq g < (n+1)e$, and then $g = ne$ or else $0 < g - ne < e$. Clearly, we have $ne < me$ if, and only if, $n < m$.   □

The following Proposition is the first step in proving Hölder's embedding theorem 7.8.

7.7 Proposition *Every Archimedean ordered group is commutative.*

Proof If G is Archimedean and not commutative, then there are positive elements $a, b \in G$ such that $a + b < b + a$. We may write $b + a = a + b + c$, and in view of 7.6 we may assume that there is a positive $d \in G$ with $2d < c$. We find non-negative integers m, n satisfying $md \leq a < (m+1)d$ and $nd \leq b < (n+1)d$, and we set $k = m + n + 2$. Then $kd < a + b + c = b + a < kd$, a contradiction.   □

The following important embedding theorem is essentially due to HÖLDER 1901 §12.

7.8 Theorem (Hölder) *Every Archimedean ordered group G is order-isomorphic to some subgroup of the ordered group $\mathbb{R}$.*

Proof We follow PRIESS-CRAMPE 1983 I §3 Satz 4; compare also BLYTH 2005 10.16. We may assume that G is not trivial. Fix an arbitrarily chosen positive element $e \in G$. For each $a \in G$, we define two subsets of $\mathbb{Q}$, the 'lower set' $L(a)$ and the 'upper set' $U(a)$ such that these two sets form a Dedekind cut of $\mathbb{Q}$. This means that $\mathbb{Q} = L(a) \cup U(a)$ and that $q \in L(a)$, $r \in U(a)$ implies $q < r$. The definition is

$$L(a) = \left\{ \tfrac{m}{n} \mid m \in \mathbb{Z}, n \in \mathbb{N}, me \leq na \right\}$$
$$U(a) = \left\{ \tfrac{u}{v} \mid u \in \mathbb{Z}, v \in \mathbb{N}, ue > va \right\} .$$

If $me \leq na$ and $ue > va$, then $mve \leq nva < nue$ and $mv < nu$. This proves that $L(a) < U(a)$. That $\mathbb{Q}$ is the union of the two sets is a direct consequence of their definition. Moreover, $L(a)$ and $U(a)$ are not empty, since G is Archimedean. Thus we have obtained a Dedekind cut, and we define $f(a) = \sup L(a) = \inf U(a)$, the real number determined by this cut.

We show next that $L(a) + L(b) \subseteq L(a + b)$, whence $f(a) + f(b) \leq f(a + b)$. Together with a similar result for upper sets, this will imply that f is a group homomorphism. So let $me \leq na$ and $ue \leq vb$; then $(mv + nu)e \leq nva + nvb = nv(a + b)$ as desired; note that we used commutativity (7.7) for the last equation.

It remains to be shown that f is injective. So let $0 < a$ in G. By the Archimedean property, we may choose $n \in \mathbb{N}$ such that $e < na$, and then $1/n \in L(a)$ shows that $f(a) > 0$.   □

7.9 Corollary *Every non-trivial completely ordered group is isomorphic to one of the ordered groups $\mathbb{Z}$ or $\mathbb{R}$.*

Proof A completely ordered group G is Archimedean (7.5) and isomorphic to a completely ordered and, hence, closed subgroup of $\mathbb{R}$ by 7.8. On the other hand, every subgroup of $\mathbb{R}$ is either cyclic or dense (1.4), and the assertion follows.   □

7.10 Corollary *If G is an ordered group which is isomorphic to $\mathbb{R}$ as a chain, then G is isomorphic to $\mathbb{R}$ as an ordered group.*   □

In general, the connection between an ordered group and its underlying chain is much looser. For example, ALLING–KUHLMANN 1994 show that for any $\alpha > 0$, there exist $2^{\aleph_\alpha}$ non-isomorphic ordered divisible abelian groups of cardinality $\aleph_\alpha$ that are all isomorphic as chains.

The opposite of 7.10 is also far from true; namely, there are many non-isomorphic ordered groups that are isomorphic to $\mathbb{R}^+$ as groups. For example, every hyperplane of the rational vector space $\mathbb{R}$ (compare 1.11ff) is an ordered group (with the ordering induced from $\mathbb{R}$) and is not completely ordered, but Archimedean. A non-Archimedean ordering on the group $\mathbb{R}^+$ can be obtained using the isomorphism $\mathbb{R}^+ \cong \mathbb{R}^+ \oplus \mathbb{R}^+$ of 1.15 and the *lexicographic ordering* on the latter group, defined by

$$(x, y) \leq (x', y') \;\; \Leftrightarrow \;\; x < x' \text{ or } x = x' \wedge y \leq y' \,.$$

Another example is given by the non-standard rationals; see 22.4 and 22.8. See 7.14 and ALLING–KUHLMANN 1994 for refinements of these constructions.

Subgroups of the ordered group $\mathbb{R}$ abound; compare 1.15. *Quotient groups*, on the other hand, are scarce. Indeed, the kernel of an epimorphism $\varphi : \mathbb{R} \to G$ of ordered groups is an interval since φ is monotone, hence $\varphi = 0$ or φ is an isomorphism.

In contrast to the huge automorphism group of the additive group $\mathbb{R}^+$ (see 1.32), the automorphism group of the ordered group $\mathbb{R}$ is rather tame:

7.11 Theorem   *The endomorphisms of the ordered group $\mathbb{R}$ are precisely the maps $\varphi_a : x \mapsto ax$, where $0 \leq a \in \mathbb{R}$. Thus, the semigroup $\mathrm{End}(\mathbb{R}, +, <)$ is isomorphic to the multiplicative semigroup of non-negative real numbers.*

Proof Let φ be an order-preserving endomorphism. If $\varphi(r) = 0$ for some $r \neq 0$, then $\varphi[0, r] = \{0\}$ by monotonicity, and then $\varphi = 0 = \varphi_0$. Therefore, we may assume that φ is injective. We set $a = \varphi(1) > 0$ and define an order-preserving endomorphism $\psi = \varphi_{a^{-1}}\varphi$; then $\psi(1) = 1$, and we have to show that $\psi = \mathrm{id}$. Unique divisibility of $\mathbb{R}^+$ (see 1.8) implies that $\psi(q) = q$ for all $q \in \mathbb{Q}$. If $\psi(r) \neq r$ for some $r \in \mathbb{R}$, then there is a rational number between $\psi(r)$ and r (see 3.1), and we obtain a contradiction to monotonicity of ψ. □

The preceding result is a special case of the next one. Nevertheless, we include two independent proofs, as both of them are instructive.

7.12 Theorem   *For any subgroup $G \leq \mathbb{R}^+$, the order-preserving group homomorphisms $\varphi : G \to \mathbb{R}^+$ are precisely the maps $\varphi_a : g \mapsto ag$, where $0 \leq a \in \mathbb{R}$.*

Proof As in the previous proof, we only have to consider injective homomorphisms φ. For any two positive elements $g, h \in G$ and $m, n \in \mathbb{N}$ we have the equivalence

$$\frac{m}{n} \leq \frac{h}{g} \iff mg \leq nh \iff m\varphi(g) \leq n\varphi(h) \iff \frac{m}{n} \leq \frac{\varphi(h)}{\varphi(g)} .$$

In other words, the real numbers h/g and $\varphi(h)/\varphi(g)$ are both equal to the supremum of the same set of rational numbers. Hence, the two numbers are equal, and we infer that $\varphi(g)/g = \varphi(h)/h$ is a positive constant $a \in \mathbb{R}$. This shows that $\varphi(g) = \varphi_a(g)$ for $0 < g$, and it follows that this equality holds for all $g \in G$. □

The exponential map is a group isomorphism $\mathbb{R}^+ \cong \mathbb{R}^\times_{\mathrm{pos}}$ (see 2.2), and is monotone. Hence $\mathbb{R}^+$ and $\mathbb{R}^\times_{\mathrm{pos}}$ are isomorphic ordered groups, and we rephrase the previous result in terms of multiplication as follows.

7.13 Corollary *For any subgroup $H \leq \mathbb{R}_{\text{pos}}^{\times}$, the order-preserving group homomorphisms $H \to \mathbb{R}_{\text{pos}}^{\times}$ are precisely the maps $h \mapsto h^a$, where $0 \leq a \in \mathbb{R}$.* □

The following result on the number of orderings of $(\mathbb{R}, +)$ will be used later for constructions of real closed fields and of involutions in $\text{Aut}\,\mathbb{C}$; see 12.14 and 14.16.

7.14 Theorem *Let G be the group $\mathbb{R}^+$, let $\aleph = 2^{\aleph_0} = \text{card}\,G$, and let $c \in \{\aleph_0, \aleph, 2^{\aleph}\}$. Then there exist precisely $2^{\aleph}$ isomorphism types of ordered groups $(G, <)$ such that $\text{card}\,\text{Aut}(G, <) = c$.*

Proof There exist only $2^{\aleph}$ (ordering) relations on the set $\mathbb{R}$, hence it suffices to find at least $2^{\aleph}$ isomorphism types of ordered groups $(G, <)$ as required. We construct rational vector spaces G of cardinality $\aleph$ with suitable orderings, rather than taking $G = \mathbb{R}^+$ directly; compare 1.17; the constructions use the axiom of choice.

First we treat the case $c = \aleph_0$. The rational vector space $\mathbb{R}$ contains $2^{\aleph}$ hyperplanes (see Section 1, Exercise 4). Take for G any of these hyperplanes, and endow it with the ordering induced from the natural ordering of $\mathbb{R}$. By 7.12, each hyperplane is order-isomorphic to at most $\aleph$ other hyperplanes. Therefore we obtain $2^{\aleph}$ isomorphism types of ordered groups $(G, <)$; compare 61.13a. Again by 7.12, the automorphism group $\text{Aut}(G, <)$ consists of all maps $x \mapsto rx$ with $0 < r \in \mathbb{R}$ and $rG = G$. Clearly $F := \{r \in \mathbb{R} \mid rG = G\} \cup \{0\}$ is a subfield of $\mathbb{R}$, and G is a vector space over F, hence also the quotient group $\mathbb{R}/G \cong \mathbb{Q}$ is a vector space over F. Thus $\text{card}\,F = \aleph_0 = \text{card}\,\text{Aut}(G, <)$.

Next we consider the case $c = \aleph$. Choose any subfield F of $\mathbb{R}$ such that $\text{card}\,F = \aleph = [\mathbb{R} : F]$; such subfields can be obtained by splitting a transcendency basis of $\mathbb{R}$ over $\mathbb{Q}$ into two parts of cardinality $\aleph$ (see 64.20). There exist $2^{\aleph}$ subspaces G of $\mathbb{R}$ considered as a vector space over F. As in the previous paragraph, we use 7.12 to conclude that we obtain $2^{\aleph}$ isomorphism types of non-trivial ordered groups $(G, <)$. Here we have $\text{card}\,\text{Aut}(G, <) = \aleph$; see 7.12 and observe that each map $x \mapsto rx$ with $0 < r \in F$ is an automorphism of $(G, <)$.

Now we deal with the case $c = 2^{\aleph}$. Let $(G, <)$ be one of the ordered groups constructed so far, with $\text{card}\,\text{Aut}(G, <) \in \{\aleph_0, \aleph\}$. We denote by $G^{(\mathbb{R})}$ the direct sum $\bigoplus_{r \in \mathbb{R}} G$ endowed with the lexicographic ordering. Then $\text{card}\,G^{(\mathbb{R})} = \aleph$ by 61.14 and 61.13. The automorphism group of the ordered group $G^{(\mathbb{R})}$ contains the Cartesian power $\bigtimes_{r \in \mathbb{R}} \text{Aut}(G, <)$, which has cardinality $2^{\aleph}$, hence $\text{card}\,\text{Aut}\,G^{(\mathbb{R})} = 2^{\aleph}$.

We claim that distinct isomorphism types of ordered groups $(G, <)$ give distinct isomorphism types of ordered groups $G^{(\mathbb{R})}$; this implies that there are enough isomorphism types as required. For each positive element $g \in G^{(\mathbb{R})}$, the two sets $A_g := \bigcup_{n \in \mathbb{N}} \{x \in G^{(\mathbb{R})} \mid -ng \leq x \leq ng\}$ and $B_g := \bigcap_{n \in \mathbb{N}} \{x \in G^{(\mathbb{R})} \mid -g < nx < g\}$ are convex subgroups. The ordered quotient group A_g / B_g is called a *jump* of $G^{(\mathbb{R})}$. Now g has finite support, and we denote by r be the smallest real number with $g_r > 0$. We obtain $A_g = \{x \in G^{(\mathbb{R})} \mid x_s = 0 \text{ for } s < r\}$ and $B_g = \{x \in G^{(\mathbb{R})} \mid x_s = 0 \text{ for } s \leq r\}$, since $(G, <)$ is an Archimedean ordered group. Therefore each jump determined by a positive element g of $G^{(\mathbb{R})}$ is isomorphic to $(G, <)$.   □

Exercises

(1) The vector group $\mathbb{R}^2$ becomes a non-Archimedean ordered group when endowed with the lexicographic ordering defined by $(a, b) < (a', b') \leftrightharpoons a < a'$, or $a = a'$ and $b < b'$.

(2) A non-commutative, non-Archimedean ordered group is obtained by taking on $\mathbb{R}^2$ the lexicographic ordering and the group operation $(a, b) + (c, d) = (a + c, b + e^a d)$ of the semidirect product $\mathbb{R} \ltimes \mathbb{R}$; compare 9.4f. Generalize this to arbitrary semidirect products of ordered groups.

(3) The set $\{2^m 3^n \mid m, n \in \mathbb{Z}\}$ is dense in the set of positive real numbers.

(4) Let $c \in \mathbb{R} \setminus \mathbb{Q}$ and $A = \mathbb{Z} + \mathbb{Z}c$. Describe the structure of the group of all those automorphisms of the group A which preserve the ordering inherited from $\mathbb{R}$.

8   The real numbers as a topological group

A topological group should be a group $(G, +)$ carrying a topology compatible with the group structure. We could model the compatibility conditions after the conditions for ordered groups as expressed in the remark following 7.1. In other words, we would require that left or right multiplication with any fixed element defines a homeomorphism of G. This leads to the notion of a semi-topological group, which is too weak for most purposes; compare 62.1 and the subsequent remarks. (But at least, this shows that the topological space underlying a topological group is homogeneous.) Instead, we define topological groups as follows.

8.1 Definition   Let $(G, +)$ be a group and let τ be a topology on G. Then $(G, +, \tau)$ is called a *topological group* if addition $(g, h) \mapsto g + h$ is a continuous map $G \times G \to G$ and inversion $g \mapsto -g$ is a continuous map $G \to G$. Usually, it will be assumed that τ is a Hausdorff topology.

If we equip $\mathbb{R}^+$ with the cofinite topology (a proper subset is closed if, and only if, it is finite), then we obtain a semi-topological group with continuous inversion, but addition $\mathbb{R} \times \mathbb{R} \to \mathbb{R}$ is discontinuous. On the other hand, $\mathbb{R}^+$ with the Sorgenfrey topology (5.73) is a para-topological group, which means that addition is continuous in two variables, but inversion need not be (and in our example, is not) continuous.

8.2 Proposition $\mathbb{R}^+$ *is a topological group with respect to the usual topology on* $\mathbb{R}$.

Proof If $x, x', y, y' \in \mathbb{R}$ and $|x - x'| < \varepsilon$, $|y - y'| < \varepsilon$, then the triangle inequality gives $|x + y - (x' + y')| < 2\varepsilon$ and $|-x - (-x')| < \varepsilon$.

The same proof works for the additive group of any (skew) field with absolute value (as defined in 55.1). □

We remark that local compactness of the group $\mathbb{R}^+$ is the most important of its topological properties in the context of topological groups. Other relevant properties are connectedness, separability, and the fact that the topology is induced by a metric.

We shall see shortly that every ordered group is a topological group (8.4), and this will prove 8.2 once more in view of Section 7. While this proof is less trivial than the one just given, its first step, which follows next, will facilitate the construction of topological groups in general. The objective is to clarify what is needed for a semi-topological group to be a topological group.

8.3 Proposition *Let* $(G, +)$ *be a group with a topology such that*
  (i) *the left and right addition maps* $\lambda_c : x \mapsto c + x$ *and* $\rho_c : x \mapsto x + c$
      of G *are homeomorphisms,*
  (ii) *addition* $(x, y) \mapsto x + y$ *is continuous at* $(0, 0)$, *and*
  (iii) *inversion* $x \mapsto -x$ *is continuous at* 0.
Then G *is a topological group.*

Proof Writing the inversion map i as $i = \lambda_{-c} i \rho_{-c}$, we see that it is continuous at c. Similarly, the decomposition $(x, y) \mapsto (-c + x, y - d) \mapsto -c + x + y - d \mapsto x + y$ of the addition shows that it is continuous at (c, d). □

8.4 Proposition *Every ordered group* $(G, +, <)$ *is a topological group with respect to the topology induced by the ordering.*

Proof We verify the conditions of 8.3. According to the remark following 7.1, the maps λ_c and ρ_c are order-preserving bijections and, hence, homeomorphisms with respect to the topology induced by the ordering. Likewise, inversion is an order-reversing bijection and, again, a homeomorphism.

Given a neighbourhood $]a, b[$ of 0, we define $c = \min\{-a, b\}$; then $I :=]-c, c[\subseteq]a, b[$. We are looking for an interval $J =]-d, d[$ such that $J + J \subseteq I$, thus proving continuity of addition at $(0, 0)$. If c is a smallest positive element, then $J := I$ will do. If not, choose d such that $0 < 2d < c$, applying 7.6. Again, $J + J \subseteq I$ holds and G is a topological group by 8.3.   □

Subgroups and quotients

As pointed out in Section 62, the adequate notion of substructure for topological groups is that of a closed subgroup. This includes all open subgroups (62.7), and the assumption of closedness implies that coset spaces inherit a natural Hausdorff topology from the group (62.8).

The topological group of real numbers has very few closed subgroups in spite of the abundance of subgroups in general (1.15). We record this fact in the following proposition, which is merely a restatement of Theorem 1.4.

8.5 Proposition   *The closed subgroups of the topological group $\mathbb{R}^+$ are precisely $\{0\}$, the cyclic groups $r\mathbb{Z} \cong \mathbb{Z}$, $0 \neq r \in \mathbb{R}$, and $\mathbb{R}$ itself.*   □

In particular, the topological group $\mathbb{R}$ does not have any non-trivial compact subgroups, nor any 'small' subgroups (that are contained in a preassigned neighbourhood of the neutral element). These properties play an important role in the theory of locally compact groups. We shall show in 8.19 that the property 8.5 characterizes $\mathbb{R}$ among all locally compact, connected groups.

We complement the last result by its generalization to the vector groups $\mathbb{R}^n$, which are topological groups as well (same proof as 8.2).

8.6 Theorem   *Let H be a closed subgroup of a vector group $\mathbb{R}^n$. Then there is a basis $v_1, \ldots, v_n$ of $\mathbb{R}^n$ such that H consists of all linear combinations $\sum_{i=1}^{n} r_i v_i$ satisfying $r_i \in \mathbb{Z}$ for $i > i_1$ and $r_i = 0$ for $i > i_2$, where $i_1 \leq i_2 \leq n$. In particular, H is isomorphic to the direct sum $\mathbb{R}^{i_1} \oplus \mathbb{Z}^{i_2 - i_1}$.*

Proof By 8.5, the intersection of H with a one-dimensional subspace $\mathbb{R}v$ of $\mathbb{R}^n$ is either all of $\mathbb{R}v$ or infinite cyclic or trivial. Let $U \leq \mathbb{R}^n$ be the largest vector space contained in H and choose a vector space V

complementary to U. Now H is the direct sum of U and $H \cap V$, hence the problem is reduced to the case that H contains no vector space $\neq 0$. In this case, H is discrete, or else we find non-zero elements $h_k \in H$ tending to 0, and we could assume that the vectors $w_k := h_k \|h_k\|^{-1}$ converge to some vector w with $\|w\| = 1$. As in 8.5, an argument similar to the proof of 1.4 would then show that the set $\{\, z\|h_k\| \mid k \in \mathbb{N}, z \in \mathbb{Z} \,\}$ is dense in $\mathbb{R}$, and this implies that $\mathbb{R}w \subseteq H$, a contradiction.

Take an element $u \in H \smallsetminus \{0\}$ such that $\|u\|$ is minimal, and consider the image $G \cong H/\langle u \rangle$ of H in $\mathbb{R}^n/\mathbb{R}u$. This is again a discrete group. Otherwise, there are elements $g_k \in H \smallsetminus \mathbb{R}u$ and real numbers r_k such that $g_k - r_k u \to 0$. Adding suitable integers to r_k, we could arrange that the set $\{\, r_k \mid k \in \mathbb{N} \,\}$ is bounded, contradicting the discreteness of H. Using induction over n, we may assume that G consists of all integral combinations of some independent vectors $\tilde{v}_2, \ldots, \tilde{v}_s \in \mathbb{R}^n/\mathbb{R}u$. If we choose an inverse image $v_i \in H$ for each $\tilde{v}_i$, we get a set $v_1 = u, v_2, \ldots, v_s$ having the properties asserted in the theorem. $\qquad\square$

Next we turn to quotients of the topological group $\mathbb{R}^+$. Only quotients modulo closed subgroups inherit a reasonable structure, and 8.5 says that there is only one possibility up to isomorphism.

8.7 The circle group as a topological quotient group of $\mathbb{R}^+$ The circle $\mathbb{S}_1$ has been introduced in Section 5 as the set of all vectors of length 1 in $\mathbb{R}^2$. Identifying $\mathbb{R}^2$ with the field $\mathbb{C}$ of complex numbers, we see that $\mathbb{S}_1$ is a closed subgroup of the multiplicative group $\mathbb{C}^\times$. The formulae $(x + iy)(u + iv) = xu - yv + i(xv + yu)$ and $(x + iy)^{-1} = (x^2 + y^2)^{-1}(x - iy)$ show that multiplication and inversion in $\mathbb{C}^\times$ are continuous, and $\mathbb{C}^\times$ is a topological group containing $\mathbb{S}_1$ as a closed subgroup.

Now remember the surjective map

$$p : \mathbb{R} \to \mathbb{S}_1 : t \to e^{2\pi i t}$$

introduced in 5.16. It is a group homomorphism by the exponential law $e^{z+w} = e^z e^w$, and the topology of $\mathbb{S}_1$ is the quotient topology with respect to this map; see 5.16. It follows that $\mathbb{S}_1$ is isomorphic to the topological quotient group of $\mathbb{R}$ modulo $\ker p = \mathbb{Z}$, as introduced in 62.11.

As an abstract group, the quotient $\mathbb{R}/\mathbb{Z}$ appeared first in 1.20 and was denoted $\mathbb{T}$. The symbol abbreviates the name *torus group*, which is primarily used for the topological group $\mathbb{S}_1$ and for the higher dimensional tori $\mathbb{S}_1 \times \cdots \times \mathbb{S}_1 = \mathbb{T}^n$ (where n is the number of factors); compare 5.64.

Originally, the name torus refers to the topological space $\mathbb{S}_1 \times \mathbb{S}_1$, the doughnut surface. Because of the isomorphism

$$\mathbb{S}_1 \cong \mathbb{T} = \mathbb{R}^+/\mathbb{Z},$$

we shall not be too strict in distinguishing between the group $\mathbb{S}_1 \leq \mathbb{C}^\times$ and the factor group $\mathbb{T} = \mathbb{R}/\mathbb{Z}$, but at least for the next proposition and its proof, we shall be careful about this point.

8.8 Proposition (a) *The closed subgroups of the torus group* $\mathbb{S}_1 \cong \mathbb{T}$
   are precisely the groups $\mathbb{S}_1$ *and the finite cyclic groups generated by*
   the roots of unity $e^{2\pi i/n}$, *where* $n \in \mathbb{N}$.
 (b) *The only connected subgroups of* $\mathbb{S}_1$ *are* $\mathbb{S}_1$ *itself and the trivial*
   group.
 (c) *The open set* $\{z \in \mathbb{S}_1 \mid \operatorname{Re} z > 0\}$ *does not contain any non-trivial*
   subgroup of $\mathbb{S}_1$.

Proof (a) Let $G \leq \mathbb{S}_1$ be a closed subgroup. Then $p^{-1}(G)$ is a closed subgroup of $\mathbb{R}$; compare 8.7. Moreover, this subgroup contains the number $1 \in \ker p$. Now the assertion follows from 8.5.

  (b) If $G \leq \mathbb{S}_1$ is a non-trivial connected subgroup, then $p^{-1}(G)$ contains a non-trivial connected subset and, hence, contains some neighbourhood of 0. It follows that $p^{-1}(G) = \mathbb{R}$.

  (c) If $G \leq \mathbb{S}_1$ is a non-trivial subgroup contained in the right half plane, then the closure $\overline{G}$ is one of the groups listed in (a) and is contained in the closed right half plane. Such a group does not exist.   □

8.9 Theorem   *Let G be a non-trivial (Hausdorff) topological group. If there is a continuous epimorphism $\mathbb{R} \to G$, then G is isomorphic to the torus $\mathbb{T}$ as a topological group or isomorphic to $\mathbb{R}$ as a group. In the second case, the topology of G may be coarser than the topology of $\mathbb{R}$; compare 5.66.*

Proof The kernel of the given epimorphism φ is a closed subgroup of $\mathbb{R}$ because G is a Hausdorff space. By 8.5, it follows that either φ is injective, which implies that $G \cong \mathbb{R}$ as a group, or after rescaling (using $\varphi(rt)$ instead of $\varphi(t)$) we may assume that $\ker \varphi = \mathbb{Z}$. Then φ induces a group isomorphism $\psi : \mathbb{R}/\mathbb{Z} \to G$ such that $\varphi = \psi p$, where p is the canonical epimorphism $\mathbb{R} \to \mathbb{R}/\mathbb{Z}$. Moreover, it follows from the definition of the quotient topology that ψ is continuous. Indeed, the preimage $\psi^{-1}(U)$ of an open set $U \subseteq G$ has open preimage under p because $\varphi = \psi p$ is continuous, and this property characterizes $\psi^{-1}(U)$ as an open set.

Now ψ is a continuous bijection of the compact space $\mathbb{R}/\mathbb{Z} \approx \mathbb{S}_1$ onto the Hausdorff space G, hence ψ is a homeomorphism, and an isomorphism of topological groups. □

Characterizations

The following result shows that the topology of $\mathbb{R}$ determines the group structure. We shall return to this point later (8.16).

8.10 Theorem *Let G be a topological group whose underlying topological space is homeomorphic to $\mathbb{R}$. Then G is isomorphic to $\mathbb{R}$ as a topological group. In particular, G is abelian.*

Proof We show that there is an ordering on G inducing the given topology and making G an ordered group. Since G is connected by assumption, the ordering is complete (3.3), and the assertion is a consequence of 7.9.

We define the ordering on G such that the given homeomorphism $G \to \mathbb{R}$ becomes an isomorphism of chains. We have to show that the left and right addition maps λ_c and ρ_c of G are monotone; compare 7.1 and the subsequent remarks. Since G is a topological group, we know that these maps are homeomorphisms. Hence, they are either monotone or antitone; compare 5.51. Any antitone homeomorphism f of $\mathbb{R}$ has a fixed point (apply the intermediate value theorem to $f - \mathrm{id}$). For $c \neq 0$ it follows that λ_c and ρ_c are indeed monotone. □

We shall push the last result as far as possible in 8.21 and 8.22. Here we want to prove a similar result for the torus group. This will be an easy application of some standard results about covering maps, which we present first. A good reference is GREENBERG 1967 §§5 and 6.

8.11 Definition (a) A surjective continuous map $p : E \to B$ of topological spaces is called a *covering map* if every point $b \in B$ has an open neighbourhood U such that $p^{-1}(U)$ is a disjoint union of open subsets U_i, $i \in I$, each of which is mapped homeomorphically onto U by p. Neighbourhoods like U will be called *special neighbourhoods*, and the U_i are called the *sheets* over U.

(b) A topological space X is said to be *simply connected* if X is pathwise connected and if every loop can be contracted in X. This means that, given a continuous map $f : [0,1] \to X$ satisfying $f(0) = f(1) = x_0$, there is a continuous map $F : [0,1] \times [0,1] \to X$ such that $F(s,0) = f(s)$ for all s and $F(s,t) = x_0$ whenever $s \in \{0,1\}$ or $t = 1$.

The standard example of a simply connected space is $\mathbb{R}^n$, where the map F is most easily defined by $F(s,t) = tx_0 + (1 - t)f(s)$. The most prominent example of a covering map is the quotient map $p : \mathbb{R} \to \mathbb{S}_1$ of 8.7. Every proper open subset $U \subseteq \mathbb{S}_1$ is a special neighbourhood; this follows from the existence of the complex logarithm function.

Instead of going into the details of the theory of covering spaces, we shall just quote the one result that we need. For a proof, see GREENBERG 1967 loc. cit.

8.12 Theorem Let $p : E \to B$ be a covering map, and $h : X \to B$ a continuous map with X simply connected. If $x \in X$ and $e \in E$ are points such that $h(x) = p(e)$, then there is a unique continuous map $\widetilde{h} : X \to E$ sending x to e and satisfying $h = p\widetilde{h}$.

The map $\widetilde{h}$ is called a *lift* of h over p. We apply this result to topological groups:

8.13 Theorem (a) Let $(G, +)$ be a topological group with neutral element 0 and $p : E \to G$ a covering map with E simply connected. Given $e \in E$ such that $p(e) = 0$, there is a unique group structure on E with neutral element e such that E is a topological group and p is a group homomorphism.

(b) If H is a simply connected topological group, then every continuous group homomorphism $\psi : H \to G$ lifts to a unique continuous group homomorphism $\widetilde{\psi} : H \to E$, that is, $\psi = p\widetilde{\psi}$.

(c) Every continuous group endomorphism $\varphi : G \to G$ lifts to a unique continuous group endomorphism $\widetilde{\varphi} : E \to E$, that is, $p\widetilde{\varphi} = \varphi p$.

$$
\begin{array}{ccc}
E & \xrightarrow{\widetilde{\varphi}} & E \\
\downarrow{\scriptstyle p} & & \downarrow{\scriptstyle p} \\
G & \xrightarrow{\varphi} & G
\end{array}
$$

Proof (a) Let $h : E \times E \to E$ be given by $h(a, b) = p(a) + p(b)$. It is easy to verify that $E \times E$ is simply connected, hence there is a unique lift $\widetilde{h} : E \times E \to E$ sending (e, e) to e. This map is going to be the addition in E, that is, $a + b = \widetilde{h}(a, b)$. Similarly, inversion in E is defined as a lift of inversion in G. The group axioms are all verified by the same method, exploiting the uniqueness part of 8.12. For example, consider the associative law. It says that the two maps $E \times E \times E \to E$ defined by $f_1(a, b, c) = (a + b) + c$ and by $f_2(a, b, c) = a + (b + c)$ agree. Now the corresponding maps in G do agree, and f_1 is a lift of $(a, b, c) \mapsto (p(a) + p(b)) + p(c)$, similarly for f_2. Thus we have two lifts of the same map, which agree at the point (e, e, e), so they are equal.

(b) The existence of $\widetilde{\psi}$ is immediate from 8.12 (for x and e, take the neutral elements). The homomorphism property is again translated into equality of two lifts, namely of $(a, b) \mapsto \widetilde{\psi}(a) + \widetilde{\psi}(b)$, which lifts the map $(a, b) \mapsto \psi(a) + \psi(b)$, and $(a, b) \mapsto \widetilde{\psi}(a + b)$, which lifts $(a, b) \mapsto \psi(a + b)$. Again, the two lifts are equal because they agree at (e, e).

(c) This follows if we apply (b) to $\psi = \varphi p$. □

Now we return to the torus group.

8.14 Theorem *Every topological group homeomorphic to $\mathbb{S}_1$ is isomorphic as a topological group to $\mathbb{T}$.*

Proof Using the covering map $p : \mathbb{R} \to \mathbb{S}_1$ and 8.13, we may define a topological group structure on $\mathbb{R}$ such that p becomes a continuous group epimorphism. The result follows from 8.9; note that p is not injective, so we are dealing with the first case of 8.9. □

8.15 Theorem *The only topological groups whose underlying spaces are connected 1-manifolds are $\mathbb{R}$ and $\mathbb{T}$.*

Proof The essential step is to show that the space of such a group G is separable; then we can apply the classification of connected, separable 1-manifolds (5.17) and our results 8.10 and 8.14.

Let U be an open neighbourhood of 0 homeomorphic to $\mathbb{R}$. Then $V := U \cap (-U) = -V$ is a separable open set. The union W of the sets $nV = V + \cdots + V = \{ v_1 + \cdots + v_n \mid v_i \in V \}$ with $n \in \mathbb{N}$ is an open subgroup, hence also closed (62.7, all cosets are open), and $W = G$ by connectedness. Now W is separable: first, V contains a countable dense set A, giving rise to a countable dense set $A^n = A \times \cdots \times A \subseteq V^n$; under addition, A^n is mapped to nA, hence nA is dense in nV, and the union of all sets nA, $n \in \mathbb{N}$, is still countable by 61.13. □

The last result can be obtained as a corollary of deep results in the theory of locally compact groups. It has been proved by Montgomery, Zippin, and Gleason (see MONTGOMERY–ZIPPIN 1955 4.10 p. 184) that every locally Euclidean topological group is a Lie group, that is, it admits an analytic structure such that the group operations (addition and inversion) are analytic maps. Lie groups can be classified using their Lie algebras, which are linearizations of the group structures. Then 8.15 reduces to the rather easy classification of 1-dimensional Lie groups.

8.16 Independence of group structure and topology We have just seen that the topology of $\mathbb{R}$ and of $\mathbb{T}$ determines the group structure up to isomorphism. The converse is far from true, and we give a number of examples of non-homeomorphic topological groups whose underlying groups are isomorphic to $\mathbb{T}$ or to $\mathbb{R}$.

(1) The group $\mathbb{R}^+$ is isomorphic to any vector group $\mathbb{R}^n$ (see 1.15), and $\mathbb{R}^n$ is a topological group. Unlike $\mathbb{R}$, the topological space $\mathbb{R}^n$, $n > 1$, remains connected after deletion of a point, hence the two spaces are not homeomorphic.

(2) $\mathbb{R}$ with the torus topology 5.65 is a (non-closed) subgroup of the topological group $\mathbb{T}^2$ (compare 5.66), and hence is again a topological group. Proposition 5.67 amply demonstrates that this topology is not homeomorphic to the usual one.

(3) Let A be a subspace of the rational vector space $\mathbb{R}$ such that A has the same (transfinite) dimension as $\mathbb{R}$ itself. Then the additive group A^+ is isomorphic to $\mathbb{R}^+$; compare 1.15. For example, A could be a hyperplane. In any case, A is a (non-closed) subgroup of the topological group $\mathbb{R}$, hence A is a topological group. The topology of A is totally disconnected and not locally compact, because both A and its complement are dense in $\mathbb{R}$.

(4) By 1.17, the group $\mathbb{R}^+$ is a direct sum of $\aleph$ copies of $\mathbb{Q}^+$. The direct sum is a subgroup of the direct product, which is a topological group with respect to the product topology, and the direct sum is a topological group with respect to the induced topology.

(5) The solenoid $\mathbb{Q}_d^*$ introduced in 8.29 is compact, connected, and isomorphic to $\mathbb{R}$ as a group; see 8.32 and 8.33.

(6) It was shown in 1.25 that $\mathbb{T} \cong \mathbb{R}^n \oplus \mathbb{T}$ as groups. Of course, both are topological groups, but one of the spaces ($\mathbb{T}$) is compact, and the other is not.

(7) A systematic investigation of group topologies on $\mathbb{R}$ is presented in CH'ÜAN–LIU 1981, leading to the result that there are exactly $2^{2^{\aleph}}$ non-isomorphic topological groups ($\aleph = \operatorname{card}\mathbb{R}$) that are algebraically isomorphic to $\mathbb{R}$. Among these, countably many are compact, and similar results are given for topologies with other special properties.

By contrast, the topology on the additive group of integers is uniquely determined under mild additional conditions, as we show now.

8.17 Theorem *The discrete topology is the only locally compact Hausdorff topology making the additive group $\mathbb{Z}$ of integers (or, in fact, any countable group) into a topological group.*

Proof Our group is a union of countably many singletons $\{x\}$, which are closed by assumption. At least one of the singletons must contain a non-empty open subset. This follows from the Baire category theorem, which holds in locally compact spaces, see DUGUNDJI 1966 XI.10.3 p. 250. We conclude that some (and, hence, every) singleton is open. □

Theorem 5.15 characterizes the topological group $\mathbb{R}$, and we shall continue by proving a number of other characterization theorems. The first one (8.18) is due, in the form presented here, to MORRIS 1986. The proof uses the theorem of Mal'cev–Iwasawa, and also the theorem of Peter–Weyl, which says that each compact group has enough (63.11) continuous unitary representations. A complex matrix A is called unitary, if $A\bar{A}^{\mathrm{T}} = 1$; thus a one-dimensional continuous unitary representation is the same thing as a character.

8.18 Theorem *Let G be a non-discrete, locally compact Hausdorff topological group, and suppose that all proper closed subgroups of G are discrete. Then the topological group G is isomorphic to the additive group $\mathbb{R}$ of real numbers or to the torus group $\mathbb{T}$.*

Proof We denote by G^1 the connected component of G, that is, the largest connected subgroup; note that G^1 is closed, because the closure of a connected set is connected (compare 62.13). We infer that either $G^1 = G$, or G^1 is discrete and therefore trivial. In the second case, G has a neighbourhood basis at 1 consisting of compact open proper subgroups (see 62.13 and Exercise 3 of Section 63); these subgroups are discrete, so G is discrete, which is a contradiction. Hence G is connected. The theorem of Mal'cev–Iwasawa (62.14) then shows in view of our hypothesis that G is either compact or isomorphic to $\mathbb{R}$. It remains to consider the case where G is connected and compact.

We claim that G contains an element g of infinite order. The Peter–Weyl theorem yields a non-trivial continuous unitary representation $\varphi : G \to \mathrm{GL}_n\mathbb{C}$; see HOFMANN–MORRIS 1998 2.27, 2.28 p. 44, STROPPEL 2006 14.33 (compare also HEWITT–ROSS 1963 22.12, 22.13, PONTRYAGIN 1986 §32, §33, ADAMS 1969 3.39). The corresponding trace map $G \to \mathbb{C} : g \mapsto \mathrm{tr}\,\varphi(g)$ is continuous and not constant, since $\mathrm{tr}\,\varphi(g) = n$

implies $\varphi(g) = \mathrm{id}$ (note that $\varphi(g)$ can be diagonalized and has eigen-values of absolute value 1). If all elements $g \in G$ have finite order, then the eigenvalues of $\varphi(g)$ are roots of unity. Hence the connected set $\{\,\mathrm{tr}\,\varphi(g) \mid g \in G\,\} \ne \{n\}$ consists of finite sums of roots of unity and is therefore countable, which is a contradiction.

An element $g \in G$ of infinite order generates a dense subgroup of G; otherwise the closure of $\langle g \rangle$ would be a proper compact subgroup, hence discrete by assumption, and thus finite, which is absurd. We conclude that G is abelian.

Thus by Theorem 63.13, there exists a non-trivial character $\chi : G \to \mathbb{T}$ (in fact, the representation φ can be chosen to be irreducible; then φ is one-dimensional, hence $\chi = \varphi$ is such a character). The connected image $\chi(G)$ is all of $\mathbb{T}$, and the kernel of χ is a proper closed subgroup, hence discrete and finite. It is straightforward to show that G is locally homeomorphic to $\mathbb{T}$, hence a compact 1-manifold. Now 5.17 and 8.14 imply that $G \cong \mathbb{T}$ as topological groups. $\square$

As a consequence, we show that the property 8.5 (all non-trivial proper closed subgroups are infinite cyclic) is in fact a characterizing property of $\mathbb{R}$.

8.19 Corollary   *Let G be a non-discrete, locally compact Hausdorff topological group, and suppose that all proper closed subgroups of G are cyclic. Then the topological group G is isomorphic to the additive group $\mathbb{R}$ of real numbers or to the torus group $\mathbb{T}$. If all non-trivial proper closed subgroups of G are infinite cyclic, then $G \cong \mathbb{R}$ is the only possibility.*

Proof The hypothesis of 8.19 implies that of 8.18, because a closed cyclic subgroup of G discrete by 8.17. The second statement is easily deduced since the torus contains non-trivial finite subgroups (in fact, we obtain a proof for the second statement in 8.19 simply by omitting the second paragraph of the proof for 8.18). $\square$

The next result is due to MONTGOMERY 1948. We need the notion of topological dimension. By definition, a topological space X has (cover-ing) dimension $\dim X \le n$ if every finite open cover of X has a refine-ment containing at most $n+1$ sets with non-empty intersection. A good reference on dimension is NAGAMI 1970; compare also SALZMANN *et al.* 1995 Section 92 and Theorems 93.5 to 93.7.

8.20 Theorem   *Let G be a locally compact, connected topological group. If G is one-dimensional and not compact, then G is isomorphic to the topological group $\mathbb{R}$.*

Proof The theorem of Mal'cev–Iwasawa (62.14) shows that G is homeomorphic to $C \times \mathbb{R}^k$, where $C \leq G$ is a (maximal) compact subgroup. This implies that $k = \dim \mathbb{R}^k \leq \dim G$, hence $k = 1$ since G is not compact. It follows that $\dim G = \dim C + 1$; this is not as trivial as it seems, see NAGAMI 1970 42-3 for a proof; compare also SALZMANN *et al.* 1995 92.11. Now $\dim C = 0$ implies that C is totally disconnected. On the other hand, connectedness of $G \approx C \times \mathbb{R}$ implies that C is connected. Finally, we conclude that $C = \{0\}$, and then 8.10 or the theorem of Mal'cev–Iwasawa shows that $G \cong \mathbb{R}$ as a topological group. $\qquad \square$

The next result should be compared with 8.10. Instead of $G \approx \mathbb{R}$ we make only weak assumptions concerning connectivity. Yet we obtain a rather strong assertion.

8.21 Theorem *Let G be a Hausdorff topological group. If G is connected but $G \smallsetminus \{0\}$ is not, then there is a continuous group isomorphism $\varphi : G \to \mathbb{R}$. Thus, G can be obtained from the topological group $\mathbb{R}$ by refining the topology.*

Proof (1) We claim that $H = G \smallsetminus \{0\}$ has precisely two connected components (later we shall use this fact to define an ordering on G). Let $H = U \cup V$ be a separation as defined in 5.7, that is, the sets U and V are non-empty, open and disjoint. If both U and V are connected, then there is nothing to prove. If not, then one of the two sets admits a separation, and $H = U \cup V \cup W$ with non-empty, disjoint open sets U, V, W. Applying 5.8c to U and $V \cup W$, we obtain that $\overline{U} = U \cup \{0\}$ is connected. Select connected components C_U, C_V, C_W of U, V, W, respectively. Then the product $C_V \times \overline{U}$ is connected, and continuity of addition and inversion implies that $C_V + (-\overline{U})$ is connected, too. This set is contained in H and contains the connected component C_V, hence $C_V + (-C_U) \subseteq C_V$. By symmetry, we also have $C_U + (-C_V) \subseteq C_U$ and, after taking inverses on both sides, $C_V + (-C_U) \subseteq -C_U$. These results show that the intersection $C_V \cap (-C_U)$ is not empty. Both these sets are connected components of H, hence they are equal. The same holds with W in place of V, which is a contradiction.

(2) Repeating the argument (1) with two sets instead of three, we see that the connected components of H are interchanged by inversion, so we denote them by C and $-C$. We define $x \leq y \rightleftharpoons y - x \in \overline{C} = C \cup \{0\}$, and we claim that this turns G into an ordered group. First we have to show that the relation $\leq$ transitive, i.e., that $x \leq y$ and $y \leq z$ together imply $x \leq z$. The connected set $C + \overline{C} \subseteq H$ contains C, hence $C = C + \overline{C}$

and $\overline{C} = \overline{C} + \overline{C}$, and transitivity follows. From $\overline{C} \cap -\overline{C} = \{0\}$ we infer that $x \leq y$ and $y \leq x$ together imply $x = y$, and $\leq$ is an ordering relation.

Suppose that $a \leq b$ and $x \neq 0$. Then $b + x - (a + x) = b - a$ shows that $a + x \leq b + x$. In order to prove that $x + a \leq x + b$, i.e., that $x + (b - a) - x \in \overline{C}$, we need to know that $x + \overline{C} - x \subseteq \overline{C}$. Now $x + C - x$ is a connected subset of H, and contains an element of C (namely, x or $-x$, whichever belongs to C). Therefore, $x + C - x \subseteq C$ and $x + \overline{C} - x \subseteq \overline{C}$. Thus, we have defined an ordered group.

(3) We compare topologies: the sets C, $C + a$ and $-C + b$ are open in the given topology of G, hence the intersection $]a, b[= (C+a) \cap (-C+b)$ is open, too. Therefore, the identity mapping from G with the given topology to G with the order topology is continuous, and the order topology is connected. By 3.3, it follows that the ordering is complete, and 7.9 shows that the ordered group G is isomorphic to $\mathbb{Z}$ or to $\mathbb{R}$; however, the first possibility is excluded by connectedness.   □

Groups G as described in 8.21 do exist; we give an example in 8.23. One additional hypothesis, however, suffices to exclude these possibilities, as we prove next.

8.22 Theorem   *If G satisfies the hypothesis of 8.21 and is locally compact, then G is isomorphic to the topological group $\mathbb{R}$.*

Proof Any compact subgroup $C \leq G$ is mapped by φ to a compact subgroup of $\mathbb{R}$. Using injectivity of φ together with 8.5, we infer that $C = \{0\}$. Now the theorem of Mal'cev–Iwasawa (62.14) implies that G is homeomorphic to $\mathbb{R}^k$ for some k. For $k > 1$, Euclidean space $\mathbb{R}^k$ is not separated by any point, hence $k = 1$ by our hypothesis. In this case, the theorem of Mal'cev–Iwasawa (or 8.10) asserts that $G \cong \mathbb{R}$ as a topological group.   □

A counter-example

We show that Theorem 8.21 cannot be improved in the sense that G is isomorphic to $\mathbb{R}$ as a topological group. The example given in the proof of 8.23 below is due to LIVENSON 1937. See also 14.9 for an even stronger result.

8.23 Theorem   *There is a non-closed subgroup $G \subseteq \mathbb{R}^2$ such that G, but not $G \smallsetminus \{0\}$, is connected in the topology induced by $\mathbb{R}^2$, and such that G as a topological group is not isomorphic to $\mathbb{R}$.*

Proof We obtain G as the graph of a certain discontinuous group homomorphism $\psi : \mathbb{R} \to \mathbb{R}$, i.e., $G = \{(x, \psi(x)) \mid x \in \mathbb{R}\}$. The homomorphism is chosen in such a way that G meets the boundary of almost every open set of $\mathbb{R}^2$.

There is a countable basis for the topology of $\mathbb{R}^2$, hence there exist (at most) $\aleph = 2^{\aleph_0}$ open subsets of $\mathbb{R}^2$. A Hamel basis B of $\mathbb{R}$ as a vector space over $\mathbb{Q}$ also has cardinality $\aleph$ (see Theorem 1.12), hence there is a surjective map $f : B \to \mathcal{U}$ onto the following collection $\mathcal{U}$ of open sets in $\mathbb{R}^2$.

For an open set $U \subseteq \mathbb{R}^2$, define $U^b = p(U) \cap p(\mathbb{R}^2 \setminus \overline{U})$, where $p : \mathbb{R}^2 \to \mathbb{R}$ is the projection onto the first factor, $p(x, y) = x$. The set U belongs to $\mathcal{U}$ if $U^b \neq \emptyset$. In other words, the condition for U is that some line $x = \text{const}$ (a parallel to the y-axis) meets both U and $\mathbb{R}^2 \setminus U$ and, hence, meets the boundary ∂U. In fact, this happens for all x in the open set U^b.

Now assume that $b \in B$ and $U = f(b) \in \mathcal{U}$. Then $b\mathbb{Q}$ contains a non-zero element of U^b, and we may choose $\psi(b) \in \mathbb{R}$ such that $\mathbb{Q}(b, \psi(b))$ intersects ∂U. In this way, we obtain a homomorphism $\psi : \mathbb{R} \to \mathbb{R}$ whose graph G meets the boundary of every $U \in \mathcal{U}$. An immediate consequence is that G is dense in $\mathbb{R}^2$.

The y-axis ($x = 0$) separates $\mathbb{R}^2$ and meets G only in the point $0 = (0, 0)$, hence $G \setminus \{0\}$ is disconnected.

In order to show that G is connected, we consider the intersection $C = G \cap H$ with the half plane $H = \{(x, y) \mid x > 0\}$. We shall show that C is connected; then also $C \cup \{0\} \subseteq \overline{C}$ and $G = C \cup \{0\} \cup -C$ are connected.

Suppose that C is disconnected. Then there exist open subsets U, V of H such that $C \subseteq U \cup V$ and $U \cap V \cap C = \emptyset$. The latter condition implies that $U \cap V = \emptyset$, because C is dense and $U \cap V$ is open. Then moreover $\overline{U} \cap V = \emptyset$ because U is contained in the closed set $H \setminus V$. Now consider the case $U \in \mathcal{U}$. Then ∂U contains an element of C; this element is not contained in U, hence it belongs to $V \cap \overline{U}$, which is a contradiction. On the other hand, if $U \notin \mathcal{U}$, then $p(U) \cap p(V) \subseteq U^b = \emptyset$. Moreover, $p(U)$ and $p(V)$ are open sets, but their union $p(U) \cup p(V) = p(C) = \{(x, 0) \mid x > 0\}$ is connected, which is a contradiction.

It remains to be shown that G is not isomorphic to $\mathbb{R}$ as a topological group, and this is very easy. In the rational vector space $\mathbb{R}$, every one-dimensional subspace is dense with respect to the usual topology. Since G is dense in $\mathbb{R}^2$, this is far from true in G.   □

Automorphisms and endomorphisms

To formulate the appropriate notion of *automorphisms of a topological group* G, we require that automorphisms are at the same time homeomorphisms of the underlying topological space and group homomorphisms. Automorphisms in this sense form a group denoted $\mathrm{Aut}_c\, G$.

8.24 Proposition *The automorphisms of the topological group $\mathbb{R}$ are precisely the multiplication maps $\varphi_a : x \mapsto ax$, where $0 \neq a \in \mathbb{R}$. In particular, the automorphism group $\mathrm{Aut}_c\, \mathbb{R}$ is isomorphic to $\mathbb{R}^\times$.*

Proof Everything is similar to 7.11, only $a = \varphi(1)$ may be any non-zero number, and φ may be antitone; compare 5.51. $\square$

8.25 Definition The endomorphisms of an abelian group G form a ring $\mathrm{End}\, G$ called the *endomorphism ring* of G. Multiplication of this ring is the composition of maps and addition is defined by $(\varphi + \psi)(x) = \varphi(x) + \psi(x)$.

Note that commutativity of G is indispensable; the homomorphism property of $\mathrm{id} + \mathrm{id}$ requires that $x + x + y + y = x + y + x + y$ holds as an identity in G.

If G is a topological group, then the continuous endomorphisms form a subring, denoted $\mathrm{End}_c\, G$, of $\mathrm{End}\, G$.

8.26 Theorem *The continuous endomorphisms of the topological group $\mathbb{R}$ are precisely the maps φ_a for arbitrary $a \in \mathbb{R}$. In particular, the ring $\mathrm{End}_c\, \mathbb{R}$ of continuous endomorphisms is isomorphic to the field $\mathbb{R}$.*

Proof Each endomorphism φ of $\mathbb{R}^+$ is $\mathbb{Q}$-linear, by the unique divisibility of $\mathbb{R}^+$ (1.8, 1.9). Let $a := \varphi(1)$. Then $\varphi(x) = \varphi(1 \cdot x) = a \cdot x$ for $x \in \mathbb{Q}$. As $\mathbb{Q}$ is dense in $\mathbb{R}$, the continuity of φ implies that this equation holds for all $x \in \mathbb{R}$. Thus $\varphi = \varphi_a$. $\square$

For a stronger statement, see Exercise 4.

8.27 Theorem *The ring $\mathrm{End}_c\, \mathbb{T}$ of continuous endomorphisms of $\mathbb{T}$ is isomorphic to the ring $\mathbb{Z}$ of integers, and $\mathrm{Aut}_c\, \mathbb{T} \cong C_2$ is cyclic of order 2.*

Proof By 8.13c, every continuous endomorphism φ of $\mathbb{T}$ lifts to a continuous endomorphism $\widetilde{\varphi}$ of $\mathbb{R}$, that is, $\varphi p = p\widetilde{\varphi}$. From this equation it follows that $\widetilde{\varphi}$ maps $\ker p = \mathbb{Z}$ into itself. From 8.26, we know that $\widetilde{\varphi}$ has the form φ_a for some $a \in \mathbb{R}$, and $a = \varphi_a(1)$ must be an integer.

Conversely, if $a \in \mathbb{Z}$ then φ_a can be 'pushed down' to an endomorphism of $\mathbb{T}$, sending $x \in \mathbb{T}$ to x^a. This gives the isomorphism

$\operatorname{End}_c \mathbb{T} \cong \mathbb{Z}$, and the automorphisms are the invertible elements in this ring, that is, the maps φ_a for $a \in \{1, -1\}$. □

In the remainder of this section, we consider topological groups whose endomorphism rings are skew fields.

Groups having an endomorphism field

Let G be a locally compact, connected abelian topological group all of whose non-trivial continuous endomorphisms are invertible, so that $\operatorname{End}_c G$ is a skew field. We shall show that there are exactly two such groups G. To warm up, we treat the analogous question for groups without topology (or with discrete topology); the result 8.28 is due to SZELE 1949; see also FUCHS 1973 111.1. In fact, this is more than an exercise: the connected case will be reduced to the discrete case using character theory. Thus the reader should absorb the definition of character groups from Section 63, and then refer to that section as need arises.

8.28 Theorem *Let G be an (abstract) abelian group. The ring $\operatorname{End} G$ is a skew field if, and only if, $G \cong C_p$ is cyclic of prime order or $G \cong \mathbb{Q}^+$ is the additive group of rationals. We have field isomorphisms $\operatorname{End} C_p \cong \mathbb{F}_p$ if p is a prime, and $\operatorname{End} \mathbb{Q}^+ \cong \mathbb{Q}$.*

Proof (1) Suppose that $\operatorname{End} G$ is a skew field, and let F be its (commutative) prime field, i.e., the smallest subfield of $\operatorname{End} G$. Then $F \cong \mathbb{F}_p$ is of prime order, or $F \cong \mathbb{Q}$. The group G is a vector space over $\operatorname{End} G$ and, hence, over F; the product of a scalar $\varphi \in \operatorname{End} G$ and a vector $g \in G$ is defined by applying φ to the group element: $\varphi g = \varphi(g)$. Every vector space of dimension more than 1 has plenty of non-invertible endomorphisms, which are also endomorphisms of the underlying abelian group. (Here we used the existence of bases and, hence, the Axiom of Choice.) Thus G is isomorphic to the additive group of F.

(2) Conversely, let φ be an endomorphism of C_p. The order of a subgroup divides the order of the group, hence the only subgroups of C_p are $\{0\}$ and C_p itself. Either $\ker \varphi = \{0\}$ and φ is bijective, or $\ker \varphi = C_p$ and $\varphi = 0$.

Now let φ be an endomorphism of $\mathbb{Q}^+$. If $n \in \mathbb{N}$ and $\varphi(nq) = 0$, then $n\varphi(q) = 0$ and $\varphi(q) = 0$. Thus, $\ker \varphi$ is divisible. The only divisible subgroups of $\mathbb{Q}^+$ are $\{0\}$ and $\mathbb{Q}^+$ itself, hence $\varphi = 0$ or φ is injective. In the latter case, $\varphi(\mathbb{Q})$ is a non-trivial divisible subgroup of $\mathbb{Q}^+$, and φ is bijective. □

8.29 Theorem *There are only two connected, locally compact, Hausdorff abelian groups G such that the ring $\mathrm{End}_c\, G$ of continuous endomorphisms is a skew field, namely the topological group $\mathbb{R}$ and the character group $\mathbb{Q}_d^*$, where $\mathbb{Q}_d$ denotes the additive group of rationals with discrete topology. We have field isomorphisms $\mathrm{End}_c\, \mathbb{R} \cong \mathbb{R}$ and $\mathrm{End}_c(\mathbb{Q}_d^*) \cong \mathbb{Q}$.*

The group $\mathbb{Q}_d^*$ belongs to the class of so-called *solenoids*. It will be featured in 8.32 below.

Proof Suppose that G is a group satisfying the assumptions of the theorem and that $\mathrm{End}_c\, G$ is a skew field. In our abelian case, the theorem of Mal'cev–Iwasawa asserts that $G \cong C \oplus \mathbb{R}^n$, where C is a compact connected (abelian) group; see 63.14. There are only two cases where the existence of non-invertible endomorphisms $\varphi \neq 0$ is not obvious, namely, $C = \{0\}$ and $n = 1$, i.e., $G \cong \mathbb{R}$, and $n = 0$, i.e., G is compact. In the first case, we know that $\mathrm{End}_c\, G \cong \mathbb{R}$ is a field (8.26).

In the remaining case, G compact, we shall apply character theory. It suffices to determine the character group G^*, because the Pontryagin duality theorem asserts that $G \cong G^{**}$; see 63.27. By 8.30 below, $\mathrm{End}_c\, G$ is anti-isomorphic to $\mathrm{End}_c(G^*)$ as a ring, and G^* is a discrete group according to 63.5. Thus, all endomorphisms of G^* are continuous and we may apply 8.28; we find that $G^* \cong C_p$ (a cyclic group of prime order) or $G^* \cong \mathbb{Q}_d$. The first possibility entails that $G = C_p^*$ is a finite Hausdorff space, hence disconnected.

The second possibility leads to $G = \mathbb{Q}_d^*$, which is a connected group because $\mathbb{Q}_d$ has no compact (i.e., finite) subgroups other than $\{0\}$; compare 63.30. Moreover, $\mathbb{Q}_d^*$ is compact according to 63.5, and $\mathrm{End}_c(\mathbb{Q}_d^*) \cong \mathrm{End}_c(\mathbb{Q}_d) = \mathrm{End}\,\mathbb{Q}^+ \cong \mathbb{Q}$ by 8.28; note that an anti-isomorphism is an isomorphism in this case. $\square$

8.30 Theorem *Let G be a locally compact abelian group and G^* its character group. Then the ring $\mathrm{End}_c\, G$ is anti-isomorphic to the ring $\mathrm{End}_c(G^*)$.*

In fact, let γ^* denote the adjoint of $\gamma \in \mathrm{End}_c\, G$ as defined in 63.3. Then $\gamma \mapsto \gamma^*$ is bijective and satisfies $(\gamma\delta)^* = \delta^*\gamma^*$ and $(\gamma+\delta)^* = \gamma^*+\delta^*$.

Proof The identities follow directly from the definition of the adjoint. Our notation differs from 63.3, where γ acts from the right and γ^* acts from the left. Here, all maps act from the left, which enforces the reversal of factors in the first identity.

We have to show that $\gamma \mapsto \gamma^*$ is bijective. We use the duality theorem 63.27 and identify A^{**} with A via the duality isomorphism. This means that we define γ^* for $\gamma \in \mathrm{End}_c G$ by $\langle \gamma(g), \chi \rangle = \langle g, \gamma^*(\chi) \rangle$ and σ^* for $\sigma \in \mathrm{End}_c(G^*)$ by $\langle g, \sigma(\chi) \rangle = \langle \sigma^*(g), \chi \rangle$; here, $g \in G$ and $\chi \in G^*$. Now we have $\langle \gamma(a), \chi \rangle = \langle a, \gamma^*(\chi) \rangle = \langle \gamma^{**}(a), \chi \rangle$ for every $\chi \in G^*$ (and for fixed a). Since G has enough characters (compare 63.12), it follows that $\gamma = \gamma^{**}$, which proves bijectivity. □

To end this section, we compute the character groups of $\mathbb{R}$ and of some related groups (see also Exercise 5 of Section 52), and finally we give a description of $\mathbb{Q}_d^*$.

8.31 Proposition (a) *The topological group $\mathbb{R}$ is self-dual, $\mathbb{R}^* \cong \mathbb{R}$.*

(b) *The character group $\mathbb{T}^*$ is isomorphic to $\mathbb{Z}$, the discrete group of integers. An isomorphism $\mathbb{Z} \to \mathbb{T}^*$ is given by $n \mapsto \chi_n$, where $\chi_n : \mathbb{T} \to \mathbb{T}$ sends $z \in \mathbb{T}$ to its power z^n.*

(c) *The character group $\mathbb{Z}^*$ is isomorphic to the torus group $\mathbb{T}$. An isomorphism $\mathbb{T} \to \mathbb{Z}^*$ is given by $z \mapsto \chi_z$, where $\chi_z : \mathbb{Z} \to \mathbb{T}$ sends the integer n to z^n.*

Proof (1) We begin with the simpler parts. As a group, $\mathbb{T}^*$ is simply the additive group of the ring $\mathrm{End}_c \mathbb{T}$, which is isomorphic to the ring of integers (8.27). The topology of $\mathbb{T}^*$ is discrete by 63.5. Of course, this can be seen directly: the endomorphisms of $\mathbb{T}$ are of the form $z \mapsto z^n$, $n \in \mathbb{Z}$, and if a sequence of such maps converges (pointwise convergence suffices), then it is finally constant. This proves assertion (b), and (c) follows by the duality theorem 63.27. Again, this can be verified directly, a character χ of $\mathbb{Z}$ is determined by the image $\chi(1) \in \mathbb{T}$, and this defines the isomorphism $\mathbb{Z}^* \to \mathbb{T}$.

(2) Theorem 18.3b shows that for every character $\chi : \mathbb{R} \to \mathbb{T}$ there is a unique continuous endomorphism $\widetilde{\chi} : \mathbb{R} \to \mathbb{R}$ such that $\chi = p\widetilde{\chi}$, where $p : \mathbb{R} \to \mathbb{T}$ is the quotient map, $p(t) = e^{2\pi i t}$; note that we are treating $\mathbb{T}$ as a multiplicative group once more. According to 8.26, we have $\widetilde{\chi}(t) = rt$ for some real number r, and $\chi(t) = e^{2\pi i r t} =: \chi_r(t)$. Clearly, $\chi_{r+s} = \chi_r \chi_s$, and only the topology of $\mathbb{R}^* \cong \mathbb{R}^+$ remains to be determined.

It suffices to check the continuity of the group homomorphism $r \mapsto \chi_r$ and of its inverse at the neutral elements $0 \in \mathbb{R}$ and $1 \in \mathbb{T}$, respectively. So let $C \subseteq \mathbb{R}$ be compact and let $U \subseteq \mathbb{T}$ be a neighbourhood of 1. We want to determine $\varepsilon > 0$ such that $|r| < \varepsilon$ implies that $e^{2\pi i r c} \in U$ for all $c \in C$. This is possible because $(r, t) \mapsto e^{2\pi i r t}$ is continuous. Conversely, let $\varepsilon > 0$ be given. We have to specify a compact set $C \subseteq \mathbb{R}$ and a neighbourhood U of 1 in $\mathbb{T}$ such that $\chi_r(C) \subseteq U$ implies $|r| < \varepsilon$. We let $C = [0,1] \subseteq \mathbb{R}$ and choose a sufficiently small U and a local inverse $\mathrm{Log} : U \to \mathbb{R}$ of p such that $\mathrm{Log}(1) = 0$ (a modification of the complex logarithm). If $\chi_r(C) \subseteq U$, then $\tilde{\chi}_r = \mathrm{Log}\,\chi_r$ holds on C. Continuity of Log implies that $rC = \tilde{\chi}_r(C) \subseteq \,]{-}\varepsilon, \varepsilon[$ if U is small enough. This implies that $|r| < \varepsilon$. $\square$

8.32 The solenoid $\mathbb{Q}_d^*$ We shall describe the character group of the rationals in various respects. For proofs and for further information, we refer to HEWITT–ROSS 1963 10.12-10.13 and 25.4-25.5. Being the character group of a discrete group, the solenoid $\mathbb{Q}_d^*$ is compact (63.5). Since $\mathbb{Q}$ does not have any non-trivial compact subgroups, the solenoid is connected (63.30).

An explicit description can be derived from the following description of $\mathbb{Q}$: the set of rationals can be obtained as the union of the sets $(n!)^{-1}\mathbb{Z}$, $n \in \mathbb{N}$. More abstractly, this can be expressed by saying that $\mathbb{Q}$ is the inductive limit of the sequence

$$\mathbb{Z} \xrightarrow{\;2\;} \mathbb{Z} \xrightarrow{\;3\;} \mathbb{Z} \xrightarrow{\;4\;} \mathbb{Z} \xrightarrow{\;5\;} \mathbb{Z} \longrightarrow \ldots,$$

where the homomorphism marked n is given by $z \mapsto nz$. Via Pontryagin duality, the inductive limit is transformed into the projective limit of the dual sequence; see HOFMANN–MORRIS 1998 7.11(iv). The dual sequence is given by

$$\mathbb{T} \xleftarrow{\;2\;} \mathbb{T} \xleftarrow{\;3\;} \mathbb{T} \xleftarrow{\;4\;} \mathbb{T} \xleftarrow{\;5\;} \mathbb{T} \longleftarrow \ldots,$$

where the arrow marked n maps $t \in \mathbb{T}$ to t^n if we think of $\mathbb{T}$ as a multiplicative group. The projective limit can be described explicitly: in the product group $\mathbb{T}^{\mathbb{N}}$, the projective limit is the subgroup

$$\mathbb{Q}_d^* = \{\,(t_n)_{n \in \mathbb{N}} \mid t_n \in \mathbb{T},\; t_n^n = t_{n-1}\,\} \subseteq \mathbb{T}^{\mathbb{N}}\,.$$

There are other solenoids whose construction differs from the one above by the sequence of numbers assigned to the arrows. Some of the other solenoids have elements of finite order (see Exercise 6 of Section 52), but $\mathbb{Q}_d^*$ does not, because $\mathbb{Q}$ is divisible; compare 63.31.

All solenoids have in common that they are important examples of compact non-Lie groups. They contain compact, totally disconnected subgroups with quotient groups isomorphic to $\mathbb{T}$. Locally, a solenoid is homeomorphic to the product of an arc with the Cantor set; see HEWITT–ROSS 1963 10.15 or HOFMANN–MORRIS 1998 8.23 (equivalence of (3) and (8)).

The following fact was observed by HALMOS 1944.

8.33 Proposition *As a group, the solenoid $\mathbb{Q}_d^*$ is isomorphic to $\mathbb{R}^+$.*

Proof Since $\mathbb{Q}$ is uniquely divisible (in other words, divisible and torsion free), the character group $\mathbb{Q}_d^*$ has the same properties; see 63.31 and 63.32. Thus, $\mathbb{Q}_d^*$ is a rational vector space, see 1.9, and it remains to determine the cardinality of a basis or of $\mathbb{Q}_d^*$ itself (1.14).

There are at most $\aleph$ characters $\chi : \mathbb{Q}_d \to \mathbb{T}$, as χ is determined by the sequence of values $\chi(1/n) \in \mathbb{T}$, $n \in \mathbb{N}$. Since $\operatorname{card}\mathbb{T} = \operatorname{card}\mathbb{R} = 2^{\aleph_0}$ (see 1.10), the set of these sequences has cardinality $(2^{\aleph_0})^{\aleph_0} = 2^{\aleph_0} = \operatorname{card}\mathbb{R}$.

Conversely, the standard character $\mathbb{Q}_d \to \mathbb{Q}/\mathbb{Z} \leq \mathbb{R}/\mathbb{Z} = \mathbb{T}$ may be followed by any automorphism of $\mathbb{Q}/\mathbb{Z}$, and it is a consequence of 1.28 together with 1.27 that there are $\aleph$ such automorphisms. □

Exercises

(1) It is known that for any two sets $A, B \subseteq \mathbb{R}$ of positive Lebesgue measure, the sum $A + B$ contains a non-trivial interval; compare 10.7. Prove that there exists even a meagre set S of Lebesgue measure zero such that $S + S = \mathbb{R}$.

(2) With respect to Sorgenfrey's topology on $\mathbb{R}$ (5.73), addition is continuous, but not the map $x \mapsto -x$.

(3) Fill in the details for examples 8.16(3) and (4). In particular, show that A is not locally compact and that the direct product of (any number of) copies of $\mathbb{Q}^+$ is a topological group.

(4) Endow the ring of continuous endomorphisms $\operatorname{End}_c \mathbb{R}$ with the compact-open topology (see 63.2). Show that $\operatorname{End}_c \mathbb{R} \cong \mathbb{R}$ as a topological field.

9 Multiplication and topology of the real numbers

First we verify that the multiplicative group $\mathbb{R}^\times = \mathbb{R} \smallsetminus \{0\}$ is a topological group. In fact, we show slightly more, so that the result will combine with 8.2 to prove that $\mathbb{R}$ is a topological field.

9.1 Theorem *Multiplication $(x, y) \mapsto xy$ is continuous on $\mathbb{R} \times \mathbb{R}$, and inversion $\mathbb{R}^\times \to \mathbb{R}^\times$ is continuous. In particular, $\mathbb{R}^\times$ is a topological group.*

Proof Consider $(x + g)(y + h) = xy + gy + hx + gh$. Given $\varepsilon \in \,]0, 1]$, set $m = \max\{|x|, |y|, 1\}$ and $\delta = \varepsilon/(3m)$. If $|g|, |h| < \delta$, then $|gy + hx + gh| < m\delta + m\delta + \delta^2 < \varepsilon$. This yields continuity of multiplication. For $x \neq 0$ and $\varepsilon > 0$, choose $\delta = \min\{\varepsilon|x|^2, |x|\}/2$. If $|x - y| < \delta$, then $|y| \geq |x|/2$ and $|y - x| < \varepsilon|xy|$. Therefore, $|1/x - 1/y| < \varepsilon$, and inversion is continuous.

The same proof works for the additive group of any (skew) field with absolute value (as defined in 55.1). □

9.2 Proposition *The multiplicative group* $\mathbb{R}_{\mathrm{pos}}$ *of positive real numbers is isomorphic to* $\mathbb{R}^+$ *as a topological group, and* $\mathbb{R}^\times$ *is isomorphic to the direct product* $\mathbb{R}^+ \times C_2$ *as a topological group.*

Proof In 2.2 we described a group isomorphism $\varphi : \mathbb{R}^\times \to \mathbb{R}^+ \oplus C_2 = \mathbb{R}^+ \times C_2$ using the exponential function. Explicitly, φ is given by $\varphi(r) = (\ln|r|, r|r|^{-1})$. This map is a homeomorphism. $\square$

Our goal in this section is to demonstrate that the topology of $\mathbb{R}^\times$ does not determine the multiplication; in fact, there is precisely one (non-commutative) group structure other than the usual one which makes $\mathbb{R}^\times$ a topological group. Instead of constructing this group in an abstract way, we prefer to show that it arises naturally as the isometry group of the metric space of real numbers.

9.3 Definition: The affine group of the real line We consider the mappings $\mathbb{R} \to \mathbb{R}$ defined by

$$\tau_{a,b} : x \mapsto a + bx, \quad a, b \in \mathbb{R}, \ b \neq 0 .$$

They are bijective and form a group $\mathrm{Aff}\,\mathbb{R}$, the *affine group* of the real line. Multiplication in this group is composition of maps, and is given by

$$\tau_{a,b}\tau_{c,d} = \tau_{a+bc,\,bd} .$$

The map $\tau_{0,1}$ is the identity, and inversion is given by $\tau_{a,b}^{-1} = \tau_{-a/b,1/b}$. Using the parameters a, b, we may consider $\mathrm{Aff}\,\mathbb{R}$ as a subset of $\mathbb{R}^2$, and the topology induced by $\mathbb{R}^2$ makes $\mathrm{Aff}\,\mathbb{R}$ a topological group. We might add that the map $\mathrm{Aff}\,\mathbb{R} \times \mathbb{R} \to \mathbb{R}$ given by $(\tau, r) \mapsto \tau(r)$ is continuous, so $\mathrm{Aff}\,\mathbb{R}$ is a *topological transformation group* acting continuously on $\mathbb{R}$.

Sending the map $\tau_{a,b}$ to the matrix $\begin{pmatrix} 1 & 0 \\ a & b \end{pmatrix}$, we may represent $\mathrm{Aff}\,\mathbb{R}$ as a group of matrices, and the action on $\mathbb{R}$ corresponds to

$$\begin{pmatrix} 1 & 0 \\ a & b \end{pmatrix}\begin{pmatrix} 1 \\ x \end{pmatrix} = \begin{pmatrix} 1 \\ a + bx \end{pmatrix} .$$

From a purely group theoretic point of view, $\mathrm{Aff}\,\mathbb{R}$ is best described as a semidirect product.

9.4 Semidirect products Suppose that the group $(G, \cdot)$ contains a normal subgroup A and a (not necessarily normal) subgroup B such that $A \cap B = \{1\}$ and $A \cdot B = G$. For $b \in B$, let ψ_b be the automorphism of A induced by the inner automorphism $\vartheta_b : g \mapsto bgb^{-1}$ of G. Then $b \mapsto \psi_b$

is a homomorphism $B \to \operatorname{Aut} A$, and the map $ab \mapsto (a,b) \in A \times B$ is an isomorphism of G onto the *semidirect product* $A \rtimes_\psi B$, defined by

$$A \rtimes_\psi B := A \times B \text{ as a set, and } (a,b) \cdot (c,d) := (a \cdot \psi_b(c), b \cdot d) .$$

To verify the isomorphism, it suffices to note that $(ab)(cd) = a(bcb^{-1})bd$.

Conversely, given two groups A, B and a homomorphism $\psi : B \to \operatorname{Aut} A$, the above formula defines a semidirect product group. The semidirect product reduces to a direct product if, and only if, $\psi_b = \operatorname{id}$ for all b. Another equivalent condition is that also B is a normal subgroup of G.

Comparison with 9.3 shows that $\operatorname{Aff} \mathbb{R}$ is a semidirect product $R \rtimes S$ of the normal subgroup $R \cong \mathbb{R}^+$ of *translations*, given by $b = 1$, with the subgroup $S \cong \mathbb{R}^\times$ defined by $a = 0$,

$$\operatorname{Aff} \mathbb{R} \cong R \rtimes S \cong \mathbb{R}^+ \rtimes \mathbb{R}^\times .$$

Note that the group operation in R appears as addition in 9.3, and that $\psi_b(c) = bc$. Hence, the multiplication in the semidirect product has to be written as $(a,b)(c,d) = (a + bc, bd)$. The action of S on R by conjugation (which determines the structure of the semidirect product) is precisely the action of $\operatorname{Aut} \mathbb{R}$ on $\mathbb{R}$; compare 8.24.

9.5 Definition: The motion group $\mathbb{M}$ of the real line We single out a subgroup of $\operatorname{Aff} \mathbb{R}$ by restricting b in 9.3 to $\{1, -1\}$. We call this group $\mathbb{M}$ the *motion group* of $\mathbb{R}$, that is, $\mathbb{M} = \{ x \mapsto a \pm x \mid a \in \mathbb{R} \}$. This is a semidirect product,

$$\mathbb{M} \cong \mathbb{R}^+ \rtimes_\psi C_2,$$

where C_2 is identified with the subgroup $\{1, -1\} \leq S$ and -1 acts on $R \cong \mathbb{R}^+$ by inversion, that is, $\psi_{-1}(r) = -r$. Other examples of abelian groups extended by inversion are the finite dihedral groups $D_n \cong C_n \rtimes C_2$ (the symmetry groups of the regular n-gons, $n \geq 3$) and the orthogonal group of the plane, $O_2\mathbb{R} \cong (SO_2\mathbb{R}) \rtimes C_2$. The proof is left as an exercise.

We note that every element of $\mathbb{M}$ is either a translation $\tau_{a,1}$ or a reflection $\tau_{a,-1}$. More precisely, the involution $\tau_{a,-1}$ reflects the real line $\mathbb{R}$ about the point $a/2$. We leave it as an exercise to prove that $\mathbb{M}$ consists precisely of the isometries of the standard metric on $\mathbb{R}$. This is what the name motion group expresses.

Obviously, $\mathbb{M}$ is homeomorphic to $\mathbb{R}^\times$, and $\mathbb{M}$ is not abelian, so its group structure differs from that of $\mathbb{R}^\times$. It is easier to compare the two groups if we write $\mathbb{R}^\times \cong \mathbb{R}_{\mathrm{pos}} \times C_2$ and $\mathbb{M} \cong \mathbb{R}_{\mathrm{pos}} \rtimes C_2$. Then the multiplication in the semidirect product becomes

$$(r, e)(s, f) = (rs^e, ef) \ .$$

Our next result is that these two topological groups are the only ones that can live on the topological space $\mathbb{R} \smallsetminus \{0\}$.

9.6 Theorem *Every topological group $(G, \cdot)$ homeomorphic to $\mathbb{R}^\times$ is isomorphic as a topological group to $\mathbb{R}^\times \cong \mathbb{R}^+ \times C_2$ or to the motion group $\mathbb{M} \cong \mathbb{R}^+ \rtimes C_2$.*

Proof (1) Let $R \approx \mathbb{R}$ be the connected component of G containing the neutral element 1. Then R is a normal subgroup of G (see 62.13), and $R \cong \mathbb{R}^+ \cong \mathbb{R}_{\mathrm{pos}}$ as a topological group by 8.10 and 9.2. We may assume that in fact $(R, \cdot) = (\mathbb{R}_{\mathrm{pos}}, \cdot)$. The only other connected component of G must be a coset Rs.

(2) The factor group G/R has order 2, hence $s^2 \in R$. Consider the inner automorphism $\vartheta_s : g \mapsto sgs^{-1}$ induced by s. The square ϑ_s^2 is the inner automorphism induced by $s^2 \in R$; it induces the identity on the abelian group R. Using 8.24, we infer that there are precisely two possibilities for the automorphism of R induced by ϑ_s: it is either the identity or inversion.

(3) If ϑ_s induces inversion, then the equation $s^2 = \vartheta_s(s^2) = s^{-2}$ shows that $s^2 = 1$ and $\langle s \rangle \cong C_2$. Now 9.4 shows that we obtain an isomorphism $G \to \mathbb{R}_{\mathrm{pos}} \rtimes C_2$ of topological groups by sending $r \in R$ to $(r, 1)$ and $rs \in Rs$ to $(r, -1)$.

(4) Suppose that ϑ_s induces the identity on R. Choose an element $r \in R$ such that $r^2 = s^2$. We set $s' = sr^{-1}$; then $(s')^2 = sr^{-1}sr^{-1} = \vartheta_s(r^{-1})s^2r^{-1} = 1$, and $\vartheta_{s'} = \vartheta_s\vartheta_{r^{-1}}$ induces the identity on R. Then the same map as in step 3 is an isomorphism with respect to the direct product multiplication on $\mathbb{R}_{\mathrm{pos}} \times C_2$. $\square$

Exercises

(1) Show that all automorphisms of the topological group $\mathbb{R}^\times$ have the form $x \mapsto x^a$, where $a \in \mathbb{R}^\times$, and deduce that $\mathrm{Aut}_c\, \mathbb{R}^\times \cong \mathbb{R}^\times$. (How should x^a be defined for $x < 0$?)

(2) Prove that the motion group $\mathbb{M}$ introduced in 9.5 consists precisely of the isometries, i.e., of the maps preserving the standard metric $d(x, y) = |x - y|$ on $\mathbb{R}$.

(3) Verify the semidirect product decompositions of dihedral groups and of the orthogonal group stated in 9.5.

10   The real numbers as a measure space

A measure μ on a (metric) space M associates with some subsets of M
non-negative real numbers, measuring their 'sizes', in such a way that
μ is additive on finite or countable families of pairwise disjoint sets.
If M carries a group structure, μ is usually required to be translation
invariant. Generally, it is not possible to extend μ consistently to the
class of all subsets of M; see 10.9–11. The members of the largest
domain $\mathfrak{M} \subseteq 2^M$ to which μ can be extended are called measurable.
In the real case, the pair $(\mathfrak{M}, \mu)$ is unique up to a scalar factor. The
very complicated nature of $\mathfrak{M}$ is illustrated by results 10.15ff. For an
introduction to measure theory see HALMOS 1950 or COHN 1993.

10.1 Definition   A *σ-field* (or *σ-algebra*) on a space M is a subset
$\mathfrak{S} \subseteq 2^M$ such that:

(a) $\emptyset \in \mathfrak{S}$ and $(S \in \mathfrak{S} \Rightarrow M \smallsetminus S \in \mathfrak{S})$

(b) If $S_\nu \in \mathfrak{S}$ for $\nu \in \mathbb{N}$, then $\bigcup_{\nu \in \mathbb{N}} S_\nu \in \mathfrak{S}$ (and $\bigcap_{\nu \in \mathbb{N}} S_\nu \in \mathfrak{S}$ by (a)).
The σ-field $\langle \mathfrak{D} \rangle_\sigma$ generated by a set $\mathfrak{D} \subseteq 2^M$ is the intersection of all
σ-fields in 2^M containing $\mathfrak{D}$. (Note that 2^M itself is a σ-field.)

In a *metric* space M, the open sets and the closed sets generate the
same σ-field $\mathfrak{B}$; it is called the *Borel field* of M, and each element of $\mathfrak{B}$
is called a *Borel set*.

In this section, M will usually be the space $\mathbb{R}$ with its ordinary metric.
Since each open set in $\mathbb{R}$ is a union of countably many intervals, the Borel
field of $\mathbb{R}$ is generated by the open or the closed intervals or the compact
sets as well.

10.2 Definition   A *measure space* $(M, \mathfrak{S}, \mu)$ consists of a metric space
M, a σ-field $\mathfrak{S}$ containing the Borel field of M, and a σ-additive function
$\mu : \mathfrak{S} \to [0, \infty]$ satisfying $\mu(\emptyset) = 0$ and $\mu(S) < \infty$ for each bounded set
$S \in \mathfrak{S}$. Precisely, μ is called σ-additive if

$$\mu\left(\bigcup_{\nu \in \mathbb{N}} S_\nu\right) = \sum_{\nu \in \mathbb{N}} \mu(S_\nu) \qquad (*)$$

for any union of pairwise disjoint sets $S_\nu \in \mathfrak{S}$, $\nu \in \mathbb{N}$.

The measure μ is said to be *regular* if

$$\mu(S) = \inf\{\,\mu(O) \mid S \subseteq O \wedge O \text{ is open}\,\}$$

for each $S \in \mathfrak{S}$, and μ is *complete* if $T \subset S \in \mathfrak{S}$ and $\mu(S) = 0$ imply
$T \in \mathfrak{S}$.

If $(\mathbb{R}, \mathfrak{S}, \mu)$ is a measure space, then μ is called *(translation) invariant*,
if $\mu(S + t) = \mu(S)$ for all $S \in \mathfrak{S}$ and $t \in \mathbb{R}$.

10.3 Lebesgue measure There is a unique regular, complete, and translation invariant measure λ on $\mathbb{R}$ such that $\lambda([0,1]) = 1$. This measure is known as *Lebesgue measure*, and the sets in the corresponding σ-field $\mathfrak{M}$ are the *measurable sets*.

We sketch a proof for this assertion and characterize the sets in $\mathfrak{M}$. Since λ is assumed to be additive, regular, and invariant, $\lambda([a,b]) = b - a$ for each interval of rational length and then for all intervals (by regularity). Each open subset O of $\mathbb{R}$ is the disjoint union of its connected components, and these are (at most countably many) open intervals; compare 5.6. Thus, $\lambda(O)$ is necessarily the sum of the lengths of the connected components of O. Since each compact set C is a difference of a closed interval and an open set, also $\lambda(C)$ is well-defined. By

$$\overline{\lambda}(X) = \inf\big\{\,\lambda(O) \mid X \subseteq O \wedge O \text{ is open in } \mathbb{R}\,\big\}$$

we can define an *outer measure* $\overline{\lambda} : 2^{\mathbb{R}} \to [0, \infty]$, and $\overline{\lambda}$ is obviously translation invariant. Moreover, $\overline{\lambda}$ is easily seen to be σ-*subadditive*: $\overline{\lambda}(\bigcup_{\nu \in \mathbb{N}} X_\nu) \leq \sum_{\nu \in \mathbb{N}} \overline{\lambda}(X_\nu)$. The *inner measure* $\underline{\lambda}$ is defined dually, approximating X from within by closed or compact sets. It satisfies $\underline{\lambda}(\bigcup_\nu X_\nu) \geq \sum_\nu \underline{\lambda}(X_\nu)$ for any countable family of pairwise disjoint sets X_ν. By the very definition, $\underline{\lambda}(X) \leq \overline{\lambda}(X)$ for all $X \subseteq M$.

The only way to enforce regularity consists in choosing the right σ-algebra. We put

$$X \in \mathfrak{M} \;\rightleftharpoons\; 0 = \inf\big\{\,\overline{\lambda}(O \smallsetminus X) \mid X \subseteq O \wedge O \text{ is open in } \mathbb{R}\,\big\}$$

and $\lambda(X) = \overline{\lambda}(X)$ for $X \in \mathfrak{M}$. By construction, λ is translation invariant.

10.4 Theorem $(\mathbb{R}, \mathfrak{M}, \lambda)$ *is a regular complete measure space.*

A *proof* consists of several steps, most of which are easy consequences of the definitions:

(a) If $X_\nu \in \mathfrak{M}$ for $\nu \in \mathbb{N}$, then $X = \bigcup_\nu X_\nu \in \mathfrak{M}$. In fact, for each $\varepsilon > 0$ there are open sets O_ν and U_ν such that $X_\nu \subseteq O_\nu \subseteq X_\nu \cup U_\nu$ and $\lambda(U_\nu) < \varepsilon 2^{-\nu}$. Put $U = \bigcup_\nu U_\nu$. Then $X \subseteq \bigcup O_\nu \subseteq X \cup U$, and $\lambda(U) < 2\varepsilon$ is arbitrarily small.

(b) If $X \in \mathfrak{M}$ and O is open in $\mathbb{R}$, then $X \cap O \in \mathfrak{M}$.

(c) If $X \in \mathfrak{M}$ and $\lambda(X) < \infty$, then $\inf_C \overline{\lambda}(X \smallsetminus C) = 0$, where C varies over all compact subsets of X. In particular, $X \in \mathfrak{M}$ implies $\mathbb{R} \smallsetminus X \in \mathfrak{M}$ for each bounded set X. Together with (a) and (b) this shows that $\mathfrak{M}$ is a σ-field.

(c′) Step (c) can be rephrased as follows: if $\overline{\lambda}(X) < \infty$, then $X \in \mathfrak{M}$ if, and only if, $\underline{\lambda}(X) = \overline{\lambda}(X)$.

(d) If $X = \bigcup_\nu X_\nu$ is a countable union of pairwise disjoint measurable sets X_ν, then $\lambda(X) < \infty$ if, and only if, $\sum_\nu \lambda(X_\nu) < \infty$. In this case $\overline{\lambda}(X) \leq \sum_\nu \lambda(X_\nu) \leq \underline{\lambda}(X)$, and (c′) shows that $X \in \mathfrak{M}$ and $\lambda(X) = \sum_\nu \lambda(X_\nu)$. Thus, λ is σ-additive on $\mathfrak{M}$.

(e) By definition, λ is regular. If $\overline{\lambda}(X) = 0$ and $Y \subseteq X$, then $Y \in \mathfrak{M}$. Therefore, λ is also complete. $\qquad\square$

10.5 Uniqueness If $(\mathbb{R}, \mathfrak{S}, \mu)$ is a measure space, and if μ satisfies the conditions imposed upon λ in 10.3, then $\mathfrak{S} = \mathfrak{M}$ and $\lambda = \mu$.

Indeed, λ and μ agree on intervals and hence on the Borel field $\mathfrak{B}$. Since μ is regular, $\mu(S) = \overline{\lambda}(S)$ for each $S \in \mathfrak{S}$. Moreover, S is contained in a Borel set $B \in \mathfrak{B}$ (in fact, an intersection of countably many open sets) such that $\mu(S) = \mu(B)$ and $\overline{\lambda}(B \setminus S) = 0$. This implies $S \in \mathfrak{M}$. We will show that $\mathfrak{S} = \mathfrak{M}$. Any set N with $\lambda(N) = 0$ is contained in a Borel set of measure 0. Because μ is complete, it follows that $N \in \mathfrak{S}$. An arbitrary set $X \in \mathfrak{M}$ can be written as $X = B \cup N$ with $B \in \mathfrak{B}$ and $\lambda(N) = 0$. Hence $X \in \mathfrak{S}$. $\qquad\square$

10.6 Examples (a) The countable field $\mathbb{R}_{\mathrm{alg}}$ of real algebraic numbers has measure $\lambda(\mathbb{R}_{\mathrm{alg}}) = 0$. Argument (a) of the proof of 10.4 shows that $\mathbb{R}_{\mathrm{alg}}$ has open neighbourhoods O of arbitrarily small measure $\lambda(O)$, and the complement $\mathbb{R} \setminus O$ consists entirely of transcendental numbers.

(b) The Cantor set $\mathcal{C} = \{\sum_{\nu=1}^{\infty} c_\nu 3^{-\nu} \mid c_\nu \in \{0, 2\}\}$ has the measure $\lambda(\mathcal{C}) = 0$. In fact, $\mathcal{C}$ is obtained from the interval $[0, 1]$ by repeatedly removing 'middle thirds'; see 5.35. Thus, $\lambda(\mathcal{C}) \leq (2/3)^n$ for any $n \in \mathbb{N}$. Since $\operatorname{card} \mathcal{C} = \aleph = 2^{\aleph_0}$, it follows from completeness that $\operatorname{card} \mathfrak{M} = 2^{\aleph}$. Note that $\mathcal{C} + \mathcal{C} = [0, 2]$ has positive measure (NYMANN 1993).

(c) A modified construction leads to a set which is homeomorphic to $\mathcal{C}$ but has positive measure: one starts again with $[0, 1]$, instead of middle thirds, however, in the ν-th step one removes an open interval of length $\nu^{-2}\lambda(J)$ from the middle of each remaining closed interval J. This yields a compact set $\mathcal{C}_2$ of measure $\prod_{\nu=2}^{\infty}(1 - \nu^{-2}) = 1/2$, the value of the infinite product being obtained as follows:

$$\prod_{\nu=2}^{n} (\nu^2 - 1)/\nu^2 = (n-1)! \, (n+1)! \, /2(n!)^2 = (n+1)/2n \ .$$

10.7 Differences *If $\underline{\lambda}(S) > 0$, then $S - S = \{s - t \mid s, t \in S\}$ contains some open neighbourhood of 0.*

Proof Since S contains a compact subset of positive measure, we may assume that S itself is compact. There is an open neighbourhood O of S such that $\lambda(S) \geq \frac{5}{6} \cdot \lambda(O)$. The set O is a union of disjoint intervals. Therefore, O contains an interval J with $\lambda(S \cap J) \geq \frac{4}{5} \cdot \lambda(J)$. Put $T = S \cap J$ and $\lambda(J) = 5\varepsilon$, and assume that $|x| < \varepsilon$ and $x \notin T - T$. Then $x + T$ and T are both contained in an ε-neighbourhood of J, hence in an interval K of length 7ε, and $(x + T) \cap T = \emptyset$, but $\lambda(x + T) = \lambda(T) \geq 4\varepsilon$. This contradicts additivity and shows that $]-\varepsilon, \varepsilon[\subseteq S - S$.   □

The fact that $\mathbb{R}$ is a vector space over $\mathbb{Q}$ yields many examples of *non-measurable* sets. For any Hamel basis B (see 1.13), the *lattice* $L = L_B$ consists of all numbers $\sum_{b \in B} n_b \cdot b$ with $n_b \in \mathbb{Z}$ and $n_b = 0$ for all but finitely many $b \in B$. We will also consider an arbitrary hyperplane H of $\mathbb{R}$ over $\mathbb{Q}$.

10.8 Inner measure *Each lattice L and each hyperplane H has inner measure 0. More generally, $\underline{\lambda}(G) = 0$ for any proper additive subgroup $G < \mathbb{R}$.*

Proof Since G is a group, $G - G = G$. If $\underline{\lambda}(G) > 0$, then $G - G$ contains an open neighbourhood of 0 by 10.7, but such a neighbourhood generates the full group $\mathbb{R}$.   □

10.9 Proposition *A hyperplane H intersects an arbitrary open interval J in a set of positive outer measure. In particular, $H \cap J \notin \mathfrak{M}$.*

Proof Assume that $\overline{\lambda}(H \cap J) = 0$. Then $\lambda(H \cap J) = 0$ by 10.8. There are countably many elements $h_\nu \in H$ such that $\bigcup_\nu (J + h_\nu) = \mathbb{R}$. Note that $(H \cap J) + h_\nu = H \cap (J + h_\nu)$. Thus, $\lambda(H \cap J) = 0$ implies $\lambda(H) = 0$. Because $\mathbb{R}/H \cong \mathbb{Q}$, the set $\mathbb{R}$ is a union of countably many cosets of H, and it would follow that $\lambda(\mathbb{R}) = 0$.   □

10.10 Proposition *Any lattice $L = L_B$ in the rational vector space $\mathbb{R}$ has outer measure $\overline{\lambda}(L) = \infty$. Therefore, $L \notin \mathfrak{M}$.*

Proof The elements $\frac{1}{2}b$ with $b \in B$ represent uncountably many distinct cosets of L in $\mathbb{R}$. Hence one cannot argue as in the previous proof. The definition of $\overline{\lambda}$ shows that $\overline{\lambda}(r \cdot S) = r \cdot \overline{\lambda}(S)$ for any positive number r. Obviously, $\mathbb{R} = \bigcup_{n \in \mathbb{N}} (n!)^{-1} L$, and hence $\infty \leq \overline{\lambda}(\mathbb{R}) \leq \sum_n (n!)^{-1} \cdot \overline{\lambda}(L) = e \cdot \overline{\lambda}(L)$.   □

The following analysis shows that a lattice L is non-measurable in a very strong sense.

10.11 Theorem *If L is a lattice in $\mathbb{R}$ and U is open, then $\overline{\lambda}(L \cap U) = \lambda(U)$, while $\underline{\lambda}(L \cap U) = 0$.*

Proof By 10.8, it suffices to prove the first claim. Two distinct elements of a basis of $\mathbb{R}$ over $\mathbb{Q}$ generate a dense subgroup of $\mathbb{R}$; see 1.6b. In particular, $\overline{L} = \mathbb{R}$.

(a) We show that for each $\vartheta \in \,]0,1[$ and for each $\varepsilon > 0$, there is some open interval J of length $\lambda(J) < \varepsilon$ such that $\overline{\lambda}(L \cap J) > \vartheta \cdot \lambda(J)$.

In fact, $\overline{\lambda}(L) = \infty$ and subadditivity of $\overline{\lambda}$ imply that there is a compact interval C with $\overline{\lambda}(L \cap C) := m > 0$. By definition of $\overline{\lambda}$, we can find an open neighbourhood O of $L \cap C$ satisfying $\lambda(O) < m + r$ with an arbitrarily small number $r > 0$. Let $2r < (1 - \vartheta) \cdot m$. The open set O can be covered by intervals J_ν of length less that ε such that $\lambda(O) \leq \sum_\nu \lambda(J_\nu) < m + 2r$.

Suppose that $\overline{\lambda}(L \cap J_\nu) \leq \vartheta \cdot \lambda(J_\nu)$ for all ν; then $m \leq \sum_\nu \overline{\lambda}(L \cap J_\nu) \leq \vartheta \cdot (m + 2r) < m$, a contradiction.

(b) We may assume that U is an open interval. Let J be chosen according to step (a). Because L is dense in $\mathbb{R}$, there are finitely many elements $t_1, \ldots, t_k \in L$ such that the intervals $J + t_\kappa \subseteq U$ are pairwise disjoint and that $\lambda(U) - k \cdot \lambda(J) < \varepsilon$. It follows that

$$\overline{\lambda}(L \cap U) \geq \overline{\lambda}\left(\bigcup_{\kappa=1}^{k} L \cap (J + t_\kappa)\right) = \sum_{\kappa=1}^{k} \overline{\lambda}\left((L \cap J) + t_\kappa\right)$$
$$= k \cdot \overline{\lambda}(L \cap J) > k \cdot \vartheta \cdot \lambda(J) \,,$$

hence $\overline{\lambda}(L \cap U) > \vartheta \cdot (\lambda(U) - \varepsilon)$ is arbitrarily close to $\lambda(U)$. □

Remark With the same arguments, an analogous result can be proved for hyperplanes.

Intuitively, both *meagre sets* (i.e., countable unions of nowhere dense sets) and sets of Lebesgue measure zero (*null sets*) are small. The two notions are quite independent of each other. This is illustrated by the following astonishing result:

10.12 Proposition $\mathbb{R}$ *is a union of a meagre set M and a null set N.*

Proof Let $\mathbb{Q} = \{ r_\nu \mid \nu \in \mathbb{N} \}$ and put $J_{\kappa\nu} = \{ x \in \mathbb{R} \mid |x - r_\nu| < 2^{-\kappa-\nu} \}$ for $\kappa \geq 1$. Then $O_\kappa = \bigcup_{\nu=1}^{\infty} J_{\kappa\nu}$ is an open neighbourhood of the dense set $\mathbb{Q}$, and $\mathbb{R} \smallsetminus O_\kappa$ is nowhere dense. The set $N = \bigcap_\kappa O_\kappa$ is a null set (since $\lambda(O_\kappa) \leq 2^{1-\kappa}$) and $\mathbb{R} \smallsetminus N$ is meagre. □

We mention a kind of duality between meagre sets and null sets (for a *proof* see OXTOBY 1971 §19):

10.13 Theorem (Erdös)   *Under the continuum hypothesis* $2^{\aleph_0} = \aleph_1$
there exists an involutory mapping $\varphi : \mathbb{R} \to \mathbb{R}$ *such that* $\lambda(X) = 0$ *if,*
and only if, $\varphi(X)$ *is meagre.*

Traditionally, any union of countably many closed sets is called an
F_σ-set.

10.14 Measurable sets   *A set* $S \subseteq \mathbb{R}$ *is measurable if, and only if,*
there is an F_σ-*set* F *and a null set* N *with* $S = F \cup N$.

Proof If S is measurable, so is $X = \mathbb{R} \smallsetminus S$. By definition of $\mathfrak{M}$, the set
X is contained in an intersection D of countably many open sets such
that $\overline{\lambda}(D \smallsetminus X) = 0$. Hence $F = \mathbb{R} \smallsetminus D$ is an F_σ-set and $S \smallsetminus F = D \smallsetminus X$
has measure 0.   □

In order to illustrate the high degree of complexity of the σ-field $\mathfrak{M}$
of the measurable sets, we introduce the notions of Borel classes and of
Souslin sets. Starting with the compact or the open sets, all Borel sets
can be obtained by alternatingly forming countable unions and countable
intersections. For $\mathfrak{X} \subseteq 2^{\mathbb{R}}$, let $\mathfrak{X}_\sigma$ be the set of all unions and $\mathfrak{X}_\delta$ the set
of all intersections of countable subfamilies of $\mathfrak{X}$.

We shall now use ordinal numbers; refer to Section 61 for notions and
notation.

10.15 Borel hierarchy   Denote by $\mathfrak{O} = \mathfrak{O}_0$ the set of all open subsets
of $\mathbb{R}$, and put $\mathfrak{O}_\nu = \bigcup_{\mu \in \nu} (\mathfrak{O}_\mu)_{\delta\sigma}$.

By definition, $\mathfrak{O}_\nu \subseteq \mathfrak{B}$ for each ordinal ν. Because any countable
subset of the least uncountable ordinal number Ω has an upper bound
in Ω (see 61.5), we have $(\mathfrak{O}_\Omega)_{\delta\sigma} = \mathfrak{O}_\Omega$. Since closed sets are contained in
$\mathfrak{O}_\delta$, it follows inductively that the complement of any set in $\mathfrak{O}_\mu$ belongs
to $\mathfrak{O}_{\mu+1}$. Hence $\mathfrak{B} = \mathfrak{O}_\Omega$.

10.16 Borel sets   *We have* card $\mathfrak{B} = \aleph = 2^{\aleph_0} <$ card $\mathfrak{M}$.

Proof By 61.10, we have card $\mathfrak{X}_\sigma$, card $\mathfrak{X}_\delta \leq (\text{card } \mathfrak{X})^{\aleph_0}$. Since $\mathbb{R}$ has
a countable basis, there are only $\aleph_0^{\aleph_0} = \aleph$ open sets. By induction it
follows that card $\mathfrak{O}_\nu = \aleph$ for $\nu \in \Omega$. Now $2^{\aleph_\alpha} > \aleph_\alpha$ by 61.11, hence
card $\Omega = \aleph_1 \leq \aleph$, and with 61.13(a) we obtain card $\mathfrak{O}_\Omega = \aleph$.

Because each subset of the Cantor set $\mathcal{C}$ (see 10.6(b)) is a null set,
card $\mathfrak{M} = 2^{\aleph} > \aleph$.   □

10.17 Theorem   *For* $\nu \leq \Omega$, *the Borel classes* $\mathfrak{O}_\nu$ *are strictly increasing.*

A *proof* can be found in HAUSDORFF 1935 §33 I; compare also MILLER
1979.

10.18 The Souslin operation Let $\mathfrak{F}$ be the set of all finite sequences of elements of $\mathbb{N}$. For any sequence $\nu \in \mathbb{N}^{\mathbb{N}}$ of natural numbers, put $\nu|\kappa = (\nu_\iota)_{\iota \le \kappa} \in \mathfrak{F}$. Consider an arbitrary family $\mathfrak{T} \subseteq 2^{\mathbb{R}}$. Any mapping $\mathfrak{F} \to \mathfrak{T}$, i.e., any choice of sets $T_{\nu|\kappa} \in \mathfrak{T}$ determines a *Souslin set* $\bigcup_{\nu \in \mathbb{N}^{\mathbb{N}}} \bigcap_{\kappa \in \mathbb{N}} T_{\nu|\kappa}$ over $\mathfrak{T}$; note that the union is to be taken over uncountably many sequences ν.

The family of all Souslin sets over $\mathfrak{T}$ is denoted by $\mathfrak{T}_S$. The choice $T_{\nu|\kappa} = T_\kappa \in \mathfrak{T}$ shows that $\mathfrak{T}_\delta \subseteq \mathfrak{T}_S$, similarly $T_{\nu|\kappa} = T_{\nu_1} \in \mathfrak{T}$ gives $\mathfrak{T}_\sigma \subseteq \mathfrak{T}_S$. An argument related to general distributivity shows that $(\mathfrak{T}_S)_S = \mathfrak{T}_S$; see HAUSDORFF 1935 §19 I, JACOBS 1978 Chapter XIII Proposition 1.3, or ROGERS–JAYNE 1980 Theorem 2.3.1.

Let $\mathfrak{S} = \mathfrak{O}_S$ denote the collection of the Souslin sets over the open subsets of $\mathbb{R}$, and write $\mathfrak{C}$ for the class of all compact sets.

10.19 Proposition *The family $\mathfrak{S}$ contains all Borel subsets of $\mathbb{R}$. Moreover, $\mathfrak{C}_S = \mathfrak{S}$.*

Proof By 10.15, the Borel field $\mathfrak{B}$ is the smallest family of subsets of $\mathbb{R}$ such that $\mathfrak{O} \subseteq \mathfrak{B}$ and $\mathfrak{B}_\delta = \mathfrak{B}_\sigma = \mathfrak{B}$, and 10.18 shows that $\mathfrak{S}_\delta, \mathfrak{S}_\sigma \subseteq \mathfrak{S}_S = \mathfrak{S}$. Hence $\mathfrak{B} \subseteq \mathfrak{S}$. Moreover, $\mathfrak{C} \subseteq \mathfrak{O}_\delta$ and $\mathfrak{O} \subseteq \mathfrak{C}_\sigma$ imply $\mathfrak{C}_S \subseteq \mathfrak{O}_S$ and $\mathfrak{S} \subseteq (\mathfrak{C}_\sigma)_S \subseteq (\mathfrak{C}_S)_S = \mathfrak{C}_S$. □

10.20 Analytic sets *By definition, an analytic set is the image of some continuous map $\varphi : \mathbb{N}^{\mathbb{N}} \to \mathbb{R}$. Each analytic set is contained in $\mathfrak{S}$.*

Proof It is tacitly understood that $\mathbb{N}^{\mathbb{N}}$ carries the product topology. A typical neighbourhood $U_\kappa(\nu)$ of $\nu \in \mathbb{N}^{\mathbb{N}}$ consists of all sequences $\mu \in \mathbb{N}^{\mathbb{N}}$ such that $\mu|\kappa = \nu|\kappa$. For a fixed ν, continuity of φ implies that $\varphi U_\kappa(\nu)$ is contained in some ε-neighbourhood $V_\varepsilon(\varphi(\nu))$, where the $\varepsilon = \varepsilon_\kappa$ converge to 0 as κ increases. Choose $O_{\nu|\kappa}$ as an ε_κ-neighbourhood of $\varphi U_\kappa(\nu)$. Then $O_{\nu|\kappa} \subseteq V_{2\varepsilon}(\varphi(\nu))$ and $\bigcap_{\kappa \in \mathbb{N}} O_{\nu|\kappa} = \{\varphi(\nu)\}$. The proof is completed by taking the union over all $\nu \in \mathbb{N}^{\mathbb{N}}$. □

10.21 Theorem *Each Borel set and even each Souslin set is analytic.*

Proof By 4.10, the space $\mathfrak{J}$ of irrational real numbers may be identified with $\mathbb{N}^{\mathbb{N}}$.
(a) $\mathfrak{J}_\nu = \mathfrak{J} \cap [\nu, \nu+1]$ is homeomorphic to $\mathfrak{J}$. If $\varphi_\nu : \mathfrak{J}_\nu \to \mathbb{R}$ is continuous, then there is a continuous map $\varphi : \mathfrak{J} \to \mathbb{R}$ with $\varphi|_{[\nu,\nu+1]} = \varphi_\nu$. Obviously $\varphi(\mathfrak{J}) = \bigcup_\nu \varphi_\nu(\mathfrak{J}_\nu)$. Hence a union of countably many analytic sets is analytic.
(b) The set $\mathfrak{J} = \mathbb{N}^{\mathbb{N}}$ contains a copy of the Cantor set $\mathcal{C} = \{0,1\}^{\mathbb{N}}$ and

$((c_\nu)_\nu \mapsto \sum_\nu c_\nu 2^{-\nu}) : \mathcal{C} \to [0,1]$ is a continuous surjection. It extends
to a continuous map $\mathfrak{J} \to [0,1]$ by normality of $\mathfrak{J}$. Hence each closed
interval is analytic, and (a) implies that open sets are analytic.

(c) Let $\varphi_\nu : \mathfrak{J} \to \mathbb{R}$ be continuous, and assume that $\bigcap_{\nu \in \mathbb{N}} \varphi_\nu(\mathfrak{J}) = E \neq \emptyset$.
The set

$$D = \left\{ x = (x_\nu)_\nu \in \mathfrak{J}^{\mathbb{N}} \mid \forall_\nu \; \varphi_\nu(x_\nu) = \varphi_1(x_1) \right\}$$

is closed in $\mathfrak{J}^{\mathbb{N}}$, and $(x \mapsto \varphi_1(x_1)) : D \to E$ is a continuous surjection.
Now D is a retract of $\mathfrak{J}^{\mathbb{N}}$, because $\mathfrak{J}^{\mathbb{N}} = \mathbb{N}^{\mathbb{N} \times \mathbb{N}} \approx \mathbb{N}^{\mathbb{N}} = \mathfrak{J}$ and each closed
subset of $\mathbb{N}^{\mathbb{N}}$ is a retract (see Exercise 7 or ENGELKING 1969). Hence,
$(x \mapsto \varphi_1(x_1))$ extends to $\mathfrak{J}^{\mathbb{N}}$, and the intersection E of countably many
analytic sets is analytic. 10.15 proves the first claim.

(d) By the last part of 10.19, Souslin sets can be obtained from compact
sets instead of open sets, hence each Souslin set is of the form $A = \bigcup_{\nu \in \mathbb{N}^{\mathbb{N}}} \bigcap_{\kappa \in \mathbb{N}} C_{\nu|\kappa}$ with compact sets $C_{\nu|\kappa}$. We may assume, moreover,
that $\kappa < \lambda$ implies $C_{\nu|\kappa} \supseteq C_{\nu|\lambda}$. Write $C_{\nu|\kappa} = \bigcup_{i \in \mathbb{N}} C_{\nu|\kappa}^i$, where each
$C_{\nu|\kappa}^i$ is contained in some interval of length κ^{-1}. Infinite distributivity
gives $\bigcap_{\kappa \in \mathbb{N}} C_{\nu|\kappa} = \bigcup_{\iota \in \mathbb{N}^{\mathbb{N}}} \bigcap_{\kappa \in \mathbb{N}} C_{\nu|\kappa}^{\iota_\kappa}$. Therefore the $C_{\nu|\kappa}$ may be chosen
of diameter at most κ^{-1}. We define a set D by $\nu \in D \leftrightharpoons \bigcap_{\kappa \in \mathbb{N}} C_{\nu|\kappa} = \{x_\nu\} \neq \emptyset$. The map $\varphi : D \to A : \nu \mapsto x_\nu$ is a continuous surjection.
Moreover, D is closed in $\mathbb{N}^{\mathbb{N}}$, since $\nu \notin D$ implies $C_{\nu|\kappa} = \emptyset$ for some
$\kappa \in \mathbb{N}$, and then $\mu \notin D$ for all μ with $\mu|\kappa = \nu|\kappa$. Again, D is a retract
of $\mathbb{N}^{\mathbb{N}}$ and φ extends to a continuous map $\mathbb{N}^{\mathbb{N}} \to A$. $\square$

10.22 Proposition *The family of all Souslin sets in $\mathbb{R}$ has cardinality*
card $\mathfrak{S} = \aleph$.

Proof Since $\mathbb{R}$ has a countable basis for the topology, card $\mathfrak{O} = \aleph = 2^{\aleph_0}$.
The set $\mathfrak{F}$ of all finite sequences $\nu|\kappa$ is countable by 61.14. Each Souslin
set is determined by a choice of open sets $O_{\nu|\kappa}$, that is, by a map $\mathfrak{F} \to \mathfrak{O}$.
There are $\aleph^{\aleph_0} = \aleph$ such maps (61.13b). $\square$

10.23 Theorem *There is a Souslin set X in $\mathbb{R}$ such that $\mathbb{R} \smallsetminus X \notin \mathfrak{S}$.*
In particular, X is not a Borel set.

Proof By 61.14, the set $\mathfrak{F}$ of all finite sequences $\nu|\kappa$ is countable, hence
there is a bijection $\mathfrak{F} \to \mathbb{N}$ denoted by $(\nu|\kappa \mapsto \overline{\nu}(\kappa))$. The map $(\nu \mapsto \overline{\nu})$
is injective and defines a bijection of $\mathbb{N}^{\mathbb{N}}$ onto its image $\mathcal{N} \subseteq \mathbb{N}^{\mathbb{N}}$. Each
Souslin set is now of the form $\bigcup_{n \in \mathcal{N}} \bigcap_{\kappa \in \mathbb{N}} O_{n(\kappa)}$. Choose a countable
basis $\mathcal{B} = (B_\mu)_{\mu \in \mathbb{N}}$ for the open sets and note that $\mathcal{B}_\sigma = \mathfrak{O}$. For $\xi \in \mathbb{N}^{\mathbb{N}}$
put $S(\xi) = \bigcup_{n \in \mathcal{N}} \bigcap_{\kappa \in \mathbb{N}} B_{\xi_{n(\kappa)}}$. Then $\{ S(\xi) \mid \xi \in \mathbb{N}^{\mathbb{N}} \} = \mathcal{B}_S = \mathfrak{S}$.

We identify $\mathbb{N}^{\mathbb{N}}$ with the space $\mathfrak{I}$ of all irrational numbers in $\mathbb{R}$ as in 4.10. Let $X \subseteq \mathbb{R}$ be defined by $\xi \in X \rightleftharpoons \xi \in S(\xi)$. Assume that $\mathbb{R} \smallsetminus X = S(\xi)$ for some $\xi \in \mathbb{N}^{\mathbb{N}}$. Then $\xi \in \mathbb{R} \smallsetminus X \Leftrightarrow \xi \in S(\xi) \Leftrightarrow \xi \in X$, a contradiction. Therefore, $\mathbb{R} \smallsetminus X$ is not a Souslin set.

In order to show that $X \in \mathfrak{S}$, let $T_\kappa = \{\, \xi \in \mathfrak{I} \mid \xi \in B_{\xi_\kappa} \,\}$, and note that $\xi \in B_{\xi_\kappa}$ if, and only if, there is some $\mu \in \mathbb{N}$ such that $\xi \in B_\mu$ and $\mu = \xi_\kappa$. Now $\{\, \xi \mid \xi_\kappa = \mu \,\}$ is open in $\mathbb{N}^{\mathbb{N}}$, hence T_κ is open in $\mathfrak{I}$, and therefore $T_\kappa \in \mathfrak{O}_\delta$. Thus $X = \bigcup_{n \in \mathcal{N}} \bigcap_{\kappa \in \mathbb{N}} T_{n(\kappa)} \in \mathfrak{O}_S$. $\square$

10.24 Souslin sets are measurable *The set $\mathfrak{M}$ of all Lebesgue measurable sets satisfies $\mathfrak{M}_S = \mathfrak{M}$. Hence $\mathfrak{S} \subseteq \mathfrak{M}$, even $\mathfrak{S} \subset \mathfrak{M}$, since* $\operatorname{card} \mathfrak{S} = \aleph < \operatorname{card} \mathfrak{M}$.

Proofs of this standard result can be found in SAKS 1937 Chapter II Theorem 5.5, or ROGERS–JAYNE 1980 Corollary 2.9.3. They are somewhat technical and will be omitted here.

Finally we mention the following result by LACZKOVICH 1998:

10.25 *Every analytic proper subgroup of the reals can be covered by an F_σ null set.*

Exercises

(1) In the notation of 10.15, show that $\mathfrak{O}_1 \neq \mathfrak{O}$.

(2) Represent $\mathbb{Q}$ as a Souslin set.

(3) Any sublattice S in the intersection of finitely many hyperplanes in the rational vector space $\mathbb{R}$ has infinite outer measure.

(4) There are many σ-fields between $\mathfrak{S}$ and $\mathfrak{M}$.

(5) The modified Cantor set $\mathcal{C}_2$ (defined in 10.6c) has non-measurable subsets.

(6) A map $\varphi : \mathbb{R} \to \mathbb{R}$ is said to be *measurable* if, and only if, each open set O has an inverse image $\varphi^{-1}(O) \in \mathfrak{M}$. If φ is measurable, then the inverse image of each Borel set is measurable, but not necessarily the inverse image of each set $X \in \mathfrak{M}$.

(7) If A is closed in $\mathbb{N}^{\mathbb{N}}$, then there is a continuous map $\rho : \mathbb{N}^{\mathbb{N}} \to A$ with $\rho|_A = \operatorname{id}_A$. (Such a map is called a *retraction*.)

11   The real numbers as an ordered field

In this section we show that the field $\mathbb{R}$ can be made into an ordered field in a unique way. Conversely, ordering and addition of the real numbers determine the multiplication up to isomorphism. After a brief survey on exponential functions in ordered fields we characterize those ordered fields which are isomorphic to subfields of $\mathbb{R}$.

11.1 Definition   An *ordered skew field* is a skew field F equipped with the additional structure of a chain $(F, <)$ such that the following *monotonicity laws* hold. For all $a, b, c \in F$ such that $a < b$, we stipulate that

$$a + c < b + c$$
$$ac < bc \qquad \text{if} \quad 0 < c .$$

From these axioms it follows that $bc < ac$ if $c < 0$; indeed, from $c < 0$ we obtain $0 < -c$ by the first monotonicity law, and the second one yields $a(-c) < b(-c)$. This implies $0 < ac - bc$, and the claim follows.

The first law says that the additive group of F becomes an ordered group in the sense of 7.1. Note that $(F, +)$ is commutative by 6.1, hence we do not need to stipulate the monotonicity of left addition. There is a similar lack of symmetry in the second law although at this moment the multiplication in F is not assumed to be commutative. We shall show in 11.5 that the missing law comes for free. If F happens to be commutative, then the second law alone implies that the *positive* elements $a > 0$ form an ordered group under multiplication.

In any ordered skew field F, we define the *absolute value* of $a \in F$ by

$$|a| = \max\{a, -a\} .$$

This absolute value satisfies the *triangle inequality*

$$|a + b| \le |a| + |b| ,$$

which is proved by adding the inequalities $\varepsilon a \le |a|$ and $\varepsilon b \le |b|$, where $\varepsilon \in \{1, -1\}$, to obtain $\varepsilon(a + b) \le |a| + |b|$.

11.2 Remark   The monotonicity law for multiplication implies that every square a^2, $0 \ne a \in F$, is positive: $a^2 > 0$. In particular, it is not possible to make the complex numbers into an ordered field, because both $-1 = i^2$ and $1 = 1^2$ would be positive, resulting in the contradiction $0 = -1 + 1 > 0$. Moreover, in every ordered skew field, we have $1 = 1^2 > 0$ and $1 + \cdots + 1 > 0$, hence the characteristic of an ordered skew field is always zero.

An ordering on a field F is introduced most conveniently by specifying the set of positive elements. The properties that this set has to satisfy are collected in the following definition.

11.3 Definition   A *domain of positivity* in a skew field F is a subgroup $P \le F^\times$ of index 2 in the multiplicative group which is closed under addition. In other words, P does not contain zero and forms a group under multiplication, and we have $P + P \subseteq P$ and $F^\times = P \cup aP$ for each $a \notin P \cup \{0\}$.

11.4 Proposition (a) *Every ordered skew field $(F, <)$ defines a domain of positivity by*

$$P = \{x \in F \mid 0 < x\}.$$

(b) *Conversely, every domain of positivity P in a skew field F defines an ordering which makes F an ordered skew field by the rule*

$$a < b \Leftrightarrow b - a \in P.$$

(c) *The operations in (a) and (b) are inverses of each other.*

Proof (a) The monotonicity laws imply that $P \leq F^\times$ is a subgroup which is closed under addition. Moreover, the product of any two negative elements is positive, hence the index of this subgroup is at most 2. Finally, -1 is negative, so the index is exactly 2.

(b) Being a group under multiplication, P contains 1 but not 0, hence it does not contain -1 since P is closed under addition. It follows that $F^\times = P \cup -P$ (disjoint union), and this implies that for distinct elements $a, b \in F$, precisely one of the relations $a < b$, $b < a$ holds. Moreover, the relation $<$ is transitive: $a < b < c$ implies that $c - a = (c - b) + (b - a) \in P + P$. Thus, we have an ordering relation.

Monotonicity of addition holds because $(b + c) - (a + c) = b - a$. If $a < b$ and $0 < c$, then $bc - ac = (b - a)c \in PP = P$. If $a < b$ and $c < 0$, then $bc - ac \in P \cdot (-P) = -(PP) = -P$.

(c) This assertion is easily verified. $\qquad\square$

11.5 Remark We can now explain the lack of symmetry in the monotonicity laws for multiplication: the domain of positivity is a subgroup of $F^\times$ of index two, hence a normal subgroup (the unique non-trivial right coset must coincide with the unique non-trivial left coset). This implies that $cb - ca = c(b - a) \in cP = Pc$ if $a < b$, hence $ca < cb \Leftrightarrow ac < bc$; in other words, monotonicity of left multiplication holds automatically.

We take it for granted that the field of real numbers is an ordered field with respect to the usual ordering (see 42.11 or Section 23), and we proceed to show that this is the only way of making $\mathbb{R}$ into an ordered field. We need a fact known from elementary analysis.

11.6 Lemma *Every positive real number has a square root in $\mathbb{R}$.*

Proof It suffices to find a root of $a > 1$ since $c^2 = a$ implies $(c^{-1})^2 = a^{-1}$. Then the non-empty set $B = \{\, b \in \mathbb{R} \mid 1 \le b \wedge b^2 < a \,\}$ is bounded above by a, hence it has a supremum c. For every $n \in \mathbb{N}$, there exists $b \in B$ such that $c - n^{-1} < b$, hence $(c - n^{-1})^2 < a \le (c + n^{-1})^2$. Passing to the limit as $n \to \infty$, we obtain that $c^2 = a$. $\square$

11.7 Corollary   *There is only one way of ordering the field of real numbers. The same is true of the rational number field.*

Proof Every domain of positivity $Q \subseteq \mathbb{R}$ contains the set S of non-zero squares, and the usual domain of positivity P is contained in S, as we have just seen (11.6). Now Q and P have the same finite index in $\mathbb{R}^{\times}$, hence they are equal.

The reasoning is different for the rational numbers. There, one sees that $1 = 1^2 > 0$, hence all $n \in \mathbb{N}$ are positive by monotonicity, and then $1/n$ and, finally, m/n for $m \in \mathbb{N}$ are positive. $\square$

Not only is the ordering of $\mathbb{R}$ uniquely determined by the field structure, as we have just seen, but also the multiplication is determined (up to isomorphism) by the ordered additive group. This follows from the characterization of the ordered field $\mathbb{R}$ given in 42.9 and 42.11 ($\mathbb{R}$ is the only completely ordered field). Here we give a direct proof.

11.8 Proposition   *If $\mathbb{R}$ is an ordered skew field with the usual addition and ordering and some multiplication $*$, then this ordered skew field is isomorphic to $\mathbb{R}$ with the usual structure.*

Proof For $a > 0$, the distributive law together with the monotonicity law tells us that left multiplication $x \mapsto a * x$ is an automorphism of the ordered additive group. According to 7.11 or 7.12, this implies that there exists some (necessarily positive) $b \in \mathbb{R}$ such that $a * x = bx$ for every x. Taking $x = 1$, we see that $b = a * 1$, and we have

$$a * x = (a * 1)x$$

for all x. We proved this only for $a > 0$, but the distributive law gives $(-a) * x = -(a * x)$, hence the above equation holds for all a. Taking $x = e$, the neutral element of the multiplication $*$, we obtain that $a = a * e = (a * 1)e$ and $a * 1 = ae^{-1}$, the inverse being taken with respect to the usual multiplication. Finally, we get

$$a * x = ae^{-1}x \,.$$

It follows that $e = (1 * 1)^{-1}$ is positive. We have $(ae) * (xe) = (ax)e$, and the map $a \mapsto ae$ is an isomorphism of $\mathbb{R}$ with the usual structure onto the given ordered skew field. □

Property 11.7 of the real numbers is special: in general, two ordered fields can be non-isomorphic as ordered fields although they are isomorphic as fields; they can even be isomorphic as chains at the same time. In other words, a given field structure and a given chain structure can sometimes be put together in essentially different ways. This is shown by the following examples.

11.9 Examples We construct three ordered fields such that no two of them are isomorphic as ordered fields, but their underlying fields and chains are separately all isomorphic. Two of the ordered fields will be Archimedean (this refers to the ordered additive group; compare 7.4), but not the third one.

The fields will all be isomorphic to the simple transcendental extension $\mathbb{Q}(t)$ of the field of rational numbers; compare 64.19. Taking t to be a transcendental real number, we obtain a subfield $\mathbb{Q}(t) \subseteq \mathbb{R}$, and this subfield inherits an Archimedean ordering from $\mathbb{R}$. The field $\mathbb{Q}(t^2) \subseteq \mathbb{R}$ is again a simple transcendental extension and inherits an Archimedean ordering, but a hypothetical order isomorphism $\varphi : \mathbb{Q}(t^2) \to \mathbb{Q}(t)$ has to fix 1 and, hence, all rational numbers (remember that the additive group of $\mathbb{R}$ is uniquely divisible, 1.8). Since φ preserves the ordering and $\mathbb{Q} \subseteq \mathbb{R}$ is dense, φ would be the identity by 11.13 below, and $\mathbb{Q}(t) = \mathbb{Q}(t^2)$. But t does not belong to $\mathbb{Q}(t^2)$, because the substitution $t \mapsto -t$ induces the identity on $\mathbb{Q}(t^2)$.

The third ordered field is obtained by taking the usual ordering on the subfield $\mathbb{Q} \subseteq \mathbb{Q}(t)$ and putting $\mathbb{Q} < t$. The extension to an ordering on $\mathbb{Q}(t)$ is enforced: $t < t^2 < t^3 < \ldots$ implies that a polynomial $p(t)$ is positive if, and only if, the coefficient of the highest power of t is positive, and a quotient of two positive or two negative polynomials is positive, etc. We do get an ordered field, which is not Archimedean since $\mathbb{N} < t$. See also Exercises 1 and 3 of Section 12.

It remains to explain why the three ordered fields have isomorphic underlying chains. In fact, the chains are all countable (by 61.14) and strongly dense in themselves, as shown by an easy verification. The isomorphism follows from these properties by Theorem 3.4.

11.10 Exponential fields This is a variation of the theme considered in Section 2, adapted to ordered fields. We have seen in 2.14 that there are many ordered fields whose additive group is isomorphic to the multiplicative group of positive elements. The real numbers have a much stronger property, however. Namely, the group $\mathbb{R}^+$ is isomorphic *as an ordered group* to the multiplicative group $\mathbb{R}^\times_{\text{pos}}$ of positive elements through the exponential function. An ordered field F is called an *exponential field* if there is an isomorphism of ordered groups $e : F^+ \to F^\times_{\text{pos}}$, and any such isomorphism is called an *exponential function*. Sometimes additional 'growth conditions' are imposed in order to make the function e more similar to the real exponential function; compare 56.16.

The Archimedean exponential fields (i.e., those whose additive ordered group is Archimedean) are comparatively simple. First of all, every Archimedean ordered field is isomorphic to an ordered subfield of the real numbers; see 11.14 below. Then the homomorphism property together with unique divisibility shows that $e(x) = e(1)^x$ for every rational number x. Since $\mathbb{Q} \subseteq \mathbb{R}$ is dense and e preserves the ordering, this equation holds for every x. It follows that the rational numbers, for example, are not an exponential field; compare Section 1, Exercise 6. Next, it is not hard to see that every subfield of $\mathbb{R}$ is contained in a unique minimal exponential subfield of $\mathbb{R}$; compare Exercise 1.

The study of non-Archimedean exponential fields, initiated by ALLING 1962, is much harder. The monograph KUHLMANN 2000 gives an impressive view of its present state, focusing on extensions of the real number field which admit an extension of the ordinary exponential. A natural candidate for such an exponential field appears to be a Hahn power series field $\mathbb{R}((\Gamma))$ with exponents taken from a divisible ordered abelian group Γ, as described in 64.25. Such fields can be used, like the Puiseux fields in 2.10, to obtain examples with a multiplicative group isomorphic to the additive group of some other field. However, it was shown in KUHLMANN *et al.* 1997 that such a field is never exponential unless Γ is trivial. (These fields do, however, admit a non-surjective logarithm; see KUHLMANN 2000.) On the other hand, ALLING 1962 and KUHLMANN 2000 give many examples of non-Archimedean exponential fields; such examples can be obtained as countable unions of Laurent series fields. Other examples of non-Archimedean exponential fields include the non-standard real numbers (see 24.3) and a large class of examples given by ALLING 1962, which we describe next.

11.11 Non-Archimedean exponential fields constructed from rings of continuous functions Let X be a completely regular topological space (for example, $X = \mathbb{N}$ or $X = \mathbb{R}$), and consider the ring $C(X)$ of continuous real-valued functions on X with pointwise addition and multiplication of functions. If we set $f < g$ whenever $f(x) < g(x)$

for all $x \in X$, then $C(X)$ becomes a partially ordered ring. We want to form the quotient of $C(X)$ modulo one of its maximal ideals M (compare 64.3). It can be shown that M is *convex*, that is, $0 \leq f \leq u \in M$ implies $f \in M$, whence the ordering on $C(X)$ carries over to the field $F = C(X)/M$ and turns it into an ordered field; see GILLMAN–JERISON 1960 Theorem 5.5. The field F contains the real numbers as a subfield (namely, the residue classes of the constant functions). If F is Archimedean, then it follows that it is isomorphic to $\mathbb{R}$; compare our discussion of Archimedean exponential fields above. The question is how we can obtain non-Archimedean examples.

We should avoid using the ideals $\{\, f \mid f(p) = 0\,\}$ where $p \in X$, because they are the kernels of the evaluation maps $f \mapsto f(p)$ and their quotient fields are isomorphic to $\mathbb{R}$. If X is not compact, then any maximal ideal containing the ideal of all functions with compact support is not of this type. Now we assume moreover that $C(X)$ contains an unbounded function u; this excludes pseudo-compact spaces like the long line $\mathbb{L}$ introduced in 5.25. It is plausible that then for some ideal M the residue class $u + M$ is an upper bound for all classes of constant functions, hence $C(X)/M$ is non-Archimedean. A proof is given in GILLMAN–JERISON 1960 5.7.

For the construction of an exponential we also need the fact that the zero set $f^{-1}(0)$ of a function determines the maximal ideals M containing f. More precisely, f belongs to M if, and only if, its zero set meets the zero set of every member of M; see GILLMAN–JERISON 1960 2.6.

Using the ordinary exponential function exp, we obtain a mapping $e :$ $f \mapsto \exp \circ f$ of $C(X)$ into itself, which is a homomorphism $(C(X), +) \to$ $(C(X), \cdot)$. For $u \in M$, also $e(u) - 1$ belongs to M because it has the same zero set. It follows that $e(f + u) - e(f) = e(f)(e(u) - 1) \in M$, and e induces a map on the quotient field $F = C(X)/M$, again denoted by e. On constant functions, e agrees with the ordinary exponential. The map e preserves the ordering on $C(X)$; in particular, the functions $\exp \circ f$ are all positive, hence we have constructed an order-preserving homomorphism $(F, +) \to (F_{\mathrm{pos}}, \cdot)$ extending the real exponential. This is in fact an isomorphism, because for $f > 0$ we have $\exp \circ \log \circ f = f$.

We remark that according to GILLMAN–JERISON 1960 13.4, the exponential fields constructed here are all real closed; compare Section 12.

Our next goal is a characterization of the ordered subfields of $\mathbb{R}$; compare 11.14. We need the following notions in order to express the characteristic properties.

11.12 Definition: Properties of pairs of ordered fields Let $F \subseteq G$
be a *pair of ordered fields*, that is, G is an ordered field and F is a subfield,
endowed with the induced ordering.

We say that the pair is *Archimedean* or *cofinal* if every $g \in G$ is
dominated by some $f \in F$, that is, $g \leq f$. (In particular, G is an
Archimedean ordered field if the pair $\mathbb{Q} \subseteq G$ is Archimedean, where $\mathbb{Q}$
is the prime field of G; compare 11.7.)

We say that the pair is *dense* if F is dense in the chain G. Note
that an ordered field is strongly dense in itself in the sense of 3.1, hence
it does not matter how we define density of chains, and it suffices to
require that for every pair $a < b$ in G there is an element $f \in F$ such
that $a \leq f \leq b$.

The pair $F \subseteq G$ is called *order-rigid* if every order automorphism of
G that fixes every $f \in F$ is the identity.

Finally, we say that the pair is *algebraic* if G is an algebraic extension
of the field F; compare 64.5.

The following result about the relationships among these properties
is taken from BAER 1970a. We remind the reader that our fields are
commutative.

11.13 Proposition *Let $F \subseteq G$ be a pair of ordered fields. Then the
following implications hold among the properties defined in 11.12.*

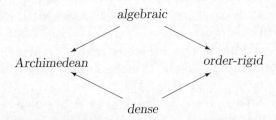

There is no implication between 'dense' and 'algebraic' in either direction. Consequently, none of the four implications shown is reversible.

About the mutual relationship of the properties 'Archimedean' and
'order-rigid' we know nothing.

Proof (1) We show that an algebraic pair is Archimedean. Let $g \in G$
be given; if $g \leq 1$, we have nothing to show, hence we assume that
$g > 1$. The pair is algebraic, hence g satisfies an equation $a_0 + a_1 g +
\cdots + a_{n-1} g^{n-1} + g^n = 0$, where $a_k \in F$. After dividing by g^{n-1}, we
obtain $g = -(a_{n-1} + \cdots + a_1 g^{2-n} + a_0 g^{1-n})$, and the triangle inequality
yields $|g| \leq |a_{n-1}| + \cdots + |a_0| \in F$.

(2) Suppose that the pair is algebraic; we show that it is order-rigid. Let α be an automorphism of G fixing each element of F. As before, every $g \in G$ satisfies a polynomial equation $p(g) = 0$ with coefficients in F, and α has to permute the roots of p in G. There are at most $n = \deg p$ roots, and α has to respect their ordering. This is only possible if α fixes all roots.

(3) If the pair is dense, then it is order-rigid. Indeed, if an automorphism α of the ordered field G fixes all elements of F and moves $g \in G$, then there is some $f \in F$ lying between g and $\alpha(g)$, and α fixes f; this is impossible.

(4) Again if the pair is dense, let $g \in G$ be given; there is $f \in F$ such that $g \leq f \leq g + 1$, and this shows that the pair is Archimedean.

(5) The pair $\mathbb{Q} \subseteq \mathbb{R}$ is dense, hence also Archimedean and order-rigid, but not algebraic.

(6) We describe a pair which is algebraic but not dense, namely, the pair $\mathbb{R}((t^2)) \subseteq \mathbb{R}((t))$ of Laurent series fields; the elements of $\mathbb{R}((t))$ are Laurent series $\sum_{k \geq n} r_k t^k$, where $n \in \mathbb{Z}$ and $r_k \in \mathbb{R}$; see 64.23. The subfield $\mathbb{R}((t^2))$ consists of those series where $r_k = 0$ whenever k is odd. This pair of fields is algebraic; indeed, the variable t (viewed as the power series with $r_1 = 1$ and $r_k = 0$ in all other cases) satisfies the polynomial equation $x^2 - t^2 = 0$ over $\mathbb{R}((t^2))$. Every element of $\mathbb{R}((t))$ can be split into odd and even powers, and thus can be written in the form $\sum_{k \geq n} r_k t^{2k} + t \cdot \sum_{k \geq n} s_k t^{2k}$. This shows that $\mathbb{R}((t))$ is a quadratic extension field of $\mathbb{R}((t^2))$.

It remains to define an ordering on the field $\mathbb{R}((t))$ in such a way that it becomes an ordered field and that the subfield $\mathbb{R}((t^2))$ is not dense. This is done by specifying $P = \{ \sum_{k \geq n} r_k t^k \mid n \in \mathbb{Z}, r_n > 0 \}$ as the domain of positivity. For every positive $r \in \mathbb{R}$, the power series t and $r - t$ are positive, hence $0 < t < r$. We have $t < 2t$, but no element of $\mathbb{R}((t^2))$ lies between t and $2t$ (hence the pair of fields is not dense). Indeed, from $t \leq g = \sum_{k \geq n} r_k t^{2k}$ and $r_n \neq 0$ it follows that $n \leq 0$, while $g \leq 2t$ implies that $n \geq 1$. $\qquad \square$

11.14 Theorem *Let F be an ordered field. Then the following assertions are mutually equivalent.*

(a) *F is a subfield of $\mathbb{R}$ endowed with the induced ordering.*

(b) *The prime field $\mathbb{Q} \subseteq F$ is dense.*

(c) *For every subfield S of F, the pair $\mathbb{Q} \subseteq S$ is order-rigid.*

(d) *F is Archimedean, i.e., the pair $\mathbb{Q} \subseteq F$ is cofinal.*

Proof (1) We prove the implications (a) $\Rightarrow$ (b) $\Rightarrow$ (c) $\Rightarrow$ (d) $\Rightarrow$ (a). The first implication is obvious, and the second one has been proved in 11.13.

(2) Assume that F contains an element $t > \mathbb{Q}$. We claim that the pair of subfields $\mathbb{Q} \subseteq \mathbb{Q}(t)$ of F is not order-rigid. We know from 11.13 that $\mathbb{Q}(t)$ is a transcendental extension of $\mathbb{Q}$. The map sending $p(t)/q(t)$ to $p(t+1)/q(t+1)$ is an automorphism of $\mathbb{Q}(t)$ which fixes every element of $\mathbb{Q}$; compare 64.19. Moreover, the map preserves the ordering. Indeed, we have $1 < t < t^2 < t^3 < \ldots$, hence a polynomial in t is positive if, and only if, the coefficient of the highest power of t is positive. This coefficient remains unchanged if we substitute $t + 1$ for t.

(3) If F is Archimedean as a field, then its additive group is an Archimedean ordered group, hence there is an injective homomorphism of ordered groups $\alpha : F^+ \to \mathbb{R}^+$. Composing α with an automorphism of $\mathbb{R}^+$ we may arrange that $\alpha(1) = 1$, and then α induces the identity on the prime field $\mathbb{Q} \subseteq F$. We claim that α is a homomorphism with respect to multiplication (and hence an isomorphism of ordered fields $F \cong \alpha(F) \subseteq \mathbb{R}$). Let $\lambda_f : F \to F$ be left multiplication by $f \in F^\times$, that is, $\lambda_f(g) = fg$. Similarly, we have $\lambda_{\alpha(f)} : \alpha(F) \to \mathbb{R}$. The maps $\lambda_{\alpha(f)}$ and $\alpha \circ \lambda_f \circ \alpha^{-1}$ coincide on the dense subset $\mathbb{Q} \subseteq \alpha(F)$ because the group $\mathbb{R}^+$ is uniquely divisible. Both maps are order-preserving if $f > 0$ and order-reversing if $f < 0$. This implies that the maps coincide, and the proof is complete.

The implication (d) $\Rightarrow$ (a) is proved again in 42.12, using the concept of completion. $\square$

11.15 Corollary *Every completely ordered field is isomorphic to* $\mathbb{R}$.

Proof By 7.5, such a field F is Archimedean, and then F is isomorphic to a subfield of $\mathbb{R}$ by 11.14. This subfield is closed by completeness. Compare also 42.9. $\square$

11.16 Transcendental extensions   Part (2) of the proof of 11.14 suggests a general question. If F is any field and $F(t)$ is a simple transcendental extension, then the group of all automorphisms of $F(t)$ fixing F elementwise is $\mathrm{Aut}_F F(t) = \mathrm{PGL}_2 F$; see 64.19. This group consists of the maps defined by

$$t \mapsto \frac{at + b}{ct + d},$$

where $a, b, c, d \in F$ and $ad - bc \neq 0$. If F is an ordered field, we ask which of these maps preserve the ordering on $F(t)$ defined by $F < t$.

The answer is that precisely the automorphisms defined by $t \mapsto at + b$ with $0 < a$ are order-preserving. We leave the proof as Exercise 2.

Exercises

(1) Construct a countable subfield F of $\mathbb{R}$ such that exp induces an isomorphism of $(F, +)$ onto the multiplicative group of positive elements of F.

(2) Prove the claim in 11.16 on order-preserving automorphisms of simple transcendental extensions.

(3) Use Theorem 11.14 in order to give a quick proof of Proposition 11.8.

(4) Show that (the additive group of) an ordered field F is Archimedean if, and only if, the multiplicative group of positive elements is Archimedean.

12 Formally real and real closed fields

In this section we study abstract fields which can be made into ordered fields (as defined in 11.1). Real closed fields are maximal (in a suitable sense; see 12.6) with respect to this property; they are characterized by certain algebraic properties, which are familiar from the field $\mathbb{R}$ of real numbers. We prove some fundamental results due to Artin and Schreier on real closed fields.

More information on formally real and real closed fields can be found in PRESTEL 1984 (including model-theoretic aspects), SCHARLAU 1985 Chapter 3, JACOBSON 1989 Chapter 11, LANG 1993 Chapter XI, PFISTER 1995, BOCHNAK *et al.* 1998, PRESTEL–DELZELL 2001 Chapter 1 and LAM 2005 Chapter VIII.

12.1 Definition A field F is said to be *formally real*, if -1 is not a sum of squares in F.

Each ordered field F is formally real, since $a^2 \geq 0 > -1$ for every $a \in F$. We prepare the proof of the converse assertion (12.3).

12.2 Proposition *Let F be a field, let $a \in F^\times$ and let $S \subseteq F$ be a subset with the following properties:*

$$S + S \subseteq S, \ S \cdot S \subseteq S, \ x^2 \in S \text{ for every } x \in F, \text{ and } S \cap \{-1, a\} = \emptyset.$$

Then there exists an ordering $<$ on F such that $a < 0 \leq S$.

Proof Let $\mathcal{M} = \{T \mid S \subseteq T \subseteq F, T + T \subseteq T, T \cdot T \subseteq T, -1 \notin T\}$; the elements of $\mathcal{M}$ are so-called preorderings of F. If $T \in \mathcal{M}$ and $b \in F \smallsetminus T$, then $T - bT \in \mathcal{M}$; indeed, $T - bT$ contains S and is closed under addition and multiplication (since $b^2 \in S \subseteq T$), and the assumption

$-1 \in T - bT$ implies $-1 = t' - bt$ with $t', t \in T$, hence $t \neq 0$ and $b = t^{-2}t(1 + t') \in ST(1 + T) \subseteq T$, a contradiction. In particular, $S - aS \in \mathcal{M}$.

The union of any chain in $\mathcal{M}$ belongs to $\mathcal{M}$, hence the same is true for the non-empty subset $\mathcal{M}_a := \{T \in \mathcal{M} \mid S - aS \subseteq T\}$. Zorn's Lemma implies the existence of a maximal element $P \in \mathcal{M}_a$. By construction, P contains S and the element $0 - a \cdot 1 = -a$. In view of 11.4b it suffices to show that $P \smallsetminus \{0\}$ is a domain of positivity in F. This task reduces to proving that $P \cup -P = F$.

Let $b \in F \smallsetminus P$. Then $P \subseteq P - bP \in \mathcal{M}_a$, hence $P = P - bP$ by the maximality of P. In particular $-b \in P$, and therefore $b \in -P$.   □

12.3 Corollary (Artin–Schreier) *Each formally real field F can be made into an ordered field $(F, <)$.*

Proof Let S be the set of all sums of squares of F. Then $S \cdot S \subseteq S$, since $(\sum_i a_i^2)(\sum_j b_j^2) = \sum_{i,j}(a_i b_j)^2$, and $-1 \notin S$. Thus 12.2, applied with $a = -1$, yields the assertion.   □

12.4 Corollary *Let F be a field with characteristic distinct from 2. An element $a \in F^\times$ is a sum of squares in F if, and only if, $0 < a$ for every ordering $<$ of F.*

In other words: the intersection of all domains of positivity of F consists precisely of all non-zero sums of squares of F.

Proof Each non-zero sum of squares is positive with respect to any ordering of F. For the converse, we consider two cases.

If F is formally real, and if $a \in F$ does not belong to the set S of all sums of squares in F, then 12.2 implies that $a < 0$ for some ordering $<$ of F.

If F is not formally real, then -1 is a sum of squares in F, and the equation $4a = (a + 1)^2 + (-1)(a - 1)^2$ shows that every $a \in F$ is a sum of squares; furthermore F has no ordering in this case.   □

12.5 Corollary *A field F has a unique ordering if, and only if, the non-zero sums of squares form a domain of positivity of F.*

Proof If these sums of squares form a domain of positivity P, then P is the only domain of positivity of F. Conversely, if F has a unique ordering $<$, then 12.4 implies that $\{a \in F \mid 0 < a\}$ consists of all non-zero sums of squares.   □

Now we consider formally real fields which are maximal in the following sense.

12.6 Definition A field F is said to be *real closed*, if F is formally real, but no proper algebraic extension of F is formally real.

The field $\mathbb{R}$ and the field of all real algebraic numbers are examples of real closed fields (by 12.13 and 12.11). By 24.6, the field $^*\mathbb{R}$ of nonstandard real numbers is a real closed field with non-Archimedean ordering (12.7); compare also Exercise 4.

Let Γ be a divisible ordered abelian group. Then the field $\mathbb{C}((\Gamma))$ of Hahn power series is algebraically closed (see 64.25), hence $\mathbb{R}((\Gamma))$ is real closed by 12.10 below. Different isomorphism types of ordered abelian groups Γ give different isomorphism types of fields $\mathbb{R}((\Gamma))$, because the ordered group Γ is the value group of the natural ordering valuation (56.16, 64.25) determined by the unique ordering (12.7) of $\mathbb{R}((\Gamma))$.

See 12.14, 12.18 and RIBENBOIM 1992, 1993 for more examples. We mention that every real-closed field is isomorphic to an initial subfield of Conway's surreal numbers; see EHRLICH 2001.

12.7 Lemma *Each real closed field F has a unique ordering $<$, and this ordering is given by $0 < a \Leftrightarrow a = b^2$ for some $b \in F^\times$.*

Proof By 12.3 the field F has some ordering $<$. It suffices to show that each element $a \in F$ with $0 < a$ is a square in F. Otherwise we could consider the quadratic field extension $F(\sqrt{a})$, which is not formally real (as F is real closed). Then there exist elements $x_j, y_j \in F$ such that

$$-1 = \textstyle\sum_j (x_j + y_j \sqrt{a})^2 = \sum_j (x_j^2 + ay_j^2) + (\sum_j 2x_j y_j)\sqrt{a} \ .$$

As $\sqrt{a} \notin F$ we infer that $-1 = \sum_j (x_j^2 + ay_j^2) \geq 0$, which is absurd. $\square$

12.8 Euclidean and Pythagorean fields A field F is said to be *Euclidean* if the non-zero squares of F form a domain of positivity of F (equivalently, F is formally real and for each $a \in F^\times$, either a or $-a$ is a square); then F has a unique ordering, which is defined as in 12.7.

The field $\mathbb{R}$ is Euclidean; see Lemma 11.6. The *Euclidean closure* E of $\mathbb{Q}$ in $\mathbb{R}$ consists of all real algebraic numbers that can be reached by iterated quadratic extensions $F(\sqrt{c})$ with $0 < c \in F$, starting from $F = \mathbb{Q}$. A point (x, y) of the real plane $\mathbb{R}^2$ can be constructed with ruler and compass, starting from the the unit square, if, and only if, $(x, y) \in E^2$; see MARTIN 1998 Chapter 2.

We mention a related concept: a field F is called *Pythagorean*, if each sum of squares of F is a square in F. The *Pythagorean closure*

P of $\mathbb{Q}$ in $\mathbb{R}$ consists of all real algebraic numbers that can be reached by iterated quadratic extensions $F(\sqrt{1+c^2})$ with $c \in F$, starting from $F = \mathbb{Q}$ (compare SCHARLAU 1985 p. 52 or LAM 2005 VIII.4). A point $(x, y) \in \mathbb{R}^2$ can be constructed with ruler and dividers, starting from the unit square, if, and only if, $(x, y) \in P^2$; see MARTIN 1998 Chapter 5 or HILBERT 1903 §§36f (constructions by 'Lineal und Eichmaß').

We remark that the field P is the splitting field in $\mathbb{R}$ of the set of all rational polynomials which are solvable by real radicals; see PAMBUCCIAN 1990.

A formally real field F is Euclidean if, and only if, F has no formally real quadratic extension; see BECKER 1974 Satz 1 or LAM 2005 VIII.1.7. By a result of Whaples, which was rediscovered by Diller and Dress (see LAM 2005 VIII.5.5 p. 269), a field F with characteristic not 2 is Pythagorean precisely if F has no cyclic Galois extension of degree 4. See also Exercise 9.

The field P is a proper subfield of E; see Exercise 5. Each real closed field is Euclidean, by 12.7, but the fields E and P are not real closed (since they are proper subfields of the ordered field of all real algebraic numbers). See RIBENBOIM 1992 Section 6 or LAM 2005 VIII for more examples of Euclidean or Pythagorean fields.

A real closed field F is isomorphic to a subfield of $\mathbb{R}$ if and only if the unique ordering of F is Archimedean; see 11.14. The following result generalizes the rigidity theorem 6.4 for $\mathbb{R}$.

12.9 Corollary (i) *Let E be an algebraic extension field of a field F. If E is real closed, then $\mathrm{Aut}_F\, E = \{\mathrm{id}\}$, i.e., each field automorphism of E which fixes each element of F is trivial.*

(ii) *Each real closed subfield F of $\mathbb{R}$ is rigid, i.e., $\mathrm{End}\, F = \{\mathrm{id}\}$.*

Proof (i) By 12.7 the field E has a unique ordering, hence every element $\varphi \in \mathrm{Aut}_F\, E$ preserves this ordering. Furthermore φ leaves invariant the finite set $f^{-1}(0)$ of roots of each polynomial $f \in F[x]$, whence $\varphi = \mathrm{id}$; compare step (2) of the proof of 11.13.

(ii) By 12.7 the field F has a unique ordering; this ordering coincides with the ordering induced by $\mathbb{R}$. Each endomorphism φ of F preserves this unique ordering. Furthermore φ fixes each element of the prime field $\mathbb{Q}$. Since $\mathbb{Q}$ is dense in $\mathbb{R}$ and in F, we infer that $\varphi = \mathrm{id}$. □

Concerning 12.9(ii), we remark that SHELAH 1983 constructs rigid real closed fields which are not subfields of $\mathbb{R}$ (using suitable set-theoretic assumptions).

Now we show that real closed fields are characterized by certain algebraic properties of $\mathbb{R}$; see 12.10, 12.12 and 12.15.

12.10 Theorem (Artin–Schreier)   *The following conditions for a field F are equivalent.*

(i) *F is real closed.*

(ii) *F is Euclidean, and each polynomial in $F[x]$ of odd degree has a root in F.*

(iii) *The element -1 is not a square in F, and the field $F(\sqrt{-1})$ is algebraically closed.*

Proof (i) implies (ii): By 12.7, F is Euclidean. We show that each polynomial $f \in F[x]$ of odd degree n has a root in F. It suffices to consider irreducible polynomials f. We proceed indirectly and choose a counter-example f of minimal degree $n = \deg f$. Then $n > 1$ and f has no root in F. Let c be a root of f in an algebraic closure of F. Then $F(c)$ is a proper algebraic extension of the real closed field F, hence $F(c) = \{\, g(c) \mid g \in F[x],\ \deg g < n \,\}$ is not formally real. Therefore

$$-1 = \sum_j g_j(c)^2$$

for suitable polynomials $g_j \in F[x]$ with $\deg g_j < n$, and we infer that c is a root of the polynomial $h := 1 + \sum_j g_j^2$. Hence the minimal polynomial f of c divides h. The degree of h is even (consider leading coefficients and use the ordering) and strictly smaller than $2n$, hence the quotient h/f has odd degree $\deg h - n < n$. Since n was chosen to be minimal, h has a root b in F, hence $-1 = \sum_j g_j(b)^2$. This means that F is not formally real, a contradiction.

(ii) implies (iii): By (ii), the element -1 is not a square in F, hence $F(i) := F(\sqrt{-1})$ is a quadratic field extension of F; it remains to show that $F(i)$ is algebraically closed. We consider an irreducible polynomial $f \in F(i)[x]$. Let E be the field obtained from $F(i)$ by adjoining all roots of f (in an algebraic closure of $F(i)$). Then E is a Galois extension of F (note that F has characteristic zero, hence $E|F$ is separable by 64.11), with a finite Galois group G (see 64.8, 64.10). The field of all fixed elements of a Sylow 2-subgroup H of G is an extension of F of odd degree $k = [G : H]$; see 64.18. By (ii) and by the existence of a primitive element (64.12) we have $k = 1$, hence G is a 2-group. The Galois group G_i of the Galois extension $E|F(i)$ is a subgroup of G, hence also a 2-group. In the following paragraph we show that G_i is trivial; then $E = F(i)$, thus $F(i)$ has no proper algebraic extension and is therefore algebraically closed.

If G_i is not trivial, then G_i has a subgroup H of index 2 (because the composition factors of any finite 2-group are cyclic of order 2; see also Sylow's theorem in the form of JACOBSON 1985 1.13 p. 80). The fixed elements of H form a quadratic extension of $F(i)$. However, by (ii) each element of $F = \{a \in F \mid a > 0\} \cup \{0\} \cup \{a \in F \mid -a > 0\}$ is a square in $F(i)$, and we show now that each element of $F(i) \smallsetminus F$ is a square in $F(i)$; indeed, if $a, b \in F$ with $b \neq 0$, then $a^2 + b^2 > 0$ and hence $a^2 + b^2 = c^2$ with $0 < c \in F^\times$, which gives $c > a$; thus $2(c - a) = d^2$ with $d \in F^\times$, and then $a + ib = (b/d + id/2)^2$. Hence $F(i)$ has no quadratic extension (note that $-1 \neq 1$ in F), which is a contradiction.

(iii) implies (i): Since $F(i) = F(\sqrt{-1})$ is algebraically closed, each element $a + ib$ with $a, b \in F$ is of the form $(c + id)^2$ with $c, d \in F$. Thus $a = c^2 - d^2$ and $b = 2cd$, which implies $a^2 + b^2 = (c^2 + d^2)^2$. We conclude that each sum of squares in F is in fact a square in F. Now -1 is not a square, hence not a sum of squares, whence F is formally real. Furthermore the only proper algebraic extension $F(i)$ is not formally real, thus F is real closed. $\qquad\square$

12.11 Corollary *Let F be a subfield of a real closed field E. Then the field $F^{\natural E}$ of all elements of E which are algebraic over F is a real closed subfield of E.*

Proof Let $a \in F$ be positive with respect to the unique ordering (12.7) of E. Then the polynomial $x^2 - a$ has a root in E, and this root belongs to $F^{\natural E}$. Hence the non-zero squares of $F^{\natural E}$ form a domain of positivity of $F^{\natural E}$, whence $F^{\natural E}$ is Euclidean. Furthermore, if $f \in F^{\natural E}[x]$ is a polynomial of odd degree, then f has a root in E by 12.10; again this root belongs to $F^{\natural E}$. Thus $F^{\natural E}$ is real closed by 12.10. $\qquad\square$

12.12 Corollary *A field F is real closed if, and only if, F admits an ordering $<$ such that the intermediate value theorem holds for polynomials $f \in F[x]$, i.e., $a < b$ and $f(a)f(b) < 0$ imply that $f(c) = 0$ for some $c \in F$ with $a < c < b$.*

Proof Let F be real closed. By 12.7, the field F has a unique ordering and each positive element is a square. We may assume that the leading coefficient of the given polynomial f is 1. By 12.10(iii), f decomposes into factors $x - c$ of degree 1 and quadratic factors $x^2 - 2dx + e = (x - d)^2 + e - d^2$ with $e - d^2 > 0$. Each irreducible quadratic factor $x^2 - 2dx + e$ assumes only positive values on F. Therefore the change of the sign of f between a and b occurs also at some factor $x - c$ of f, whence c is a root of f with $a < c < b$.

Conversely, assume that the intermediate value theorem holds for polynomials. Let $a \in F$ with $a > 0$. The intermediate value theorem for the polynomial $x^2 - a$ implies that a is a square of some element of F (between 0 and $\max\{a, 1\}$). Hence F is Euclidean. Furthermore each polynomial $f = \sum_{j=0}^{n} a_j x^j \in F[x]$ of odd degree n assumes positive and negative values, as we show now. We may assume that $a_n = 1$. Choose $b \in F$ larger than $1 + \sum_j |a_j|$, where $|a| := \max\{a, -a\}$. Then $\sum_{j=0}^{n-1} \pm a_j b^j \leq \sum_{j=0}^{n-1} |a_j| b^{n-1} < b^n$ for any combination of signs $\pm$, hence $f(b) > 0 > f(-b)$. Thus f has a root in F by the intermediate value theorem, and 12.10 implies that F is real closed.   □

12.13 Corollary *The field $\mathbb{R}$ of real numbers is real closed and the field $\mathbb{C} = \mathbb{R}(\sqrt{-1})$ of complex numbers is algebraically closed.*

Proof The intermediate value theorem holds for real polynomials, by 5.4 and the continuity of polynomials. Hence $\mathbb{R}$ is real closed by 12.12 (or by 12.10). Now 12.10 shows that $\mathbb{C}$ is algebraically closed.   □

The fact that $\mathbb{C}$ is algebraically closed is traditionally called the Fundamental Theorem of Algebra. The proof given above is due to Artin; see FINE–ROSENBERGER 1997 or EBBINGHAUS *et al.* 1991 Chapter 4, for other proofs.

Now we construct plenty of real closed fields (to be used in Theorem 14.16).

12.14 Theorem Let $\aleph = 2^{\aleph_0} = \operatorname{card} \mathbb{R}$.

  (i) *There exist precisely $2^{\aleph}$ isomorphism types of real closed fields (of cardinality $\aleph$) with Archimedean ordering (12.7). All these fields are rigid.*

 (ii) *There exist precisely $2^{\aleph}$ isomorphism types of real closed fields F with $\operatorname{card} F = \aleph$ and $\operatorname{card} \operatorname{Aut} F = 2^{\aleph}$.*

Proof (i) Real closed subfields of $\mathbb{R}$ are rigid by 12.9. According to 11.14, each Archimedean ordered field is isomorphic to a subfield of $\mathbb{R}$, hence there exist at most $2^{\aleph}$ isomorphism types as in (i). Now we construct $2^{\aleph}$ isomorphism types.

Using Zorn's Lemma, choose a transcendency basis T of $\mathbb{R}$ over $\mathbb{Q}$. Then $\mathbb{R}$ is algebraic over the purely transcendental field $\mathbb{Q}(T)$, hence $\operatorname{card} \mathbb{R} = \operatorname{card} \mathbb{Q}(T)$ by 64.5. The set T is infinite, as $\mathbb{R}$ is uncountable. Hence $\operatorname{card} T = \operatorname{card} \mathbb{Q}(T) = \operatorname{card} \mathbb{R} = \aleph$; see 64.20. For each subset

$S \subseteq T$ we write $\mathbb{Q}(S)^{\natural}$ for the algebraic closure in $\mathbb{C}$ of $\mathbb{Q}(S)$, and we denote by $F_S := \mathbb{R} \cap \mathbb{Q}(S)^{\natural}$ the real part of $\mathbb{Q}(S)^{\natural}$, as in 6.2d. Then $\mathbb{Q}(S)^{\natural}$ has degree 2 over F_S, since $i \in \mathbb{Q}^{\natural} \subseteq \mathbb{Q}(S)^{\natural}$ and F_S is the field of fixed elements of the complex conjugation map γ, restricted to $\mathbb{Q}(S)^{\natural}$. Note that F_S is a real closed field by 12.11, hence it has a unique ordering (defined by the squares of F_S); see 12.7.

We claim that the fields F_S with $S \subseteq T$ are mutually non-isomorphic. To prove this, we observe that the unique ordering of F_S is the restriction of the ordering of $\mathbb{R}$. Each field isomorphism $\varphi : F_S \to F_{S'}$, where $S, S' \subseteq T$, preserves the unique ordering and fixes each element of $\mathbb{Q}$. The strong density (3.1) of $\mathbb{Q}$ in $\mathbb{R}$ implies that φ is the identity, whence $F_S = F_{S'}$ and $S = T \cap F_S = T \cap F_{S'} = S'$.

This shows that there are $2^{\aleph}$ isomorphism types of real closed fields with Archimedean ordering. If we admit only subsets S with $\operatorname{card} S = \operatorname{card} T = \aleph$, then $\operatorname{card} F_S = \aleph$, and we still obtain $2^{\aleph}$ isomorphism types, since $\operatorname{card}\{\, S \mid S \subseteq T, \operatorname{card} S = \operatorname{card} T \,\} = 2^{\aleph}$ by 61.16.

(ii) The additive group of any field F as in (ii) is isomorphic to $(\mathbb{R}, +)$; see 1.14. Hence there exist at most $2^{\aleph}$ isomorphism types of such fields (note that $\operatorname{card}(\mathbb{R}^{\mathbb{R} \times \mathbb{R}}) = \aleph^{\aleph^2} = \aleph^{\aleph} = 2^{\aleph}$ by 61.12 and 61.13b). Now we construct that many isomorphism types.

For every divisible ordered abelian group Γ, the real Hahn power series with countable support form a real closed field $\mathbb{R}((\Gamma))_1$; see 64.25. If $\operatorname{card} \Gamma \leq \aleph$, then $\operatorname{card} \mathbb{R}((\Gamma))_1 = \aleph$, since $\operatorname{card} \mathbb{R}((\Gamma))_1 \leq \operatorname{card}(\mathbb{R} \times \Gamma)^{\mathbb{N}} = \aleph^{\aleph_0} = \aleph$; see 61.13b. Different isomorphism types of ordered abelian groups Γ give different isomorphism types of fields $\mathbb{R}((\Gamma))_1$, because the ordered group Γ is the value group of the natural ordering valuation (56.16) of $\mathbb{R}((\Gamma))_1$. (See ALLING–KUHLMANN 1994 Corollary 4.1 for a related result.)

Now let Γ be the ordered group $(\mathbb{R}, +, <)$ where $<$ is an ordering of $(\mathbb{R}, +)$ such that $\operatorname{card} \operatorname{Aut} \Gamma = 2^{\aleph}$. By 7.14 there are $2^{\aleph}$ isomorphism types for Γ, hence we have that many isomorphism types of fields $\mathbb{R}((\Gamma))_1$ with cardinality $\aleph$. Moreover, each (non-trivial) automorphism of the ordered group Γ induces a (non-trivial) field automorphism of $\mathbb{R}((\Gamma))_1$, and this implies that $\mathbb{R}((\Gamma))_1$ has precisely $2^{\aleph}$ field automorphisms (not more, since $\aleph^{\aleph} = 2^{\aleph}$ by 61.13b). $\qquad\square$

There exist many isomorphism types of skew fields which have dimension 4 over a subfield isomorphic $\mathbb{R}$; this amazing remark is due to GERSTENHABER–YANG 1960 p. 462; see also DESCHAMPS 2001. Indeed, for any field F as in 12.14, the quaternion skew field $H_F =$

$F + Fi + Fj + Fij$ with centre F can be defined in the usual fashion; compare 34.17. Then $F(i) = F + Fi \cong \mathbb{C} = \mathbb{R} + i\mathbb{R}$ by 12.10 and 64.21, and $H_F = F(i) + F(i)j$. Hence H_F has dimension 4 over a subfield isomorphic to $\mathbb{R}$.

These copies of $\mathbb{R}$ are not contained in the centre F of H_F, hence there is no contradiction to the classical result of Frobenius, which implies that Hamilton's quaternion skew field $\mathbb{H}$ is the only skew field of dimension 4 over its centre $\mathbb{R}$ (see 58.11 or the remarks after 13.8).

The following characterization result is a remarkable achievement of Artin and Schreier; the proof given below uses simplifications due to Leicht 1966 and Waterhouse 1985 (see also Lam 2005 VIII.2.21 for the special case of fields F of characteristic 0).

12.15 Theorem (Artin–Schreier) *Let F be a field. Then the algebraic closure $F^\natural$ of F has finite dimension n over F if, and only if, F is real closed (then $n = 2$) or algebraically closed (then $n = 1$).*

Proof If F is real closed, then $F^\natural$ has degree 2 over F by 12.10. Now we assume that $F^\natural \neq F$ has finite degree over F. We have to show that F is real closed; this is achieved by verifying condition 12.10(iii), see steps (3) and (4). Let p denote the characteristic of F.

(1) We claim that $F^\natural | F$ is a Galois extension.

Clearly $F^\natural | F$ is normal (64.9), so by 64.17 we have to show that $F^\natural | F$ is separable. This holds for $p = 0$ (see 64.11), so let $p > 0$. Then $x \mapsto x^p$ is a field endomorphism, hence $F^p = \{\, a^p \mid a \in F \,\}$ is a subfield of F, and $(F^\natural)^p = F^\natural$ since $F^\natural$ is algebraically closed. Furthermore the degree $[F^\natural : F^p] = [(F^\natural)^p : F^p] \leq [F^\natural : F]$ is finite, hence $F = F^p$.

The equation $F = F^p$ means, by definition, that F is perfect. This implies that every algebraic extension of F is separable; indeed, if an irreducible polynomial $f \in F[x]$ has multiple roots, then f and its derivative f' have a common divisor. Thus $f' = 0$, hence $f = \sum_i a_i x^{pi}$ with $a_i \in F$. Since F is perfect, we have $a_i = b_i^p$ with suitable $b_i \in F$, hence $f = \sum_i b_i^p x^{pi} = (\sum_i b_i x_i)^p$, a contradiction to the irreducibility of f.

(2) We claim that the characteristic p does not divide the order of the Galois group G of $F^\natural | F$.

Otherwise G contains an element g of order $p > 0$ (by Sylow's theorem, compare Jacobson 1985 1.13 p. 80), and $F^\natural$ is a (cyclic Galois) extension of degree p of the field $E = \operatorname{Fix} g$ of fixed points of g. The trace map $\operatorname{tr} = 1 + g + \cdots + g^{p-1}$ is an E-linear endomorphism of $F^\natural$ into E. By Dedekind's independence lemma (see Jacobson 1985 4.14

p. 291 or COHN 2003a 7.5.1 or LANG 1993 VI.4), the powers g^j with $0 \le j < p$ are linearly independent. In particular, tr is not the trivial endomorphism. Thus $\mathrm{tr}(F^{\natural}) = E$, and we can choose $a \in F^{\natural}$ such that $\mathrm{tr}(a) = -1$. The element

$$b := \sum_{j=1}^{p-1} jg^j(a)$$

satisfies $g(b) = b - \mathrm{tr}(a) = b + 1$, hence $g^j(b) = b + j$ for each $j \in \mathbb{N}$. Thus $b, b+1, \dots, b+p-1$ is the orbit of b under $\langle g \rangle$. The equation $\prod_{0 \le j < p}(x - j) = x^p - x$ holds in $E[x]$, since both sides have the same roots (as a consequence of Fermat's little theorem). Substituting $x - b$ for x we infer that

$$f(x) := \prod_{0 \le j < p} (x - b - j) = (x - b)^p - (x - b) = x^p - x + (-b)^p + b$$

is the minimal polynomial of b over E (compare Section 64, Exercise 2). Its constant term $(-b)^p + b \in E$ can be written as $(-b)^p + b = \mathrm{tr}(c)$ with $c \in F^{\natural}$, and $c = d - d^p$ for some $d \in F^{\natural}$, as $F^{\natural}$ is algebraically closed. We infer that $(-b)^p + b = \mathrm{tr}(d - d^p) = \mathrm{tr}(d) - \mathrm{tr}(d)^p$, hence $\mathrm{tr}(d) \in E$ is one of the roots $b + j$ of $f(x)$. This is a contradiction, as $b \notin E$.

(3) We claim that G is a 2-group (hence $p \ne 2$ by step 2) and that $i = \sqrt{-1} \notin F$.

Let q be any prime divisor of $|G|$. Then G contains an element g of order q. Let E be the field of fixed elements of $H := \langle g \rangle$ in $F^{\natural}$. Then $F^{\natural}|E$ is a Galois extension of degree q with Galois group H (see 64.18). The irreducible polynomials in $E[x]$ have degree 1 or q (use the degree formula 64.2 and the fact that $F^{\natural}$ is algebraically closed). Hence $x^q - 1 = (x - 1)(x^{q-1} + \cdots + x + 1)$ splits over E into factors of degree 1. Since $q \ne p$ by step (2), we infer that $x^q - 1$ is separable. Thus E contains a root of unity ζ with $\zeta^q = 1 \ne \zeta$.

As a consequence, the extension $F^{\natural}|E$ can be written in the form $F^{\natural} = E(r)$ where r is a root of some irreducible binomial $f(x) = x^q - c = \prod_{h \in H}(x - h(r))$ with $c \in E$. This well known fact (see COHN 2003a 7.10.7 or LANG 1993 VI.6.2) may be proved as follows. Every element

$$r := \sum_{j=0}^{q-1} \zeta^j g^j(a)$$

with $a \in F^{\natural}$ satisfies $\zeta g(r) = r$. By Dedekind's independence lemma (see step (2) above) we can choose $a \in F^{\natural}$ such that $r \ne 0$. Then $g(r) = \zeta^{-1} r \ne r$, hence $r \notin E$ and therefore $F^{\natural} = E(r)$. Moreover $g(r^q) = g(r)^q = (\zeta^{-1} r)^q = r^q$, hence $c := r^q \in E$, and $f(x) = x^q - c$ is the (irreducible) minimal polynomial of r over E.

The algebraically closed field $F^\natural$ contains an element s with $s^q = r$. The norm $N(s) := \prod_{h \in H} h(s)$ belongs to E (as $N(s)$ is fixed by H) and satisfies the relation

$$N(s)^q = N(s^q) = N(r) = (-1)^q f(0) = (-1)^{q+1} c .$$

If q is odd, then $N(s) \in E$ is a root of f, which is a contradiction to the irreducibility of f over E. Hence $q = 2$ and $N(s)^2 = -c$. Assuming $i \in F$ we obtain that $iN(s) \in E$, and $iN(s)$ is a root of $f(x) = x^2 - c$. This contradiction shows that $i \notin F$.

(4) It remains to show that $F(i) = F^\natural$. Otherwise we could repeat the proof given above with $F(i)$ instead of F, which in step (3) would lead to the contradiction $i \notin F(i)$.   □

The following two results prepare the definition of real closures (12.18).

12.16 Proposition *Each formally real field F has an algebraic extension which is real closed.*

Each ordered field $(F, <)$ has a real closed algebraic extension E such that the unique ordering of E described in 12.7 induces on F the given ordering $<$.

Proof Let F be formally real and let $F^\natural$ be an algebraic closure of F. The set $\{E \mid F \subseteq E \subseteq F^\natural \wedge E \text{ is formally real}\}$ contains maximal elements by Zorn's Lemma; each of these maximal elements is a real closed field. This proves the first assertion.

Now let $(F, <)$ be an ordered field. We claim that the subfield $F' := F(\sqrt{a} \mid 0 < a \in F)$ of $F^\natural$ is formally real. Otherwise we consider relations of the form

$$-1 = \sum_i c_i b_i^2 ,$$

where $0 < c_i \in F$ and each b_i belongs to a subfield $F(\sqrt{a_1}, \ldots, \sqrt{a_n})$ with $0 < a_j \in F$; relations of this form exist, since we may take $c_i = 1$. We choose such a relation with minimal n, and we write $b_i = x_i + y_i \sqrt{a_n}$ with $x_i, y_i \in F(\sqrt{a_1}, \ldots, \sqrt{a_{n-1}})$; then

$$-1 = \sum_i c_i(x_i^2 + a_n y_i^2) + (\sum_i 2c_i x_i y_i)\sqrt{a_n} .$$

Since $\sqrt{a_n} \notin F(\sqrt{a_1}, \ldots, \sqrt{a_{n-1}})$ by the minimality of n, we infer that $-1 = \sum_i c_i x_i^2 + \sum_i c_i a_n y_i^2$, a contradiction to the minimality of n.

By the first assertion, F' has an algebraic extension E which is real closed. Each positive element $a \in F$ is a square in F', hence the unique ordering of E induces on F the given ordering $<$.   □

12.17 Theorem (Artin–Schreier) *Let F_1, F_2 be ordered fields, and let E_i be a real closed algebraic extension of F_i such that the unique ordering of E_i induces on F_i the given ordering, for $i = 1, 2$. Then each order-preserving isomorphism $\varphi : F_1 \to F_2$ has a unique extension to an isomorphism $\overline{\varphi} : E_1 \to E_2$, and $\overline{\varphi}$ is order-preserving.*

The uniqueness statement follows from 12.9(i) (consider the quotient of two isomorphisms $\overline{\varphi}$). For the more difficult existence of an extension $\overline{\varphi}$, see PRESTEL 1984 3.10, SCHARLAU 1985 Chapter 3 §2, LANG 1993 XI.2.9, JACOBSON 1989 Theorem 11.4 p. 656, BOCHNAK *et al.* 1998 1.3, COHN 2003a 8.8.13, PRESTEL–DELZELL 2001 1.3, or LAM 2005 VIII.2. We shall refer to 12.17 only in the following remarks and in 24.25.

12.18 Remarks on real closures A real closure of an ordered field $(F, <)$ is defined to be a real closed algebraic extension field E such that the unique ordering of E induces on F the given ordering $<$. By 12.16, each ordered field $(F, <)$ has a real closure E, and 12.17 says that E is uniquely determined up to isomorphisms of $(F, <)$.

For example, the field $\mathbb{R}_{\mathrm{alg}}$ of all real algebraic numbers is the real closure of $\mathbb{Q}$. The real closure of the Laurent series field $\mathbb{Q}((t))$, ordered such that $0 < t < q$ for each positive $q \in \mathbb{Q}$, is the field $\mathbb{R}_{\mathrm{alg}}((t^{1/\infty}))$ of all Puiseux series over $\mathbb{R}_{\mathrm{alg}}$; this is a consequence of 12.10, since adjoining $i = \sqrt{-1}$ to $\mathbb{R}_{\mathrm{alg}}((t^{1/\infty}))$ gives the field $\mathbb{C}_{\mathrm{alg}}((t^{1/\infty}))$, which is algebraically closed (see 64.24). The field $\mathbb{R}(t)$, taken with its ordering $P_{0,+}$ as in Exercise 3, has as real closure the field $\mathbb{R}((t^{1/\infty}))_{\mathrm{alg}}$ of all real Puiseux series which are algebraic over $\mathbb{R}(t)$ (by Exercise 4 and 12.11). More examples can be found in RIBENBOIM 1992, 1993.

Despite the uniqueness statement above, an algebraic closure $F^{\natural}$ of F may contain several copies of E. We describe an example of this phenomenon (see also 14.15 for a plethora of embeddings of $\mathbb{R}$ into $\mathbb{C}$). Let $F_1 = \mathbb{Q}(\sqrt[3]{2}) \subseteq \mathbb{R}$. Then $F_1 \cong F_2 := \mathbb{Q}(\zeta\sqrt[3]{2})$, where $\zeta \neq 1 = \zeta^3$ is a root of unity of order 3. The field isomorphism $\alpha : F_1 \to F_2$ with $\alpha(\sqrt[3]{2}) = \zeta\sqrt[3]{2}$ extends to an automorphism α of the algebraic closure $\mathbb{Q}^{\natural}$ of $\mathbb{Q}$ (see 64.15). By 12.11, the field $E = \mathbb{Q}^{\natural} \cap \mathbb{R} = \mathbb{R}_{\mathrm{alg}}$ is real closed, hence a real closure of $F = \mathbb{Q}$, and the same holds for the isomorphic field $\alpha(E)$. Since ζ is not real, $\zeta\sqrt[3]{2} \in \alpha(E) \smallsetminus E$, hence E and $\alpha(E)$ are distinct copies of E in $\mathbb{Q}^{\natural}$.

The existence and the uniqueness of the real closure of an ordered field (i.e. a field with a given ordering) can be proved without using Zorn's Lemma (which appears in the proof of 12.16) or the axiom of choice; see SANDER 1991, LOMBARDI–ROY 1991 and ZASSENHAUS 1970. In contrast, the existence of an ordering on every formally real field (12.3, 12.2) cannot be proved without Zorn's Lemma or one of its equivalents; see LÄUCHLI 1962.

12.19 Hilbert's 17th problem   At the International Congress of Mathematicians in 1900 at Paris, Hilbert proposed a famous list of 23 open problems. In his 17th problem, he asked whether each polynomial $f \in \mathbb{R}[x_1, \ldots, x_n]$ with $f(x) \geq 0$ for all $x \in \mathbb{R}^n$ is a sum of squares of rational functions. This was proved by Artin in 1927, using the theory of ordered fields and real closures, and Pfister proved in 1967 that f is actually a sum of 2^n squares of rational functions. For details see JA-COBSON 1989 11.4, SCHARLAU 1985 Chapter 3 §3, PRESTEL 1984 Theorem 5.7, PFISTER 1995 Chapter 6, BOCHNAK *et al.* 1998 6.1.1, 6.4.18 or PRESTEL–DELZELL 2001. These results triggered the development of real algebra and real algebraic geometry; compare also POWERS 1996.

We remark that rational functions are unavoidable in this context. In fact, the polynomial $x^4y^2 + x^2y^4 - 3x^2y^2 + 1$ is non-negative on $\mathbb{R}^2$, but not a sum of squares of polynomials; compare CHOI–LAM 1977, BOCHNAK *et al.* 1998 6.3.6, 6.6.1 or LAM 2005 p. 519 for more examples.

Exercises

(1) Let $(F, <)$ be an ordered field, and let t be transcendental over F. Show that $<$ has precisely one extension to an ordering of $F(t)$ such that $a < t$ for each $a \in F$.

(2) The field $\mathbb{Q}(\sqrt{2})$ has precisely two orderings.

(3) Show that the field $\mathbb{R}(t)$ of rational functions over $\mathbb{R}$ has domains of positivity $P_{r,+}$ and $P_{r,-}$ for $r \in \mathbb{R} \cup \{\infty\}$ with the following properties: for each positive real ε, we have $r < t < r + \varepsilon$ with respect to $P_{r,+}$, and $r - \varepsilon < t < r$ with respect to $P_{r,-}$; moreover $\mathbb{R} < t$ with respect to $P_{\infty,+}$, and $t < \mathbb{R}$ with respect to $P_{\infty,-}$. Show that $\mathbb{R}(t)$ has no other domains of positivity.

(4) The field $\mathbb{R}((t^{1/\infty}))$ of all real Puiseux series is a real closed field with non-Archimedean ordering.

(5) Show that $\sqrt{1 + \sqrt{2}} \in E \smallsetminus P$ and $\sqrt[4]{2} \in E \smallsetminus P$ in the notation of 12.8. More generally, show that $a \in E$ belongs to P if, and only if, a is totally real (that is, all roots of the minimal polynomial of a over $\mathbb{Q}$ are real).

(6) The field $\mathbb{R}$ has no maximal subfield.

(7) The field $\mathbb{R}$ contains an uncountable, strictly increasing family of real closed subfields.

(8) Let F be a field such that the degrees of the irreducible polynomials in $F[x]$ are bounded. Show that F is real closed or algebraically closed (hence these degrees are bounded by 2).

(9) A field is Euclidean if, and only if, it is Pythagorean with a unique ordering.

13   The real numbers as a topological field

In this section we introduce the concept of topological fields, which means that we add a compatible topology to the field structure. We analyse the interplay between topology and field structure, and in 13.8 we obtain a characterization of $\mathbb{R}$ and $\mathbb{C}$ due to Pontryagin.

13.1 Definition   A *topological ring* is a ring R with a topology such that
  (i) the additive group R^+ is a topological group, and
  (ii) the multiplication $R^2 \to R : (x, y) \mapsto xy$ is continuous.
Then the topology is called a *ring topology* of R. A *topological (skew) field* is a (skew) field R with a ring topology such that
(iii) the inversion $R^\times \to R^\times : x \mapsto x^{-1}$ is continuous.
Then $F^\times$ is a topological group, and the topology is called a *(skew) field topology* of F (see 13.5 for a slightly different definition).

13.2 Examples   (a) The real field $\mathbb{R}$, endowed with its usual topology, is a topological field. This fundamental fact is proved in 8.2 and 9.1.

More generally, every ordered field $(F, <)$ as defined in 11.1 is a topological field with respect to its order topology. Indeed, the additive group is easily seen to be a topological group (compare 8.2), and from the identity $xy - ab = (x - a)y + a(y - b)$ we infer that the multiplication is continuous (as in 9.1); now in view of $x^{-1} = a^{-1}(xa^{-1})^{-1}$ (see also 8.3) it suffices to prove the continuity of inversion at 1, and for any given $r \in F$ with $0 < r \leq 1$ the inequalities $-r/2 < 1 - x < r/2$ imply $0 < x^{-1} < 2$, hence $-r < x^{-1} - 1 < r$.

(b) Let F be a (skew) field with an absolute value $|\ | : F \to \mathbb{R}$, as defined in 55.1. Then the definition $d(x, y) = |x - y|$ yields a metric d on F which describes a (skew) field topology of F. This can be proved by similar considerations as in (a) and in 8.2, 9.1 (only the properties of an absolute value are needed). In particular, the field $\mathbb{C}$ of complex numbers is a topological field with its natural topology, since that topology is derived from an absolute value (see Section 14). The same remark applies to the skew field $\mathbb{H}$ of Hamiltons's quaternions; see Exercise 6.

Let F be a (skew) field with a valuation $v : F \to \Gamma \cup \{\infty\}$, as defined in 56.1; here Γ is an ordered abelian group. Then we obtain a (skew) field topology of F by taking the sets $\{\, x \in F \mid v(x - a) > \gamma \,\}$ with $\gamma \in \Gamma$ as a neighbourhood base at $a \in F$; the required arguments are again similar to the arguments in (a), see WARNER 1989 Theorem 20.16 or SHELL 1990 2.1 p. 15. We show only that inversion is continuous at 1: if $0 \leq \gamma \in \Gamma$, then $v(x-1) > \gamma$ entails $v(x) = v(x-1+1) = \min\{v(x-1), 0\} = 0$ and $v(1-x^{-1}) = v((x-1)x^{-1}) = v(x-1) - v(x) > \gamma$.

This valuation topology is always totally disconnected, because the sets $\{x \in F \mid v(x) > \gamma\}$ are subgroups of F^+ (since v is a valuation) which are open (by definition) and closed (by 62.7).

(c) Let F be a topological field, and let E be a (skew) field extension of F of finite (left or right) degree n. We endow E with the product topology with respect to any basis of E, considered as a (left or right) vector space over F. Then E is also a topological (skew) field, as we show now.

The extension E is isomorphic (via right or left regular representation $x \mapsto xa$ or $x \mapsto ax$) to a subfield of the ring $F^{n \times n}$ of all $n \times n$ matrices over F. This matrix ring $F^{n \times n}$, endowed with the product topology of F^{n^2}, is a topological ring with continuous inversion; indeed, matrix multiplication is bilinear in each matrix entry and therefore continuous; the continuity of matrix inversion may be inferred from Cramer's rule, which gives the matrix entries of A^{-1} as rational functions of the entries of A; compare SHELL 1990 7.2.

In particular, this shows again that $\mathbb{C}$ and $\mathbb{H}$ are topological (skew) fields with their natural topologies.

Several elementary properties of topological fields depend on the following easy observation. (Results 13.3–13.7 hold also for skew fields.)

13.3 Homogeneity Lemma *If F is a field with a ring topology, then the affine maps $x \mapsto ax + b$ with $a \neq 0$ form a doubly transitive group of homeomorphisms of F.*

13.4 Corollary *Let F be a field with a ring topology which is not indiscrete. Then F is a completely regular space, in particular a Hausdorff space, and F is either connected or totally disconnected.*

Proof Since the topology is not the indiscrete one, we find an open set O and points $a \notin O$, $b \in O$. By 13.3 we can cover $F \smallsetminus \{a\}$ by images of O under homeomorphisms fixing a, hence $F \smallsetminus \{a\}$ is open in F. Thus F is a T_1-space, and 62.4 says that F is completely regular. The second assertion follows from 13.3 and 5.46. □

According to SHAKHMATOV 1983, 1987, there exist field topologies which are not normal.

13.5 Lemma *Let F be a field with a T_1-topology which renders F^+ and $F^\times$ topological groups. Then F is a topological field.*

Proof By assumption, $F^\times$ is open in F. Therefore the identity $xy = (x+1)y - y$ implies that the multiplication is continuous on $F \times F^\times$, by similar identities also on $F^\times \times F$ and at $(0,0)$, hence everywhere.   □

13.6 Lemma *Let F be field with a compact ring topology which is not indiscrete. Then F is finite and discrete.*

Proof Using 13.3 we find a proper open subset W of F with $0 \in W$. Now $x \cdot 0 = 0$ for every $x \in F$, hence there exist open neighbourhoods U_x, V_x of x and 0 respectively such that $U_x \cdot V_x \subseteq W$. By compactness, F is a union of finitely many sets U_x. The intersection V of the corresponding sets V_x is a neighbourhood of 0 with $F \cdot V \subseteq W$. As F is a field and $W \neq F$, we infer that $V = \{0\}$. Hence the topology is discrete, and F is finite by compactness.   □

The usual topology of $\mathbb{R}$ does determine the field structure, as the following result shows.

13.7 Proposition *Let F be a field with a ring topology such that F is homeomorphic to $\mathbb{R}$. Then F and $\mathbb{R}$ are isomorphic as topological fields.*

Proof By 8.10 there exists an isomorphism $\varphi : \mathbb{R}^+ \to F^+$ of topological groups, and we may assume that $\varphi(1) = 1$; compare 8.24. Then the unique divisibility of $\mathbb{R}^+$ implies that $\varphi(\mathbb{Q})$ is the prime field of F, and that $\varphi|_\mathbb{Q}$ is multiplicative. Since $\mathbb{Q}$ is dense in $\mathbb{R}$, we infer that φ is multiplicative, hence an isomorphism of (topological) fields.   □

The result of Kiltinen quoted in 13.10 below shows that the usual field topology of $\mathbb{R}$ is far from being uniquely determined by the abstract field $\mathbb{R}$. However, additional assumptions on the topology do give strong uniqueness results:

13.8 Theorem (Pontryagin) *Let F be a field with a Hausdorff ring topology which is locally compact and connected. Then F is isomorphic as a topological field to $\mathbb{R}$ or to $\mathbb{C}$, endowed with their usual topologies.*

Proof According to the structure theorem 63.14 for locally compact abelian groups, F^+ is isomorphic as a topological group to a direct sum $\mathbb{R}^n \oplus C$, where C is a compact (connected) group. In fact, C is the largest compact subgroup of F^+, because $\mathbb{R}^n$ has only one (trivial) compact subgroup; see 8.6. The multiplications $x \mapsto ax$ with $a \in F^\times$ are automorphisms of F^+, and these automorphisms act transitively on $F \smallsetminus \{0\}$, leaving C invariant. In view of 13.6 this implies that $C = \{0\}$, hence F^+ is isomorphic as a topological group to the vector group $\mathbb{R}^n$.

In particular, F^+ is torsion free, whence F contains a copy $\mathbb{Q}$ of the field of rational numbers. The topological closure $\overline{\mathbb{Q}}$ of $\mathbb{Q}$ is the one-dimensional subspace spanned by 1, hence $\overline{\mathbb{Q}}$ is isomorphic to $\mathbb{R}$, even as a topological field. Clearly F has finite dimension n over $\overline{\mathbb{Q}}$, hence F is algebraic over $\overline{\mathbb{Q}} \cong \mathbb{R}$. Now $\mathbb{C}$ is algebraically closed (see 12.13) and 2-dimensional over $\mathbb{R}$, and this implies that $F \cong \mathbb{R}$ or $F \cong \mathbb{C}$. $\square$

Actually PONTRYAGIN 1932 allowed also skew fields, and he obtained the skew field $\mathbb{H}$ of Hamilton's quaternions (see Exercise 6 for a definition) as the only further possibility. In fact, the arguments above still apply, since $\overline{\mathbb{Q}} \cong \mathbb{R}$ is contained in the centre, and then $\mathbb{H}$ is the only further possibility by a well known algebraic result of Frobenius; see 58.11 or JACOBSON 1985 7.7, EBBINGHAUS *et al.* 1991 Chapter 8, PALAIS 1968 or PETRO 1987.

13.9 Corollary   (i) *Let τ be a Hausdorff ring topology of the field $\mathbb{R}$. If τ is locally compact and connected, then τ coincides with the usual topology of $\mathbb{R}$.*

(ii) *Let τ be a Hausdorff ring topology of the field $\mathbb{C}$. If τ is locally compact and connected, then τ is the image of the usual topology of $\mathbb{C}$ under some field automorphism of $\mathbb{C}$.*

Proof Part (ii) holds by 13.8, and (i) follows from 13.8 and 6.4. $\square$

Connectedness in 13.9 may be replaced by non-discreteness in both cases, as we shall see in 58.8. The assumption of local compactness cannot be omitted, since there exist strange (even locally) connected field topologies on $\mathbb{R}$; see 14.10. Compare also Exercise 11.

13.10 Existence results Each infinite field has a field topology which is neither discrete nor indiscrete; in fact, KILTINEN 1973 proved that each infinite field F of cardinality c has 2^{2^c} distinct field topologies.

WATERMAN–BERGMAN 1966 and MUTYLIN 1966 construct for any field F an arcwise connected (metrizable) field topology on the field $F_1 = F(x_t \,|\, t \in \mathbb{R})$ of rational functions in the indeterminates x_t such that F is a discrete subfield of the connected field F_1 (see also WIĘSŁAW 1988 Chapter 10); in fact, the set of indeterminates x_t is homeomorphic to $\mathbb{R}$. This shows that there exist connected fields of prime characteristic.

This was improved on by MUTYLIN 1968, who showed that any field F is a discrete subfield of a complete, metrizable, connected field E; in particular, there exist complete, metrizable, connected fields of prime characteristic. For the proof, one defines a (rather complicated) metric on F_1 which gives a ring topology on F_1 such that the completion $\widehat{F}_1$ is a ring with an open group of units, and then E is obtained as a quotient of $\widehat{F}_1$ modulo a maximal ideal.

MUTYLIN 1968 proves also that the field $\mathbb{C}$ of complex numbers is a proper subfield of a complete, metrizable field E such that E induces on $\mathbb{C}$ the usual topology (see Section 14). He defines a suitable (complicated) ring topology on $F_1 = \mathbb{C}(x)$, and then E is again obtained as a quotient of the completion $\widehat{F}_1$.

Finally we mention a result of Gelbaum, Kalisch and Olmsted which says that every Hausdorff ring topology contains a (weaker) Hausdorff field topology; see WARNER 1989 Theorem 14.4, WIĘSŁAW 1988 I §3 Theorem 3 or SHELL 1990 3.2.

Topological fields will appear again in later chapters, especially in Section 58, which deals with the classification of all locally compact (skew) fields.

A first introduction to topological fields can be found in PONTRYAGIN 1986 Chapter 4, a fuller account is given by WARNER 1989; see also WIĘSŁAW 1988, SHELL 1990. Some arguments remain valid without the associative and distributive laws; see GRUNDHÖFER–SALZMANN 1990.

Exercises

(1) Show that the non-zero ideals of $\mathbb{Z}$ form a neighbourhood base at 0 for a ring topology on $\mathbb{Q}$ which is not a field topology. Is there a ring topology of $\mathbb{R}$ which is not a field topology?

(2) Every topological field whose underlying topological space is a 1-manifold is isomorphic to $\mathbb{R}$.

(3) For transcendental real numbers s and t, the subfields $\mathbb{Q}(s)$ and $\mathbb{Q}(t)$ of $\mathbb{R}$ are algebraically isomorphic. Find necessary and sufficient conditions on s and t for the two fields to be isomorphic as topological fields. (It is an open problem if this is the case for $s = e$, $t = \pi$.)

(4) Show that every non-empty open subset of $\mathbb{R}$ contains a transcendency basis of $\mathbb{R}$ over $\mathbb{Q}$.

(5) Let F be a field with characteristic distinct from 2. Then every group topology of F^+ which renders the inversion of $F^\times$ continuous is a field topology of F.

(6) Show that the set $\mathbb{H}$ of all matrices $\begin{pmatrix} a & -\bar{b} \\ b & \bar{a} \end{pmatrix}$ with $a, b \in \mathbb{C}$, endowed with the usual matrix operations, is a skew field; this is called the skew field of Hamilton's quaternions. Show that $\mathbb{R}$ is the centre of $\mathbb{H}$ and that $\dim_{\mathbb{R}} \mathbb{H} = 4$. Show also that $\varphi(A) = |\det A|$ defines an absolute value $\varphi : \mathbb{H} \to \mathbb{R}$ (see 55.1) which induces on $\mathbb{H}$ the natural topology (as a real subspace of $\mathbb{C}^{2\times 2}$).

(7) Define real numbers by $c_1 = c = \sqrt{2}$ and $c_{n+1} = c^{c_n}$ for $n \in \mathbb{N}$. Show that the sequence $(c_n)_n$ converges and find the limit.

(8) For any sequence $a = (a_n)_n \in \mathbb{R}^{\mathbb{N}}$, we define its Cesaro transform $Ca = (c_n)_n$ as the sequence of the arithmetic means $c_n := n^{-1} \sum_{j=1}^{n} a_j$. Show that Ca converges to s if a converges to s. Find a sequence a such that Ca does not converge, but CCa does.

(9) Let $a_n \in \mathbb{R}$ with $0 < a_n < 1$ for $n \in \mathbb{N}$. Then $\sum_{n=1}^{\infty} a_n < \infty$ if, and only if, $\prod_{n=1}^{\infty}(1 - a_n) > 0$.

(10) Show that every arcwise connected topological field is locally connected.

(11) Construct an arcwise connected Hausdorff ring topology on $\mathbb{R}$ which is distinct from the usual topology. Is there also a field topology of $\mathbb{R}$ with these properties?

14   The complex numbers

Only one section of this book deals with the field $\mathbb{C}$ of complex numbers. This is due to the fact that many properties of $\mathbb{C} = \mathbb{R}(i)$ can be inferred from those of $\mathbb{R}$, as we show below.

First we consider the (topological) additive and multiplicative group of $\mathbb{C}$, and then we focus on the abstract field $\mathbb{C}$. We also exhibit some unusual features of $\mathbb{C}$, such as the existence of many discontinuous field automorphisms and of strange subfields.

The set $\mathbb{C} = \{\, a + bi \mid a, b \in \mathbb{R} \,\}$ with the usual addition and multiplication of complex numbers (in particular, $i^2 = -1$) is a field. Indeed, since the polynomial $p = x^2 + 1$ is irreducible in the real polynomial ring $\mathbb{R}[x]$, the quotient ring $\mathbb{C} := \mathbb{R}[x]/p\mathbb{R}[x]$ is a field, and we arrive at the above description of $\mathbb{C}$ if we take i as an abbreviation for the coset $x + p\mathbb{R}[x]$ (compare 64.8).

The mapping $\gamma : \mathbb{C} \to \mathbb{C} : z = a + bi \mapsto \gamma(z) := \bar{z} := a - bi$, called the complex conjugation, is a field automorphism of $\mathbb{C}$ of order 2.

The definitions $|a + bi| = \sqrt{a^2 + b^2}$ for $a, b \in \mathbb{R}$, $d(x, y) = |x - y|$ for $x, y \in \mathbb{C}$ give a metric d on $\mathbb{C}$ that describes the usual topology of $\mathbb{C}$. In fact, $1, i$ is a basis of $\mathbb{C}$, considered as a vector space over $\mathbb{R}$, and d describes the product topology, hence $\mathbb{C}$ is homeomorphic to the product space $\mathbb{R} \times \mathbb{R}$.

With this topology, $\mathbb{C}$ is a topological field; this can be verified directly, but it is also a special case of 13.2(c), and of 13.2(b) as well, since $|z| = \sqrt{z\bar{z}}$ is an absolute value of $\mathbb{C}$ (as defined in 55.1).

Another method to introduce $\mathbb{C}$ is to define it as the set of all 2×2 matrices $\begin{pmatrix} a & -b \\ b & a \end{pmatrix}$ with $a, b \in \mathbb{R}$; compare Exercise 6 of Section 13. It is easy to verify that this ring $\mathbb{C}$ is a topological field, see 13.2(c); in this description, the square $|\ |^2$ of the absolute value is just the determinant.

The topological additive group $\mathbb{C}^+$ is isomorphic to $\mathbb{R} \times \mathbb{R}$. The topological multiplicative group $\mathbb{C}^\times$ is isomorphic to $\mathbb{R} \times \mathbb{T}$, via the isomorphism $z \mapsto (|z|, z/|z|)$, which introduces polar coordinates in the plane; compare 1.20.

For the abstract groups the following holds.

14.1 Proposition *As an abstract group, $\mathbb{C}^+$ is isomorphic to $\mathbb{R}^+$, and $\mathbb{C}^\times$ is isomorphic to the torus group $\mathbb{T}$.*

Proof The group $\mathbb{C}^+ \cong \mathbb{R}^+ \oplus \mathbb{R}^+$ is divisible and torsion free, and $\mathbb{C}$ has the same cardinality as $\mathbb{R}$ by 61.12, hence $\mathbb{C}^+ \cong \mathbb{R}^+$ by 1.14. Furthermore $\mathbb{C}^\times \cong \mathbb{R} \times \mathbb{T} \cong \mathbb{T}$ by 1.25. $\square$

The topological groups $\mathbb{C}^+$ and $\mathbb{R}^+$ are not isomorphic, because $\mathbb{Z} + \mathbb{Z}i$ is a closed subgroup of $\mathbb{C}^+$ that is not cyclic; see 1.4. Therefore the usual topology of $\mathbb{C}$ cannot be induced by an ordering of the group $\mathbb{C}^+$: otherwise 3.3 and 7.9 would imply that $\mathbb{C}^+$ and $\mathbb{R}^+$ are isomorphic as ordered groups and then also as topological groups, which is not true.

The relation $i^2 = -1$ shows that there is no ordering that would be compatible with the multiplicative group $\mathbb{C}^\times$; see 11.2. In particular, $\mathbb{C}$ is not formally real (12.1). Therefore the concept of ordering will not play any role in this section.

Now we consider the influence of the topology on the additive and on the multiplicative group of $\mathbb{C}$. The following results 14.2 and 14.4 are due to VON KERÉKJÁRTÓ 1931; we remark that the group L_2 in 14.2 is a subgroup of index 2 in the affine group $\mathrm{Aff}\,\mathbb{R}$ appearing in 9.3.

14.2 Proposition *Let G be a topological group such that the underlying topological space is homeomorphic to $\mathbb{R}^2$. Then G is isomorphic as a topological group to the abelian group $\mathbb{R} \oplus \mathbb{R}$ or to the non-abelian group $L_2 := \{ \left(\begin{smallmatrix} 1 & 0 \\ a & b \end{smallmatrix} \right) \mid a, b \in \mathbb{R}, b > 0 \}$.*

Proof This is an easy consequence of the theory of Lie groups and Lie algebras: such a group G is a Lie group by the solution of Hilbert's 5th problem; see MONTGOMERY–ZIPPIN 1955 4.10 p. 184 (for groups defined on surfaces, Hilbert's 5th problem is somewhat easier; see SALZMANN *et al.* 1995 96.24). Furthermore G is simply connected, and the simply connected Lie groups (of dimension n) are classified by the real Lie algebras (of the same dimension n).

There exist only two Lie algebras of dimension 2, hence there are only two possibilities for G; since $\mathbb{R} \oplus \mathbb{R}$ and L_2 are clearly distinct, this completes the proof. $\square$

14.3 Corollary *Let A be a topological abelian group whose underlying topological space is homeomorphic to $\mathbb{C}$. Then A is isomorphic to $\mathbb{C}^+$ as a topological group.*

Proof This is an immediate consequence of 14.2. We indicate a more direct argument, which relies on the Splitting Theorem for locally compact abelian groups (63.14). This theorem implies that A is isomorphic as a topological group to $\mathbb{R}^n \oplus C$ with a compact group C. It suffices to show that C is trivial; then the assertion follows, because in $\mathbb{R}^n$ the complement of a point is connected but not simply connected only if $n = 2$.

By a topological result of KRAMER 2000, $\mathbb{R}^2$ is homeomorphic to a product $\mathbb{R}^n \times C$ only if C is homeomorphic to some space $\mathbb{R}^k$, and then $k = 0$ by the compactness of C.

Alternatively, we infer that C is simply connected (as A is simply connected), and we show now that every compact, simply connected abelian group C is trivial. By 8.13b every character $\chi : C \to \mathbb{T}$ lifts to a continuous group homomorphism $\widetilde{\chi} : C \to \mathbb{R}$, that is, $\chi = p \circ \widetilde{\chi}$ where $p : \mathbb{R} \to \mathbb{T} = \mathbb{R}/\mathbb{Z}$ is the canonical covering map as in 8.7. The compact subgroup $\widetilde{\chi}(C)$ of $\mathbb{R}$ is trivial (see 8.5), hence $\widetilde{\chi}$ and $\chi = p \circ \widetilde{\chi}$ are trivial. Thus C admits only the trivial character; by 63.13, C is trivial.   □

Each endomorphism of the group $\mathbb{C}^+$ is $\mathbb{Q}$-linear, since $\mathbb{C}^+ \cong \mathbb{R}^2$ is uniquely divisible (1.8, 1.9). Thus each continuous endomorphism of the topological group $\mathbb{C}^+$ is $\mathbb{R}$-linear, as $\mathbb{Q}$ is dense in $\mathbb{R}$. The ring of all continuous endomorphisms of $\mathbb{C}^+$ is therefore the endomorphism ring of the real vector space $\mathbb{C} = \mathbb{R}^2$, and hence is isomorphic to the ring of all real 2×2 matrices, and the group $\mathrm{GL}_2\mathbb{R}$ of all real invertible 2×2 matrices is the automorphism group of the topological group $\mathbb{C}^+$.

The influence of the topology on the multiplicative group is even stronger than in the real case (9.6); it is not necessary to assume commutativity, as we show now.

14.4 Corollary *Let G be a topological group such that the underlying topological space is homeomorphic to $\mathbb{C} \smallsetminus \{0\}$. Then G is isomorphic to $\mathbb{C}^\times$ as a topological group.*

Proof The mapping $p : \mathbb{R}^2 \to G = \mathbb{C} \smallsetminus \{0\}$ defined by $\varphi(r, s) = e^{r+is}$ is a covering map (8.11), and $\mathbb{C}$ is simply connected. Hence we obtain a unique structure of a topological group on $\mathbb{R}^2$ with neutral element $(0,0)$ such that p becomes a group homomorphism; see 8.13a. The kernel $p^{-1}(1) = \{0\} \times 2\pi\mathbb{Z}$ is a discrete normal subgroup of that connected group $\mathbb{R}^2$, hence contained in the centre (Exercise 4). By 14.2 there are only two possible group structures on $\mathbb{R}^2$, and the centre of the group L_2 is trivial. This shows that we can identify our group operation on $\mathbb{R}^2$ with the usual addition of vectors. Doing so, we can maintain that $p^{-1}(1) = \{0\} \times 2\pi\mathbb{Z}$, because the automorphism group $GL_2\mathbb{R}$ of $\mathbb{R}^2$ acts transitively on the non-zero vectors and on the non-trivial cyclic subgroups of $\mathbb{R}^2$. Therefore $G = p(\mathbb{R}^2)$ is isomorphic to $\mathbb{R}^2/p^{-1}(1) \cong \mathbb{R} \oplus (\mathbb{R}/2\pi\mathbb{Z}) \cong \mathbb{R} \times \mathbb{T} \cong \mathbb{C}^\times$ as a topological group.   □

14.5 Proposition   *The continuous endomorphisms of the topological group $\mathbb{C}^\times$ are precisely the maps*

$$\mathbb{C}^\times \to \mathbb{C}^\times : \exp(r + is) \mapsto \exp(ra + i(rb + ns))$$

where $a, b \in \mathbb{R}$ and $n \in \mathbb{Z}$. The ring of all these endomorphisms is isomorphic to the ring of all matrices $\begin{pmatrix} a & 0 \\ b & n \end{pmatrix}$ with $a, b \in \mathbb{R}, n \in \mathbb{Z}$.

Proof Let $p : \mathbb{R}^2 \to \mathbb{C} \smallsetminus \{0\}$ be the covering map used in the proof of 14.4, and let α be a continuous endomorphism of $\mathbb{C}^\times$. By 8.13 there exists a lift $\widetilde{\alpha} : \mathbb{R}^2 \to \mathbb{R}^2$, that is, a continuous endomorphism $\widetilde{\alpha}$ of $\mathbb{R}^2$ with $\alpha \circ p = p \circ \widetilde{\alpha}$. This equation implies that the kernel $p^{-1}(1) = \{0\} \times 2\pi\mathbb{Z}$ is invariant under $\widetilde{\alpha}$. As we have mentioned above, $\widetilde{\alpha}$ is in fact $\mathbb{R}$-linear. With respect to the basis $1, i$ of $\mathbb{C} = \mathbb{R}^2$ the linear map $\widetilde{\alpha}$ is described by a matrix as in 14.5. Conversely, each map as in 14.5 is a continuous endomorphism of $\mathbb{C}^\times$.

As an alternative, one could use the isomorphism $\mathbb{R}^+ \oplus \mathbb{T} \to \mathbb{C}^\times$: $(r, s + 2\pi\mathbb{Z}) \mapsto \exp(r + is)$ of topological groups, where $\mathbb{T} = \mathbb{R}/2\pi\mathbb{Z}$. For every continuous endomorphism α of $\mathbb{R} \oplus \mathbb{T}$, the image of the restriction $\alpha|_{\{0\} \oplus \mathbb{T}}$ is contained in $\mathbb{T}$ by compactness (and 8.5), hence by 8.27 this restriction is a map of the form $(0, s + 2\pi\mathbb{Z}) \mapsto (0, ns + 2\pi\mathbb{Z})$ for some $n \in \mathbb{Z}$. The restriction $\alpha|_{\mathbb{R} \oplus \{0\}}$ consists of an endomorphism of $\mathbb{R}$ and of a character of $\mathbb{R}$, hence by 8.26 and 8.31a it is a map $(r, 0) \mapsto (ra, rb + 2\pi\mathbb{Z})$ for some $a, b \in \mathbb{R}$. Assembling the pieces we obtain $\alpha(r, s + 2\pi\mathbb{Z}) = (ra, rb + ns + 2\pi\mathbb{Z})$, hence the assertion.   □

The real exponential function relates the additive and multiplicative groups of $\mathbb{R}$ very closely; see 2.2. The complex exponential function (which was used in the proofs of 14.4 and 14.5) is a continuous group homomorphism $\exp : \mathbb{C}^+ \to \mathbb{C}^\times$, which is surjective but not injective: by the periodicity of exp, the kernel is the discrete subgroup $2\pi i\mathbb{Z}$.

Results 14.1 and 1.24 show that there exist group monomorphisms $\mathbb{C}^+ \to \mathbb{C}^\times$, but we prove now that none of these is continuous.

14.6 Proposition *The continuous group homomorphisms of $\mathbb{C}^+$ into $\mathbb{C}^\times$ are precisely the maps*

$$\mathbb{C}^+ \to \mathbb{C}^\times : r + is \mapsto \exp(ar + bs + i(cr + ds))$$

where $a, b, c, d \in \mathbb{R}$. As a consequence, there exist no continuous group monomorphisms of $\mathbb{C}^+$ into $\mathbb{C}^\times$.

Proof The complex exponential function $\exp : \mathbb{C}^+ \to \mathbb{C}^\times$ is a covering map as defined in 8.11. By 8.13 each continuous group homomorphism $\varphi : \mathbb{C}^+ \to \mathbb{C}^\times$ lifts to a continuous group endomorphism $\widetilde{\varphi}$ of $\mathbb{C}^+$, that is, $\varphi = \exp \circ \widetilde{\varphi}$. The maps $\widetilde{\varphi}$ arising here are precisely the $\mathbb{R}$-linear endomorphisms of $\mathbb{C}^+ \cong \mathbb{R}^2$, and these endomorphisms are described by the real 2×2 matrices.

Assume that φ is injective. Then the linear map $\widetilde{\varphi}$ is injective, hence bijective. Thus $(\widetilde{\varphi})^{-1}(2\pi i)$ is a non-trivial element of the kernel of $\varphi = \exp \circ \widetilde{\varphi}$, which is absurd. $\square$

In the rest of this section, we study the abstract field $\mathbb{C}$. It is a basic fact that $\mathbb{C}$ is algebraically closed; see 12.13, FINE–ROSENBERGER 1997 or EBBINGHAUS *et al.* 1991 Chapter 4 for various proofs of this result, the so-called the Fundamental Theorem of Algebra. Each algebraically closed field of characteristic 0 with the same cardinality as $\mathbb{C}$ is isomorphic to $\mathbb{C}$; this is a special case of a result of Steinitz; see 64.21.

The simple transcendental extension $\mathbb{C}(t)$ and its algebraic closure $\mathbb{C}(t)^{\natural}$ have the same cardinality as $\mathbb{C}$ (see 64.16 and 64.19), hence $\mathbb{C}$ is isomorphic to $\mathbb{C}(t)^{\natural}$; see also 64.24 and 64.25. This implies that $\mathbb{C}$ admits uncountably many field automorphisms; compare 64.19. But we can do much better; see 14.11 below (or Section 64, Exercise 4).

14.7 Addition and multiplication do not determine the field of complex numbers The structure of the field $\mathbb{C}$ is not determined by the isomorphism types of its multiplicative group and its additive group (see 6.2c, d and 34.2 for analogous questions on $\mathbb{R}$ and $\mathbb{Q}$). Indeed, we show that there exists a field F with $F^\times \cong \mathbb{C}^\times$ and $F^+ \cong \mathbb{C}^+$ as groups, but $F \not\cong \mathbb{C}$ as fields.

We call a Galois extension solvable, if its Galois group is solvable. Let $F = \mathbb{C}(t)^{\mathrm{solv}}$ be the union of all finite solvable Galois extensions of $\mathbb{C}(t)$ in some fixed algebraic closure of $\mathbb{C}(t)$; this union is actually a field; see Exercise 8. The field F is called the maximal prosolvable extension of

$\mathbb{C}(t)$, because $\mathrm{Gal}_{\mathbb{C}(t)}F$ is a projective limit of finite solvable groups, but we do not need this fact. We have $\mathrm{card}\,F = \mathrm{card}\,\mathbb{C}$ (see 64.16), hence $F^+ \cong \mathbb{C}^+$ by 1.14.

We claim that the multiplicative group $F^\times$ is divisible. Let $a \in F^\times$, $n \in \mathbb{N}$, and let $p(x) = \prod_{j=1}^{k}(x - a_j) \in \mathbb{C}(t)[x]$ be the minimal polynomial of a over $\mathbb{C}(t)$. Then each a_j belongs to F. The splitting field $E_a := \mathbb{C}(t)(a_1, \ldots, a_k) \subseteq F$ of $p(x)$ over $\mathbb{C}(t)$ is a solvable Galois extension of $\mathbb{C}(t)$, by Exercise 8. Let E be the splitting field of $p(x^n)$ over $\mathbb{C}(t)$. Then $E|\mathbb{C}(t)$ is a Galois extension (by 64.10), and E is also the splitting field of $p(x^n) = \prod_i(x^n - a_i)$ over E_a. Thus $E|E_a$ is a so-called Kummer extension, and it is easy to see that the Galois group of $E|E_a$ is abelian: each element of this Galois group permutes the n-th roots of a_j by multiplication with an n-th root of unity $\zeta_j \in \mathbb{C} \subseteq E_a$ (compare JACOBSON 1985 4.7 Lemma 2 p. 253, MORANDI 1996 11.4 or COHN 2003a p. 442). Hence the Galois extension $E|\mathbb{C}(t)$ is solvable (by the main theorem of Galois theory; see 64.18(ii)). We conclude that $E \subseteq F$, hence F contains an n-th root of a.

Now we show that $F^\times \cong \mathbb{C}^\times$. The torsion subgroups of $F^\times$ and $\mathbb{C}^\times$ coincide. Being divisible, this torsion subgroup is a direct factor of both $F^\times$ and $\mathbb{C}^\times$; see 1.22, 1.23. The complementary subgroups are divisible and torsion free of the same cardinality, hence isomorphic by 1.14.

It remains to show that the field F is not isomorphic to $\mathbb{C}$. The splitting field E of the polynomial $x^5 - 5tx + 4t \in \mathbb{C}(t)[x]$ is a Galois extension of $\mathbb{C}(t)$ with the non-solvable Galois group S_5; see MALLE–MATZAT 1999 I.9.4 (in fact, every finite group is a Galois group over $\mathbb{C}(t)$; see MALLE–MATZAT 1999 I.1.5 or VÖLKLEIN 1996 Section 2.2.2). Thus E is not contained in F, hence F is not algebraically closed.

For several constructions below in this section, the following lemma is crucial.

14.8 Basic Lemma *Every transcendency basis T of $\mathbb{C}$ over $\mathbb{Q}$ has cardinality $\aleph = 2^{\aleph_0} = \mathrm{card}\,\mathbb{C}$.*

Proof $\mathbb{C}$ is algebraic over the purely transcendental field $\mathbb{Q}(T)$, hence $\mathrm{card}\,\mathbb{C} = \mathrm{card}\,\mathbb{Q}(T)$ by 64.5. The set T is infinite, as $\mathbb{C}$ is uncountable. Hence T has the same cardinality as $\mathbb{Q}(T)$ and as $\mathbb{C}$; see 64.20. □

14.9 Corollary *The field $\mathbb{C}$ admits precisely $2^\aleph$ proper field endomorphisms with mutually distinct images; thus $\mathbb{C}$ has precisely $2^\aleph$ proper subfields which are isomorphic to $\mathbb{C}$.*

Proof Using Zorn's Lemma we choose a transcendency basis T of $\mathbb{C}$ over $\mathbb{Q}$; see 64.20. For each $S \subset T$ with card $S =$ card T, we have a field isomorphism $\mathbb{Q}(T) \to \mathbb{Q}(S)$ which maps T bijectively onto S. Since $\mathbb{C}$ is an algebraic closure of $\mathbb{Q}(T)$, this field isomorphism extends by 64.15 to a field endomorphism φ_S of $\mathbb{C}$ whose image $\varphi_S(\mathbb{C})$ is algebraic over $\mathbb{Q}(S)$. We have $T \cap \varphi_S(\mathbb{C}) = S$, because T is algebraically independent. Hence the images of endomorphisms φ_S with distinct sets S as above are distinct. Furthermore card$\{\, S \mid S \subset T, \text{card}\, S = \text{card}\, T \,\} = 2^{\aleph}$ by 14.8 and 61.16. For an upper bound, we remark that $2^{\aleph}$ is the cardinality of the set $\mathbb{C}^{\mathbb{C}}$; see 61.10 and 61.15. □

Of course, none of the subfields in 14.9 is topologically closed in $\mathbb{C}$. Indeed, $\mathbb{R}$ is the only proper subfield of $\mathbb{C}$ which is closed with respect to the usual topology (since the prime field $\mathbb{Q}$ is dense in $\mathbb{R}$).

14.10 Dense connected subfields DIEUDONNÉ 1945 constructed a set $S \subseteq \mathbb{C}$ which is algebraically independent over $\mathbb{Q}$ such that the field $\mathbb{Q}(S)$ is locally connected (hence S is uncountable); the construction is related to the proof of Theorem 8.23; see SHELL 1990 Chapter 9, WIĘSŁAW 1988 Chapter 9, WARNER 1989 Exercise 27.7, BOURBAKI 1972 VI §9 Exercise 2 p. 470 for details.

Then $\mathbb{Q}(S)$ is connected (compare 13.4) and as an abstract field not isomorphic to, hence distinct from, $\mathbb{R}$ and $\mathbb{C}$ (see Exercise 5). In fact, $\mathbb{Q}(S)$ is not contained in $\mathbb{R}$, as $\mathbb{R}$ has no proper connected subfield; see 5.4. Hence $\mathbb{Q}(S)$ is a proper, dense, connected and locally connected subfield of $\mathbb{C}$.

The algebraic closure $\mathbb{Q}(S)^{\natural}$ of $\mathbb{Q}(S)$ in $\mathbb{C}$ is isomorphic to $\mathbb{C}$ by 64.21, hence $\mathbb{Q}(S)^{\natural}$ has a subfield R which is isomorphic to $\mathbb{R}$ and contains $\mathbb{Q}(S)$. This field R is connected, locally connected (as $\mathbb{Q}(S)$ is locally connected and dense in R) and dense in $\mathbb{C}$, but not complete (as R is not closed in $\mathbb{C}$).

This shows that the field $\mathbb{R}$, hence also $\mathbb{C}$, has a strange connected and locally connected field topology; compare also Exercise 11 of Section 13. (We remark that KAPUANO 1946 constructed an algebraically closed subfield F of $\mathbb{C}$ with $F \cap \mathbb{R} = \mathbb{Q}^{\natural}$ such that F has small inductive dimension 1.)

In contrast, BAER–HASSE 1932 Satz 6 (see also WARNER 1989 Exercise 27.8) showed that $\mathbb{R}$ and $\mathbb{C}$ are the only arcwise connected subfields of $\mathbb{C}$. This implies that $\mathbb{C}$ contains no arc consisting of elements which are algebraically independent over $\mathbb{Q}$ (adjoin such an arc to $\mathbb{Q}$ and use Exercise 5).

14.11 Theorem   *The group* $\operatorname{Aut}\mathbb{C}$ *of all field automorphisms of* $\mathbb{C}$ *has cardinality* $2^{\aleph} = 2^{2^{\aleph_0}}$. *The group* $\operatorname{Sym}\mathbb{C}$ *of all permutations of* $\mathbb{C}$ *is involved in* $\operatorname{Aut}\mathbb{C}$, *i.e.,* $\operatorname{Sym}\mathbb{C}$ *is isomorphic to a quotient of a subgroup of* $\operatorname{Aut}\mathbb{C}$. *Hence every group of cardinality at most* $\aleph$ *is involved in* $\operatorname{Aut}\mathbb{C}$. *Furthermore* $\operatorname{Aut}\mathbb{C}$ *has a subgroup which is free with a free generating set of cardinality* $\aleph$.

Proof   Using Zorn's Lemma we choose a transcendency basis T of $\mathbb{C}$ over $\mathbb{Q}$; see 64.20. Each permutation of T extends (uniquely) to a field automorphism of the purely transcendental field $\mathbb{Q}(T)$. Since $\mathbb{C}$ is an algebraic closure of $\mathbb{Q}(T)$, every automorphism of $\mathbb{Q}(T)$ extends to an automorphism of $\mathbb{C}$; see 64.15. Thus the symmetric group $\operatorname{Sym}T$ is involved in $\operatorname{Aut}\mathbb{C}$, and $\operatorname{Sym}T$ is isomorphic to $\operatorname{Sym}\mathbb{C}$ by 14.8. This implies that $\operatorname{card}\operatorname{Aut}\mathbb{C} = \operatorname{card}\operatorname{Sym}\mathbb{C} = 2^{\aleph}$; see 61.16. (We observe that $\operatorname{Sym}\mathbb{C}$ is not isomorphic to a subgroup of $\operatorname{Aut}\mathbb{C}$, as $\operatorname{Aut}\mathbb{C}$ contains no element of order 3; see 14.13(i)).

Every group G is isomorphic to the subgroup $\{\, x \mapsto xg \mid g \in G \,\}$ of the symmetric group $\operatorname{Sym}G$. If $\operatorname{card}G \le \aleph$, then by 61.8 we find an injective map $G \to \mathbb{C}$, hence $\operatorname{Sym}G$ is isomorphic to a subgroup of $\operatorname{Sym}\mathbb{C}$. This shows that G is involved in $\operatorname{Aut}\mathbb{C}$.

Now we take G to be a free group with a free generating set of cardinality $\aleph$. Then G has cardinality $\aleph$; see 61.14. As G is involved in $\operatorname{Aut}\mathbb{C}$, we find a subset S of $\operatorname{Aut}\mathbb{C}$ with $\operatorname{card}S = \aleph$ such that S is a free generating set modulo some subgroup of $\operatorname{Aut}\mathbb{C}$. Hence S is a free generating set for the subgroup $\langle S \rangle$ of $\operatorname{Aut}\mathbb{C}$.   □

The following proposition says that continuity is a rare phenomenon in $\operatorname{Aut}\mathbb{C}$.

14.12 Proposition   *The identity and the complex conjugation* $z \mapsto \bar{z}$ *are the only field endomorphisms of* $\mathbb{C}$ *which are continuous with respect to the usual topology of* $\mathbb{C}$. *Furthermore these are also the only field endomorphisms of* $\mathbb{C}$ *which leave* $\mathbb{R}$ *invariant.*

Proof   Each field endomorphism α fixes 1 (by definition), hence α fixes each element of $\mathbb{Q}$. Continuity implies that α fixes each real number, as $\mathbb{Q}$ is dense in $\mathbb{R}$; the same conclusion holds if $\alpha(\mathbb{R}) \subseteq \mathbb{R}$; see 6.4. From the equation $\alpha(i)^2 = \alpha(i^2) = -1$ we infer that $\alpha(i) \in \{i, -i\}$, hence α is the identity or complex conjugation.   □

Every discontinuous field automorphism α of $\mathbb{C}$ is extremely discontinuous, in the following sense: for any arc $J \subseteq \mathbb{C}$, the image $\alpha(J)$ is dense in $\mathbb{C}$; for proofs see SALZMANN 1969, KALLMAN–SIMMONS 1985 and KESTELMAN 1951. The weaker assertion that $\alpha(\mathbb{R})$ is dense in $\mathbb{C}$ is easy to prove: otherwise the topological closure of $\alpha(\mathbb{R})$ is a proper closed subfield of $\mathbb{C}$, hence equal to $\mathbb{R}$; then 6.4 implies that α is $\mathbb{R}$-linear, and hence continuous.

Now we study maximal subfields of the field $\mathbb{C}$ and involutions in the group $\operatorname{Aut} \mathbb{C}$.

14.13 Theorem (i) *The non-trivial elements in* $\operatorname{Aut} \mathbb{C}$ *are either involutions or elements of infinite order.*

(ii) *If* $\alpha \in \operatorname{Aut} \mathbb{C}$ *is an involution and* $F = \operatorname{Fix} \alpha := \{ z \in \mathbb{C} \mid \alpha(z) = z \}$, *then* F *is a real closed field,* $\mathbb{C} = F(i)$ *and* $\alpha(i) = -i$.

(iii) *The maximal subfields of* $\mathbb{C}$ *are precisely the subfields* F *such that* $[\mathbb{C} : F] = 2$. *Assigning to an involution* $\alpha \in \operatorname{Aut} \mathbb{C}$ *the corresponding field* $\operatorname{Fix} \alpha$ *of fixed elements gives a bijection between the set of all involutions in* $\operatorname{Aut} \mathbb{C}$ *and the set of all maximal subfields of* $\mathbb{C}$.

(iv) *Two involutions in* $\operatorname{Aut} \mathbb{C}$ *are conjugate if, and only if, the corresponding fields of fixed elements are isomorphic (as fields).*

(v) *Distinct involutions in* $\operatorname{Aut} \mathbb{C}$ *do not commute, their product has infinite order.*

Proof (i) If $\alpha \in \operatorname{Aut} \mathbb{C}$ has finite order n, then $n = [\mathbb{C} : \operatorname{Fix} \alpha]$ by 64.18; hence (i) follows from 12.15 (we require this result only for the field $\mathbb{C}$; this special case of 12.15 has a shorter proof, as $\mathbb{C}$ has characteristic 0).

(ii) By 64.18 we have $[\mathbb{C} : F] = 2$, hence F is real closed by 12.15, and 12.10 yields the assertion.

(iii) If F is a maximal subfield of $\mathbb{C}$, then $\mathbb{C} = F(a)$ for every $a \in \mathbb{C} \smallsetminus F$. If a were transcendental over F, then $F < F(a^2) < \mathbb{C}$, a contradiction. Thus the degree $[\mathbb{C} : F]$ is finite, in fact equal to 2 by 12.15.

This shows that the maximal subfields of $\mathbb{C}$ are precisely the subfields F with $[\mathbb{C} : F] = 2$. By Galois theory (64.18), each of these subfields is the field of fixed elements of a unique involution in $\operatorname{Aut} \mathbb{C}$, and each involution in $\operatorname{Aut} \mathbb{C}$ gives such a subfield.

(iv) If two such involutions α, β are conjugate under $\varphi \in \operatorname{Aut} \mathbb{C}$, then φ induces an isomorphism between $\operatorname{Fix} \alpha$ and $\operatorname{Fix} \beta$. Conversely, a given isomorphism $\varphi : \operatorname{Fix} \alpha \to \operatorname{Fix} \beta$ extends to an automorphism φ of $\mathbb{C}$ (by 64.15 or more directly by (ii)), and $\varphi \circ \alpha \circ \varphi^{-1}$ is then an involution fixing the elements of $\varphi(\operatorname{Fix} \alpha) = \operatorname{Fix} \beta$, and hence equal to β by (iii).

(v) If $\alpha, \beta \in \operatorname{Aut} \mathbb{C}$ are involutions, then $\alpha(\beta(i)) = \alpha(-i) = i$ by (ii). Thus $\alpha\beta$ is not an involution, again by (ii), hence it is trivial or of infinite order by (i). $\square$

14.14 Proposition *Let $\alpha \in \operatorname{Aut}\mathbb{C}$ be an involution and let $F = \operatorname{Fix}\alpha$ be the field of fixed elements of α.*

(i) *The centralizer $\operatorname{Cs}\alpha := \{\beta \in \operatorname{Aut}\mathbb{C} \mid \alpha\beta = \beta\alpha\}$ of α is isomorphic to the direct product $\langle\alpha\rangle \times \operatorname{Aut}F$.*

(ii) *The group $\operatorname{Aut}F$ is torsion free, and it is trivial if F is isomorphic to a subfield of $\mathbb{R}$.*

Proof By 14.13(ii) we have $\mathbb{C} = F(i)$ and $\alpha(i) = -i$. Each $\varphi \in \operatorname{Aut}F$ extends uniquely to an automorphism $\overline{\varphi} \in \operatorname{Aut}\mathbb{C}$ that fixes i; indeed, $\overline{\varphi}$ is given by $\overline{\varphi}(a + bi) = \varphi(a) + \varphi(b)i$ for $a, b \in F$. Clearly the group $K := \{\overline{\varphi} \mid \varphi \in \operatorname{Aut}F\}$ is a subgroup of $\operatorname{Cs}\alpha$.

Each $\beta \in \operatorname{Cs}\alpha$ acts on F and on $\{i, -i\}$, and $\beta(i) = i$ implies that $\beta = \overline{\beta|_F}$. Hence K is precisely the kernel of the action of $\operatorname{Cs}\alpha$ on the set $\{i, -i\} = \{i, \alpha(i)\}$, whence $\operatorname{Cs}\alpha = \langle\alpha\rangle \times K \cong \langle\alpha\rangle \times \operatorname{Aut}F$. This proves part (i).

Concerning (ii), we observe that the group $\operatorname{Aut}F$ is torsion free by 14.13(i, v). Furthermore F is real closed by 14.13(ii). Hence by 12.9(ii) $\operatorname{Aut}F$ is trivial if F is isomorphic to a subfield of $\mathbb{R}$. $\square$

14.15 Corollary *The complex conjugation γ defined by $\gamma(z) = \bar{z}$ has $2^{\aleph}$ conjugates in $\operatorname{Aut}\mathbb{C}$, and $\mathbb{C}$ contains $2^{\aleph}$ maximal subfields that are isomorphic to $\mathbb{R}$. The centre of the group $\operatorname{Aut}\mathbb{C}$ is trivial.*

Proof Since $\operatorname{Aut}\mathbb{R}$ is trivial (by 6.4), we infer from 14.14 that γ generates its own centralizer. Hence γ has $2^{\aleph}$ conjugates in $\operatorname{Aut}\mathbb{C}$ (see 14.11), and the centre of $\operatorname{Aut}\mathbb{C}$ is trivial. By 14.13(iii) the field $\mathbb{C}$ contains $2^{\aleph}$ maximal subfields that are isomorphic to $\mathbb{R}$. $\square$

Actually, we can do much better (or worse), as we show now.

14.16 Theorem (i) *The group $\operatorname{Aut}\mathbb{C}$ contains $2^{\aleph} = 2^{2^{\aleph_0}}$ conjugacy classes of involutions α with $\operatorname{Cs}\alpha = \langle\alpha\rangle$; each of these conjugacy classes has cardinality $2^{\aleph}$. The field $\mathbb{C}$ contains $2^{\aleph}$ mutually non-isomorphic maximal subfields that are isomorphic to a subfield of $\mathbb{R}$; in fact $2^{\aleph}$ copies of each of these maximal subfields.*

(ii) *The group $\operatorname{Aut}\mathbb{C}$ contains $2^{\aleph}$ conjugacy classes of involutions α with $\operatorname{card}\operatorname{Cs}\alpha = 2^{\aleph}$. The corresponding fields of fixed elements are maximal subfields of $\mathbb{C}$ that are not isomorphic to a subfield of $\mathbb{R}$; in fact, such a field is real closed with a non-Archimedean ordering.*

Proof Recall from 14.11 that $\operatorname{card}\operatorname{Aut}\mathbb{C} = 2^{\aleph}$.

By invoking Theorem 12.14, we obtain many isomorphism types of real closed fields F of cardinality $\aleph$. The algebraic closure $F^{\natural}$ of such a field F is isomorphic to $\mathbb{C}$; see 64.21. Hence $\mathbb{C}$ contains a maximal subfield isomorphic to F, and our results 14.13, 14.14 apply. In particular, distinct isomorphism types of fields F lead to distinct conjugacy classes of involutions in $\operatorname{Aut}\mathbb{C}$, and the cardinality of such a conjugacy class is also the number of copies of maximal subfields of $\mathbb{C}$ that are isomorphic to F.

(i) By 12.14(i) there exist $2^{\aleph}$ isomorphism types for F such that the unique ordering (12.7) of F is Archimedean. This leads to $2^{\aleph}$ conjugacy classes of involutions $\alpha \in \operatorname{Aut}\mathbb{C}$ with $\operatorname{Cs}\alpha = \langle\alpha\rangle$, and each of these conjugacy classes has cardinality $|\operatorname{Aut}\mathbb{C} : \langle\alpha\rangle| = \operatorname{card}\operatorname{Aut}\mathbb{C} = 2^{\aleph}$.

(ii) By 12.14(ii) there exist $2^{\aleph}$ isomorphism types of real closed fields F of cardinality $\aleph$ such that $\operatorname{card}\operatorname{Aut}F = 2^{\aleph}$. The maximal subfields obtained here are not isomorphic to a subfield of $\mathbb{R}$, because their unique orderings are not Archimedean; see 11.14. $\qquad\Box$

The field $^*\mathbb{R}$ of non-standard real numbers is isomorphic to a maximal subfield of $\mathbb{C}$, since $^*\mathbb{R}(i) = {}^*\mathbb{C}$ is algebraically closed by 21.7, and $^*\mathbb{C} \cong \mathbb{C}$ by 64.21 and 24.1. For the cardinality of $\operatorname{Aut}(^*\mathbb{R})$ see 24.27.

More information on (involutions in) automorphism groups of algebraically closed fields can be found in BAER 1970b, SOUNDARARAJAN 1991, SCHNOR 1992. Assuming the continuum hypothesis (61.17), EVANS–LASCAR 1997 prove that every automorphism of $\operatorname{Aut}\mathbb{C}$ is an inner automorphism.

The automorphisms of $\mathbb{C}$ which fix i form a subgroup of index 2 in $\operatorname{Aut}\mathbb{C}$. More generally, for each Galois extension E of $\mathbb{Q}$ there is a finite quotient group of $\operatorname{Aut}\mathbb{C}$ (and of $\operatorname{Aut}\mathbb{Q}^{\natural}$) which is isomorphic to the Galois group of $E|\mathbb{Q}$; see 64.18. It is an old conjecture that every finite group arises in this fashion. Inverse Galois theory is working towards a proof of this conjecture, which has been verified in many special instances, for example, for all solvable groups and for many simple groups; see MALLE–MATZAT 1999, VÖLKLEIN 1996.

The group $\operatorname{Aut}\mathbb{C}$ acts on the field $\mathbb{Q}^{\natural}$ of all algebraic numbers, inducing (by 64.15) on $\mathbb{Q}^{\natural}$ the group $\operatorname{Aut}\mathbb{Q}^{\natural}$, which is a (profinite) group of cardinality $\aleph$. The kernel of this action is the normal subgroup of all automorphisms of $\mathbb{C}$ which fix each algebraic number. According to LASCAR 1992, 1997 this kernel is a simple group (of cardinality $2^{\aleph}$, see 14.11). This unusual simple group is torsion free by 14.13(i, ii). From this point of view, $\operatorname{Aut}\mathbb{C}$ consists of a large simple subgroup, with the comparatively small group $\operatorname{Aut}\mathbb{Q}^{\natural}$ (the absolute Galois group of $\mathbb{Q}$) on top.

At this inconspicuous place of the book, we dare to indulge briefly in geometric applications.

14.17 The Euclidean plane   If we identify the plane $\mathbb{R}^2$ with $\mathbb{C}$ as indicated at the beginning of this section, then the absolute value $|z|$ of a point $z \in \mathbb{C}$ is the Euclidean distance of z from 0. The multiplicativity of the absolute value $|z| = |\bar{z}|$ implies immediately that all maps $z \mapsto az+b$ and $z \mapsto a\bar{z} + b$ with $a, b \in \mathbb{C}$ and $|a| = 1$ are Euclidean isometries; for example, the map $z \mapsto \bar{z}$ is the reflection at the axis $\mathbb{R}$, and $z \mapsto az$ is a rotation with centre 0; compare the proof of 1.20.

In fact, we have described all Euclidean isometries, and the complex affine group

$$\operatorname{Aff}\mathbb{C} = \{\, z \mapsto az + b \mid a \in \mathbb{C}^\times,\, b \in \mathbb{C} \,\} \cong \mathbb{C}^+ \rtimes \mathbb{C}^\times$$

(see 9.3, 9.4) is a subgroup of index 2 in the group $\operatorname{Aff}\mathbb{C} \rtimes \langle z \mapsto \bar{z} \rangle$ of all similarity transformations of the Euclidean affine plane $\mathbb{R}^2$.

This close connection between the field $\mathbb{C}$ and the Euclidean plane $\mathbb{R}^2$ allows to deal with plane Euclidean geometry in an algebraic fashion, using complex numbers. Many examples for this approach can be found in HAHN 1994, EBBINGHAUS *et al.* 1991 Chapter 3 §4, and SCHWERDT-FEGER 1979; see also CONNES 1998, GEIGES 2001, SCHÖNHARDT 1963, HOFMANN 1958.

14.18 Complex affine spaces   Let $n \in \mathbb{N}$. The complex affine space of dimension n is the lattice of all affine subspaces of the complex vector space $\mathbb{C}^n$. If $A \in \operatorname{GL}_n\mathbb{C}$ and $\alpha \in \operatorname{Aut}\mathbb{C}$, then the semi-linear bijection $(A, \alpha) : \mathbb{C}^n \to \mathbb{C}^n$ defined by

$$(A, \alpha)\begin{pmatrix} z_1 \\ \vdots \\ z_n \end{pmatrix} = A\begin{pmatrix} \alpha(z_1) \\ \vdots \\ \alpha(z_n) \end{pmatrix}$$

for $z_1, \ldots, z_n \in \mathbb{C}$ is clearly an automorphism of this lattice, or in other terminology: a collineation of the affine space $\mathbb{C}^n$. In fact, for $n \geq 2$ the semi-linear group $\Gamma\mathrm{L}_n\mathbb{C} := \{\,(A, \alpha) \mid A \in \operatorname{GL}_n\mathbb{C},\ \alpha \in \operatorname{Aut}\mathbb{C}\,\} = \operatorname{GL}_n\mathbb{C} \rtimes \operatorname{Aut}\mathbb{C}$ is the group of all those collineations of this affine space which fix the vector 0, by the fundamental theorem of affine geometry, but we do not need this fact.

Our aim here is to prove the following amazing result.

14.19 Theorem   *For each $n \in \mathbb{N}$ the semi-linear group $\Gamma\mathrm{L}_n\mathbb{C}$ contains a subgroup G with the following properties.*

(i) G *is transitive on the set of all ordered bases of* $\mathbb{C}^n$.

(ii) *The identity is the only element of* G *that is continuous.*

(iii) G *is a free group with a free generating set of cardinality* $\aleph = 2^{\aleph_0}$.

Proof According to 14.11, $\mathrm{Aut}\,\mathbb{C}$ contains a free subgroup $H = \langle S \rangle$, where S is a free generating set of cardinality $\aleph$. Choose any surjection $S \to \mathrm{GL}_n\mathbb{C} : s \mapsto A_s$, and define $f(s) := (A_s, s) \in \Gamma\mathrm{L}_n\mathbb{C}$. The mapping $f : S \to \Gamma\mathrm{L}_n\mathbb{C}$ extends to a group homomorphism $f : H \to \Gamma\mathrm{L}_n\mathbb{C}$, since S is a free generating set of H. We show that the image $G := f(H)$ has all the required properties.

Applying $f(s) = (A_s, s)$ to a basis of the rational vector space $\mathbb{Q}^n$ has the same effect as applying only A_s. Since we have chosen a surjection $s \mapsto A_s$, we obtain all bases of $\mathbb{C}^n$ by varying $s \in S$, hence (i) holds. (By the subsequent paragraph, the set $f(S)$ is a free generating set of G, so G is never sharply transitive on the bases).

Let $1 \neq h \in H$. Because the free group H contains no involution (see Exercise 9), h and (id, h) are discontinuous by 14.12. Furthermore $f(h) = (A, h)$ for some $A \in \mathrm{GL}_n\mathbb{C}$, and the equation $f(h) = (A, h) = (A, 1)(\mathrm{id}, h)$ shows that $f(h)$ is discontinuous, hence (ii) holds. The discontinuous map $f(h)$ cannot be the identity. Hence the group epimorphism $f : H \to G$ is injective, and G is isomorphic to the free group H, whence (iii). $\qquad\square$

In the proof above, the structure of the group $\mathrm{GL}_n\mathbb{C}$ did not matter at all; we have used only the fact that $\mathrm{GL}_n\mathbb{C}$ is a group of cardinality at most $\aleph$ consisting of homeomorphisms. This means that we obtain similar examples by substituting for $\mathrm{GL}_n\mathbb{C}$ any group of homeomorphisms of $\mathbb{C}^n$ with a reasonable transitivity property (note that the homeomorphism group of $\mathbb{C}^n$ has cardinality $\aleph$, since $\mathbb{C}^n$ has the countable dense subset $\mathbb{Q}(i)^n$).

In particular, we can consider subgroups of the full affine group of all collineations $x \mapsto \gamma(x) + a$ with $\gamma \in \Gamma\mathrm{L}_n\mathbb{C}$, $a \in \mathbb{C}^n$; by choosing the subgroup $\{ x \mapsto x + a \mid a \in \mathbb{Z}^n \} \cong \mathbb{Z}^n$, we obtain a free group of discontinuous collineations which induces on $\mathbb{Q}^n$ precisely the translation group $\mathbb{Z}^n$.

GLATTHAAR 1971 considered the complex projective plane and the group $\mathrm{GL}_3\mathbb{C}$, which is transitive on ordered quadrangles. We remark that TITS 1974 11.14 uses a similar construction of free groups in the context of algebraic groups over fields that are purely transcendental over a subfield.

Exercises

(1) Show that each non-cyclic discrete subgroup of $\mathbb{C}^+$ is of the form $\mathbb{Z}a + \mathbb{Z}b$, where $a, b \in \mathbb{C}$ are $\mathbb{R}$-linearly independent.

(2) The circle $\mathbb{S}_1$ is the largest compact subgroup of $\mathbb{C}^\times$.

(3) Show that each continuous group homomorphism $\varphi : \mathbb{C}^\times \to \mathbb{C}^+$ is of the form $\varphi(z) = a \log |z|$ with a fixed complex number $a \in \mathbb{C}$.

(4) Each discrete normal subgroup N of a connected topological group G is contained in the centre of G.

(5) Neither $\mathbb{R}$ nor $\mathbb{C}$ is purely transcendental over any proper subfield.

(6) Show that $a \in \mathbb{C}$ is contained in some maximal subfield of $\mathbb{C}$ if, and only if, the field $\mathbb{Q}(a)$ if formally real.

(7) Show that $\{\, (a+b, a+\zeta b, a+\zeta^2 b) \mid a \in \mathbb{C}, b \in \mathbb{C}^\times \,\}$ is the set of all equilateral triangles in the Euclidean plane $\mathbb{R}^2 = \mathbb{C}$, where $\zeta \in \mathbb{C}$ is a root of unity of order 3. Use the existence of discontinuous field automorphisms of $\mathbb{C}$ to prove that the following geometric notions cannot be expressed just by equilateral triangles: collinearity of points, parallelity, orthogonality and congruence of segments determined by two points (see also BETH–TARSKI 1956).

(8) Let $F \subseteq E_i \subseteq F^\natural$ be fields such that $E_i | F$ is a finite solvable Galois extension (as defined in 14.7), where $i = 1, 2$. Show that the composite $E_1 E_2 := F(E_1 \cup E_2)$ is a finite solvable Galois extension of F.

(9) Show that a free group contains no element of order 2.

2

Non-standard numbers

This chapter has a threefold purpose: to introduce ultraproducts and construct the real numbers via an ultrapower of $\mathbb{Q}$, to study the structure of an ultrapower $^*\mathbb{R}$ of the reals in a similar way as this has been done for $\mathbb{R}$ in Chapter I, and to illustrate how the non-standard extension $^*\mathbb{R}$ may be used to prove results on $\mathbb{R}$ in an easy way.

We will exclude, however, the many deeper results on non-standard numbers which require a discussion of a hierarchy of formal languages necessary for a development beyond the simplest notions of non-standard analysis; see, e.g., HURD–LOEB 1985, LINDSTRØM 1988, KEISLER 1994, or GOLDBLATT 1998. We need the axiom of choice.

21   Ultraproducts

Ultraproducts can be defined for any infinite index set. Later on, only natural numbers will be used as indices. The notion of an *ultrafilter* is essential for the construction.

21.1 Filters For any set N write $2^N = \{\, S \mid S \subseteq N \,\}$. A subset $\Phi \subseteq 2^N$ is called a *filter on* (or *over*) N if the following three conditions hold:

$$\emptyset \notin \Phi, \quad A, B \in \Phi \Rightarrow A \cap B \in \Phi, \quad (A \in \Phi \wedge A \subseteq B \subseteq N) \Rightarrow B \in \Phi \,.$$

The union of any chain of filters is again a filter. Therefore, by Zorn's Lemma, each filter on N is contained in a maximal one.

21.2 Ultrafilters A filter Ψ over N is called an *ultrafilter*, if Ψ has one of the following equivalent properties:
(a) Ψ is a maximal filter on N,
(b) for each $A \subseteq N$ either A or its complement A^{c} belongs to Ψ,
(c) whenever $A \cup B \in \Psi$, then $A \in \Psi$ or $B \in \Psi$.

Proof Obviously, (b) $\Rightarrow$ (a), and (b) is a special case of (c) because $N \in \Psi$. If $A \cup B \in \Psi$, then $(A \cup B)^{\complement} = A^{\complement} \cap B^{\complement} \notin \Psi$, and (b) $\Rightarrow$ (c).

Assume that (a) holds; we prove (b). If $A \cap X = \emptyset$ for some $X \in \Psi$, then $X \subseteq A^{\complement} \in \Psi$; if $A \cap X \neq \emptyset$ for each $X \in \Psi$, then

$$\langle \Psi, A \rangle = \{ Y \subseteq N \mid \exists_{X \in \Psi} \, A \cap X \subseteq Y \}$$

is the filter generated by Ψ and A, and $A \in \Psi$ by maximality.   □

Equivalently, an ultrafilter Ψ over N may be considered as a finitely additive probability measure $\mu : 2^N \to \{0, 1\}$, the *null* sets being exactly the sets $A \notin \Psi$. An extensive treatment of ultrafilters can be found in COMFORT–NEGREPONTIS 1974.

21.3 Free ultrafilters Obviously, each one-element set $\{a\}$ generates an ultrafilter of the form $\langle a \rangle = \{ X \subseteq N \mid a \in X \}$. These so-called *principal* ultrafilters are uninteresting.

The non-principal or *free* ultrafilters Ψ can be characterized by each of the following equivalent properties:

(a) $\bigcap \Psi = \emptyset$ (if $\bigcap \Psi = A \neq \emptyset$, then $\Psi = \bigcap_{a \in A} \langle a \rangle$ is either principal or not maximal),

(b) Ψ does not contain any finite set (if there is a finite set in Ψ, then $\bigcap \Psi \neq \emptyset$),

(c) Ψ contains the *cofinite* filter on N (consisting of all sets with finite complement).

21.4 Ultraproducts Let $P = \mathsf{X}_{\nu \in N} \, S_\nu = \{ (x_\nu)_\nu \mid \forall_{\nu \in N} \, x_\nu \in S_\nu \}$ be the product of a family of non-empty sets S_ν. If Ψ is any ultrafilter on N, two sequences are called equivalent modulo Ψ if they agree on some set of the filter, formally

$$(x_\nu)_\nu \Psi (y_\nu)_\nu \;\; \rightleftharpoons \;\; \{ \nu \in N \mid x_\nu = y_\nu \} \in \Psi \, .$$

The equivalence class of $x = (x_\nu)_\nu$ will be denoted by x / Ψ, sometimes simply by $\overline{x}$. The quotient P / Ψ is called an ultraproduct of the S_ν; generally, its structure depends on the choice of Ψ.

If $S_\nu = S$ for $\nu \in N$, we denote the ultraproduct by S^Ψ and speak of an *ultrapower*. There is a canonical injection $s \mapsto (s)_\nu / \Psi$ of S into S^Ψ.

Remarks   (1) It is often convenient to express a statement of the form $\{ \nu \in N \mid A(\nu) \} \in \Psi$ by a phrase like 'condition $A(\nu)$ holds *for almost all* $\nu \in N$'. Since Ψ contains all cofinite sets, this use of the expression 'for almost all' extends the traditional one. Property 21.2(b) shows that $\overline{x} \neq \overline{y}$ if, and only if, $x_\nu \neq y_\nu$ for almost all ν.

(2) If $\Psi = \langle n \rangle$ is a principal ultrafilter, then the projection $\pi_n : P \to S_n$ induces an isomorphism $P/\Psi \cong S_n$. For this reason, it will be tacitly assumed in all constructions that Ψ is a free ultrafilter.

(3) If S is a finite set, then condition 21.2(c), inductively applied to a union of card S sets, implies that the canonical injection $S \to S^\Psi$ is surjective.

21.5 Chains An ultraproduct of chains (i.e. linearly ordered sets) is again a chain. In fact, if $\overline{x} \neq \overline{y}$, then either $x_\nu < y_\nu$ or $x_\nu > y_\nu$ for almost all ν. (Use 21.2.)

21.6 Fields Assume that in the definition 21.4 each S_ν is a field. By the properties of an (ultra-)filter, the operations $\overline{x} + \overline{y} = \overline{x+y}$ and $\overline{x} \cdot \overline{y} = \overline{xy}$ are well-defined. Obviously, $(P/\Psi, +, \cdot)$ is a ring. If $\overline{x} \neq \overline{0}$, we may assume that $x_\nu \neq 0$ for *each* ν. Then $(x_\nu^{-1})_\nu/\Psi$ is a multiplicative inverse of $\overline{x}$, and P/Ψ is even a field. Analogous assertions hold for other algebraic structures, because all first-order statements (i.e., statements which do not involve quantifiers for set variables, compare CHANG–KEISLER 1990 1.3) carry over to ultraproducts.

21.7 Theorem: Algebraically closed fields *An ultraproduct of algebraically closed fields is algebraically closed.*

Proof Let $P = \times_{\nu \in N} S_\nu$ be a product of algebraically closed fields. For $k > 0$ and $\kappa = 0, 1, \ldots, k$ consider coefficients $\overline{c}_\kappa = (c_{\kappa\nu})_\nu/\Psi \in P/\Psi$. We may assume that $c_{k\nu} \neq 0$ for each $\nu \in N$. Since S_ν is algebraically closed, there is some $x_\nu \in S_\nu$ such that $\sum_{\kappa=0}^k c_{\kappa\nu} x_\nu^\kappa = 0$. Then $\overline{x} = (x_\nu)_\nu/\Psi$ satisfies $\sum_{\kappa=0}^k \overline{c}_\kappa \overline{x}^\kappa = 0$. $\qquad\square$

An analogous result holds for real closed fields, since these fields can be characterized by a first-order property; see 12.10.

21.8 Example An ultraproduct $F = \times_{p \in \mathbb{P}} \mathbb{F}_p/\Psi$ of all Galois fields of prime order has the following properties:

(a) char $F = 0$,

(b) each element in F is a sum of two squares,

(c) the element $-\overline{1}$ is a square in F if, and only if, the set
   $\{p \in \mathbb{P} \mid p \equiv 1 \bmod 4\}$ belongs to Ψ,

(d) F has cardinality $\aleph = 2^{\aleph_0}$,

(e) the algebraic closure $F^\natural$ of F is isomorphic to $\mathbb{C}$,

(f) $F^\natural$ is a proper subfield of $\times_{p \in \mathbb{P}} \mathbb{F}_p^\natural/\Psi$.

Proof (a) One has $n \cdot 1 \neq 0$ in $\mathbb{F}_p$ for almost all $p \in \mathbb{P}$, hence $n \cdot \overline{1} \neq \overline{0}$ for all $n \in \mathbb{N}$.

(b) Put $Q = \{\, x^2 \mid x \in \mathbb{F}_p \,\}$. If p is odd, then Q has $(p+1)/2$ elements, and so has $c - Q$ for any $c \in \mathbb{F}_p$. Therefore, Q and $c - Q$ intersect, and hence there are $a, b \in \mathbb{F}_p$ with $a^2 + b^2 = c$ (see also 34.14). By the construction of F, each $\bar{c} \in F$ is a sum of two squares.

(c) Recall that $\mathbb{F}_p^\times$ is a cyclic group of order $p - 1$ (Section 64, Exercise 1). Hence -1 is a square in $\mathbb{F}_p$ if, and only if, $p - 1 \in 4\mathbb{N}$. The claim is now an immediate consequence of the definition of an ultraproduct.

(d) is a special case of the following Theorem 21.9.

(e) is true for any field satisfying (a) and (d); see 64.21 and 64.20.

(f) Choose $c_p \in \mathbb{F}_p^\natural$ such that c_p has degree s_p over $\mathbb{F}_p$ and the s_p are unbounded. Then $\bar{c} = (c_p)_p / \Psi$ is not algebraic over F: if $\sum_{\mu=0}^m \bar{a}_\mu \bar{c}^\mu = 0$ for $\bar{a}_\mu \in F$, then $\sum_{\mu=0}^m a_{\mu p} c_p^\mu = 0$ and $s_p \le m$ for almost all $p \in \mathbb{P}$.   □

21.9 Theorem: Cardinality If Ψ *is a free ultrafilter over a countable set, say over* $\mathbb{N}$, *and if the cardinalities of the sets* S_ν *satisfy the conditions* $\nu < \operatorname{card} S_\nu \le \aleph$, *where* $\aleph := 2^{\aleph_0}$, *then* $\operatorname{card}\left(\bigtimes_{\nu \in \mathbb{N}} S_\nu / \Psi \right) = \aleph$.

Proof For each of the $\aleph$ functions $f = (f_\nu)_\nu \in \{0,1\}^\mathbb{N}$ define a map $f^\circ : \mathbb{N} \to \mathbb{N}$ by $f^\circ(n) = \sum_{\nu < n} f_\nu 2^\nu < 2^n$. Choose $s(\mu, \nu) \in S_\nu$ for $\mu = 0, 1, \dots, \nu$ such that $s(\mu, \nu) \ne s(\mu', \nu)$ whenever $\mu \ne \mu'$. Put $\widehat{f}_\nu = s(f^\circ(h_\nu), \nu)$, where h_ν is the largest integer such that $2^{h_\nu} - 1 \le \nu$. (This condition implies that $f^\circ(h_\nu) \le \nu$, thus it guarantees that $\widehat{f}_\nu$ is defined.)

If $f \ne g$, then $f_\kappa \ne g_\kappa$ for some κ and $f^\circ(n) \ne g^\circ(n)$ for all $n > \kappa$. Now $\nu \ge 2^{\kappa+1}$ implies $h_\nu > \kappa$, $f^\circ(h_\nu) \ne g^\circ(h_\nu)$ and $\widehat{f}_\nu \ne \widehat{g}_\nu$. Hence $\widehat{f}$ and $\widehat{g}$ represent different elements of the ultraproduct P/Ψ, and $\aleph \le \operatorname{card}(P/\Psi) \le \operatorname{card} P \le \prod_\nu \operatorname{card} S_\nu \le \aleph^{\aleph_0} = 2^{\aleph_0}$ by 61.13b.   □

21.10 Extension of functions Let an ultrafilter Ψ over N be given. Any map $h : S \to T$ has a canonical extension $h^\Psi : S^\Psi \to T^\Psi$ defined by $h^\Psi((s_\nu)_\nu / \Psi) = (h(s_\nu)_\nu)/\Psi$ or, in short, $h^\Psi(\bar{s}) = \overline{h(s)}$. The extension will often also be denoted by h without danger of confusion.

Exercises

(1) Discuss an ultraproduct $E = \bigtimes_{\nu \in \mathbb{N}} \mathbb{F}_{p^\nu} / \Psi$ in a similar way as the field F in example 21.8.

(2) Put $h(x) = x^3$. Show that h^Ψ is an automorphism of the chain $\mathbb{R}^\Psi$.

(3) Determine the image of $\sin^\Psi : \mathbb{R}^\Psi \to \mathbb{R}^\Psi$.

22   Non-standard rationals

In this section, the structure of the ultrapower $\mathbb{Q}^\Psi$ with respect to a given free ultrafilter Ψ over $\mathbb{N}$ will be studied in a similar (but less detailed) manner as the structure of $\mathbb{R}$ has been discussed in Chapter 1. The special choice of Ψ will not play any role (compare 24.28), and $\mathbb{Q}^\Psi$ will also be denoted by the customary name $^*\mathbb{Q}$. The properties of $^*\mathbb{Q}$ will be compared with those of $\mathbb{Q}$ or $\mathbb{R}$.

Relevant results on $\mathbb{Q}$ can be found in Chapter 3, most of them are quite familiar. By $\mathbb{P}$ we denote the set of all primes in $\mathbb{N}$, as usual. The inclusions $\mathbb{P} \hookrightarrow \mathbb{N} \hookrightarrow \mathbb{Z} \hookrightarrow \mathbb{Q}$ give rise to natural embeddings $^*\mathbb{P} \hookrightarrow {}^*\mathbb{N} \hookrightarrow {}^*\mathbb{Z} \hookrightarrow {}^*\mathbb{Q}$. Statements for which no proof is given follow directly from the definition 21.4.

22.1 Fact $^*\mathbb{Q}$ *is an ordered field of cardinality* $\aleph$, *and* $^*\mathbb{Q}$ *is the field of fractions of its subring* $^*\mathbb{Z}$. *Fractions can be added and multiplied in the same way as in* $\mathbb{Q}$.

The assertion that card $^*\mathbb{Q} = \aleph$ is a special case of Theorem 21.9. From Lagrange's theorem (34.18) we infer the following result.

22.2   *Each positive element in* $^*\mathbb{Z}$ *and hence also in* $^*\mathbb{Q}$ *is a sum of at most four squares.*

22.3 Corollary $^*\mathbb{Q}$ *has only one ordering which satisfies the monotonicity laws.*

Notation   For positive numbers, we write $a \ll b \rightleftharpoons \forall_{n \in \mathbb{N}} \, n \cdot a < b$. The absolute value $|c|$ of c is defined by the condition $0 \leq |c| \in \{c, -c\}$, as usual, and c is said to be *infinitely small* or *infinitesimal* if $|c| \ll 1$.

22.4 Additive group $^*\mathbb{Q}^+$ *is a vector space of dimension* $\aleph$ *over the subfield* $\mathbb{Q}$, *hence* $^*\mathbb{Q}^+ \cong \mathbb{R}^+$, *and* $^*\mathbb{Q}^+$ *is not locally cyclic.*

Proof The dimension is the cardinality of a basis $\mathfrak{b}$ of $^*\mathbb{Q}$ over $\mathbb{Q}$. Since $^*\mathbb{Q}$ is the union of all subspaces spanned by finitely many elements in the basis, $\aleph = \text{card} \, ^*\mathbb{Q} = \text{card} \, \mathfrak{b}$. In detail, any finite-dimensional vector space over $\mathbb{Q}$ is countable. Hence $\mathfrak{b}$ is infinite. By 61.14, the basis $\mathfrak{b}$ has exactly card $\mathfrak{b}$ finite subsets. Thus, $^*\mathbb{Q}$ is covered by card $\mathfrak{b}$ countable subspaces, and card $^*\mathbb{Q} = \text{card} \, \mathfrak{b}$. $\square$

22.5 Primes   Let $\bar{1} \neq \bar{p} = (p_\nu)_\nu/\Psi$. *Then* $\bar{p}$ *is a prime element in* $^*\mathbb{N}$ *if, and only if,* p_ν *is a prime in* $\mathbb{N}$ *for almost all* ν. *Hence* $^*\mathbb{P}$ *is the set of all prime elements in* $^*\mathbb{N}$. $\square$

22.6 Multiplication   *$^*\mathbb{P}$ generates a proper subgroup of the multiplicative group $^*\mathbb{Q}^\times_{\mathrm{pos}}$ of the positive elements in $^*\mathbb{Q}$.*

Proof Consider $\mathbb{P} = \{p_\kappa\}_\kappa$ in its natural ordering. Put $q_\nu = \prod_{\kappa \le \nu} p_\kappa$ and $\overline{q} = (q_\nu)_\nu/\Psi$. If $\overline{s} = (s_\nu)_\nu/\Psi$ is a product of m elements in $^*\mathbb{P}$, then, for almost all ν, the component s_ν has only m prime factors, and $\overline{s} \ne \overline{q}$.   □

22.7 Proposition   *$^*\mathbb{Q}^\times_{\mathrm{pos}}$ is a non-Archimedean ordered group.*

Proof If $0 < c \ll 1$, then $(1 + c)^n \le (1 + \frac{1}{n})^n < \sum_\nu 1/\nu! < 3$.   □

22.8 Ordering   *The group $^*\mathbb{Z}^+$ is a non-Archimedean ordered group of cardinality $\aleph$. The shift $\overline{z} \mapsto \overline{z} + 1$ is an automorphism of the chain $^*\mathbb{Z}$, it maps each element onto its immediate successor. There is an embedding φ of the first uncountable ordinal $\Omega = \omega_1$ into $^*\mathbb{N}$, but $^*\mathbb{N}$ is not well-ordered.*

Proof For an inductive definition of φ, let $\varphi(0) = \overline{0}$ and define $\varphi(\nu + 1)$ as the successor of $\varphi(\nu)$; if $\sigma \in \Omega$ is not of the form $\nu + 1$, the countable set $\{\,\varphi(\nu) \mid \nu < \sigma\,\}$ has an upper bound $\overline{s} \in {}^*\mathbb{N}$ (see 24.11), and $\varphi(\sigma)$ may be chosen as $\overline{s} + 1$. The shift $\overline{z} \mapsto \overline{z} + 1$ maps $\mathbb{N}$ into itself and induces an automorphism of the chain $M = {}^*\mathbb{N} \smallsetminus \mathbb{N}$. Hence M has no smallest element.   □

Exercises

(1) Let $\mathbb{Q}^\natural$ and $\mathbb{Q}^\Psi$ be embedded into $(\mathbb{Q}^\natural)^\Psi$ in a natural way. Show that $\mathbb{Q}^\natural \cap \mathbb{Q}^\Psi = \mathbb{Q}$.

(2) The continued fraction for $\sqrt{2}$ gives a sequence of rational numbers s_ν converging to $\sqrt{2}$ such that $s_{2\nu} < \sqrt{2} < s_{2\nu+1}$. Let $\overline{s} = (s_\nu)_\nu/\Psi \in \mathbb{Q}^\Psi$. Show that $\overline{s}^2 < 2$ or $\overline{s}^2 > 2$ depending on the choice of Ψ.

(3) The number of primes $p \le n$ is denoted by $\pi(n)$. The prime number theorem says that $n^{-1}\pi(n)\log n$ converges to 1 for $n \to \infty$; see, e.g., HARDY–WRIGHT 1971 Theorem 6 p. 9 or ZAGIER 1997. Write p_ν for the νth prime in $\mathbb{N}$ and use the prime number theorem in order to show that $(\nu)_\nu/\Psi \ll (p_\nu)_\nu/\Psi \ll (\nu^2)_\nu/\Psi$.

23   A construction of the real numbers

Axiomatically, the real numbers are characterized as a completely ordered field (as defined in 42.7). There are several ways in which $\mathbb{R}$ can be obtained by completion of $\mathbb{Q}$. Most familiar are the following two.

(I) First the ordered set $\mathbb{Q}$ is completed, say by Dedekind cuts, and then the algebraic operations are extended. The crucial step is to show that the resulting ring has no zero divisors.

(II) All sequences converging to 0 form a maximal ideal M in the ring C of all rational Cauchy sequences. The ordering of $\mathbb{Q}$ is extended to C/M, and one has to show that this quotient is order complete (see 23.7). Details and related methods can be found in Chapter 4.

The procedure of MEISTERS–MONK 1973 to be described in this section is particularly transparent. Compared to (I) and (II), it has the advantage that the algebraic operations and the ordering relation are extended simultaneously. If R is the ring of all finite elements (see 23.1) in $^*\mathbb{Q}$ and N is the (maximal) ideal of all infinitesimals (Section 22), then R/N turns out to be a completely ordered field, hence $\mathbb{R} \cong R/N$. The proof is independent of Section 22, it uses only 21.4 and 21.5.

A proper subring R of a field F is said to be a *valuation ring*, if $a \in F \smallsetminus R$ implies $a^{-1} \in R$; compare 56.4.

23.1   The set $R = \{\, a \in {}^*\mathbb{Q} \mid \exists_{n \in \mathbb{N}} \, |a| < n \,\}$, consisting of all 'finite' elements, is a valuation ring.

Proof Sum and product of finite elements are obviously finite again, so that R is a ring. If $a \notin R$, then $1 \ll |a|$ and a^{-1} is infinitesimal, hence $a^{-1} \in R$. □

23.2   The infinitesimals form a maximal ideal N of R, and $R \smallsetminus N = R^{\times}$ is the group of units in R.

Proof By definition, N is an ideal in R. If $a^{-1} \notin R$, then $|a|^{-1} \gg 1$, and $a \in N$. Therefore, each element in $R \smallsetminus N$ has an inverse in R, and any proper ideal in R is contained in N. □

23.3   With the ordering induced by $^*\mathbb{Q}$, the ring R is an ordered ring, and the ideal N is convex (that is, if $a, b \in N$ and $a < c < b$, then $c \in N$). □

23.4 Corollary   The residue field $F := R/N$ is an ordered field.

Proof The canonical projection $R \to R/N$ preserves the ordering: in fact, $0 < c \notin N$ implies $N < c + N$ by the convexity of N. □

23.5 Lemma   For a given $k \in \mathbb{N}$, any element $b \in R$ has a representation $b = (b_\nu)_\nu / \Psi$ such that $|b_\nu - b_\mu| < k^{-1}$ for all $\mu, \nu \in \mathbb{N}$.

Proof Since b is finite, there is some $m \in \mathbb{N}$ such that $|b| < m$ and hence $|b_\nu| < m$ for almost all ν. Divide the interval $[-m, m]$ into finitely many subintervals of length less than k^{-1}. By property 21.2(c) of ultrafilters, there is one among these subintervals, say J, such that $K = \{\nu \in \mathbb{N} \mid b_\nu \in J\} \in \Psi$. According to the definition of the equivalence relation Ψ, the b_μ with $\mu \notin K$ may be chosen arbitrarily in the interval J without changing the element b. □

23.6 Proposition *The canonical inclusion* $\kappa : \mathbb{Q} \hookrightarrow {}^*\mathbb{Q}$ *(by constant sequences) induces an embedding* $\mathbb{Q} \hookrightarrow F$ *with strongly dense image.*

Proof Since $\kappa : \mathbb{Q} \hookrightarrow R$ and $\mathbb{Q}^\kappa \cap N = \{0\}$, the map κ induces an embedding $\mathbb{Q} \hookrightarrow F$. Consider elements $x, y \in R$ with $x < y$ and $y - x \notin N$. Then there exists $k \in \mathbb{N}$ such that $y - x > 3k^{-1}$. By 23.5, we may assume that $x = (x_\nu)_\nu/\Psi$ with $|x_\nu - x_\mu| < k^{-1}$ for all $\mu, \nu \in \mathbb{N}$ and that y satisfies an analogous condition. Choose μ such that $y_\mu - x_\mu > 3k^{-1}$ and put $c = (x_\mu + y_\mu)/2 \in \mathbb{Q}$. Then $x + N < c < y + N$. □

23.7 Theorem *As an ordered field, F is complete.*

Proof Let $\emptyset \neq M \subset F$, and assume that M has an upper bound in F. Then $T = \{t \in \mathbb{Q} \mid M < t\}$ and $S = \{s \in \mathbb{Q} \mid s < T\}$ are not empty, and (S, T) is a Dedekind cut: in fact, if $z \in \mathbb{Q}$ and $z \notin S$, then $t \leq z$ for some $t \in T$, hence $M < z$ and $z \in T$.

There is a least integer $t_0 \in T$, and $s_0 = t_0 - 1 \in S$. Define inductively $s_\nu \in S$ and $t_\nu \in T$ as follows: put $m_\nu = (s_\nu + t_\nu)/2$ and let

$$s_{\nu+1} = \begin{cases} m_\nu & \text{if } m_\nu \in S \\ s_\nu & \text{if } m_\nu \in T \end{cases}, \quad t_{\nu+1} = \begin{cases} t_\nu & \text{if } m_\nu \in S \\ m_\nu & \text{if } m_\nu \in T \end{cases}.$$

Then $t_\nu - s_\nu = 2^{-\nu}$. Consider in R the elements $\overline{s} = (s_\nu)_\nu/\Psi$ and $\overline{t} = (t_\nu)_\nu/\Psi$, and note that by construction $\overline{s} < T$ and $S < \overline{t}$. From $t_\nu - s_\nu = 2^{-\nu}$ it follows that $\overline{t} - \overline{s} \in N$. Thus, $\overline{s}$ and $\overline{t}$ represent the same element $r \in F$, and $S \leq r \leq T$. The fact that $\mathbb{Q}$ is strongly dense in F implies that r is a least upper bound of M in F: if $r < x$ for some $x \in M$, there exists $t \in \mathbb{Q}$ with $r < t < x$, but then $t \in T$ and $x < t$. This contradiction shows $M \leq r$. If $M \leq u < r$, then $M < t < r$ for some $t \in \mathbb{Q}$. This implies $t \in T$ and $r \leq t$, again a contradiction. □

Remark Suppose that $\mathbb{R}$ is given. By mapping each rational sequence $c = (c_\nu)_\nu$ which converges to $r \in \mathbb{R}$ to the element $\overline{c} + N$, the identity mapping of $\mathbb{Q}$ extends to an isomorphism $\mathbb{R} \cong F$.

There is an analogous construction starting with the order complete real field $\mathbb{R}$ instead of $\mathbb{Q}$. The finite elements in the ultrapower $^*\mathbb{R} = \mathbb{R}^\Psi$ also form a valuation ring

$$\mathfrak{R} = \left\{ x \in {}^*\mathbb{R} \mid \exists_{n \in \mathbb{N}} \, |x| < n \right\} ,$$

and its maximal ideal $\mathfrak{n} = \mathfrak{R} \smallsetminus \mathfrak{R}^\times$ consists of the *infinitesimals*: $\mathfrak{n} = \{ x \in \mathfrak{R} \mid |x| \ll 1 \}$.

23.8 Proposition *The canonical injection* $\kappa : \mathbb{R} \hookrightarrow {}^*\mathbb{R}$ *(via constant sequences) induces an isomorphism* $\mathbb{R} \cong \mathfrak{R}/\mathfrak{n}$, *and* $\mathfrak{R} = \mathbb{R} + \mathfrak{n}$.

Proof By definition, $\mathbb{R} = \mathbb{R}^\kappa$ is cofinal in $\mathfrak{R}$ and $\mathbb{R} \cap \mathfrak{n} = \{0\}$. Therefore, κ induces an inclusion of $\mathbb{R}$ into the ordered field $\mathfrak{R}/\mathfrak{n}$. Assume that there is an element $a \in \mathfrak{R}$ such that $a + \mathfrak{n} \notin \mathbb{R}$. Since $\mathbb{R}$ is complete and cofinal in $\mathfrak{R}$, the set $\{ t \in \mathbb{R} \mid t < a \}$ has a least upper bound $s \in \mathbb{R}$. This implies $s - \varepsilon < a < s + \varepsilon$ for all positive ε in $\mathbb{R}$, hence $a - s \in \mathfrak{n}$ in contradiction to the choice of a. $\square$

23.9 Terminology By 23.8, each finite element b belongs to a unique residue class $r + \mathfrak{n}$ or *monad* with $r = {}^\circ b \in \mathbb{R}$; the element ${}^\circ b$ is known as the *standard part* of b. The equivalence relation $x \approx y \rightleftharpoons {}^\circ x = {}^\circ y \Leftrightarrow y - x \in \mathfrak{n}$ will play an important role.

Exercises

(1) Show directly that R/N is an Archimedean ordered field.

(2) A *Euclidean* field is an ordered field in which each positive element is a square, and the Euclidean closure E of $\mathbb{Q}$ is the smallest Euclidean subfield of $\mathbb{R}$; see 12.8. Compare $^*E$ and the Euclidean closure K of $^*\mathbb{Q}$ in $^*\mathbb{R}$.

(3) A *Pythagorean* field is a field in which the sum of two squares is again a square. By Exercise 5 of Section 12, the Pythagorean closure P of $\mathbb{Q}$ is properly contained in E. What can be said about $^*P$?

24   Non-standard reals

In several respects, the field $\mathbb{R}$ is simpler than $\mathbb{Q}$. A similar observation can be made with regard to ultrapowers of $\mathbb{R}$ and $\mathbb{Q}$. Usually, an ultrapower of $\mathbb{R}$ is denoted by $^*\mathbb{R}$, the specific ultrafilter used in the construction is not explicitly referred to. This is partially justified by Theorem 24.26 below. The properties of the algebraical and topological structures underlying $^*\mathbb{R}$ will be discussed according to the scheme of

Chapter 1. Algebraically, $\mathbb{R}$ and $^*\mathbb{R}$ share many features, the topology of $^*\mathbb{R}$, however, is not so nice.

24.1 Fact $^*\mathbb{R}$ *is an ordered field of cardinality* $\mathrm{card}\,^*\mathbb{R} = \mathrm{card}\,\mathbb{R} = \aleph$.

This follows immediately from 21.5 and $\mathbb{R} \hookrightarrow {}^*\mathbb{R} = \mathbb{R}^{\mathbb{N}}/\Psi$ without using Theorem 21.9.   □

24.2 Addition *The additive group of* $^*\mathbb{R}$ *is a rational vector space of dimension* $\aleph$, *hence it is isomorphic to* $\mathbb{R}^+$.

Remark By 22.4 we have $^*\mathbb{Q}^+ \cong {}^*\mathbb{R}^+$, but $^*\mathbb{Q}$ is a *proper* subfield of $^*\mathbb{R}$ because $\mathbb{R} < {}^*\mathbb{R}$ and 2 is not a square in $^*\mathbb{Q}$.

24.3 Exponentiation By 21.10, there is a function $\exp : {}^*\mathbb{R} \to {}^*\mathbb{R}$, and $\exp$ is an isomorphism of $^*\mathbb{R}^+$ onto $^*\mathbb{R}^{\times}_{\mathrm{pos}}$ (which preserves the ordering). Its inverse is the extended logarithm. The assertions are immediate consequences of the definitions. Together with 24.2 we get

24.4 Corollary *The multiplicative group* $^*\mathbb{R}^{\times}$ *is isomorphic to* $\mathbb{R}^{\times}$.

In spite of 24.2 and 24.4, however, the fields $^*\mathbb{R}$ and $\mathbb{R}$ are not isomorphic (since $^*\mathbb{R}$ has a non-Archimedean ordering).

24.5 Ordering In the natural (non-Archimedean) ordering of $^*\mathbb{R}$ (the one inherited from $\mathbb{R}$), the positive elements are exactly the squares, i.e., $^*\mathbb{R}$ is a *Euclidean* field. Hence this is the only ordering relation on $^*\mathbb{R}$ which satisfies the monotonicity laws.

Proof According to 21.10, the absolute value on $\mathbb{R}$ extends to $^*\mathbb{R}$ as follows: if $x = (x_\nu)_\nu/\Psi$, then $|x| = (|x_\nu|)_\nu/\Psi$. Obviously, the latter term is a square. The uniqueness of the ordering is also a consequence of the next theorem.   □

Remember from 12.15 that a field F of characteristic 0 is real closed if a quadratic extension of F is algebraically closed.

24.6 Theorem $^*\mathbb{R}$ *is a real closed field, and* $^*\mathbb{C}$ *is isomorphic to* $\mathbb{C}$.

Proof The field $^*\mathbb{C}$ is algebraically closed by 21.7. Its elements can be written in the form $(a_\nu + ib_\nu)_\nu/\Psi = \bar{a} + i\bar{b}$. Hence $^*\mathbb{C} = {}^*\mathbb{R} + {}^*\mathbb{R}i$. The first result follows also from the fact that being a real closed field is a first-order property; compare Section 21. By 64.20 and 64.21, any algebraically closed field of characteristic 0 and cardinality $\aleph$ is isomorphic to $\mathbb{C}$.   □

Ordering and topology

Remember notation and results discussed at the end of Section 23. In particular, $\mathfrak{R} = \mathbb{R} + \mathfrak{n}$ is a valuation ring in $^*\mathbb{R}$. Associated with $\mathfrak{R}$ is the valuation $v : {}^*\mathbb{R}^\times \to \Gamma^+ \cong {}^*\mathbb{R}^\times / \mathfrak{R}^\times$. The relation $v(x) \leq v(y) \rightleftharpoons y\mathfrak{R} \subseteq x\mathfrak{R}$ turns the value group Γ into an ordered group; compare 56.5. For convenience, put $v(0) = \infty > \Gamma$.

The sets $V_\gamma = \{ x \in {}^*\mathbb{R} \mid v(x) > \gamma \}$ with $\gamma \in \Gamma$ can be taken as neighbourhoods of 0, as in 13.2. On the other hand, the open intervals form a basis of a natural topology τ on $^*\mathbb{R}$. It is easy to show that the topology defined by the V_γ coincides with the order topology τ (see Exercise 7 below).

In the sequel, some properties of the topological space $\mathfrak{T} := ({}^*\mathbb{R}, \tau)$ will be described. Like any ordered space, $\mathfrak{T}$ is normal; compare BIRKHOFF 1948 Chapter III Theorem 6, BOURBAKI 1966 Chapter IX §4 Exercise 5 or LUTZER 1980. Whereas each homeomorphism of $\mathbb{R}$ preserves or reverses the ordering, the topology of $\mathfrak{T}$ does in no way determine the ordering of $^*\mathbb{R}$; see 24.14. The ordering of $^*\mathbb{R}$ is discussed in DI NASSO–FORTI 2002.

24.7 Proposition: Homogeneity *The space $\mathfrak{T}$ is doubly homogeneous, i.e., the homeomorphism group is doubly transitive on $\mathfrak{T}$; see also 33.13.*

Proof Since $^*\mathbb{R}$ is a topological field by 13.2(a), each non-constant map $x \mapsto ax + b$ is a homeomorphism.   □

24.8 *The space $\mathfrak{T}$ is totally disconnected: the only non-empty connected sets are the points.*

Proof If some connected set would contain more than one point, then any two points would lie in a connected set by 24.7, and $\mathfrak{T}$ would be connected (DUGUNDJI 1966 Theorem V.1.5), but $\mathfrak{R}$ and each of its cosets is open, hence also closed.   □

24.9 *The (canonically embedded) subspace $\mathbb{R}$ is discrete, and $\mathbb{R}$ is closed in $\mathfrak{T}$.*

Proof The infinitesimal ideal $\mathfrak{n}$ is open and $(r + \mathfrak{n}) \cap \mathbb{R} = \{r\}$ for each $r \in \mathbb{R}$. Each monad without its standard point is open and so is the complement of $\mathfrak{R}$. The union of these sets is the complement of $\mathbb{R}$.   □

24.10 $\mathfrak{T}$ *is a union of uncountably many pairwise disjoint open sets. Hence $\mathfrak{T}$ is not separable and not a Lindelöf space.*

Proof In fact, $\mathfrak{T}$ is the union of all cosets of the open set $\mathfrak{n}$.   □

24.11 Proposition *Any countable subset of $^*\mathbb{R}$ has an upper bound in $(^*\mathbb{N}, <)$.*

Proof Consider a sequence $\overline{x}_\kappa = (x_{\kappa\nu})_\nu/\Psi \in {}^*\mathbb{R}$, choose $z_\nu \in \mathbb{N}$ such that $x_{\kappa\nu} \leq z_\nu$ for $\kappa \leq \nu$, and put $\overline{z} = (z_\nu)_\nu/\Psi$. Then $\overline{x}_\kappa \leq \overline{z}$ for each κ: in fact, $z_\nu \geq x_{\kappa\nu}$ for all $\nu \geq \kappa$.   □

24.12 Corollary *The space $\mathfrak{T}$ is not first countable, i.e., there are no countable neighbourhood bases. In particular, $\mathfrak{T}$ is not metrizable. Each convergent sequence in $\mathfrak{T}$ is finally constant.*

Proof Suppose that the open intervals $]c - a_\kappa, c + a_\kappa[$ with $\kappa \in \mathbb{N}$ form a neighbourhood base at $c \in \mathfrak{T}$. If $b = d^{-1} > a_\kappa^{-1}$ for all $\kappa \in \mathbb{N}$, then the interval $]a - d, a + d[$ is properly contained in all the given neighbourhoods.

If a sequence of elements $c_n \in \mathfrak{T} \smallsetminus \{0\}$ converges to 0, then the $|c_n|^{-1}$ are bounded by some $s \in \mathfrak{T}$ and $|c_n| \geq s^{-1} > 0$, a contradiction.   □

24.13 Proposition *$\mathfrak{T}$ is not locally compact, not even locally pseudo-compact.*

Proof A space X is called pseudo-compact, if each continuous function $f : X \to \mathbb{R}$ has a bounded and then even a compact image. If X is locally pseudo-compact, then each point has a pseudo-compact *closed* neighbourhood (since $f(\overline{A}) \subseteq \overline{f(A)}$ by continuity). Because each continuous function f on a closed *interval* in $\mathfrak{T}$ has a continuous extension $\widehat{f} : \mathfrak{T} \to \mathbb{R}$ with the same range as f, local pseudo-compactness of $\mathfrak{T}$ would imply the existence of a pseudo-compact interval $[a, b]$ and then each closed interval would be pseudo-compact by 24.7. Consider now an infinite element c such that $\mathbb{R} \subset [-c, c]$. Since $\mathfrak{T}$ is normal, each continuous function on the closed subspace $\mathbb{R}$ can be extended to $[-c, c]$, but continuous functions on $\mathbb{R}$ need not be bounded.   □

24.14 Homeomorphisms *Each permutation of the subset $\mathbb{R}$ of $\mathfrak{T}$ is induced by a homeomorphism of $\mathfrak{T}$.*

Proof The open and closed monads can be permuted arbitrarily by homeomorphisms of $\mathfrak{R}$, and $(r + \mathfrak{n}) \cap \mathbb{R} = \{r\}$.   □

24.15 Proposition *The value group Γ is a non-Archimedean group.*

The *proof* follows easily from 24.11: let $\mathfrak{R} < x^{-1}$ and choose z such that $x^{-n} < z^{-1}$ for all $n \in \mathbb{N}$. Then $0 < x^{-n}z < 1$, $z \in x^n\mathfrak{R}$, and $n \cdot v(x) = v(x^n) \le v(z)$. □

Remark In contrast with 24.5, the topology of an infinite field is never uniquely determined: a field of cardinality c admits 2^{2^c} different field topologies (KILTINEN 1973). Only additional assumptions restrict the number of topologies as well as the structure of the field. The most prominent among such assumptions is local compactness. In fact, up to finite extensions, all locally compact topological fields can be described quite explicitly; see 58.7 or WEIL 1967 Chapter I. The following result shows that the field $^*\mathbb{R}$ has a different structure than any locally compact field.

24.16 *There is no locally compact field topology on $^*\mathbb{R}$.*

Proof By 58.8, every formally real, locally compact field is isomorphic to $\mathbb{R}$. □

The *Sorgenfrey topology* on $\mathbb{R}$ has all half-open intervals $]a, b]$ as a basis; it is a source of interesting examples; see 5.73. The Sorgenfrey topology σ on $^*\mathbb{R}$ is defined analogously. It is strictly finer (stronger) than τ, since $]a, b[= \bigcup_{x \in]a,b[}]a, x] \in \sigma$ but $]a, b] \notin \tau$. The following statements, for which no proofs are given, can easily be verified.

24.17 *Addition in $(^*\mathbb{R}, +, \sigma)$ is continuous, but not the map $x \mapsto -x$.*

24.18 Homogeneity *The homeomorphism group of $(^*\mathbb{R}, \sigma)$ is transitive on ordered pairs.*

24.19 *The monads are open and closed in $(^*\mathbb{R}, \sigma)$. Hence assertions 24.8–12 are true for $(^*\mathbb{R}, \sigma)$ as well as for $\mathfrak{T}$.*

η_1-*fields*

According to HAUSDORFF 1914 p. 180/1, a chain $K_<$ is called an η_1-*set*, if for any two subsets A, B of K such that $A < B$ and $A \cup B$ is at most countable, there is an element $c \in K$ with $A < c < B$. (Here, A or B may be empty.) Chains with this property, and η_1-fields in particular, have been studied a good deal; compare PRIESS-CRAMPE 1983 Chapter IV and DALES–WOODIN 1996. Noteworthy are the uniqueness results on η_1-structures of cardinality $\aleph_1$. The following result provides examples.

24.20 Theorem *If Ψ is an ultrafilter over $\mathbb{N}$ and if F is any subfield of $\mathbb{R}$, then $(F^{\Psi}, <)$ is an η_1-field.*

This can be proved by a slight modification of the argument in 24.11: write $A = \{\bar{a}_\kappa\}_\kappa$ with $\bar{a}_\kappa = (a_{\kappa\nu})_\nu / \Psi$ and analogously $B = \{\bar{b}_\lambda\}_\lambda$. The assumption $A < B$ implies that $\forall_{\kappa,\lambda\leq\mu}\, a_{\kappa\nu} < b_{\lambda\nu}$ for almost all ν. An easy induction over μ shows that we may assume $a_{\kappa\nu} < b_{\lambda\nu}$ for all κ, λ and ν. Let $\max_{\kappa\leq\nu} a_{\kappa\nu} < c_\nu < \min_{\lambda\leq\nu} b_{\lambda\nu}$. Then $c = (c_\nu)_\nu / \Psi$ satisfies the assertion. $\square$

24.21 Cardinality *Any η_1-chain $K_<$ has cardinality at least $2^{\aleph_0}$.*

Proof Since K is dense in itself, $\mathbb{Q}$ can be embedded as a subchain into K; see the remark following 3.4. Each of the $\aleph$ irrational numbers $s \in \mathbb{R} \setminus \mathbb{Q}$ determines a Dedekind cut (A_s, B_s) in $\mathbb{Q}$. By assumption, there is an element $c_s \in K$ with $A_s < c_s < B_s$. If $s < t$, then $A_t \cap B_s$ contains some element d, and $c_s < d < c_t$. Hence $(s \mapsto c_s) : \mathbb{R} \setminus \mathbb{Q} \to K$ is an injective map. $\square$

24.22 *A strongly dense subchain D of an η_1-chain K is itself an η_1-chain.*

Proof If A and B are countable subsets of D with $A < B$, then repeated application of the definition shows that there is more than one element and hence an interval J in K between A and B. Since D is dense, J contains a point of D. $\square$

The following uniqueness theorem has already been proved by HAUS-DORFF 1914 p. 181; see also PRIESS-CRAMPE 1983 Chapter IV §2.

24.23 *Any two η_1-chains K and L of cardinality $\aleph_1$ are isomorphic as chains.*

This result will not be used in the sequel. It can be proved by transfinite induction, mapping alternately an element of K to L and vice versa. Details are left to the reader. 24.21 shows that η_1-chains of cardinality $\aleph_1$ exist only if the continuum hypothesis $2^{\aleph_0} = \aleph_1$ is assumed, i.e., if $2^{\aleph_0}$ is the smallest uncountable cardinal; compare 61.17.

Combining the idea of the proof with Artin–Schreier Theory (see Section 12 or JACOBSON 1989 Chapter 11), one obtains (ERDÖS *et al.* 1955):

24.24 Uniqueness Theorem *Any two real closed η_1-fields F and F' of cardinality $\aleph_1$ are isomorphic as fields and hence as ordered fields.*

In the proof of the theorem the following is needed.

24.25 Lemma *Any η_1-field F contains a strongly dense set T of algebraically independent elements over $\mathbb{Q}$.*

Proof by transfinite induction. Because of 24.21, the field F has cardinality $\aleph_\alpha \geq 2^{\aleph_0}$. The set of all open intervals in F can be written as $\{ J_\kappa \mid \kappa < \omega_\alpha \}$, where ω_α is the smallest ordinal such that $\operatorname{card} \omega_\alpha = \aleph_\alpha$. Consider an ordinal number $\sigma < \omega_\alpha$, and assume that algebraically independent elements $t_\rho \in J_\rho$ have been found for $\rho < \sigma$. Let E_σ be the field of all elements in F which are algebraic over $\mathbb{Q}(t_\rho)_{\rho<\sigma}$, and note that $\operatorname{card} E_\sigma = \aleph_0 \operatorname{card} \sigma < \aleph_\alpha$. By homogeneity, $\operatorname{card} J_\sigma = \operatorname{card}[-1,1] = \aleph_\alpha$. Hence there is an element $t_\sigma \in J_\sigma \smallsetminus E_\sigma$, and $\{t_\rho\}_{\rho \leq \sigma}$ is algebraically independent by construction. □

Proof of Theorem 24.24. By 24.25, there exist strongly dense maximal algebraically independent sets (transcendency bases) $T \subset F$ and $T' \subset F'$ of cardinality $\aleph_1$. These bases shall be considered as well-ordered sets of order type ω_1 (with respect to an ordering different from the one inherited from F or F' respectively). Denote the real closure of A in F by $[A]_F$, and similarly for F'. Let $\sigma < \omega_1$ and assume that elements $t_\rho \in T$ have been selected for $\rho < \sigma$. Put $E_\sigma = [\mathbb{Q}(t_\rho)_{\rho<\sigma}]_F$ and note that E_σ is countable because σ has only countably many predecessors. Suppose that we have an isomorphism $\varphi_\sigma : E_\sigma \cong E'_\sigma$ which maps t_ρ to t'_ρ for $\rho < \sigma$. Since E_σ is real closed, φ_σ preserves the ordering, and since E_0 is the real closure of $\mathbb{Q}$, the isomorphism φ_0 exists and is uniquely determined.

We will distinguish between even and odd ordinal numbers as follows: define λ to be even if λ is a limit ordinal, i.e., if $\lambda \neq \xi + 1$ for each ordinal ξ, and let $\xi + 1$ be odd (even) if ξ is even (odd). It follows from 61.5 that each ordinal is of the form $\lambda + \mu$ with λ a limit ordinal and $\mu \in \omega$. Hence each ordinal is either even or odd. The map $\xi \mapsto \xi + 1$ is an order-preserving bijection of the even ordinals onto the odd ones, and $\{\xi \in \omega_1 \mid \xi \text{ is even}\}$ is order-isomorphic to ω_1.

If σ *is even*, let t_σ be the first element in $T \smallsetminus \{t_\rho\}_{\rho<\sigma}$. Because T is algebraically independent, $t_\sigma \notin E_\sigma$, and $E_\sigma = A \cup B$ with $A < t_\sigma < B$. Applying the map φ_σ (which preserves the ordering) to A and B, we obtain sets A' and B' such that $A' < B'$ and $A' \cup B' = E'_\sigma$ is a countable subset of $E'_\sigma \cup T'$. This set is dense in the η_1-field F'. Hence $E'_\sigma \cup T'$ is an η_1-chain by 24.22, and there exist elements $t' \in T'$ with $A' < t' < B'$. The first of these elements (in the well-ordering of T') is taken as t'_σ. By construction, $a < s := t_\sigma \Leftrightarrow a^{\varphi_\sigma} < s' := t'_\sigma$ for all $a \in E_\sigma$. This condition is necessary and sufficient for φ_σ to have an order-preserving extension $\widehat{\varphi}_\sigma : E_\sigma(t_\sigma) \cong E'_\sigma(t'_\sigma)$ mapping t_σ onto t'_σ. In fact, since E_σ is real closed, each monic polynomial $f(s) \in E_\sigma[s]$ splits uniquely into a product of linear factors $s - c_\nu$ and some quadratic factors $(s - p_\mu)^2 + q_\mu$ with $q_\mu > 0$. In the ordering induced by F on $E_\sigma(s)$ the element $f(s)$ is positive if, and only if, the number of linear factors with $s < c_\nu$ is even, and this condition is preserved by φ_σ. The isomorphism $E_\sigma[s] \cong E'_\sigma[s']$ of chains then extends in a natural way to the fields of fractions. Finally, $\widehat{\varphi}_\sigma$ can be extended to an isomorphism $\varphi_{\sigma+1}$ of the real closures; see 12.17. This concludes the inductive step in the even case.

If σ is odd, proceed analogously with the roles of F and F' interchanged and all maps reversed. Because the first remaining elements of T or T' are selected alternately, both bases will be exhausted by the induction. Thus, $\bigcup_{\sigma \in \omega_1} E_\sigma = [\mathbb{Q}(T)]_F = F$, and the common extension of the φ_σ yields an isomorphism $\varphi : F \cong F'$. □

Let $\mathbb{R}_{\mathrm{alg}} = \mathbb{Q}^\natural \cap \mathbb{R}$ denote the real closed field of all real algebraic numbers.

24.26 Corollary *If the continuum hypothesis $2^{\aleph_0} = \aleph_1$ is assumed, and if Φ and Ψ are arbitrary free ultrafilters over $\mathbb{N}$, then $\mathbb{R}_{\mathrm{alg}}^\Phi \cong \mathbb{R}^\Psi$ as ordered fields. In particular, any two ultrapowers of $\mathbb{R}$ over a countable index set are isomorphic.*

Proof By 24.20 and the argument of 24.6, both $\mathbb{R}_{\mathrm{alg}}^\Phi$ and $\mathbb{R}^\Psi$ are real closed η_1-fields of cardinality $2^{\aleph_0}$; compare also HATCHER–LAFLAMME 1983. □

In fact, the continuum hypothesis implies that every ultrapower of $\mathbb{R}$ as in 24.26 is isomorphic to the field $\mathbb{R}((\Gamma))_1$ of Hahn power series with countable support (as defined in 64.25), where $\Gamma = \mathbb{R}((S))_1$ is the Hahn group over Sierpiński's ordered set S; see DALES–WOODIN 1996 4.30 p. 88.

A slight modification at the beginning of the proof of 24.24 shows that the field $*\mathbb{R}$ has at least $\aleph_1$ automorphisms, in contrast with the fact that $\mathbb{R}$ is rigid.

24.27 Corollary *If $2^{\aleph_0} = \aleph_1$, then $\mathrm{Aut}\,*\mathbb{R}$ is uncountable.*

The *proof* will be formulated for the field $*\mathbb{R}_{\mathrm{alg}}$. If $1 \ll t \in *\mathbb{R}_{\mathrm{alg}}$, then t is transcendental over $\mathbb{R}_{\mathrm{alg}}$ because $t^n > f(t)$ for any polynomial f over $\mathbb{R}_{\mathrm{alg}}$ of degree $< n$. Fix an element $t_0 \gg 1$. For each of the $\aleph_1$ different elements $t' \gg 1$ there is an isomorphism $\tau : \mathbb{R}_{\mathrm{alg}}(t_0) \cong \mathbb{R}_{\mathrm{alg}}(t')$

which preserves the ordering and maps t_0 to t'. Exactly as before, τ can be extended to an automorphism φ of $^*\mathbb{R}_{\mathrm{alg}}$. $\qquad\square$

Regarding other algebraic structures, we mention the following by-product of results in model theory.

24.28 Uniqueness *If the continuum hypothesis* $2^{\aleph_0} = \aleph_1$ *is assumed, and if S is an algebraic structure of cardinality at most $\aleph = \aleph_1$, then any two ultrapowers of S over a countable index set are isomorphic.*

A *proof* can be found in POTTHOFF 1981 Corollary 23.6 or CHANG–KEISLER 1990 6.1.9.

Exercises

(1) The ultrapower $\mathbb{R}^\Psi$ can also be written in the form $\mathbb{R}^\mathbb{N}/M$ where M is a maximal ideal in the ring $\mathbb{R}^\mathbb{N}$. Express M by Ψ and vice versa.

(2) Show that $\exp : {}^*\mathbb{C}^+ \to {}^*\mathbb{C}^\times$ is surjective, and determine the kernel.

(3) Define b^a in $^*\mathbb{R}$ for an arbitrary positive element b.

(4) The field $\mathbb{C}$ has an involutory automorphism fixing the elements of a sub-field $F \cong {}^*\mathbb{R}$.

(5) As a chain, $^*\mathbb{R}$ is isomorphic to each of its open intervals, but not to the chain $\mathfrak{R}$.

(6) As a real vector space, $^*\mathbb{R}$ does not have a countable basis.

(7) Show that the valuation topology of $^*\mathbb{R}$ has a basis consisting of the open intervals.

25   Continuity and convergence

By embedding the real numbers $\mathbb{R}$ into $\mathfrak{R} = \mathbb{R} + \mathfrak{n}$ instead of identifying $\mathbb{R}$ with $\mathfrak{R}/\mathfrak{n}$, arguments involving infinitely small entities, which were used before the invention of 'ε and δ', could be made precise. Of course, $\mathfrak{R}$ is only a ring; if the numbers are required to form a field, one has to admit infinitely large elements also and work in the field of fractions $^*\mathbb{R} = \mathfrak{R}(\mathfrak{R} \smallsetminus \{0\})^{-1}$.

Several notions of elementary analysis can be dealt with very easily within this framework; see for example 25.1 and 25.7 for the relation of simple and uniform continuity. Our discussion will be restricted to such parts of non-standard analysis which do not presuppose some routine in formal logic and model theory.

Extensions of functions as defined in 21.10 will be used continually, often in the following form.

25.0 Definition Let Ψ be a given ultrafilter over $\mathbb{N}$. Consider any map $h : D \to \mathbb{R}$ with $D \subseteq \mathbb{R}$. Write $^*D = D^\Psi = \{\bar{c} = c/\Psi \mid c = (c_\nu)_\nu \in D^\mathbb{N}\}$ and put $^*h(\bar{c}) = (h(c_\nu))_\nu/\Psi = \overline{h(c)}$. Then $^*h$ is a well-defined map such that the diagram

$$
\begin{array}{ccc}
D & \xrightarrow{\ h\ } & \mathbb{R} \\
\downarrow & & \downarrow \\
{}^*D & \xrightarrow{\ {}^*h\ } & {}^*\mathbb{R}
\end{array}
$$

is commutative. If there is no danger of confusion, $^*h$ will also be denoted by h. Recall that we write $x \approx y$ if $x - y$ is infinitely small.

25.1 Continuity *A function $f : \mathbb{R} \to \mathbb{R}$ is continuous at $a \in \mathbb{R}$ if, and only if,*

$$\mathop{\forall}_{x \in {}^*\mathbb{R}} \left(x \approx a \Rightarrow {}^*f(x) \approx {}^*f(a) \right) .$$

Consequently, the composition of continuous functions is again continuous.

Proof (a) Put $x = (x_\nu)_\nu/\Psi$. Then $x \approx a$ if, and only if, for each $m \in \mathbb{N}$ the condition $|x_\nu - a| < m^{-1}$ holds for almost all ν. If f is continuous at a in the usual sense, then for each $n \in \mathbb{N}$ there is some m such that $|x_\nu - a| < m^{-1}$ implies $|f(x_\nu) - f(a)| < n^{-1}$. Therefore, $^*f(x) \approx {}^*f(a)$.

(b) Assume that f is not continuous at a. Then there is some $n \in \mathbb{N}$ such that for each $\nu \in \mathbb{N}$ there exists some x_ν with $|x_\nu - a| < \nu^{-1}$ and $|f(x_\nu) - f(a)| \geq n^{-1}$. In particular, $x \approx a$, but $|^*f(x) - {}^*f(a)| \notin \mathfrak{n}$. $\square$

25.2 Convergent sequences *A sequence $s : \mathbb{N} \to \mathbb{R}$ converges to $t \in \mathbb{R}$ if, and only if, $^*s(m) \approx t$ for each infinite $m \in {}^*\mathbb{N}$.*

Proof Suppose that for each $n \in \mathbb{N}$ there is some $\mu \in \mathbb{N}$ such that $|s(\nu) - t| < n^{-1}$ for all $\nu > \mu$. Consider $m = (m_\nu)_\nu/\Psi \in {}^*\mathbb{N} \smallsetminus \mathbb{N}$. Then $\{\nu \in \mathbb{N} \mid m_\nu \leq \mu\} \notin \Psi$ and $m_\nu > \mu$ for almost all ν, hence $|^*s(m) - t| < n^{-1}$.

If the s_ν do not converge to t, then there is an element $n \in \mathbb{N}$ and for each $\nu \in \mathbb{N}$ some $m_\nu > \nu$ such that $|s(m_\nu) - t| \geq n^{-1}$. It follows that $m = (m_\nu)_\nu/\Psi \in {}^*\mathbb{N} \smallsetminus \mathbb{N}$ and that $^*s(m) \not\approx t$. $\square$

25.3 Example We show that $\sqrt[n]{n}$ converges to 1: put $s(n) = \sqrt[n]{n} - 1$. Then $(s(n) + 1)^n = n$ and, by the binomial expansion, $\binom{n}{2}s(n)^2 < n - 1$ or $|s(n)| < \sqrt{2n^{-1}}$. Hence $^*s(m) \approx 0$ for each $m \in {}^*\mathbb{N} \smallsetminus \mathbb{N}$.

25.4 Corollary *A real function f is continuous at $a \in \mathbb{R}$ if, and only if, $f(a + s_\nu)$ converges to $f(a)$ whenever the s_ν converge to 0.*

Proof We may assume that $a = 0 = f(a)$, and we write $s_m = (s_{m_\nu})_\nu / \Psi$ for $m = (m_\nu)_\nu / \Psi$. If f is continuous at 0, then $s_m \approx 0$ implies $f(s_m) \approx 0$ for each infinite $m \in {}^*\mathbb{N}$. On the other hand, any element in $\mathfrak{n}$ can be written in the form s_m with $m = (\nu)_\nu / \Psi$.   □

25.5 Monotone convergence *If $s : \mathbb{N} \to \mathbb{R}$ is an increasing sequence and if ${}^*s(m) \approx t$ for some $m \in {}^*\mathbb{N} \smallsetminus \mathbb{N}$ and $t \in \mathbb{R}$, then s converges to t.*

Proof For each $n \in \mathbb{N}$, the assumption on $m = (m_\nu)_\nu / \Psi$ implies $m_\nu \geq n$ for almost all ν, hence $s(n) \leq {}^*s(m)$ and $s(n) \leq t$. If $r < t$ and $r \in \mathbb{R}$, then $r < s(m_\nu)$ for almost all ν. Since s is increasing, $r < s(n) \leq t$ for all but finitely many $n \in \mathbb{N}$.   □

25.6 Cauchy criterion *A sequence $s : \mathbb{N} \to \mathbb{R}$ converges if, and only if,*

$$m, n \in {}^*\mathbb{N} \smallsetminus \mathbb{N} \implies {}^*s(m) \approx {}^*s(n) .$$

Proof If s converges to $t \in \mathbb{R}$ in the ordinary sense, then ${}^*s(m) \approx {}^*s(n)$ by 25.2. Suppose now that s is not a Cauchy sequence in the ordinary sense. Then for some $k \in \mathbb{N}$, there are arbitrarily large m_ν and $n_\nu \in \mathbb{N}$ with $s(n_\nu) - s(m_\nu) > k^{-1}$, and ${}^*s(m) - {}^*s(n) \notin \mathfrak{n}$.   □

25.7 Uniform continuity *A real function f is uniformly continuous on $\mathbb{R}$ if, and only if,*

$$\bigvee_{x,y \in {}^*\mathbb{R}} \left(x \approx y \Rightarrow {}^*f(x) \approx {}^*f(y) \right) .$$

Proof The arguments are quite similar to those in 25.1. Assume that f is uniformly continuous. Then for each $n \in \mathbb{N}$ there is some m such that $|y_\nu - x_\nu| < m^{-1}$ implies $|f(y_\nu) - f(x_\nu)| < n^{-1}$, and the conclusion of 25.7 is true.

   If f is not uniformly continuous, then there is some $n \in \mathbb{N}$ such that for each $\nu \in \mathbb{N}$ there are elements x_ν and y_ν with $|y_\nu - x_\nu| < \nu^{-1}$ and $|f(y_\nu) - f(x_\nu)| \geq n^{-1}$. Now $x = (x_\nu)_\nu / \Psi$ and $y = (y_\nu)_\nu / \Psi$ satisfy $x \approx y$ but not ${}^*f(x) \approx {}^*f(y)$.   □

Note If one of the variables in 25.7 is restricted to $\mathbb{R}$, then, according to 25.1, the condition characterizes ordinary continuity of f.

25.8 Corollary *Any continuous function $f : [a, b] \to \mathbb{R}$ is even uniformly continuous.*

Proof For real $t \leq a$ or $t \geq b$ let f be constant. Consider $x, y \in {}^*\mathbb{R}$ with $x \approx y$. If $x, y \in [a, b]$, then x and y are finite, and $x \approx y$ implies $^{\circ}x = {}^{\circ}y = c \in \mathbb{R}$, hence ${}^*f(x) \approx f(c) \approx {}^*f(y)$ by 25.1. If $x < a \leq y$ and $x \approx y$, then $y \approx a$, as $\mathfrak{n}$ is convex, and ${}^*f(x) = f(a) \approx {}^*f(y)$ by continuity.   □

25.9 Ring of continuous functions *Sum and product of continuous functions are continuous. If f is continuous at a and $f(a) \neq 0$, then $g = 1/f$ is also continuous at a.*

Proof The first part is an immediate consequence of 25.1 and the fact that $\mathfrak{n}$ is an ideal in $\mathfrak{R}$.

For the second part, note that ${}^*f(x) \approx f(a)$ implies ${}^*f(x) \notin \mathfrak{n}$ and ${}^*g(x) \in \mathfrak{R}$. Hence ${}^*g(x) - g(a) = \big(f(a) - {}^*f(x)\big){}^*g(x)g(a) \in \mathfrak{n}$.   □

Exercises

These exercises should be dealt with in the setting of this section.

(1) Which sequences $s : \mathbb{N} \to \{0, 1\}$ are convergent?

(2) Show that $(x \mapsto x^2) : \mathbb{R} \to \mathbb{R}$ is not uniformly continuous.

(3) Let $f : \mathbb{R} \to \mathbb{R}$ be any function. If ${}^*f$ is constant on each monad, then f is constant.

26   Topology of the real numbers in non-standard terms

A few basic properties of the topological space $\mathbb{R}$ (with the usual topology) will be dealt with via the embedding $\mathbb{R} \hookrightarrow {}^*\mathbb{R}$. Remember that the topology τ defined in Section 24 induces on $\mathbb{R}$ the discrete topology (24.9), so that $\mathbb{R}$ is not to be considered as subspace of ${}^*\mathbb{R}$.

If $A \subset \mathbb{R}$, write ${}^*A = A^{\mathbb{N}}/\Psi$ and note that ${}^*(\mathbb{R} \smallsetminus A) = {}^*\mathbb{R} \smallsetminus {}^*A$ (by 21.2). For other notation see the last part of Section 23.

26.1 Open sets *A set $A \subset \mathbb{R}$ is open in $\mathbb{R}$ if, and only if, $A + \mathfrak{n} \subseteq {}^*A$.*

Proof If $a \in A$ and A is open in $\mathbb{R}$, then $(a - r, a + r) \cap \mathbb{R} \subseteq A$ for some positive real number r, and $a + \mathfrak{n} \subseteq (a - r, a + r) \subseteq {}^*A$.

If A is not open, then there is an element $a \in A$ which does not belong to the interior of A, and some sequence of numbers $c_\nu \in \mathbb{R} \smallsetminus A$ converges to a. It follows that $c = (c_\nu)_\nu/\Psi \notin {}^*A$ and $c - a \in \mathfrak{n}$.   □

26.2 Closure *The closure of a set A in $\mathbb{R}$ is given by $\overline{A} = (^*A + \mathfrak{n}) \cap \mathbb{R} = {}^\circ(^*A \cap \mathfrak{R})$.*

Proof If $t \in \mathbb{R}$ and $t \in \overline{A}$, then there is a sequence of numbers $a_\nu \in A$ which converges to t, and $\overline{a} = (a_\nu)_\nu / \Psi \in {}^*A \cap (t + \mathfrak{n})$. Conversely, $t \in \mathbb{R} \setminus \overline{A} = O$ implies $t + \mathfrak{n} \subseteq {}^*O \subseteq {}^*(\mathbb{R} \setminus A)$ by 26.1. □

26.3 Continuity *If $f : \mathbb{R} \to \mathbb{R}$ is continuous, then each open set $U \subseteq \mathbb{R}$ has an open pre-image $O = f^{\leftarrow}(U)$, moreover, $f(\overline{A}) \subseteq \overline{f(A)}$ for each $A \subseteq \mathbb{R}$.* (In fact, each of these equivalent conditions characterizes continuity of f.)

Proof By continuity, $x \approx a \in O$ implies ${}^*f(x) \approx f(a) \in U$. Since U is open, 26.1 gives ${}^*f(x) \in {}^*U$, and this equivalent with $x \in {}^*O$. Moreover, if $a \approx x \in {}^*A$, then $f(a) \approx {}^*f(x) \in {}^*(f(A))$ and hence $f(a) \in \overline{f(A)}$. □

26.4 Boundedness *A set $A \subseteq \mathbb{R}$ is bounded if, and only if, ${}^*A \subseteq \mathfrak{R}$.*

Proof By definition, ${}^*A$ contains infinite elements if, and only if, A is unbounded. □

26.5 Compactness *A is compact if, and only if, for each $x \in {}^*A$ there is an element $a \in A$ with $x \approx a$, in short, if ${}^\circ(^*A) = A$.*

This is *proved* by combining 26.2 and 26.4. □

26.6 Corollary *Compactness is preserved by continuous maps.*

For a *proof*, note that $f(^*A) = {}^*(f(A))$ and apply 25.1. □

26.7 Intersections *If $A_{\kappa+1} \subset A_\kappa$ for $\kappa \in \mathbb{N}$, and if each A_κ is compact and non-empty, then $D = \bigcap_{\kappa \in \mathbb{N}} A_\kappa \neq \emptyset$.*

Proof Choose $a_\nu \in A_\nu$ and put $\overline{a} = (a_\nu)_\nu / \Psi$. Then $\overline{a} \in {}^*A_\kappa$ for each κ (since $a_\nu \in A_\kappa$ for $\nu \geq \kappa$), and the standard part d of $\overline{a}$ is in A_κ by 26.5. Hence $d \in D$. □

26.8 Accumulation points *A real number t is an accumulation point of $A \subseteq \mathbb{R}$ if, and only if, there is an element $\overline{a} \in {}^*A$ such that $t \approx \overline{a}$ and $t \neq \overline{a}$.*

Proof According to 26.2, the condition is equivalent with $t \in \overline{A \setminus \{t\}}$, and this is the standard definition of an accumulation point. □

26.9 Corollary (Bolzano–Weierstraß) *Each bounded infinite subset A of $\mathbb{R}$ has an accumulation point.*

Proof The assumptions imply $A \subset {}^*A \subseteq \mathfrak{R}$, and then the standard part of any element $\bar{a} \in {}^*A \smallsetminus A$ is an accumulation point of A.   □

Exercises

(1) Use 26.1 in order to show that any union of open sets is open and that the intersection of two open sets is open. Show also that $\overline{A \cup B} = \overline{A} \cup \overline{B}$.

(2) Deduce from results in this chapter that $\mathbb{Q}$ is dense in $\mathbb{R}$.

(3) Show that the intersection of countably many dense open subsets of $\mathbb{R}$ is dense in $\mathbb{R}$ (but there are open neighbourhoods of $\mathbb{Q}$ in $\mathbb{R}$ of arbitrarily small Lebesgue measure).

(4) A continuous function on a closed interval assumes its maximum.

(5) If $C = [a, b] \subset \mathbb{R}$ and if $f : C \to \mathbb{R}$ is continuous, then ${}^*f$ assumes its maximum at a standard point of ${}^*C$.

27   Differentiation

Hardly any topic is better suited for non-standard methods than differentiation. Reasoning in the extension ${}^*\mathbb{R}$ of $\mathbb{R}$, several heuristic arguments dealing with infinitely small numbers can be put on a sound foundation.

For the sake of simplicity, only real-valued functions defined on an open and connected domain $D \subseteq \mathbb{R}$ will be considered.

27.1 Characterization   *A function $f : D \to \mathbb{R}$ is differentiable at $a \in D$ in the usual sense with derivative $f'(a) \in \mathbb{R}$ if, and only if,*

$$\left({}^*f(a + h) - f(a) \right) h^{-1} \approx f'(a) \text{ for each } h \approx 0, h \neq 0 .   (*)$$

Proof Assume that f is differentiable at a. This means that for each $n \in \mathbb{N}$ there is an $m \in \mathbb{N}$ such that $0 < |s| < m^{-1}$ implies

$$|(f(a + s) - f(a)) \, s^{-1} - f'(a)| < n^{-1} .$$

If $0 \neq h = (s_\nu)_\nu / \Psi \approx 0$, then $0 < |s_\nu| < m^{-1}$ for almost all $\nu \in \mathbb{N}$, and the assertion follows.

Suppose conversely that f is not differentiable at a or that $f'(a) \neq t \in \mathbb{R}$. Then there exists a sequence (s_ν) which converges to 0 such that $|(f(a + s_\nu) - f(a)) s_\nu^{-1} - t| \geq \varepsilon$ for some positive $\varepsilon \in \mathbb{R}$. Letting $h = (s_\nu)_\nu / \Psi$, we have $h \approx 0$, but $\left({}^*f(a + h) - f(a) \right) h^{-1} \not\approx t$.   □

27.2 Continuity *If f is differentiable at a, then f is also continuous at a.*

Proof From equation $(*)$ we obtain ${}^*f(a+h) - f(a) \approx f'(a) \cdot h \approx 0$ for $h \approx 0$. $\qquad\square$

27.3 Corollary: Chain Rule *If f and g are differentiable and if the composition $g \circ f$ is defined in a neighbourhood of a, then $(g \circ f)'(a) = g'(f(a)) \cdot f'(a)$.*

Proof Let $a \neq x \approx a$. Then $\big({}^*f(x) - f(a)\big)(x-a)^{-1} \approx f'(a)$ and hence ${}^*f(x) - f(a) \approx 0$. Therefore $\big({}^*(g \circ f)(x) - (g \circ f)(a)\big)\big({}^*f(x) - f(a)\big)^{-1} \approx g'(f(a))$ whenever ${}^*f(x) \neq f(a)$. $\qquad\square$

27.4 Products *If f and g are differentiable at a, then*

$$(f \cdot g)'(a) = f'(a) \cdot g(a) + f(a) \cdot g'(a) \ .$$

Proof Let $x \approx a$ and $x \neq a$. Then $g(x) \approx g(a)$ by 27.2, and the identity $(f \cdot g)(x) - (f \cdot g)(a) = (f(x) - f(a))g(x) + f(a)(g(x) - g(a))$ together with 27.1 gives the assertion. $\qquad\square$

The following kind of Cauchy criterion does not require knowledge of the derivative in advance.

27.5 Differentiability *A function $f : D \to \mathbb{R}$ is differentiable at $a \in D$ if, and only if, $x, y \approx a \neq x, y$ implies*

$$({}^*f(x) - f(a))(x-a)^{-1} \approx ({}^*f(y) - f(a))(y-a)^{-1} \ . \qquad (**)$$

Proof For $x \in \mathbb{R}$ put $q(x) = (f(x) - f(a))(x-a)^{-1}$. Assume first that ${}^*q(z)$ is infinitely large for some $z = (z_\kappa)_\kappa / \Psi \approx a$. Then for each $n \in \mathbb{N}$ there are infinitely many $\kappa \in \mathbb{N}$ with $0 < |z_\kappa - a| < n^{-1}$ and $q(z_\kappa) \geq n$. Among the z_κ define inductively a sequence of elements x_ν converging to a such that $q(x_{\nu+1}) \geq q(x_\nu) + 1$, and let $y_\nu = x_{\nu+1}$. Then $x, y \approx a$ and ${}^*q(y) \geq {}^*q(x) + 1 \not\approx {}^*q(x)$.

Hence $(**)$ implies that ${}^*q(z) \in \mathfrak{R}$. Denoting the standard part of ${}^*q(z)$ by $f'(a)$, differentiability follows from 27.1. $\qquad\square$

27.6 Continuous derivative *A function $f : D \to \mathbb{R}$ is continuously differentiable if, and only if, the following condition holds:*

$$\underset{a \in D}{\forall} \ \underset{d \in \mathbb{R}}{\exists} \ \underset{x,y \in {}^*\mathbb{R}}{\forall} \ \left[x \approx a \approx y \neq x \Rightarrow ({}^*f(y) - {}^*f(x))(y-x)^{-1} \approx d \right] \quad (\dagger)$$

Proof For simplicity, write $q(x,y) = (f(y) - f(x))(y - x)^{-1}$, and let x be the smaller one of the two distinct elements x and y. Represent $\overline{x}$ in the usual way as $\overline{x} = (x_\nu)_\nu/\Psi$ etc.

Let f be differentiable and $\overline{x}, \overline{y} \approx a$. By the mean value theorem, we find real numbers z_ν between x_ν and y_ν such that $q(x_\nu, y_\nu) = f'(z_\nu)$ and hence ${}^*q(\overline{x},\overline{y}) = {}^*f'(\overline{z})$. The convexity of monads gives $\overline{z} \approx a$, and continuity of f' shows that ${}^*f'(\overline{z}) \approx f'(a)$; see 25.1.

By 27.1 or (**), condition (†) implies that f is differentiable at each point $a \in D$. Assume that for every $\varepsilon > 0$ there is some $\nu \in \mathbb{N}$ such that $|q(x,y) - f'(a)| < \varepsilon$ for any two distinct real numbers x, y in a ν^{-1}-neighbourhood of a; then f' is continuous at a. Suppose now that this is not true. Then for some $\varepsilon > 0$ and for each $\nu \in \mathbb{N}$ there are elements x_ν and y_ν with $a - \nu^{-1} < x_\nu < y_\nu < a + \nu^{-1}$ and $|q(x_\nu, y_\nu) - f'(a)| \geq \varepsilon$. Consider $\overline{x} = (x_\nu)_\nu/\Psi$ and $\overline{y} = (y_\nu)_\nu/\Psi$. By construction, $\overline{x} < \overline{y}$ and $\overline{x} \approx \overline{y} \approx a$, but $|q(\overline{x},\overline{y}) - f'(a)| \geq \varepsilon$. This contradicts (†). □

27.7 Examples Let $p(x) = x^n$. If $x \approx a$, then $(x^n - a^n)(x - a)^{-1} = \sum_{\nu=1}^{n} x^{n-\nu}a^{\nu-1} \approx n \cdot a^{n-1} = p'(a)$.

Let $s(x) = \sin x$. The power series expansion shows immediately that $\lim_{x \to 0} x^{-1} s(x) = 1$. Hence $h \approx 0$ implies $s(h)h^{-1} \approx 1 = s'(0)$. Moreover, $\cos h \approx 1$ and $(1 - \cos h)h^{-1} = s(h)^2 h^{-1}(1 + \cos h)^{-1} \approx 0$. Therefore, $(s(x + h) - s(x))h^{-1} = (\cos x \sin h - \sin x(1 - \cos h))h^{-1} \approx \cos x = s'(x)$.

Exercises

(1) Show that $f'(a) = 0$ if f is differentiable and has a local minimum at a.

(2) If f is differentiable, $f(a) \neq 0$, and $g = 1/f$, then $g'(a) = -f'(a)/f(a)^2$.

28   Planes and fields

Any (commutative) field F coordinatizes a Pappian affine plane with point space F^2 such that the lines are described by linear equations. This correspondence between fields and planes is compatible with the ultraproduct construction. Thus, all results in this chapter have a geometric interpretation.

28.1 A *projective plane* $\mathcal{P}$ consists of a point set P and a set $\mathfrak{L}$ of lines $L \subset P$ such that any two distinct points belong to a unique line, and, dually, any two lines intersect in a point. By deleting from $\mathcal{P}$ a line and all of its points, one obtains an *affine plane*.

If the well known Pappos theorem holds, then we speak of a Pappian plane. Precisely the Pappian affine planes can be represented by a (commutative) field in the way indicated above.

The lines of an affine or projective plane $\mathcal{D} = (P, \mathfrak{L})$ are determined by the *collinearity* relation λ on P^3 which holds exactly if three points belong to the same line. Using the notion of collinearity, ultraproducts of planes can conveniently be defined as follows.

28.2 Ultraproducts of planes Given affine or projective planes $\mathcal{D}_\nu = (P_\nu, \lambda_\nu)$ and a free ultrafilter Ψ over $\mathbb{N}$, the relation λ holds for three points $\overline{p}_\kappa = (p_{\kappa\nu})_\nu / \Psi$ in the ultraproduct of the P_ν if $\lambda_\nu(p_{1\nu}, p_{2\nu}, p_{3\nu})$ holds for almost all $\nu \in \mathbb{N}$. Ultraproducts of affine or projective planes are again affine or projective planes, and the same is true for Pappian planes, because all first-order properties carry over to ultraproducts.

28.3 *If the field F coordinatizes a (Pappian) plane $\mathcal{D}$, then any ultrapower $\mathcal{D}^\Psi$ is coordinatized by F^Ψ.*

28.4 *If Ψ is a free ultrafilter over $\mathbb{N}$, then the ultrapower $\mathcal{Q}^\Psi$ of the projective plane $\mathcal{Q}$ over the rational field admits a homomorphism (i.e., a collinearity preserving mapping) onto the real projective plane.*

Proof As in Section 23, let R denote the valuation ring of all finite elements in $\mathbb{Q}^\Psi$. Each point in $\mathcal{Q}^\Psi$ has homogeneous coordinates $z_\kappa \in R$ such that at least one of the z_κ is a unit in $R^\times$. The canonical projection $R \to R/N \cong \mathbb{R}$ induces a homomorphism onto the real projective plane (because no coordinate triple is mapped to 0). $\qquad\square$

A *Euclidean plane* is an affine plane over a Euclidean field (see 12.8) with an orthogonality relation $\perp$ for lines, orthogonality having the familiar properties. (Remember that lines given by $y = ax$ and $y = bx$ are orthogonal if, and only if, $ab = -1$.)

28.5 Euclidean planes *An ultrapower $^*\mathcal{E} = \mathcal{E}^\Psi$ of the ordinary Euclidean plane $\mathcal{E}$ over the real numbers is a Euclidean plane.*

28.6 Example Let C be the parabola in $\mathcal{E}$ with the equation $y = x^2$, let $p = (a, a^2) \in C$, and put $^*C = C^\mathbb{N}/\Psi$. Then $^*C = \{ (x, x^2) \mid x \in {}^*\mathbb{R} \}$. We show that the tangent at p to C consists of the standard parts of the points on a line through p and some other point $q \in {}^*C$ infinitely close to p. Put $q = (u, u^2)$ where $u \approx a$. The line connecting p and q has the equation $y = (a + u)(x - a) + a^2$. If $s \approx x$, $t \approx y$ and $s, t \in \mathbb{R}$, then $t \approx 2a(s - a) + a^2$, hence $t = 2a(s - a) + a^2$.

3

Rational numbers

This chapter treats the different kinds of structures on the field $\mathbb{Q}$ of rational numbers (algebra, order and topology) and various combinations of them in the same way as it was done in Chapter 1 for the field $\mathbb{R}$ of real numbers.

We sometimes profit from the fact that $\mathbb{Q}$ is embedded in $\mathbb{R}$ so that we can use results from Chapter 1. In doing so, we take the field $\mathbb{R}$ for granted. Constructions of $\mathbb{R}$ from $\mathbb{Q}$ are presented in Chapter 2 (via an ultrapower of $\mathbb{Q}$) and in Chapter 4 (by completion of $\mathbb{Q}$).

31 The additive group of the rational numbers

Under the usual addition, the rational numbers form a group $(\mathbb{Q}, +)$, or briefly $\mathbb{Q}^+$. Being a subgroup of $\mathbb{R}^+$, this group has already been studied to some extent in Section 1, together with the factor group $\mathbb{Q}^+/\mathbb{Z}$. We continue the investigation of these groups, we characterize them in the class of all groups, and we study their endomorphism rings.

31.1 Definition A group G is called *locally cyclic*, if the subgroup $\langle a_1, a_2, \ldots, a_n \rangle$ generated by finitely many elements $a_1, a_2, \ldots, a_n$ of G is always cyclic, that is, this subgroup may be generated in fact by a single element of G. By induction, this is equivalent to the property that the subgroup generated by any two elements is cyclic. In particular, every locally cyclic group is abelian.

31.2 For $0 \neq b \in \mathbb{Q}$ the subgroup $\langle b \rangle$ of $\mathbb{Q}^+$ generated by b is isomorphic to $\langle 1 \rangle = \mathbb{Z}^+$, as there is an automorphism of $\mathbb{Q}^+$ mapping b to 1.

Indeed, such an automorphism is given by $r \mapsto r/b$; it induces an isomorphism of $\langle b \rangle$ onto $\mathbb{Z}$. $\qquad\square$

31.3 Characterization theorem *The group $\mathbb{Q}^+$ is torsion free, divisible and locally cyclic, and is, in fact, the only non-trivial group having these properties (up to isomorphism).*

Proof It is clear that $\mathbb{Q}^+$ is torsion free and divisible (for the definitions of these notions, the reader is referred to 1.2 and 1.7). In 1.6(c), it is shown that $\mathbb{Q}^+$ is locally cyclic.

Conversely, let A be any torsion free, divisible and locally cyclic group which is not reduced to the neutral element. Then A is abelian, and the first two properties imply that A can be made into a rational vector space; see 1.9. As A is locally cyclic, any two elements are contained in a 1-dimensional $\mathbb{Q}$-linear subspace. Thus the dimension of A over $\mathbb{Q}$ is 1, and the group A is isomorphic to $\mathbb{Q}^+$.   □

According to 31.2 and 31.3, the factor group of $\mathbb{Q}^+$ by any non-trivial finitely generated subgroup is isomorphic to the factor group $\mathbb{Q}^+/\mathbb{Z}$.

31.4 The group $\mathbb{Q}^+/\mathbb{Z}$ is a *torsion group*, which means that every element has finite order. In fact, it is the torsion subgroup of the torus group $\mathbb{R}^+/\mathbb{Z}$, i.e., the subgroup consisting of all elements of finite order. Indeed, the real numbers x having an integer multiple $nx \in \mathbb{Z}$ are precisely the rational numbers.

31.5 *The group $\mathbb{Q}^+/\mathbb{Z}$ is a divisible, locally cyclic torsion group.*

Proof The group $\mathbb{Q}^+/\mathbb{Z}$ is the homomorphic image of $\mathbb{Q}^+$ under the canonical projection $\mathbb{Q}^+ \to \mathbb{Q}^+/\mathbb{Z}$ and therefore inherits the properties of being divisible and locally cyclic from $\mathbb{Q}^+$; see 31.3.   □

In the sequel, we shall find out to what extent the properties of 31.5 suffice to characterize $\mathbb{Q}^+/\mathbb{Z}$. For this, the following description will be helpful.

31.6 Let p be a prime. The *p-primary component* of an abelian group is the subgroup consisting of all elements whose order is a power of p. The p-primary component of $\mathbb{Q}^+/\mathbb{Z}$ is the so-called *Prüfer group* (see 1.26)

$$C_{p^\infty} = \{ \tfrac{m}{p^n} + \mathbb{Z} \mid m \in \mathbb{Z}, n \in \mathbb{N} \} \subseteq \mathbb{Q}^+/\mathbb{Z} \,.$$

In 1.28 it has been shown that

$$\mathbb{Q}^+/\mathbb{Z} = \bigoplus\nolimits_{p \in \mathbb{P}} C_{p^\infty} \,,$$

where the sum is taken over the set $\mathbb{P}$ of all primes, just as every abelian torsion group is the direct sum of all its primary components; compare

Exercise 2 of Section 1. As a consequence, we may infer that each Prüfer group C_{p^∞} is divisible since $\mathbb{Q}^+/\mathbb{Z}$ is; see also 1.27 for a direct proof of this fact.

31.7 Lemma *Let A be a divisible group. If A contains a non-trivial element of finite order and if the prime p divides this order, then A contains a subgroup isomorphic to C_{p^∞}.*

Proof We use additive notation, although A is not required to be abelian. By assumption, there is an element $0 \neq x_1 \in A$ such that $px_1 = 0$. Since A is divisible, we may choose inductively elements x_n for $2 \leq n \in \mathbb{N}$ such that $px_2 = x_1$, $px_3 = x_2$, ..., $px_{n+1} = x_n$. Then $p^{n-1}x_n = x_1 \neq 0$ and $p^n x_n = 0$, so that the order of x_n is p^n. The map

$$C_{p^\infty} \to A : \tfrac{m}{p^n} + \mathbb{Z} \mapsto mx_n$$

(for $m \in \mathbb{Z}$, $n \in \mathbb{N}$) is a well-defined injective homomorphism; indeed, $mx_n = 0$ if, and only if, the order p^n of x_n divides m, in other words, if, and only if, $m/p^n \in \mathbb{Z}$. The image of this homomorphism is a subgroup of A isomorphic to C_{p^∞}.   □

31.8 Characterization Theorem *Up to isomorphism, the divisible, locally cyclic torsion groups are precisely the direct sums*

$$\bigoplus_{p \in \mathbb{P}'} C_{p^\infty} \text{ for arbitrary subsets } \mathbb{P}' \subseteq \mathbb{P} .$$

Proof (1) We show that such a direct sum has the asserted properties. The Prüfer groups are divisible torsion groups (see 31.6 and 1.27), hence the same holds for any direct sum of Prüfer groups. Moreover, each direct sum as above is a subgroup of $\mathbb{Q}^+/\mathbb{Z} = \bigoplus_{p \in \mathbb{P}} C_{p^\infty}$ and therefore inherits the property of being locally cyclic from $\mathbb{Q}^+/\mathbb{Z}$.

(2) Now let A be an arbitrary divisible, locally cyclic torsion group. Let $\mathbb{P}'$ be the set of all primes dividing the order of some group element. We claim that the group $B = \bigoplus_{p \in \mathbb{P}'} C_{p^\infty}$ is isomorphic to A. In order to prove this, we construct a homomorphism $\varphi : B \to A$ and show that it is bijective.

For each prime $p \in \mathbb{P}'$, there exists an injective homomorphism $\varphi_p : C_{p^\infty} \to A$ by 31.7, and we define φ by

$$\varphi((c_p)_{p \in \mathbb{P}'}) = \sum_{p \in \mathbb{P}'} \varphi_p(c_p) .$$

Here, we use the notation for elements of direct sums introduced in 1.16, and the fact that A is abelian (31.1). If $c = (c_p)_{p \in \mathbb{P}'}$ is distinct from 0, then some component c_p is non-zero. Multiplying the image with the product r of the orders of all other components c_q for $q \in \mathbb{P}' \setminus \{p\}$, we obtain $r\varphi(c) = r\varphi_p(c_p)$, and this is distinct from 0 since the order of $\varphi_p(c_p)$ is prime to r.

We have shown that the subgroup $\varphi(B)$ of A is isomorphic to B. Assume that an element $a \in A \setminus \varphi(B)$ exists. Then all prime factors of the order of a belong to $\mathbb{P}'$, hence $\varphi(B)$ contains an element b having the same order as a. The subgroup of A generated by a, b is cyclic and finite by assumption, but a finite cyclic group contains at most one subgroup of any given order, hence $\langle a \rangle = \langle b \rangle$. This contradiction shows that $\varphi(B) = A$.

We remark that in step (2) above, one could also use Exercise 2 of Section 1.   □

31.9 Corollary   *A group is isomorphic to $\mathbb{Q}^+/\mathbb{Z}$ if, and only if, it is a divisible, locally cyclic torsion group and if every prime p divides the order of some group element.*

Proof Since $\mathbb{Q}^+/\mathbb{Z} = \bigoplus_{p \in \mathbb{P}} C_{p^\infty}$, this follows immediately from 31.8; indeed, from the direct sum in 31.8 the set $\mathbb{P}'$ may be retrieved as being the set of primes dividing the orders of elements of the group.   □

We now turn to endomorphisms of $\mathbb{Q}^+$ and $\mathbb{Q}^+/\mathbb{Z}$. Recall that the set $\operatorname{End} A$ of endomorphisms of an abelian group A is a ring in a natural way; its addition is obtained from the operation of the group A, and multiplication is the composition of maps.

31.10   *The endomorphism ring of $\mathbb{Q}^+$ is $\operatorname{End} \mathbb{Q}^+ = \{ x \mapsto rx \mid r \in \mathbb{Q} \}$; as a ring, it is isomorphic to the field $\mathbb{Q}$.*

Remark In 8.28, we have seen that the only other abelian groups whose endomorphism rings are fields are the cyclic groups of prime order.

Proof For an endomorphism φ of $\mathbb{Q}^+$ and for $m \in \mathbb{Z}$, $n \in \mathbb{N}$ we have $n\varphi(m/n) = \varphi(n \cdot m/n) = \varphi(m) = m\varphi(1)$, hence $\varphi(m/n) = m\varphi(1)/n$, so that $\varphi(x) = rx$ where $r = \varphi(1)$. Conversely, it is immediate that for arbitrary $r \in \mathbb{Q}$ this is an endomorphism of $\mathbb{Q}^+$. It is also clear that $\varphi \mapsto \varphi(1)$ is a ring isomorphism of $\operatorname{End} \mathbb{Q}^+$ onto $\mathbb{Q}$.   □

31.11 Theorem *The endomorphism ring of $\mathbb{Q}^+/\mathbb{Z}$ is isomorphic to the following subring of the direct product $R := \bigtimes_{k \in \mathbb{N}} \mathbb{Z}/k!\mathbb{Z}$ of the factor rings $\mathbb{Z}/k!\mathbb{Z}$, with componentwise addition and multiplication:*

$$\operatorname{End}(\mathbb{Q}^+/\mathbb{Z}) \cong \left\{ (m_k + k!\mathbb{Z})_k \in R \mid \forall_{k \in \mathbb{N}}\ m_{k+1} \equiv m_k \bmod k! \right\}$$

Proof (i) A rational number m/n for $m \in \mathbb{Z}$, $n \in \mathbb{N}$ can be written as $m(n-1)!/n!$, hence $\mathbb{Q}^+/\mathbb{Z}$ is the union of the subgroups

$$C_{1/k!} := (1/k!)\mathbb{Z}/\mathbb{Z}$$

for $k \in \mathbb{N}$, which form an ascending chain of subgroups.

Every endomorphism φ of $\mathbb{Q}^+/\mathbb{Z}$ induces an endomorphism φ_k of the subgroup $C_{1/k!}$ since this subgroup consists precisely of those elements of $\mathbb{Q}^+/\mathbb{Z}$ whose $k!$-fold multiple is the neutral element, and this property is preserved by endomorphisms.

The endomorphisms φ_k of $C_{1/k!}$ induced by φ satisfy

$$\varphi_{k+1}|_{C_{1/k!}} = \varphi_k. \tag{1}$$

Conversely, if a sequence of endomorphisms $\varphi_k \in \mathrm{End}\,C_{1/k!}$ for $k \in \mathbb{N}$ satisfies this compatibility condition (1), then they can be merged into an endomorphism φ of $\mathbb{Q}^+/\mathbb{Z}$ which induces them all.

(ii) Since $C_{1/k!}$ is a cyclic group, there exists, for given $m \in \mathbb{Z}$, a unique endomorphism $\mu_{k,m}$ of $C_{1/k!}$ mapping the generator $(k!)^{-1} + \mathbb{Z}$ to $m(k!)^{-1} + \mathbb{Z}$; this endomorphism maps every element to its m-fold multiple. The surjective map

$$\mathbb{Z} \to \mathrm{End}\,C_{1/k!} : m \mapsto \mu_{k,m} \tag{2}$$

is clearly a ring homomorphism with kernel $\{m \mid m(k!)^{-1} \in \mathbb{Z}\} = k!\mathbb{Z}$, so that by factorization this ring homomorphism yields a ring isomorphism

$$\mathbb{Z}/k!\mathbb{Z} \to \mathrm{End}\,C_{1/k!} : m + k!\mathbb{Z} \mapsto \mu_{k,m}. \tag{3}$$

(iii) For a sequence $(m_k + k!\mathbb{Z})_k \in R$, the corresponding sequence of endomorphisms $\varphi_k := \mu_{k,m_k} \in \mathrm{End}\,C_{1/k!}$ satisfies the compatibility condition (1) if, and only if, $m_{k+1} - m_k$ belongs to the kernel $k!\mathbb{Z}$ of the homomorphism (2), i.e., if $m_{k+1} \equiv m_k \bmod k!$. As explained in step (i), this is equivalent to the existence of a unique endomorphism of $\mathbb{Q}^+/\mathbb{Z}$ inducing all the endomorphisms μ_{k,m_k}. Mapping such sequences to the resulting endomorphisms of $\mathbb{Q}^+/\mathbb{Z}$ we obtain a homomorphism of the ring of all such sequences onto $\mathrm{End}(\mathbb{Q}^+/\mathbb{Z})$ which is induced by the isomorphisms (3) and therefore is an isomorphism itself.   □

31.12 Remarks   (1) The additive group of $\mathrm{End}(\mathbb{Q}^+/\mathbb{Z})$ can also be described as the group $\mathrm{Hom}(\mathbb{Q}^+/\mathbb{Z}, \mathbb{T})$ of homomorphisms to the torus group $\mathbb{T} = \mathbb{R}^+/\mathbb{Z}$, since $\mathbb{Q}^+/\mathbb{Z}$ is the torsion subgroup of $\mathbb{T}$ (see 31.4) and thus is mapped into itself by every homomorphism to $\mathbb{T}$.

(2) The situation of 31.11 can be handled very efficiently by using the notion of *direct and inverse limits* of directed systems of homomorphisms. For an introduction to these, the reader may consult, for example, FUCHS 1970 Sections 11, 12 p. 53ff, ENGLER–PRESTEL 2005 5.1 or, in a more general context, JACOBSON 1989 2.5 p. 70ff.

The ring used in 31.11 to describe $\mathrm{End}(\mathbb{Q}^+/\mathbb{Z})$ is the inverse limit $\varprojlim_k \mathbb{Z}/k!\mathbb{Z}$ of the following system of homomorphisms:

$$\mathbb{Z}/k!\mathbb{Z} \to \mathbb{Z}/j!\mathbb{Z} : m + k!\mathbb{Z} \mapsto m + j!\mathbb{Z} \quad \text{for } j \leq k \in \mathbb{N}. \tag{1}$$

We shall indicate a duality between this and the fact that the group $\mathbb{Q}^+/\mathbb{Z} = \bigcup_{k \in \mathbb{N}} \frac{1}{k!}\mathbb{Z}/\mathbb{Z} = \bigcup_{k \in \mathbb{N}} C_{1/k!}$ can be considered as the direct limit $\varinjlim_k C_{1/k!}$ of the system of inclusion homomorphisms

$$C_{1/j!} \to C_{1/k!} \quad \text{for } j \leq k \in \mathbb{N}; \tag{2}$$

see FUCHS 1970 Section 11 Example 1 p. 56. The system (1) is dual to the system (2) in the sense that it is obtained from (2) by applying the functor $\mathrm{Hom}(\ ,\mathbb{T})$. Indeed, similarly as in Remark (1) and using arguments from the proof of 31.11, one sees that $\mathrm{Hom}(C_{1/k!},\mathbb{T}) = \mathrm{End}\, C_{1/k!} \cong \mathbb{Z}/k!\mathbb{Z}$. According to a general fact (see FUCHS 1970 Theorem 44.2 p. 185), one may conclude that $\mathrm{End}(\mathbb{Q}^+/\mathbb{Z}) = \mathrm{Hom}(\mathbb{Q}^+/\mathbb{Z},\mathbb{T})$ coincides with

$$\mathrm{Hom}(\varinjlim_k C_{1/k!},\mathbb{T}) = \varprojlim_k \mathrm{Hom}(C_{1/k!},\mathbb{T}) \cong \varprojlim_k \mathbb{Z}/k!\mathbb{Z},$$

thus proving 31.11 in a more categorical setting.

(3) Another description of $\mathrm{End}(\mathbb{Q}^+/\mathbb{Z}) = \mathrm{Hom}(\mathbb{Q}^+/\mathbb{Z},\mathbb{T})$ using the p-adic numbers will be given in 52.10.

31.13 Corollary *The ring* $\mathrm{End}(\mathbb{Q}^+/\mathbb{Z})$ *has the cardinality* $2^{\aleph_0}$ *of the continuum.*

Proof Rather than deriving this from the description of $\mathrm{End}(\mathbb{Q}^+/\mathbb{Z})$ in 31.11, we follow another line. We describe $\mathbb{Q}^+/\mathbb{Z}$ as the direct sum $\bigoplus_{p \in \mathbb{P}} C_{p^\infty}$ of all Prüfer groups; see 1.28. For every subset $S \subseteq \mathbb{P}$ there is an endomorphism of the direct sum mapping the element $(a_p)_p$ to $(a_p')_p$ where $a_p' = a_p$ if $p \in S$ and $a_p' = 0$ otherwise. Since there are infinitely many primes (see 32.7ff) we have thus constructed a subset of $\mathrm{End}(\mathbb{Q}^+/\mathbb{Z})$ having the cardinality $2^{\aleph_0}$ of the power set of $\mathbb{P}$. On the other hand, $\mathrm{End}(\mathbb{Q}^+/\mathbb{Z})$ is contained in the set of all maps of $\mathbb{Q}^+/\mathbb{Z}$ into itself, and since $\mathbb{Q}$ is countable, this set has cardinality $\aleph_0^{\aleph_0} = 2^{\aleph_0}$ as well; see 61.15. Thus, our assertion is proved. $\square$

Exercises

(1) No proper subgroup of $\mathbb{Q}^+$ is isomorphic to $\mathbb{Q}^+$.

(2) Show that $\mathbb{Q}^+$ has no maximal subgroup.

(3) Are the subgroups of $\mathbb{Q}^+$ generated by $2^{\mathbb{Z}}$ and by $3^{\mathbb{Z}}$ isomorphic?

(4) Show that $\mathbb{Q}^+$ has $2^{\aleph_0}$ mutually non-isomorphic subgroups.

32　The multiplication of the rational numbers

The multiplicative group of non-zero rational numbers will be denoted by $\mathbb{Q}^{\times}$. The positive rational numbers form a subgroup $\mathbb{Q}^{\times}_{\mathrm{pos}}$ of $\mathbb{Q}^{\times}$.

We show that $\mathbb{Q}^{\times}_{\mathrm{pos}}$ is a free abelian group with the prime numbers as a free set of generators. The rank of this group, that is, the cardinality of a free set of generators, is $\aleph_0$, since there are infinitely many primes; various proofs of this truly classical result will be presented at the end of this section (see 32.7ff.). Before this, the homomorphisms between $\mathbb{Q}^{\times}$ or $\mathbb{Q}^{\times}_{\mathrm{pos}}$ and $\mathbb{Q}^+$ will be determined. We shall see in particular that the additive and the multiplicative group of $\mathbb{Q}$ are radically different, contrary to the situation in the field of real numbers (compare Section 2).

32.1 Proposition *The multiplicative group $\mathbb{Q}^{\times}_{\mathrm{pos}}$ of positive rational numbers is the direct sum of the subgroups $\langle p \rangle = \{ p^m \mid m \in \mathbb{Z} \} \cong \mathbb{Z}$ generated by the prime numbers p:*

$$\mathbb{Q}^{\times}_{\mathrm{pos}} = \bigoplus_{p \in \mathbb{P}} \langle p \rangle \cong \mathbb{Z}^{(\mathbb{P})}, \quad \text{and} \quad \mathbb{Q}^{\times} = \{1, -1\} \oplus \mathbb{Q}^{\times}_{\mathrm{pos}} \cong C_2 \oplus \mathbb{Z}^{(\mathbb{P})}.$$

Proof Recall from 1.16 that the elements of the direct sum $\mathbb{Z}^{(\mathbb{P})}$ of copies of $\mathbb{Z}$ indexed by the set $\mathbb{P}$ of primes are the maps $\alpha : \mathbb{P} \to \mathbb{Z}$ with the property that the set of primes p such that $\alpha(p) \neq 0$ is finite. For such a map, the product $\prod_{p \in \mathbb{P}} p^{\alpha(p)}$ has only finitely many factors different from 1 and therefore is a well-defined element of $\mathbb{Q}^{\times}_{\mathrm{pos}}$. It is easy to verify that the map

$$\mathbb{Z}^{(\mathbb{P})} \to \mathbb{Q}^{\times}_{\mathrm{pos}} : \alpha \mapsto \prod_{p \in \mathbb{P}} p^{\alpha(p)}$$

is a homomorphism. It is bijective, since every natural number can be written as a product of prime powers in a unique way. Under this isomorphism, the summands of the direct sum $\mathbb{Z}^{(\mathbb{P})}$ correspond to the subgroups $\langle p \rangle$ generated by the primes $p \in \mathbb{P}$. Thus, $\mathbb{Q}^{\times}_{\mathrm{pos}}$ is the direct sum of these subgroups.

The map $\{1, -1\} \oplus \mathbb{Q}^{\times}_{\mathrm{pos}} \to \mathbb{Q}^{\times} : (\varepsilon, r) \mapsto \varepsilon r$ clearly is an isomorphism. $\square$

A free abelian group is a direct sum $\mathbb{Z}^{(I)}$ of copies of $\mathbb{Z}$, and the rank of $\mathbb{Z}^{(I)}$ is the cardinality of I; compare FUCHS 1970 Section 14 p. 72ff. Granting for the moment that there are infinitely many primes (which will be proved in 32.7ff.), in other words, that the cardinality of $\mathbb{P}$ is $\aleph_0$, the result of 32.1 can thus be expressed as follows.

32.2 *The group $\mathbb{Q}_{\mathrm{pos}}^{\times}$ is a free abelian group of rank $\aleph_0$.* □

In order to study homomorphisms defined on the group $\mathbb{Q}^{\times}$, which according to 32.1 may be represented as a direct sum, we will use the following general facts.

32.3 Homomorphisms of direct sums We consider a family of abelian groups A_i indexed by the elements of a set I, and a further abelian group B; here, these groups shall be written additively. Given a homomorphism $\varphi_i : A_i \to B$ for every index $i \in I$, we may define a mapping

$$\textstyle\sum_{i \in I} \varphi_i : \bigoplus_{i \in I} A_i \to B : (a_i)_{i \in I} \mapsto \sum_{i \in I} \varphi_i(a_i) \,;$$

indeed, in the sum on the right only finitely many terms are different from 0, since the same holds for the a_i, by definition of a direct sum. It is straightforward to see that this map is a homomorphism. Conversely, every homomorphism $\varphi : \bigoplus_{i \in I} A_i \to B$ is obtained in this manner if one considers the summand A_i as a subgroup of the direct sum in the obvious way and takes φ_i to be the restriction of φ to A_i. Thus, we obtain a bijection

$$\textstyle\bigtimes_{i \in I} \mathrm{Hom}(A_i, B) \to \mathrm{Hom}(\bigoplus_{i \in I} A_i, B) : (\varphi_i)_{i \in I} \mapsto \sum_{i \in I} \varphi_i \,,$$

and a straightforward verification shows that this bijection is a homomorphism and, hence, an isomorphism.

32.4 Theorem *The additive group of the endomorphism ring $\mathrm{End}\,\mathbb{Q}_{\mathrm{pos}}^{\times}$ is isomorphic to the direct product $(\mathbb{Q}_{\mathrm{pos}}^{\times})^{\mathbb{P}}$ of countably many copies of $\mathbb{Q}_{\mathrm{pos}}^{\times}$ indexed by $\mathbb{P}$:*

$$(\mathrm{End}\,\mathbb{Q}_{\mathrm{pos}}^{\times}, +) \cong (\mathbb{Q}_{\mathrm{pos}}^{\times})^{\mathbb{P}} \,, \quad and \quad (\mathrm{End}\,\mathbb{Q}^{\times}, +) \cong C_2 \times (\mathbb{Q}^{\times})^{\mathbb{P}} \,.$$

In particular, both groups have the cardinality of the continuum.

Remarks Again, we take it for granted that there are infinitely many primes.

The given isomorphisms are not suited for describing the multiplication of these endomorphism rings; see Exercise 4.

Proof By 32.1, the group $\mathbb{Q}^\times$ is the direct sum $\{1, -1\} \oplus \mathbb{Z}^{(\mathbb{P})}$ of $\{1, -1\}$ and of copies of $\mathbb{Z}$. Thus by 32.3, $\operatorname{End}\mathbb{Q}^\times$ is the direct product

$$\operatorname{Hom}(\{1, -1\}, \mathbb{Q}^\times) \times (\operatorname{Hom}(\mathbb{Z}, \mathbb{Q}^\times))^\mathbb{P} \tag{1}$$

of $\operatorname{Hom}(\{1, -1\}, \mathbb{Q}^\times)$ and of copies of $\operatorname{Hom}(\mathbb{Z}, \mathbb{Q}^\times)$.

Since the only element of $\mathbb{Q}^\times$ of order 2 is -1, the only homomorphisms of $\{1, -1\}$ to $\mathbb{Q}^\times$ are the constant homomorphism mapping both 1 and -1 to 1, and the inclusion homomorphism; thus

$$\operatorname{Hom}(\{1, -1\}, \mathbb{Q}^\times) \cong C_2 \,.$$

The additive group $\mathbb{Z}$ is generated by 1, therefore every homomorphism $\varphi : \mathbb{Z} \to \mathbb{Q}^\times$ satisfies $\varphi(m) = \varphi(m \cdot 1) = \varphi(1)^m$, and

$$\operatorname{Hom}(\mathbb{Z}, \mathbb{Q}^\times) = \{\, m \mapsto r^m \mid r \in \mathbb{Q} \smallsetminus \{0\} \,\} \cong \mathbb{Q}^\times \,;$$

recall that addition of homomorphisms is defined by multiplication of the values in $\mathbb{Q}^\times$. Analogously $\operatorname{Hom}(\mathbb{Z}, \mathbb{Q}_{\mathrm{pos}}^\times) \cong \mathbb{Q}_{\mathrm{pos}}^\times$. The assertion on $\operatorname{End}\mathbb{Q}^\times$ now follows from (1), and for $\operatorname{End}\mathbb{Q}_{\mathrm{pos}}^\times$ it is obtained in the same way by disregarding the summand $\{1, -1\}$.

The cardinality of both direct products is $\aleph_0^{\aleph_0} = 2^{\aleph_0}$, the cardinality of the continuum (compare 61.15). □

32.5 Theorem $\operatorname{Hom}(\mathbb{Q}^\times, \mathbb{Q}^+) \cong \operatorname{Hom}(\mathbb{Q}_{\mathrm{pos}}^\times, \mathbb{Q}^+) \cong (\mathbb{Q}^+)^\mathbb{P}$, *in particular, these groups have the cardinality of the continuum.*

Proof The proof is analogous to the preceding arguments: from 32.1 and 32.4 we infer that $\operatorname{Hom}(\mathbb{Q}^\times, \mathbb{Q}^+) \cong \operatorname{Hom}(\{1, -1\}, \mathbb{Q}^+) \times \operatorname{Hom}(\mathbb{Q}_{\mathrm{pos}}^\times, \mathbb{Q}^+)$ and $\operatorname{Hom}(\mathbb{Q}_{\mathrm{pos}}^\times, \mathbb{Q}^+) \cong (\operatorname{Hom}(\mathbb{Z}, \mathbb{Q}^+))^\mathbb{P}$. Because the group $\mathbb{Q}^+$ is torsion free, the set $\operatorname{Hom}(\{1, -1\}, \mathbb{Q}^+)$ contains only the trivial homomorphism, and $\operatorname{Hom}(\mathbb{Z}, \mathbb{Q}^+) = \{\, m \mapsto mr \mid r \in \mathbb{Q} \,\} \cong \mathbb{Q}^+$. □

In 2.2 we noted that the additive group of the real numbers is isomorphic to the multiplicative group of positive real numbers (via the exponential function). In this respect, the rational numbers are quite different, as can already be seen from the different cardinalities of $\operatorname{End}\mathbb{Q}^+$ and $\operatorname{End}\mathbb{Q}_{\mathrm{pos}}^\times$ (31.10 and 32.4). The difference is even more drastic.

32.6 Proposition   *The only homomorphism of $\mathbb{Q}^+$ into $\mathbb{Q}^\times$ is the constant map $x \mapsto 1$.*

Proof The idea is to use the divisibility of $\mathbb{Q}^+$ on the one hand and to show that, on the other hand, the only element of $\mathbb{Q}^\times$ having an n-th root for every $n \in \mathbb{N}$ is 1.

For a homomorphism $\varphi : \mathbb{Q}^+ \to \mathbb{Q}^\times$, for $0 \neq x \in \mathbb{Q}$ and $n \in \mathbb{N}$ one has $\varphi(x) = \varphi(n \cdot x/n) = \varphi(x/n)^n$, so that $\varphi(x)$ has an n-th root in $\mathbb{Q}^\times$ for every $n \in \mathbb{N}$.

We now show that 1 is the only element of $\mathbb{Q}^\times$ having this property. Under the isomorphism $\mathbb{Q}^\times \cong C_2 \times \mathbb{Z}^{(\mathbb{P})}$ of 32.1, an element with this property is mapped to an element of the direct sum which is divisible by every $n \in \mathbb{N}$ (the group operation of the direct sum being addition). Now in the summands C_2 and $\mathbb{Z}$, the only element which has this divisibility property clearly is 0, which proves our assertions. $\square$

Existence of infinitely many primes

As announced, we now give several proofs of the following classical result:

32.7 Theorem   *There are infinitely many primes.*

32.8 The proof from Euclid's Elements   This is the oldest proof of which we know. One shows that, given any finite list $p_1, p_2, \ldots, p_n$ of primes, there is a further prime not contained in the list. Let $m = \prod_{k=1}^n p_k$. The number $m + 1$ has a prime factor p, and p cannot be one of the primes p_k ($k = 1, \ldots, n$) of the list, else p would be a divisor of m and hence also of 1, which is impossible. $\square$

32.9 Proof by the number of non-quadratic factors   For a natural number n, we consider its unique decomposition into prime powers, and for every prime p appearing with an odd power we split off a factor p. In this way, we may represent n uniquely in the form $n = f_n \cdot s_n^2$, where $f_n, s_n \in \mathbb{N}$ and f_n is not divisible by any square, in other words, every prime is contained at most once as a factor in f_n. The numbers s_n^2 and f_n will be called the quadratic and the non-quadratic factor of n.

For a natural number $m \leq n$, we obviously have $s_m \leq \sqrt{m} \leq \sqrt{n}$, so that there are at most $\sqrt{n}$ possible different quadratic factors of numbers $\leq n$. Hence, for these numbers there must be at least $\sqrt{n}$ different non-quadratic factors f_m. Let $\pi(n)$ be the number of primes $\leq n$. We obtain that $2^{\pi(n)} \geq \sqrt{n}$, since f_m with $m \leq n$ contains each prime $\leq n$ at most once. Therefore $4^{\pi(n)} \geq n$ and $\pi(n) \geq \ln n / \ln 4$, in particular $\pi(4^k) \geq k$. Hence, there exist infinitely many primes. $\square$

The following three proofs use sequences of special numbers.

32.10 Proof by the Mersenne numbers   These are the numbers of the form $2^p - 1$, where p is a prime number; see 2.4. Let q be a prime factor of $2^p - 1$; we show that p divides $q - 1$. This then proves that for every prime number p there is a greater prime number q, so that there must be infinitely many of them.

We consider the multiplicative group $\mathbb{F}_q^\times = \mathbb{F}_q \setminus \{0\}$ of the prime field $\mathbb{F}_q$ with q elements (in other words, of the factor ring $\mathbb{Z}/q\mathbb{Z}$). The multiplicative group has $q - 1$ elements. As q divides $2^p - 1$, so that $2^p \in 1 + q\mathbb{Z}$, we infer that p is the order of the element $2 = 1 + 1 \in \mathbb{F}_q^\times$. Now the order of an element of a group divides the order of the group; thus p divides $q - 1$. □

Remark Many Mersenne numbers are prime numbers themselves; it is as yet unknown, however, if there are infinitely many prime Mersenne numbers.

For a prime $p < 5\,000$ the corresponding Mersenne number is prime if, and only if, $p = 2, 3, 5, 7, 13, 17, 19, 31, 61, 89, 107, 127, 521, 607, 1279,$ $2203, 2281, 3217, 4253, 4423$. Much larger prime Mersenne numbers have been found, such as $2^{25\,964\,951} - 1$ (Nowak 2005), $2^{30\,402\,457} - 1$ and $2^{32\,582\,657} - 1$ (Cooper and Boone 2005, 2006). A current record is kept on the internet site `www.mersenne.org`.

32.11 Proof by the Fermat numbers These are the numbers

$$F_n = 2^{(2^n)} + 1$$

for $n \in \mathbb{N} \cup \{0\}$. We shall show that any two of these numbers are relatively prime. Considering their prime factors, one may conclude that there must be infinitely many prime numbers.

More precisely, we shall prove the following recursion formula:

$$\prod_{k=0}^{n-1} F_k = F_n - 2 \,.$$

From this it follows immediately that two Fermat numbers F_k and F_n for $k < n$ are relatively prime: If d is a common divisor, then by the recursion formula d also divides 2, and since the Fermat numbers are odd, we conclude that $d = 1$.

The recursion formula is proved by induction on n. We have $F_1 - 2 = 3 = F_0$. Assuming the formula to be valid, one obtains $\prod_{k=0}^{n} F_k = (F_n - 2) \cdot F_n = (2^{(2^n)} - 1)(2^{(2^n)} + 1) = 2^{(2^{n+1})} - 1 = F_{n+1} - 2$. □

Remarks Fermat had conjectured all Fermat numbers to be prime. Now the numbers $F_0 = 3$, $F_1 = 5$, $F_2 = 17$, $F_3 = 257$, $F_4 = 65\,537$ indeed are prime; but until today no further Fermat number has been found that is prime.

Let us prove for example that F_5 is not prime: $641 = 5 \cdot 2^7 + 1 = 5^4 + 2^4$, so that mod 641 one obtains the congruences $5^4 \equiv -2^4$ and $5 \cdot 2^7 \equiv -1$. Raised to the 4-th power, the latter yields $5^4 \cdot 2^{28} \equiv 1$. Using the former congruence, one obtains $-2^4 \cdot 2^{28} \equiv 1$ or, in other words, $F_5 = 2^{32} + 1 \equiv 0$, and 641 divides F_5.

One of the instances in which the Fermat numbers are significant is the following. A regular n-gon can be constructed by ruler and compass if, and only if, in the decomposition of n into primes only 2 and prime Fermat numbers occur, and the latter occur only with exponent 1. See, for example, JACOBSON 1985 Theorem 4.18 p. 274.

32.12 Proof by the Fibonacci numbers These are the numbers f_k (for $k \in \mathbb{N} \cup \{0\}$) defined recursively by

$$f_0 = 0; \quad f_1 = 1; \quad f_{k+1} = f_k + f_{k-1}$$

for $k \in \mathbb{N}$. By induction on $k \in \mathbb{N}$ we show that

$$f_n = f_k \cdot f_{n-k+1} + f_{k-1} \cdot f_{n-k} \quad \text{for } k < n \in \mathbb{N}. \tag{1}$$

This is clear for $k = 1$. Assuming (1) to be true for a certain k and all $n > k$, we obtain for $k + 1$ instead of k and $n > k + 1$ the fact that $f_{k+1} \cdot f_{n-(k+1)+1} + f_{(k+1)-1} \cdot f_{n-(k+1)} = f_{k+1} \cdot f_{n-k} + f_k \cdot f_{n-k-1} = f_k \cdot f_{n-k} + f_{k-1} \cdot f_{n-k} + f_k \cdot f_{n-k-1} = f_k \cdot f_{n-k+1} + f_{k-1} \cdot f_{n-k} = f_n$. Also by induction on k, it is easy to see that the greatest common divisor of two consecutive Fibonacci numbers is 1:

$$\gcd(f_k, f_{k+1}) = 1 \quad \text{for } k \in \mathbb{N}. \tag{2}$$

Next we generalize this to obtain

$$\gcd(f_k, f_n) = f_d \quad \text{where} \quad d = \gcd(k, n) \quad \text{for } k, n \in \mathbb{N}. \tag{3}$$

For the proof, we may assume $k \leq n$ and proceed by induction on n. For $n = 1$ and $k = 1$, the assertion is clear. Now let $2 \leq N \in \mathbb{N}$, and assume (3) to be true for all $k \leq n \in \mathbb{N}$ such that $n < N$ and hence for all $k, n \in \mathbb{N}$ smaller than N. We prove (3) for N instead of n and for $k \leq N$. If $k = 1$ or $k = N$, the assertion is trivial. For $1 < k < N$, we obtain from (1) and (2) that $\gcd(f_k, f_N) = \gcd(f_k, f_{k-1} \cdot f_{N-k}) = \gcd(f_k, f_{N-k})$; by the induction hypothesis, it follows that $\gcd(f_k, f_N) = f_D$ where $D = \gcd(k, N - k) = \gcd(k, N)$.

We now use (3) to show that for every finite list $p_1 = 2$, $p_2 = 3$, $p_3 = 5$, $p_4 = 7$, $\ldots$, p_n of n pairwise different prime numbers arranged by their size, there is a prime number not contained in the list. We do so by considering the Fibonacci numbers

$$f_3 = 2 = p_1, \ f_4 = 3 = p_2, \ f_{p_3} = 5 = p_3, \ f_{p_4}, \ \ldots, \ f_{p_n}.$$

By assertion (3), these are relatively prime. If all the prime factors of these n numbers were from the above list containing just n prime numbers, then f_{p_k} would have to be prime for $3 \leq k \leq n$, and $f_{p_k} = p_k$ by reasons of order. But this is obviously false ($p_4 = 7 \neq 13 = f_7$).   □

Remark The Fibonacci numbers are related to the approximation of $\zeta = (\sqrt{5} + 1)/2$ by continued fractions; see Section 4. The numbers c_ν for this approximation according to 4.1 are easily found to be $c_\nu = 1$ for all $\nu \in \mathbb{N} \cup \{0\}$, and the numbers p_ν and q_ν, according to the recursion formulae (†) in 4.2, are just the Fibonacci numbers $p_\nu = f_{\nu+2}$, $q_\nu = f_{\nu+1}$. Hence, the quotients $p_\nu/q_\nu = f_{\nu+2}/f_{\nu+1}$ of two consecutive Fibonacci numbers converge to $(\sqrt{5} + 1)/2$.

32.13 Proof by the logarithm function For $1 \leq x \in \mathbb{R}$, let $\pi(x)$ be the number of primes $\leq x$. Let us number the primes in ascending order: $\mathbb{P} = \{p_1, p_2, p_3, \ldots\}$. We now estimate the following integral from above using a step function:

$$\ln x = \int_1^x \frac{1}{t} \, dt \ \leq \ \sum_{n \in \mathbb{N}, \, n \leq x} \frac{1}{n}.$$

The sum on the right-hand side may be estimated from above by $\sum n^{-1}$ where summation is extended over all $n \in \mathbb{N}$ having only prime divisors $\leq x$. This sum in turn is the Cauchy product of geometric series

$$\prod_{p \in \mathbb{P}, \, p \leq x} \sum_{k=0}^{\infty} \frac{1}{p^k} = \prod_{p \in \mathbb{P}, \, p \leq x} \frac{p}{p-1} = \prod_{n=1}^{\pi(x)} \frac{p_n}{p_n - 1}.$$

By the rough estimate $n + 1 \leq p_n$ it follows that

$$\frac{p_n}{p_n - 1} = 1 + \frac{1}{p_n - 1} \leq 1 + \frac{1}{n} = \frac{n+1}{n}$$

and hence

$$\ln x \leq \prod_{n=1}^{\pi(x)} \frac{n+1}{n} = \pi(x) + 1.$$

In particular, $\lim_{x \to \infty} \pi(x) = \infty$, since the same is known for the logarithm, so that there must be infinitely many primes.

In fact, we have obtained an estimate for $\pi(x)$ which is better than the estimate in 32.9. The growth of $\pi(x)$ has been studied intensively; the very deep prime number theorem says that $\lim_{x\to\infty} \pi(x) \ln x/x = 1$; see, for example, HARDY–WRIGHT 1971 Theorem 6 p. 9 or ZAGIER 1997. □

32.14 A topological proof We define a topology on $\mathbb{N}$ by declaring a subset open if it is the union of subsets of the form $(a + b\mathbb{Z}) \cap \mathbb{N}$ for $a, b \in \mathbb{N}$. In order to see that this is a topology indeed, we have to show that the intersection of two open sets is open; for this, it suffices to verify that for $a, b, a', b' \in \mathbb{N}$ the intersection $(a + b\mathbb{Z}) \cap (a' + b'\mathbb{Z}) \cap \mathbb{N}$ is open. For an element c of this intersection, one easily verifies that

$$c \in (c + bb'\mathbb{Z}) \cap \mathbb{N} \subseteq (a + b\mathbb{Z}) \cap (a' + b'\mathbb{Z}) \cap \mathbb{N}.$$

Thus, this intersection contains an open neighbourhood of every element and hence is open.

For $b \in \mathbb{N} \smallsetminus \{1\}$, the set $\mathbb{N} \smallsetminus b\mathbb{N} = \bigcup_{k=1}^{b-1} (k + b\mathbb{Z}) \cap \mathbb{N}$ is open in this topology, so that $b\mathbb{N}$ is closed.

Now, if the set $\mathbb{P}$ of prime numbers were finite, then $\mathbb{N} \smallsetminus \{1\} = \bigcup_{p\in\mathbb{P}} p\mathbb{N}$ would be closed and the singleton $\{1\}$ would be open. But in this topology, all non-empty open sets are infinite. Thus, we have obtained a contradiction.

We remark that in the same way we could have used the more sophisticated topology on $\mathbb{N}$ described in 5.21. For the present purpose, however, the topology used here does the trick just as well. □

Exercises

(1) The multiplicative group $\mathbb{Q}^\times$ has uncountably many subgroups which are isomorphic to $\mathbb{Q}^\times$.

(2) Which subgroups of $\mathbb{Q}^\times$ can be embedded into $\mathbb{Q}^+$?

(3) The group $\mathbb{Q}^\times$ has uncountably many factor groups isomorphic to $\mathbb{Q}^\times$.

(4) Consider the infinite matrices $(z_{pq})_{p,q\in\mathbb{P}}$ over $\mathbb{Z}$ indexed by the set $\mathbb{P}$ of primes such that for each $q \in \mathbb{P}$ the corresponding column $(z_{pq})_p$ has only finitely many non-zero entries. These matrices form a ring $\mathbb{Z}^{(\mathbb{P})\times\mathbb{P}}$ with the usual addition and multiplication of matrices. Show that $\mathbb{Z}^{(\mathbb{P})\times\mathbb{P}}$ is isomorphic to the endomorphism ring $\operatorname{End}\mathbb{Q}^\times_{\mathrm{pos}}$. Compare this with 32.4.

(5) Show that there are infinitely many primes of the form $3n-1$ and infinitely many of the form $4n - 1$.

(6) Show that there are infinitely many primes of the form $3n + 1$.

(7) Show that there are infinitely many prime elements in the ring $\mathbb{J} = \mathbb{Z} + i\mathbb{Z}$.

33   Ordering and topology of the rational numbers

As an ordered set, $\mathbb{Q}$ has already been studied in Section 3. According to Cantor's characterization 3.4, every totally ordered countable set which is strongly dense in itself is order isomorphic to $\mathbb{Q}$. This result will be used for the study of the topological space $\mathbb{Q}$. Characterizations of $\mathbb{Q}$ as a topological space will be proved. A few results are given about the fact that algebraic irrational numbers are relatively hard to approximate by rational numbers, starting with Liouville's classical result on this subject. Transitivity properties of the group $\mathrm{Aut}(\mathbb{Q}, <)$ of order-preserving bijections and of the homeomorphism group $\mathcal{H}(\mathbb{Q})$ of $\mathbb{Q}$ are presented. The section closes with references concerning normal subgroups of $\mathrm{Aut}(\mathbb{Q})$ and the simplicity of $\mathcal{H}(\mathbb{Q})$.

33.1 The topology of $\mathbb{Q}$ (a) The usual topology on $\mathbb{Q}$ is the order topology obtained from the structure of $\mathbb{Q}$ as an ordered set. Basic open sets in this topology are the open intervals $]a, b[\subseteq \mathbb{Q}$ for $a < b \in \mathbb{Q}$. An arbitrary subset is open if, and only if, it is a union of open intervals.

(b) This topology can also be described as the topology induced on $\mathbb{Q}$ by the order topology on $\mathbb{R}$. In order to see this, one has to verify that for real numbers $r < s \in \mathbb{R}$ the set $]r, s[_{\mathbb{R}} \cap \mathbb{Q}$ is open in the order topology of $\mathbb{Q}$. (The notation $]r, s[_{\mathbb{R}}$ is used to indicate that we mean an interval in $\mathbb{R}$.) Of course, this is an interesting question only if r or s are not rational. In general, for every $x \in]r, s[_{\mathbb{R}} \cap \mathbb{Q}$ we find rational numbers $a_x, b_x \in \mathbb{Q}$ such that $r < a_x < x < b_x < s$, since $\mathbb{Q}$ is dense in $\mathbb{R}$, and then $x \in]a_x, b_x[\subseteq]r, s[_{\mathbb{R}} \cap \mathbb{Q}$, which shows that the latter set is open in the order topology of $\mathbb{Q}$. The fact that $\mathbb{Q}$ is dense in $\mathbb{R}$ is essential for this argument, as the example in (d) below shows.

(c) The topology of $\mathbb{Q}$ is also induced by the metric $d(a, b) = |b - a|$. Indeed, the interval $]a, b[$ is just the ball of radius $(b - a)/2$ centred at $(a + b)/2$.

(d) We demonstrate by an example that there are subsets of $\mathbb{Q}$ which are homeomorphic to $\mathbb{Q}$ in their order topology, but not in the topology induced from $\mathbb{R}$, so that these two topologies do not coincide. Consider

$$M =]-\infty, -1[\cup \{0\} \cup]1, \infty[\subseteq \mathbb{Q}.$$

There is an order-preserving bijection

$$M \to \mathbb{Q} : x \mapsto \begin{cases} x + 1 & \text{if } x < -1 \\ 0 & \text{if } x = 0 \\ x - 1 & \text{if } x > 1, \end{cases}$$

so M endowed with the order topology is homeomorphic to $\mathbb{Q}$ via this map. But in the topology induced from $\mathbb{R}$, the point 0 is isolated in M, whereas $\mathbb{Q}$ does not have isolated points.

33.2 We consider a few topological properties of $\mathbb{Q}$ illustrating the porosity of this space.

(a) *In $\mathbb{Q}$ every point has arbitrarily small neighbourhoods with empty boundaries.*

Indeed, such neighbourhoods may be chosen of the form $U :=]a,b[:=]a,b[_\mathbb{R} \cap \mathbb{Q}$ where $a < b$ are *irrational* real numbers. This set is open (see 33.1(b)), but also closed since the complement in $\mathbb{Q}$ is $]-\infty, a[\cup]b, \infty[$ and thus is open, as well. In other words, U equals its closure $\overline{U}$ and its interior $\operatorname{int} U$. This is equivalent to saying that the boundary of U, i.e. the set $\overline{U} \smallsetminus \operatorname{int} U$ is empty.

Using the notion of (small) inductive dimension ind, Property (a) may be expressed as $\operatorname{ind} \mathbb{Q} = 0$. Furthermore, it implies:

(b) *The space $\mathbb{Q}$ is totally disconnected, i.e. the only connected subsets are the singletons.*

To see this, let A be a subset containing two different points a, b. Let U be a neighbourhood of a in $\mathbb{Q}$ having empty boundary such that $b \notin U$. Then U is open and closed at the same time, as we have seen in (a), and $A = (A \cap U) \cup (A \cap (\mathbb{Q} \smallsetminus U))$ is the union of two non-empty disjoint subsets of A which are open in A, so that A is not connected.

33.3 Theorem *For every $n \in \mathbb{N}$, the Cartesian product space $\mathbb{Q}^n$ is homeomorphic to $\mathbb{Q}$.*

This will be obtained by showing that $\mathbb{Q}$ is homeomorphic to a certain space of sequences; a homeomorphism onto Cartesian products will then be established by the technique of mixing sequences.

Specifically, we shall consider the space $\mathbb{N}^\mathbb{N}$ of all sequences of natural numbers. This is a Cartesian product of countably many copies of $\mathbb{N}$. The factors $\mathbb{N}$ of this Cartesian product will be endowed with the discrete topology (in which all subsets are open), and we consider the product topology on $\mathbb{N}^\mathbb{N}$.

Now let $\operatorname{Per} \mathbb{N}^\mathbb{N}$ be the subset consisting of the finally periodic sequences, that is, of the sequences (a_n) with the property that for certain $n_0, p \in \mathbb{N}$

$$a_{n+p} = a_n \quad \text{for all } n > n_0. \tag{1}$$

33.4 Theorem *The space $\operatorname{Per} \mathbb{N}^\mathbb{N}$ is homeomorphic to $\mathbb{Q}$.*

Proof of Theorem 33.3 *by Theorem* 33.4. Granting 33.4 for the moment, we can readily obtain 33.3. It suffices to establish that the space $\operatorname{Per} \mathbb{N}^{\mathbb{N}}$ is homeomorphic to its Cartesian square $(\operatorname{Per} \mathbb{N}^{\mathbb{N}})^2$; then Theorem 33.3 follows by induction. Now the 'mixing map'

$$((a_1, a_2, a_3, \ldots), (b_1, b_2, b_3, \ldots)) \mapsto (a_1, b_1, a_2, b_2, a_3, b_3, \ldots)$$

is a homeomorphism of $\operatorname{Per} \mathbb{N}^{\mathbb{N}} \times \operatorname{Per} \mathbb{N}^{\mathbb{N}}$ onto $\operatorname{Per} \mathbb{N}^{\mathbb{N}}$, as can be verified easily; for a period of the mixed sequence one may take twice the product of periods of the original sequences. $\qquad\square$

Proof of Theorem 33.4. (1) The set $\operatorname{Per} \mathbb{N}^{\mathbb{N}}$ is countable. Indeed, the set of sequences satisfying equation (1) of 33.3 for fixed n_0 and p is countable, so that $\operatorname{Per} \mathbb{N}^{\mathbb{N}}$ is a union of countably many countable sets; now use 61.13.

(2) We shall also consider the Cartesian product $\mathbb{Z} \times \mathbb{N}^{\mathbb{N}}$ with the product topology (starting from the discrete topology on $\mathbb{Z}$ and $\mathbb{N}$) and use the homeomorphism $\mathbb{Z} \times \mathbb{N}^{\mathbb{N}} \to \mathbb{R} \smallsetminus \mathbb{Q}$ established in 4.10:

$$\varphi : \mathbb{Z} \times \mathbb{N}^{\mathbb{N}} \to \mathbb{R} \smallsetminus \mathbb{Q} : (a_0, a_1, a_2, \ldots) \mapsto [a_0; a_1, a_2, \ldots] \qquad (1)$$

mapping a sequence to the corresponding continued fraction. We recall the definition of a continued fraction from 4.1 and 4.2. For a sequence $(a_0; a_1, a_2, \ldots) \in \mathbb{Z} \times \mathbb{N}^{\mathbb{N}}$, that is $a_0 \in \mathbb{Z}$ and $a_n \in \mathbb{N}$ for $n \in \mathbb{N}$, one may form the fractions

$$[a_0; a_1, \ldots, a_k] := a_0 + \cfrac{1}{a_1 + \cfrac{1}{a_2 + \cfrac{1}{\cdots + \cfrac{1}{a_k}}}}.$$

Then, for $k \to \infty$, the sequence $([a_0; a_1, \ldots, a_k])_{k \in \mathbb{N}}$ converges to a limit $\zeta \in \mathbb{R} \smallsetminus \mathbb{Q}$, and the limit is approached in an alternating way from the left- and the right-hand side:

$$[a_0; a_1, \ldots, a_{2k}] < [a_0; a_1, \ldots, a_{2k+2}] < \zeta <$$
$$< [a_0; a_1, \ldots, a_{2k+3}] < [a_0; a_1, \ldots, a_{2k+1}]. \qquad (2)$$

The limit ζ is denoted by

$$[a_n]_{n \in \{0\} \cup \mathbb{N}} = [a_0; a_1, a_2, \ldots] := \lim_{k \to \infty} [a_0; a_1, \ldots a_k].$$

The sequence $(a_n)_{n \in \{0\} \cup \mathbb{N}}$ is called the expansion of ζ into a continued fraction (it is unique).

(3) The topological space $\operatorname{Per} \mathbb{N}^{\mathbb{N}}$ is obviously homeomorphic to the subspace $\{0\} \times \operatorname{Per} \mathbb{N}^{\mathbb{N}} \subseteq \mathbb{Z} \times \mathbb{N}^{\mathbb{N}}$ and therefore, by the homeomorphism φ in equation (1), homeomorphic to

$$P := \varphi(\{0\} \times \operatorname{Per} \mathbb{N}^{\mathbb{N}}) \subseteq \,]0, 1[_{\mathbb{R}} \,.$$

We show that P is dense in the real interval $]0, 1[_{\mathbb{R}}$. Let $0 < x < y < 1$. We choose an irrational number $\zeta \in \mathbb{R} \smallsetminus \mathbb{Q}$ such that $x < \zeta < y$. If $[0, a_1, a_2, \dots]$ is the expansion of ζ into a continued fraction, then according to equation (2) there is $k \in \mathbb{N}$ such that

$$x < [0; a_1, \dots a_{2k}] < \zeta < [0; a_1, \dots, a_{2k+1}] < y \,.$$

The number $\eta := [0; a_1, \dots, a_{2k}, a_{2k+1}, a_{2k+1}, \dots]$ is the image of a periodic sequence of period 1 under φ, so that $\eta \in P$. Again by equation (2), we see that

$$x < [0; a_1, \dots a_{2k}] < \eta < [0; a_1, \dots, a_{2k+1}] < y \,.$$

Thus, for any two different real numbers $x, y \in \,]0, 1[_{\mathbb{R}}$, there is indeed an element $\eta \in P$ lying strictly between them.

(4) In particular, the topology of P (induced from $\mathbb{R}$) coincides with the order topology of P as an ordered set; this can be obtained by the same argument as for $\mathbb{Q}$ in 33.1(b), the relevant point being the density property proved in step (3). Furthermore, because of this density property and since P is countable by step (1), we know from Theorem 3.4 that P is isomorphic as an ordered set to $\mathbb{Q}$, so that with the order topology P is homeomorphic to $\mathbb{Q}$. □

33.5 Remarks (1) In the proof above, the set $P := \varphi(\{0\} \times \operatorname{Per} \mathbb{N}^{\mathbb{N}})$ of numbers between 0 and 1 whose expansion into a continued fraction is periodic served as a link between the topological space $\operatorname{Per} \mathbb{N}^{\mathbb{N}}$ and the ordered set $\mathbb{Q}$. The set P consists in fact of all irrational solutions between 0 and 1 of quadratic equations with rational coefficients; see the references in 4.5.

(2) The same method as in the preceding proofs shows of course that $\mathbb{R} \smallsetminus \mathbb{Q}$ is homeomorphic to its own Cartesian square. This is even simpler since instead of $\operatorname{Per} \mathbb{N}^{\mathbb{N}}$ one just uses the whole space $\mathbb{Z} \times \mathbb{N}^{\mathbb{N}}$, which is homeomorphic to $\mathbb{R} \smallsetminus \mathbb{Q}$.

33.6 Example One might think that a phenomenon as in 33.3 should only occur for spaces of (small inductive) dimension 0, since one expects a decent dimension function to be additive for Cartesian products.

However, this is not true; decent dimension functions are well-behaved only with decent spaces. The following example is discussed in ERDÖS 1940, see also ENGELKING 1978 Example 1.5.17 p. 47 or PEARS 1975 Chapter 4 Example 1.8 p. 153.

Consider the Hilbert space $\ell^2\mathbb{R}$ of square summable real sequences, and let $\ell^2\mathbb{Q}$ be the subset consisting of the sequences having rational entries only. Then $\operatorname{ind}\ell^2\mathbb{Q} = 1$. Again with the technique of mixing sequences one sees that $\ell^2\mathbb{Q}$ is homeomorphic to its Cartesian square $(\ell^2\mathbb{Q})^2$ and, hence, to $(\ell^2\mathbb{Q})^n$ for all $n \in \mathbb{N}$.

We now prove a characterization of the topological space $\mathbb{Q}$ among the subspaces of $\mathbb{R}$.

33.7 Theorem *A subset of $\mathbb{R}$ is homeomorphic to $\mathbb{Q}$ in the induced topology if, and only if, it is countably infinite and has no isolated points.*

This implies that there are many subsets of $\mathbb{R}$ that are homeomorphic to $\mathbb{Q}$, but not isomorphic to $\mathbb{Q}$ as an ordered set. As a concrete example, we mention the set $\left\{ \sum_n c_n 3^{-n} \mid c_n \in \{0,2\} \wedge \exists_k \forall_{n \geq k}\, c_n = c_k \right\} \subseteq \mathcal{C}$ of end points of the intervals used in 5.35 to define the Cantor set $\mathcal{C}$.

Proof (compare SIERPIŃSKI 1920): (1) It is clear that every topological space homeomorphic to $\mathbb{Q}$ has the asserted properties.

(2) Now let Q be a countable subset of $\mathbb{R}$ without isolated points. We may assume that Q is contained in the interval $]0,1[$, since this interval is homeomorphic to $\mathbb{R}$ (for instance, via the homeomorphism $x \mapsto (2x - 1)/(1 - |2x - 1|)$).

Since an interval is uncountable, it must meet the complement of Q. Now every open subset of $]0,1[$ contains an interval whose ends are rational, and by choosing a point not in Q from each of these intervals, one obtains a countable dense subset $D \subseteq]0,1[$ such that $D \cap Q = \emptyset$. We enumerate the elements of the countable subsets Q and D in an arbitrary way:

$$Q = \{q_1, q_2, \dots\}, \quad D = \{d_1, d_2, \dots\}. \tag{1}$$

(3) To every finite binary sequence $\mathbf{i} = (i_1, i_2, \dots i_m) \in \{0,1\}^m$, where $m \in \mathbb{N}$, we now associate an interval

$$I_\mathbf{i} = [a_\mathbf{i}, b_\mathbf{i}] \subseteq [0,1]$$

and elements $p_\mathbf{i} \in Q$, $d_\mathbf{i} \in D$ in such a way that the following conditions are satisfied:

(3a) We have $Q_{\mathbf{i}} := Q \cap I_{\mathbf{i}} \neq \emptyset$, but $a_{\mathbf{i}}, b_{\mathbf{i}} \notin Q$.

(3b) If c is the first element of D encountered in the enumeration (1) such that

$$[0, c] \cap Q \neq \emptyset \neq [c, 1] \cap Q,$$

then

$$\{I_{(0)}, I_{(1)}\} = \{[0, c], [c, 1]\};$$

more precisely, $I_{(0)}$ is to be the one of the two intervals on the right-hand side containing q_1, and the other one is $I_{(1)}$.

(3c) The element $p_{\mathbf{i}}$ is the first element of Q encountered in the enumeration (1) which is contained in $Q_{\mathbf{i}}$; in particular, by (3b) we have $p_{(0)} = q_1$.

(3d) The element $d_{\mathbf{i}}$ is the first element of D encountered in the enumeration (1) which is contained in the interior of $I_{\mathbf{i}}$ and such that

$$[a_{\mathbf{i}}, d_{\mathbf{i}}] \cap Q \neq \emptyset \neq [d_{\mathbf{i}}, b_{\mathbf{i}}] \cap Q.$$

(3e) For $\mathbf{i} = (i_1, i_2, \ldots i_m)$ we define $\mathbf{i}' = (i_1, i_2, \ldots i_m, 0)$ and $\mathbf{i}'' = (i_1, i_2, \ldots i_m, 1)$. Then

$$\{I_{\mathbf{i}'}, I_{\mathbf{i}''}\} = \{[a_{\mathbf{i}}, d_{\mathbf{i}}], [d_{\mathbf{i}}, b_{\mathbf{i}}]\};$$

more precisely, $I_{\mathbf{i}'}$ is the one of the two intervals on the right-hand side containing $p_{\mathbf{i}}$,

$$p_{\mathbf{i}} \in I_{\mathbf{i}'},$$

and the other one is $I_{\mathbf{i}''}$.

These intervals and elements are constructed by recursion on m, the recursion step being already specified by the stated properties. For the recursive construction of the elements $d_{\mathbf{i}}$ in (3d), we note that the open interval $I_{\mathbf{i}} \smallsetminus \{a_{\mathbf{i}}, b_{\mathbf{i}}\} =]a_{\mathbf{i}}, b_{\mathbf{i}}[$ intersecting Q non-trivially in fact contains infinitely many elements of Q, as Q has no isolated points.

We now establish certain properties of these elements and sets. The following is clear from the construction.

(4) For finite binary sequences $\mathbf{i} = (i_1, i_2, \ldots i_m) \in \{0, 1\}^m$ and $\mathbf{j} = (j_1, j_2, \ldots j_n) \in \{0, 1\}^n$ such that $m \leq n$ one has

$$Q_{\mathbf{i}} \cap Q_{\mathbf{j}} \neq \emptyset \iff i_k = j_k \text{ for } k = 1, \ldots m \iff Q_{\mathbf{i}} \supseteq Q_{\mathbf{j}},$$

$$p_{\mathbf{i}} = p_{\mathbf{j}} \iff i_k = j_k \text{ for } k = 1, \ldots, m, \text{ and } j_k = 0 \text{ for } k > n.$$

Now we prove:

(5) *For every infinite binary sequence* $(i_\nu)_{\nu \in \mathbb{N}} \in \{0,1\}^\mathbb{N}$ *the diameter of* $Q_{(i_1,\ldots i_m)}$ *tends to* 0 *for* $m \to \infty$.

Assume that this is not the case. Then there is $\delta > 0$ and an infinite subset $M \subseteq \mathbb{N}$ such that for $m \in M$ the set $Q_{(i_1,\ldots i_m)}$ contains elements u_m, v_m satisfying $|u_m - v_m| > \delta$. We also may assume that the sequences $(u_m)_{m \in M}$ and $(v_m)_{m \in M}$ converge to elements $u, v \in [0, 1]$. Then we have $|u - v| \geq \delta$. Let $d \in D$ such that $u < d < v$. Now, by construction, the elements $d_{(i_1,\ldots i_m)} \in D$ are pairwise different for $m \in \mathbb{N}$, hence, if m is sufficiently large, $d_{(i_1,\ldots i_m)}$ appears later than d in the enumeration of D. By the properties of $d_{(i_1,\ldots i_m)}$, this implies that $Q_{(i_1,\ldots i_m)}$ lies entirely on one side of d, but for sufficiently large $m \in M$, the elements $u_m, v_m \in Q_{(i_1,\ldots i_m)}$ converging to u and v must be on different sides of d because of $u < d < v$. This contradiction proves our claim (5).

(6) *For every* $q \in Q$ *there is a finite binary sequence* $\mathbf{i} \in \{0,1\}^m$ *such that* $q = p_\mathbf{i}$, *and the length* m *of* $\mathbf{i}$ *may be chosen arbitrarily large.*

Indeed, by a recursive construction using (3e), one may construct an infinite binary sequence $(i_\nu)_{\nu \in \mathbb{N}} \in \{0,1\}^\mathbb{N}$ such that every initial part $(i_1, \ldots, i_m)$, $m \in \mathbb{N}$, of this sequence satisfies

$$q \in I_{(i_1,\ldots,i_m)} . \tag{2}$$

According to (3c), the elements $p_{(i_1,\ldots,i_m)}$ appear not later than q in the enumeration (1) of the elements of Q. Hence, they describe only finitely many elements of Q, and there is $p \in Q$ such that $p = p_{(i_1,\ldots,i_m)} \in Q_{(i_1,\ldots,i_m)}$ for infinitely many m. In view of equation (2), it follows from assertion (5) that p and q are arbitrarily close to each other, i.e., $q = p$ as claimed in (6).

(7) We now consider another subset Q' of $\mathbb{R}$ which is countable and has no isolated points, like Q. As with Q, we may assume up to homeomorphism that $Q' \subseteq]0, 1[$. We apply the same constructions as above to Q' instead of Q and we obtain elements $p'_\mathbf{i}$ and subsets $Q'_\mathbf{i}$ having the same properties with respect to Q' as the $p_\mathbf{i}$ and $Q_\mathbf{i}$ have for Q. We show that we obtain a well-defined continuous map

$$\varphi : Q \to Q' \quad \text{by setting} \quad \varphi(p_\mathbf{i}) = p'_\mathbf{i} \text{ for } \mathbf{i} \in \{0,1\}^m, \ m \in \mathbb{N}.$$

This will suffice to finish our proof, for by exchanging the roles of Q and Q' one then obtains a continuous inverse of φ, so that φ is a homeomorphism of Q onto Q'.

Now, φ is well-defined by (4), and by (6) it is defined for all elements of Q. In order to prove continuity, let $m \in \mathbb{N}$, $\mathbf{i} = (i_1, \ldots i_m) \in \{0,1\}^m$

and $\varepsilon > 0$. By (5), we find $n \in \mathbb{N}$ with $n \geq m$ such that for the finite sequence $\mathbf{i}^n = (i_1, \ldots, i_m, 0, 0, \ldots, 0) \in \{0,1\}^n$ of length n starting with $\mathbf{i}$ and continuing with zeros, the diameter of $Q'_{\mathbf{i}^n}$ is less than ε. Now $p_{\mathbf{i}} = p_{\mathbf{i}^n}$ by 4), and $Q_{\mathbf{i}^n}$ is a neighbourhood of $p_{\mathbf{i}^n}$, hence continuity of φ in this point will be clear if we establish that φ maps $Q_{\mathbf{i}^n}$ into $Q'_{\mathbf{i}^n}$. By (6), an element $p \in Q_{\mathbf{i}^n}$ may be written as $p = p_{\mathbf{j}}$ for a finite binary sequence $\mathbf{j}$ of length at least n. By (4) the first n entries of $\mathbf{i}^n$ and of $\mathbf{j}$ coincide, so that $\varphi(p) = p'_{\mathbf{j}} \in Q'_{\mathbf{j}} \subseteq Q'_{\mathbf{i}^n}$.

(8) In particular, applying this procedure to $\mathbb{Q}$ instead of Q', one obtains a homeomorphism of every such subset Q onto $\mathbb{Q}$.   □

Via standard topological tools, 33.7 yields the following purely topological characterization of $\mathbb{Q}$.

33.8 Corollary *Every regular countable topological space satisfying the first axiom of countability and having no isolated points is homeomorphic to $\mathbb{Q}$.*

Proof Let Q be such a space. Since it is countable and satisfies the first axiom of countability, is also satisfies the second axiom of countability. Being regular, it is therefore metrizable, according to Urysohn's theorem; see DUGUNDJI 1966 IX 9.2 p. 195. Since Q is countable and since a singleton is 0-dimensional, Q has dimension 0, and hence is homeomorphic to a subset of $\mathbb{R}$; see HUREWICZ–WALLMAN 1948 Theorem II 2 p. 18 and Theorem V 3 p. 60. Now one may apply 33.7.   □

Remark Theorem 33.4, which we proved independently, and Theorem 33.3, which we obtained via 33.4, could instead be derived as easy corollaries of Theorem 33.8. Indeed, the proofs of Theorems 33.7 and 33.8 do not rely on these former results. In the case of 33.3, this is the approach of NEUMANN 1985, where variants of 33.7 and 33.8 may also be found. The proof of 33.4 given here, based on Cantor's characterization 3.4 of $\mathbb{Q}$ as an ordered set, is constructive and more concrete.

33.9 Approximation of irrational numbers by rationals Every irrational number may be approximated by rational numbers, but there are differences in how easily this can be done. These differences play an important role in number theory. In certain instances, they may be used to tell algebraic numbers from transcendental numbers.

We recall that a real (or complex) number is said to be algebraic of degree $d \in \mathbb{N}$ if it satisfies a polynomial equation of degree d with integer coefficients. A number which is not algebraic of any degree is said to be transcendental.

One of the starting points of the subject of approximation by rational numbers was the following.

Theorem (Liouville 1844) *For every irrational algebraic real number a of degree d there exists an $\varepsilon > 0$ such that for every rational number $\frac{m}{n}$, $m \in \mathbb{Z}$, $n \in \mathbb{N}$*

$$\left| a - \frac{m}{n} \right| > \frac{\varepsilon}{n^d} \, .$$

This means vaguely that irrational algebraic real numbers cannot be approximated all too closely by fractions whose denominators are relatively small.

Proof Let f be a polynomial function of degree d with integer coefficients such that $f(a) = 0$. If $a = a_1, a_2, \ldots, a_d$ are the complex roots of f and $c \in \mathbb{Z}$ is the highest coefficient of f, then $f(x) = c \cdot \prod_{\nu=1}^{d}(x - a_\nu)$. Let m/n be a rational number such that $|a - m/n| \le 1$. Then

$$\left| f\left(\tfrac{m}{n}\right) \right| = |c| \left| a - \tfrac{m}{n} \right| \prod_{\nu=2}^{d} \left| a_\nu - \tfrac{m}{n} \right| \le$$
$$\le |c| \left| a - \tfrac{m}{n} \right| \prod_{\nu=2}^{d}(|a_\nu| + 1 + |a|) = C \cdot \left| a - \tfrac{m}{n} \right| ,$$

where $C > 0$ is a constant. Now, since the coefficients of f are integers, $n^d f(m/n)$ is an integer, and hence $n^d |f(m/n)| \ge 1$ except if m/n is one of the roots a_ν. With these possible exceptions we get from these estimates that $|a - m/n| \ge C^{-1}n^{-d}$. The irrational number a is not among these exceptions. Hence, if ε is chosen small enough to exclude these finitely many exceptions and such that $\varepsilon < C^{-1}$, then the assertion is true. $\square$

The following strong version of Liouville's theorem in which the dependence on the degree of the algebraic number has disappeared can be obtained immediately from Roth's theorem 4.7.

Theorem *Let $k > 2$ be a real number and a an irrational algebraic number. Then there is $\varepsilon > 0$ such that for every rational number m/n, $m \in \mathbb{Z}$, $n \in \mathbb{N}$*

$$\left| a - \frac{m}{n} \right| > \frac{\varepsilon}{n^k} \, .$$

For more information, we refer the reader to HARDY–WRIGHT 1971 Chapter XI, BAKER 1975 Chapter 7 and BAKER 1984 Chapter 6 Sections 3ff. A more elementary introduction may be found in NIVEN 1961 Chapters Six and Seven. Results of this type also have applications to Diophantine equations; see, for example, LANG 1971 Section 11 p. 671.

The preceding discussion motivates the study of the following set of real numbers (with $k = 3$).

33.10 Example For $0 < \varepsilon < 1$, let

$$A_\varepsilon = \bigcup_{\substack{m,n \in \mathbb{N} \\ m < n}} \left] \tfrac{m}{n} - \tfrac{\varepsilon}{n^3}, \tfrac{m}{n} + \tfrac{\varepsilon}{n^3} \right[\quad \text{and} \quad A = \bigcap_{0 < \varepsilon < 1} A_\varepsilon = \bigcap_{k \in \mathbb{N}} A_{1/k}.$$

The elements of A are numbers that can be approximated rather well by fractions with relatively small denominators, the exponent 3 being a measure for the degree of approximability. More precisely, A consists of those real numbers a between 0 and 1 for which the statements of Liouville's theorem with $d = 3$ and of the second theorem in 33.9 with $k = 3$ do not hold. In particular, Liouville's theorem (which we have proved) says that any irrational elements of A cannot be algebraic of degree ≤ 3; they even must be transcendental by the second theorem in 33.9 (which we did not prove). But of course

$$\mathbb{Q} \cap \,]0, 1[\,\subseteq A \subseteq \,]0, 1[.$$

We shall establish the following properties of A.

(i) *A is uncountable and hence contains irrational numbers.*
(ii) *A has Lebesgue measure 0.*

Proof (i) We use a topological tool, the so-called Baire category argument; see DUGUNDJI 1966 XI.10.1. We assume that A is countable to obtain a contradiction, and enumerate $A = \{\, a_n \mid n \in \mathbb{N} \,\}$ in an arbitrary way. For $n \in \mathbb{N}$, the set $A_{1/n}$ is open and dense in $[0, 1]$, since it contains $\mathbb{Q} \cap \,]0, 1[$, and so is then $A_{1/n} \smallsetminus \{a_n\}$. By Baire's theorem, applied in the compact space $[0, 1]$, the intersection of these sets is dense, as well, but $\bigcap_{n \in \mathbb{N}} (A_{1/n} \smallsetminus \{a_n\}) = \bigcap_{n \in \mathbb{N}} A_{1/n} \smallsetminus \{\, a_n \mid n \in \mathbb{N} \,\} = A \smallsetminus A = \emptyset$. This contradiction proves (i).

(ii) Basic information about the Lebesgue measure μ may be found in Section 10 or in BAUER 1968 §6 pp. 31ff, HALMOS 1950 §15 pp. 62ff, GORDON 1994 Chapter 1. Since the measure of an interval is its length and since μ is σ-subadditive, the measure of A_ε can be estimated as follows:

$$\mu(A_\varepsilon) \leq \sum_{m < n \in \mathbb{N}} \mu \left] \tfrac{m}{n} - \tfrac{\varepsilon}{n^3}, \tfrac{m}{n} + \tfrac{\varepsilon}{n^3} \right[= \sum_{n \in \mathbb{N}} (n-1) \cdot \frac{2\varepsilon}{n^3} \leq 2\varepsilon \cdot \sum_{n \in \mathbb{N}} \frac{1}{n^2}.$$

As the series on the right is convergent, this can be made arbitrarily small by choosing ε small enough. From $\mu(A) \leq \mu(A_\varepsilon)$ we infer that $\mu(A) = 0$. $\qquad \square$

In particular, A provides another example of a set of real numbers which has Lebesgue measure 0 without being countable. We knew from 5.36 that the Cantor set C has these properties.

As in the proof of (i), a Baire category argument will be used to prove the following result.

33.11 Proposition   *The set $\mathbb{Q}$ cannot be obtained as the intersection of countably many open subsets of $\mathbb{R}$.*

Proof Assume that $\mathbb{Q}$ is the intersection of a countable family of open subsets O_n, $n \in \mathbb{N}$. Since $\mathbb{Q}$ is dense in $\mathbb{R}$, so are the O_n. The sets $\mathbb{R} \smallsetminus \{q\}$ for $q \in \mathbb{Q}$ are a countable family of open and dense subsets of $\mathbb{R}$, as well. By Baire's theorem, the intersection $\bigcap_{n \in \mathbb{N}} O_n \cap \bigcap_{q \in \mathbb{Q}} (\mathbb{R} \smallsetminus \{q\})$ would have to be dense in $\mathbb{R}$, as well; but this intersection is empty. Hence, our assumption is false. $\square$

Order-preserving automorphisms and homeomorphisms

33.12   Let $\mathrm{Aut}(\mathbb{Q}, <)$ denote the set of all order-preserving bijections of $\mathbb{Q}$ onto itself. Under composition, $\mathrm{Aut}(\mathbb{Q}, <)$ is a group, and the same is true for the set $\mathcal{H}(\mathbb{Q})$ of homeomorphisms of the topological space $\mathbb{Q}$ onto itself. $\mathrm{Aut}(\mathbb{Q}, <)$ is a subgroup of $\mathcal{H}(\mathbb{Q})$, since the topology of $\mathbb{Q}$ is the order topology.

33.13 Lemma   *For each $n \in \mathbb{N}$, the group $\mathrm{Aut}(\mathbb{Q}, <)$ acts transitively on the set of subsets of $\mathbb{Q}$ having n elements. (In fact, this is true for any ordered skew field instead of $\mathbb{Q}$.)*

Proof It suffices to show that for a finite strictly increasing sequence $a_1 < a_2 < \cdots < a_n$, there is an order-preserving bijection α mapping k to a_k for $k \in \{1, 2, \ldots, n\}$. In fact, α can be chosen as a piecewise affine map; compare the proof of 33.14 below. $\square$

There are analogous results for certain infinite subsets. In order to avoid technicalities in the case of non-Archimedean ordered fields, we formulate such a result just for $\mathbb{Q}$.

33.14 Lemma   *For any strictly increasing unbounded infinite sequence $a_1 < a_2 < a_3 \ldots$ of rational numbers there is an order-preserving bijection of $\mathbb{Q}$ which maps k to a_k for all $k \in \mathbb{N}$.*

Proof The definition

$$\alpha(r) = \begin{cases} a_1 + r - 1 & \text{if } r \le 1 \\ a_k + (r-k)(a_{k+1} - a_k) & \text{if } r \in [k, k+1] \text{ with } k \in \mathbb{N} \end{cases}$$

yields a bijection α with the required properties.   □

33.15 Corollary *The groups* $\text{Aut}(\mathbb{Q}, <)$ *and* $\mathcal{H}(\mathbb{Q})$ *have the cardinality* $2^{\aleph_0}$ *of the continuum.*

Proof These groups consist of mappings of $\mathbb{Q}$ into itself and hence have cardinality at most $\aleph_0^{\aleph_0} = 2^{\aleph_0}$; see 61.15. By constructing an injection $\mathbb{N}^{\mathbb{N}} \to \text{Aut}(\mathbb{Q}, <)$, we shall show that the cardinality of $\text{Aut}(\mathbb{Q}, <)$ is at least $\aleph_0^{\aleph_0} = 2^{\aleph_0}$, and the same applies to $\mathcal{H}(\mathbb{Q}) \supseteq \text{Aut}(\mathbb{Q}, <)$.

An injection $\mathbb{N}^{\mathbb{N}} \to \text{Aut}(\mathbb{Q}, <)$ is obtained by mapping a sequence $(m_\nu)_{\nu \in \mathbb{N}}$ of natural numbers to an order-preserving bijection of $\mathbb{Q}$ which sends $k \in \mathbb{N}$ to $\sum_{\nu=1}^{k} m_\nu$ according to 33.14.   □

There are other aspects under which $\text{Aut}(\mathbb{Q}, <)$ appears to be quite large; for instance, this group cannot be obtained as the union of any countable chain of proper subgroups; see GOURION 1992.

33.16 Corollary *For each integer* $n \in \mathbb{N}$*, the topological space* $\mathbb{Q}$ *is* n*-homogeneous, which means that the homeomorphism group* $\mathcal{H}(\mathbb{Q})$ *is transitive on the set of ordered* n*-tuples.*

Remark Note the difference between this result and 33.13. In this respect, $\mathbb{Q}$ contrasts with $\mathbb{R}$, which is not n-homogeneous for $n \ge 3$; see 5.27. There are far more homeomorphisms of $\mathbb{Q}$ than order-preserving or order-reversing bijections, as 33.16 shows.

Proof We have to show that for any $n \in \mathbb{N}$ and for any two n-tuples $(a_1, a_2, \ldots, a_n)$, $(b_1, b_2, \ldots, b_n)$ of distinct elements of $\mathbb{Q}$ there is a homeomorphism mapping a_ν to b_ν for $\nu = 1, \ldots, n$. Choose a rational number $\varepsilon > 0$ so small that the intervals of length $\varepsilon\sqrt{2}$ centred at the a_ν and the b_ν are mutually disjoint. These intervals are open and closed in $\mathbb{Q}$, since their end points are irrational. Hence any map which permutes these intervals, induces homeomorphisms between them and fixes all other elements of $\mathbb{Q}$ is a homeomorphism of $\mathbb{Q}$ onto itself.   □

We have just used rational intervals with irrational end points and the fact that they are both open and closed in $\mathbb{Q}$. For such subsets, of which these intervals are only the tamest examples, the following transitivity property of $\mathcal{H}(\mathbb{Q})$ can be proved quite generally.

33.17 Theorem *Let C and D be non-empty proper subsets of $\mathbb{Q}$ which are both open and closed in $\mathbb{Q}$. Then there is a homeomorphism of $\mathbb{Q}$ mapping C onto D.*

Proof Since $C \neq \emptyset$ is open, it is (countably) infinite and has no isolated points. The same is true for the complement $\mathbb{Q} \smallsetminus C$, which is open and closed as well, and for D and $\mathbb{Q} \smallsetminus D$. By 33.7, all these sets are homeomorphic to $\mathbb{Q}$, so that there are homeomorphisms $\gamma : C \to D$ and $\gamma' : \mathbb{Q} \smallsetminus C \to \mathbb{Q} \smallsetminus D$. As their domains of definition are both open, they may be put together to yield a homeomorphism $\varphi : \mathbb{Q} \to \mathbb{Q}$ whose restrictions to C and to $\mathbb{Q} \smallsetminus C$ are γ and γ', respectively.   □

33.18 Notes More on $\mathcal{H}(\mathbb{Q})$ as a permutation group may be found in NEUMANN 1985. Among other things he shows that there are homeomorphisms of $\mathbb{Q}$ which permute $\mathbb{Q}$ in one single infinite cycle (this had already been proved by ABEL 1982), and even that the conjugacy classes of such homeomorphisms form a set which has the cardinality of the continuum. Furthermore, the cycle types of homeomorphisms of $\mathbb{Q}$ are investigated. TRUSS 1997 determines the conjugacy classes in $\mathcal{H}(\mathbb{Q})$ of elements of certain cycle types. MEKLER 1986 characterizes the countable subgroups of $\mathcal{H}(\mathbb{Q})$ as permutation groups.

Normal subgroups of the automorphism groups

The property proved in 33.13 that $\mathrm{Aut}(\mathbb{Q}, <)$ is transitive on subsets having two elements is quite important structurally. In Chapter 2 of GLASS 1981 automorphism groups of totally ordered sets with this property are studied systematically. Under further assumptions, such automorphism groups have only few normal subgroups, which may be determined explicitly (loc. cit. Theorem 2.3.2 p. 65ff). We formulate the result in the special case of $\mathbb{Q}$. Let

$$\Lambda(\mathbb{Q}) \quad \text{and} \quad \mathsf{P}(\mathbb{Q})$$

be the subgroups of $\mathrm{Aut}(\mathbb{Q}, <)$ consisting of all order-preserving bijections α of $\mathbb{Q}$ onto itself which fix every element of some interval $]-\infty, r]$ (for $\alpha \in \Lambda(\mathbb{Q})$) or $[r, \infty[$ (for $\alpha \in \mathsf{P}(\mathbb{Q})$), respectively. It is easy to see that these are normal subgroups of $\mathrm{Aut}(\mathbb{Q}, <)$.

33.19 Theorem (GLASS 1981) *The only non-trivial proper normal subgroups of $\mathrm{Aut}(\mathbb{Q}, <)$ are $\Lambda(\mathbb{Q})$ and $\mathsf{P}(\mathbb{Q})$ and their intersection, and the latter is the only non-trivial proper normal subgroup of both $\Lambda(\mathbb{Q})$ and $\mathsf{P}(\mathbb{Q})$.*   □

33.20 Simplicity of $\mathcal{H}(\mathbb{Q})$ and other groups of homeomorphisms
ANDERSON 1958 proved that the following groups of homeomorphisms
are simple, which means that they have no non-trivial proper normal
subgroups:

- the group of all homeomorphisms of $\mathbb{Q}$, of $\mathbb{R} \setminus \mathbb{Q}$, and of the Cantor
  set $\mathcal{C}$ (5.35)
- the group of orientation preserving homeomorphisms of the 2-sphere
  and the 3-sphere

and some more. In the case of spheres, the restriction to the group of
orientation-preserving homeomorphisms has to be made since this is a
normal subgroup of the group of all homeomorphisms.

Although the spaces considered are rather different, Anderson suc-
ceeds in developing a unified method working for them all. For each of
them, he considers a specific family of subsets which are chosen in such
a way that these families share certain properties with regard to homeomorphisms, and then he works with these properties, only. In the case
of $\mathbb{Q}$, this family consists of all subsets that are both open and closed.
As we have noted in the proof of 33.17, all these subsets are homeomorphic to each other. This, together with the ensuing transitivity property
stated in 33.17, are the basic facts about these subsets that are needed
to establish the properties used by Anderson.

Anderson then shows that for each of these groups and any two non-
trivial elements γ, δ there are six elements $\gamma_1, \ldots, \gamma_6$ such that $\gamma = \prod_{\nu=1}^{6} \gamma_\nu^{-1} \delta \gamma_\nu$. Then, if N is a normal subgroup and $\delta \in N \setminus \{\mathrm{id}\}$, it
follows that $\gamma \in N$. Since γ was arbitrary, N is the entire group.

Exercises

(1) The group $(\mathbb{Q}, +)$ has a proper subgroup S such that the chain $(S, <)$ is
isomorphic to the chain $(\mathbb{Q}, <)$.

(2) The chain $(\mathbb{Q}, <)$ has $2^{\aleph_0}$ automorphisms.

(3) Are the sets $\mathbb{Q}^2$ and $\mathbb{Q}^3$, taken with their respective lexicographic orderings,
isomorphic as chains?

(4) Let $\Gamma = \mathrm{Aut}(\mathbb{Q}, <)$ and $H = \mathcal{H}(\mathbb{Q})$ as in 33.12. Is Γ a normal subgroup of
H ? Determine the index of Γ in H.

(5) A countable metric space is not complete.

(6) Compare the Sorgenfrey topology σ on $\mathbb{Q}$ (see 5.73) with the ordinary
topology. Show that σ has a countable basis and that $(\mathbb{Q}, \sigma)$ is metrizable (in
contrast to the Sorgenfrey topology on $\mathbb{R}$).

34   The rational numbers as a field

The field $\mathbb{Q}$ is the unique infinite prime field. We discuss some of its distinctive features, in particular some results on sums of squares.

Let F be any field, and consider the ring homomorphism $\varphi : \mathbb{Z} \to F$ with $n^\varphi = n \cdot 1$. Either $\mathbb{Z}^\varphi$ is finite, in which case $\mathbb{Z}^\varphi \cong \mathbb{Z}/p\mathbb{Z}$ for some natural number p, and p is a prime (because F has no zero-divisors and $(hk) \cdot 1 = (h \cdot 1)(k \cdot 1)$ by distributivity), or φ is injective and we may identify $n \cdot 1$ with the integer n. In this case, the smallest subfield of F (the prime field) consists of all fractions m/n with $m, n \in \mathbb{Z}$ and $n \neq 0$, and we have

34.1 *If the smallest subfield F_0 of an arbitrary field is infinite, then F_0 is isomorphic to $\mathbb{Q}$.*

Without referring to a given field, $\mathbb{Q}$ can be described as the *field of fractions* of $\mathbb{Z}$ as follows: define an equivalence relation $\sim$ on $\mathbb{Z} \times \mathbb{N}$ by $(a, b) \sim (c, d) \rightleftharpoons ad = bc$, write a/b for the equivalence class of (a, b), and define addition and multiplication in the familiar way. Then $a/b = ab^{-1}$ and $\mathbb{Q} = \{ ab^{-1} \mid a \in \mathbb{Z} \wedge b \in \mathbb{N} \}$.

34.2 Remarks on addition and multiplication   The field $\mathbb{Q}$ is essentially determined by its additive group: if F is a field with $F^+ \cong \mathbb{Q}^+$, then by 31.10 there is an isomorphism $\sigma : \mathbb{Q}^+ \to F^+$ which maps 1 to the unit element e of $F = (F, +, *)$; now the distributive law implies that $\sigma(nq) = n\sigma(q) = (ne) * \sigma(q) = \sigma(n) * \sigma(q)$ for $n \in \mathbb{N}$, $q \in \mathbb{Q}$, whence σ is an isomorphism of fields.

In contrast, $\mathbb{Q}$ is not determined by its multiplicative group: we denote by $F = \mathbb{F}_3(x)$ the field of fractions of the polynomial ring $\mathbb{F}_3[x]$. Then $\mathbb{Q}^\times \cong F^\times$ by 32.1, because $F^\times$ is the direct product of $\{\pm 1\}$ with the direct sum of the cyclic groups generated by the countably many monic irreducible polynomials in $\mathbb{F}_3[x]$. Another example is the field $F = \mathbb{Q}(x)$; replacing x by a finite or countable set of indeterminates gives more examples. If a is any real algebraic number, then $\mathbb{Q}(a)^\times \cong \mathbb{Q}^\times$; this is a consequence of a result of Skolem; see FUCHS 1973 Theorem 127.2.

Similar remarks apply to the ring $\mathbb{Z}$: the additive group does essentially determine the multiplication; indeed, upon choosing one of the two generators of $\mathbb{Z}^+$ as the unit element, the multiplication is determined via the distributive law. On the other hand, there are many rings R such that the multiplicative semigroup $(R, \cdot)$ is isomorphic to $(\mathbb{Z}, \cdot)$, like the polynomial rings $\mathbb{F}_3[x]$ or $\mathbb{Z}[x]$.

See 6.2c, d and 14.7 for analogous remarks on the fields $\mathbb{R}$ and $\mathbb{C}$.

We write $\mathbb{Q}^{\square} = \{\, x^2 \mid x \in \mathbb{Q}^{\times} \,\}$ for the multiplicative group of non-zero squares of $\mathbb{Q}$. Then 32.2 implies immediately

34.3 *Written additively, the factor group* $\mathbb{Q}^{\times}/\mathbb{Q}^{\square}$ *is a vector space over* $\mathbb{F}_2$ *of dimension* $\aleph_0$.

This contrasts sharply with the real case, where $\mathbb{R}^{\times}/\mathbb{R}^{\square}$ has order 2. Next, we study the set

$$S = \mathbb{Q}^{\square} + \mathbb{Q}^{\square}$$

consisting of all sums of two non-zero squares.

34.4 Lemma *The following identity of Diophant holds in every commutative ring:*

$$(a^2 + b^2)(c^2 + d^2) = (ac \pm bd)^2 + (ad \mp bc)^2 \ . \qquad \square$$

34.5 Proposition *The set S is a subgroup of $\mathbb{Q}^{\times}$, and $\mathbb{Q}^{\square} < S < \mathbb{Q}^{\times}$.*

Proof (a) As $1 = (\frac{3}{5})^2 + (\frac{4}{5})^2$, we have $1 \in S$ and hence $\mathbb{Q}^{\square} = 1 \cdot \mathbb{Q}^{\square} \subseteq S$. The well known fact that $2 \notin \mathbb{Q}^{\square}$ gives $\mathbb{Q}^{\square} \neq S$.

(b) Lemma 34.4 implies that $S \cdot S \subseteq S \cup \mathbb{Q}^{\square} = S$. Since $s^{-1} = s(s^{-1})^2$ for $s \in S$, the set S is indeed a subgroup of $\mathbb{Q}^{\times}$.

(c) Assume that $3 \in S$. Then there are numbers $a, b, c \in \mathbb{N}$ with $a^2 + b^2 = 3c^2$ and $a, b \not\equiv 0 \bmod 3$. This would imply $1 + 1 \equiv 0 \bmod 3$, which is a contradiction. Consequently, $S \neq \mathbb{Q}^{\times}$. $\qquad \square$

A rational number q belongs to S if, and only if, it can be written in the form $q = (a^2 + b^2)c^{-2}$ with $a, b, c \in \mathbb{N}$. In order to describe S it suffices therefore to determine the set $\{\, a^2 + b^2 \mid a, b \in \mathbb{N} \,\}$. This will be done in the next steps; see also Exercise 5.

34.6 Lemma *If $a^2 + b^2 = m$ with $a, b \in \mathbb{N}$, and if p is an odd prime dividing m, then $p \equiv 1 \bmod 4$ or $a \equiv b \equiv 0 \bmod p$.*

Proof The numbers a and b may be considered as elements of the field $\mathbb{F}_p = \mathbb{Z}/p\mathbb{Z}$. If $b \in \mathbb{F}_p^{\times}$ and $c = ab^{-1}$, then $c^2 = -1 \in \mathbb{F}_p^{\times}$. It follows that the order $p - 1$ of $\mathbb{F}_p^{\times}$ is a multiple of 4. $\qquad \square$

34.7 Theorem *For an odd prime number p the following conditions are equivalent:*

(a) $p \equiv 1 \bmod 4$
(b) $p = x^2 + y^2$ *for some rational numbers x and y*
(c) $p = a^2 + b^2$ *for a unique (unordered) pair of natural numbers a, b.*

Proof (b) is weaker than (c), and (a) follows from (b) by Lemma 34.6. We show that (a) implies (c).

If (a) holds, then $\mathbb{F}_p^\times$ is a cyclic group (see Section 64, Exercise 1) of order $4k$. Hence there is an integer c with $c^2 \equiv -1 \bmod p$ and c can be chosen between 0 and $p/2$. Consider the continued fraction $p/c = [q_0; q_1, \ldots, q_\ell]$ as described in Section 4. Starting with $r_{-1} = p$ and $r_0 = c$, the q_ν are given by the Euclidean algorithm $r_{\nu-1} = q_\nu r_\nu + r_{\nu+1}$, ending with $r_\ell = (p, c) = 1$. If $s_{-1} = 1$ and if s_ν denotes the numerator of $[q_0; q_1, \ldots, q_\nu]$, then $s_0 = q_0$ and $s_{\nu+1} = q_{\nu+1} s_\nu + s_{\nu-1}$; see 4.1. By induction we show that

$$p = r_\nu s_\nu + r_{\nu+1} s_{\nu-1} \quad \text{and} \quad r_\nu^2 + s_{\nu-1}^2 \equiv 0 \bmod p;$$

indeed, $p = r_0 s_0 + r_1 s_{-1}$, and the assertion for ν implies that $p = (q_{\nu+1} r_{\nu+1} + r_{\nu+2}) s_\nu + r_{\nu+1} s_{\nu-1} = r_{\nu+1}(q_{\nu+1} s_\nu + s_{\nu-1}) + r_{\nu+2} s_\nu = r_{\nu+1} s_{\nu+1} + r_{\nu+2} s_\nu$. Thus the first claim is proved; in particular we have $r_\nu^2 s_\nu^2 \equiv r_{\nu+1}^2 s_{\nu-1}^2 \bmod p$. Now $r_\nu^2 + s_{\nu-1}^2 \equiv 0 \bmod p$ implies that $r_\nu^2 s_\nu^2 \equiv -r_\nu^2 r_{\nu+1}^2 \bmod p$ and then $r_{\nu+1}^2 + s_\nu^2 \equiv 0 \bmod p$.

There is an index κ such that $r_\kappa < \sqrt{p} < r_{\kappa-1}$, and $s_{\kappa-1} < \sqrt{p}$ in view of $r_{\kappa-1} s_{\kappa-1} < p$. Since $r_\kappa^2 + s_{\kappa-1}^2 \equiv 0 \bmod p$ and $0 < r_\kappa^2 + s_{\kappa-1}^2 < 2p$, existence of a and b follows, and only uniqueness remains to be shown. Assume that $p = u^2 + v^2$ for another pair of numbers $u, v \in \mathbb{N}$. Then $(au + bv)^2 + (av - bu)^2 = p^2$ by Diophant's identity 34.4. Since $c^{-1} \equiv -c \bmod p$ and $p = a^2 + b^2$, notation can be chosen in such a way that $a/b \equiv c \equiv u/v \bmod p$. This implies $av - bu \equiv 0 \equiv au + bv \bmod p$. Because $au + bv > 0$, it follows that $au + bv = p$ and $bu - av = 0$. Eliminating v gives $(a^2 + b^2)u = ap$ and therefore $u = a$ and $v = b$. □

34.8 Remarks Note that s_ν is the determinant $|q_0; q_1, \ldots, q_\nu|$ as in 4.2. Utilizing the formal properties of these determinants, a related proof for the existence of a and b has been given by SMITH 1855; compare also WAGON 1990. A completely different, nice and short proof is due to ZAGIER 1990; see ELSHOLTZ 2003 for a detailed discussion. The equivalence of (a) and (c) has already been proved by Fermat.

34.9 Corollary *A natural number $n = \prod_p p^{v_p}$ is a sum of two squares of integers if, and only if, v_p is even whenever $p \equiv 3 \bmod 4$.*

Proof The condition on the v_p is sufficient. This follows from $2 = 1^2 + 1^2$ and Theorem 34.7, together with Diophant's identity 34.4.

The condition is also necessary. If a prime $p \equiv 3 \bmod 4$ divides $n = a^2 + b^2$, then p divides a and b by 34.6, and p^2 can be cancelled. □

34.10 Remarks In general, Diophant's identity 34.5 yields different representations, for example, $13 \cdot 29 = 11^2 + 16^2 = 19^2 + 4^2$.

As will be shown in 34.22, the number of distinct representations of any given $n \in \mathbb{N}$ can be determined. It is thus possible to count the lattice points in the interior of a circle. If the radius of the circle is sufficiently large, the number of lattice points approximates the area. This will lead to the famous series $\pi/4 = 1 - 1/3 + 1/5 - 1/7 + - \ldots$; see 34.23.

For a characterization of the sums of two *positive* squares of integers see Exercise 5.

We put $\mathbb{P}_- = \{\, p \in \mathbb{P} \mid p \equiv -1 \bmod 4 \,\}$ and $\mathbb{P}_+ = \mathbb{P} \smallsetminus \mathbb{P}_-$.

34.11 Proposition *Both sets $\mathbb{P}_+$ and $\mathbb{P}_-$ are infinite.*

Proof A slight variation of Euclid's proof 32.8 applies to $\mathbb{P}_-$: if m is any product of finitely many primes in $\mathbb{P}_-$, then $2m + 1 \equiv -1 \bmod 4$. Hence $2m + 1$ has at least one prime divisor $q \equiv -1 \bmod 4$, and q is different from each factor of m.

Assume that $P := \prod_{p \in \mathbb{P}_+} p$ is finite. By 34.6, any odd prime factor p of $P^2 + 1$ satisfies $p \equiv 1 \bmod 4$ and hence divides P, a contradiction. □

34.12 *Written additively, both groups $\mathbb{Q}^\times/S$ and $S/\mathbb{Q}^\square$ are countably infinite vector spaces over $\mathbb{F}_2$.*

Proof Remember from 34.5 that $\mathbb{Q}^\square < S$. Therefore each coset in one of the groups has an integer representative of the form $z = \pm \prod_\nu p_\nu^{v_\nu}$ with $p_\nu \in \mathbb{P}$ and $v_\nu \in \{0,1\}$, $v_\nu = 0$ for almost all ν. Note that $2 \in S$. In the first case, it follows from 34.7 that z can be chosen as $\pm \prod_{p \in Q} p$ for some finite set $Q \subset \mathbb{P}_-$; in the second case, 34.9 implies that $z = \prod_{p \in R} p$ with $R \subset \mathbb{P}_+$. Moreover, these representatives are unique (use Lemma 34.6 in the first case). By 34.11 both groups are infinite. □

34.13 Pythagorean triples Let $a, b, c \in \mathbb{N}$ such that $\gcd(a,b) = 1$ and a is odd. Then $a^2 + b^2 = c^2$ if, and only if, $a = u^2 - v^2$, $b = 2uv$, and $c = u^2 + v^2$ for some $u, v \in \mathbb{N}$.

Proof Assume that $a^2 + b^2 = c^2$. Then b is even and c is odd (or else $a^2 \equiv b^2 \equiv 1 \bmod 4$ and $c^2 \equiv 2 \bmod 4$ which is impossible). Hence the numbers $r = b/2$, $s = (c + a)/2$, $t = (c - a)/2$ belong to $\mathbb{N}$. Moreover, $\gcd(s,t) = \gcd(c,a) = \gcd(a,b) = 1$ and $st = r^2$. Consequently, there are $u, v \in \mathbb{N}$ with $s = u^2$ and $t = v^2$, and u and v have the required properties. The converse is obvious. □

For comparison and later use we show the following.

34.14 Theorem   *In any finite field* $\mathbb{F}_q$ *each element is a sum of two squares* (*including* 0).

Proof (a) If q is even, then $x \mapsto x^2$ is a homomorphism of the additive group $\mathbb{F}_q^+$ and hence a field automorphism.

(b) In the odd case, put $Q = \{\, x^2 \mid x \in \mathbb{F}_q \,\}$, $S = Q + Q$, and $S^\times = S \smallsetminus \{0\}$. Then $Q^\times \le S^\times \le \mathbb{F}_q^\times$ since $S^\times$ is contained in the finite group $\mathbb{F}_q^\times$ and $S^\times$ is a semigroup by Diophant's identity 34.4. Note that $|\mathbb{F}_q^\times : Q^\times| = 2$ because the map $x \mapsto x^2$ has a kernel of order 2. If $S^\times = Q^\times$, then S is a subgroup of order $(q+1)/2$ in $\mathbb{F}_q^+$, a contradiction. Therefore $S^\times = \mathbb{F}_q^\times$.

Alternative proof: For each $c \in \mathbb{F}_q$ the set $c - Q$ has $(q+1)/2$ elements. Consequently, $Q \cap (c - Q) \ne \emptyset$, and there are elements $x, y \in \mathbb{F}_q$ such that $x^2 = c - y^2$. □

34.15 Remark   If $a, b \in \mathbb{F}_q^\times$, then $\{\, ax^2 + by^2 \mid x, y \in \mathbb{F}_q \,\} = \mathbb{F}_q$ by an obvious modification of the last proof.

Fermat claimed and Lagrange proved that each natural number is a sum of at most four integer squares. First we note

34.16   *A natural number of the form* $4^e n$ *with* $n \equiv 7 \bmod 8$ *cannot be written as a sum of at most three squares.*

Proof If x is odd, then $x = 4k \pm 1$ and $x^2 \equiv 1 \bmod 8$. Hence n cannot be obtained as a sum of three odd squares or of one odd and a few even squares. If $4^e n = a^2 + b^2 + c^2$, we may assume that a and b are odd and c is even, but then $a^2 + b^2 + c^2 \equiv 2 \bmod 4$. □

Quaternions are a useful tool for dealing with sums of four squares; compare EBBINGHAUS *et al.* 1991 Chapter 7, SALZMANN *et al.* 1995 §11, CONWAY–SMITH 2003, and Section 13, Exercise 6.

34.17 Review: Quaternions   The (real) *quaternions* form a skew field $\mathbb{H}$ with centre $\mathbb{R}$ and basis $1, i, j, k$ over $\mathbb{R}$ such that $i^2 = j^2 = -1$ and $k = ij = -ji$. There is an *anti-automorphism* of $\mathbb{H}$ mapping $x = x_0 + x_1 i + x_2 j + x_3 k$ onto the *conjugate* $\overline{x} = 2x_0 - x$, and $Nx = x\overline{x} = \overline{x}x = \sum_\nu x_\nu^2 \in \mathbb{R}$. Obviously, $N(xy) = (Nx)(Ny)$. This identity expresses a product of sums of four squares again as a sum of four squares.

Note that the quaternions with integer coefficients form a subring $\mathbb{G}$ of $\mathbb{H}$. If $x, y \in \mathbb{G}$ and if the coordinates satisfy $x_\nu - y_\nu \in m\mathbb{Z}$, we write

$x \equiv y \bmod m$ for convenience. This implies $xz \equiv yz \bmod m$ for any $z \in \mathbb{G}$, also $\overline{x} \equiv \overline{y} \bmod m$ and $Nx \equiv Ny \bmod m$.

34.18 Theorem (Lagrange 1770) *Each natural number n is a sum of four integer squares (including 0).*

Proof By 34.7 and the identity in 34.17, it suffices to prove the assertion for prime numbers $p \equiv 3 \bmod 4$. Theorem 34.14 implies that $a^2+b^2+1 \equiv 0 \bmod p$ for some integers a, b with $0 \leq a, b < p/2$. Hence there is a smallest number $m \in \mathbb{N}$ such that mp is a sum of four squares, i.e. that $mp = Nx$ for some $x \in \mathbb{G}$. From $|a|, |b| < p/2$ it follows that $m < p/2$. We have to show that $m = 1$.

Assume that $m \geq 2$. There is a quaternion $c \in \mathbb{G}$ with $c \equiv x \bmod m$ and $|c_\nu| \leq m/2$ for all coordinates. 34.17 gives

$$\overline{c}x \equiv Nc \equiv Nx \equiv 0 \bmod m.$$

Consequently, $Nc = hm$ for some $h \in \mathbb{N}$ and $h \leq m$ (as $|c_\nu| \leq m/2$). Moreover, since $\overline{c}x \equiv 0 \bmod m$, there exists a quaternion $z \in \mathbb{G}$ such that $\overline{c}x = m \cdot z$. Taking norms and dividing by m^2, we find $Nz = hp$, and minimality of m implies $h = m$. This is only possible if $|c_\nu| = m/2 = k \in \mathbb{N}$ for all ν, and then $x \equiv 0 \bmod k$. On the other hand, by minimality of m and because $0 < m < p/2$, the x_ν are relatively prime, and $m = 2$. Hence $Nx = 2p \equiv 2 \bmod 4$ and exactly two of the x_ν are odd, say x_1 and x_3. We put

$$y_0 = \tfrac{1}{2}(x_2 + x_0), \ y_1 = \tfrac{1}{2}(x_3 + x_1), \ y_2 = \tfrac{1}{2}(x_2 - x_0), \ y_3 = \tfrac{1}{2}(x_3 - x_1)$$

and we form a quaternion $y \in \mathbb{G}$ with coordinates y_ν. Then we have $Ny = \tfrac{1}{2}Nx = p$. $\square$

34.19 Remarks (a) If we start with $a^2 + 1 \equiv 0 \bmod p$ and restrict x, c and z to numbers in $\mathbb{G} \cap \mathbb{C}$, then the assumption $m > 1$ yields $h \leq m/2$. This contradicts minimality of m and gives a new proof of 34.7.

(b) Note that 41 is not a sum of four positive squares, and that 103 is not a sum of four distinct squares.

(c) Result 34.18 is an immediate consequence of the following more difficult result, which is proved in GROSSWALD 1985 Chapter 4 and SMALL 1986; compare also HALTER-KOCH 1982 §1, GREENFIELD 1983 and CONWAY 1997 p. 137/9.

Theorem (Legendre 1798) *Any natural number which is not of the form $4^e n$ with $n \equiv 7 \bmod 8$ can be represented as a sum of at most three squares.*

Most other proofs of 34.18 are similar to the one given above; see, for example, GROSSWALD 1985 Chapter 3 §3, SMALL 1982, or ROUSSEAU 1987.

Conway and Schneeberger have found the following remarkable generalization of 34.18; see CONWAY 2000 and BHARGAVA 2000:

Theorem *If a positive definite quadratic form given by a symmetric integral matrix (in any number of variables) represents each of the nine numbers $1, 2, 3, 5, 6, 7, 10, 14, 15$, then it is universal, i.e., it represents every natural number.*

Write $r_4(n) = \text{card}\{x \in \mathbb{G} \mid Nx = n\}$ for the number of distinct ordered quadruples of integers such that the sum of their squares is equal to n. We mention a formula for $r_4(n)$.

Theorem (Jacobi) $r_4(n) = 8 \sum \{d \in \mathbb{N} \mid d \not\equiv 0 \bmod 4 \wedge d \mid n\}$.

For a *proof* see HARDY–WRIGHT 1971 Theorem 386, ANDREWS *et al.* 1993 or HIRSCHHORN 1987.

34.20 Corollary   *Each positive rational number is a sum of four non-zero squares in $\mathbb{Q}$. Some numbers in $\mathbb{Q}_{\text{pos}}$ cannot be represented as a sum of at most three rational squares.*

Proof Write $m/n = (mn)n^{-2}$ and apply Lagrange's theorem to mn. Using any Pythagorean triple, a sum of fewer than four squares can also be written as a sum of exactly four positive squares. The second claim follows from 34.16; for example, 7 is not a sum of three squares.   □

The the subring $\mathbb{Z}(i)$ of $\mathbb{C}$ will help to make the proof of Theorem 34.22 more transparent.

34.21 Gaussian integers   *The ring $\mathbb{J} := \mathbb{Z}(i) = \mathbb{Z} + \mathbb{Z}i$ is a unique factorization domain. If $c \in \mathbb{J}$ and $c\bar{c} = p$ is a prime in $\mathbb{N}$, then c and $\bar{c}$ are prime in $\mathbb{J}$.*

These well known facts can be *proved* by means of the Euclidean algorithm with respect to the ordinary norm. The first part is a standard result and can be found in most algebra texts (e.g. HERSTEIN 1975 3.8). Assume now that $c = uv$ with $u, v \in \mathbb{J}$ and $u\bar{u} > 1$. Then $u\bar{u}$ divides the prime p, hence $v\bar{v} = 1$ and v is a unit.   □

The number $r_2(n) = \text{card}\{(x, y) \in \mathbb{Z} \times \mathbb{Z} \mid x^2 + y^2 = n\}$ of lattice points on a circle of radius $\sqrt{n}$ is expressed in the next result in terms of $\delta^{\pm}(n) = \text{card}\{t \in \mathbb{N} \mid t \equiv \pm 1 \bmod 4 \wedge t \mid n\}$.

34.22 Theorem Put $\delta(n) = \delta^+(n) - \delta^-(n)$. Then $r_2(n) = 4\delta(n)$.

Proof Consider $\rho(n) = \text{card}\{\,(a,b) \in \mathbb{N} \times (\mathbb{N}_0) \mid a^2 + b^2 = n\,\}$. Obviously, $r_2(n) = 4\rho(n)$, and we have to show that $\rho = \delta$. This will be done in several small steps.

(a) Note that $n = a^2 + b^2 \Leftrightarrow 2n = (a+b)^2 + (a-b)^2$. Hence $\rho(2n) = \rho(n) = \rho(2^e n)$. Because δ depends only on the odd divisors of n, it suffices to study odd numbers n.

(b) The assertion is true for primes: in fact, if $p \equiv 1 \bmod 4$, then $p = a^2 + b^2$ for exactly two *ordered* pairs of natural numbers by Theorem 34.7, and $\delta(p) = \delta^+(p) = 2$. If $p \equiv -1 \bmod 4$, then p is not a sum of two squares (34.6), and $\delta^+(p) = \delta^-(p) = 1$.

(c) If $a^2 + b^2 = n = mq$ for a prime $q \equiv -1 \bmod 4$ and some $m \in \mathbb{N}$, then $q \mid a, b$ by 34.6. Consequently, $\rho(mq^2) = \rho(m)$. On the other hand, $\delta(m) = -\delta(mq) = \delta(mq^2)$. In the case $q^2 \nmid m$, it follows that $\rho(n) = 0$, and one has $\delta^+(n) = \delta^-(n)$.

We may assume, therefore, that n is a product of primes $p_\nu \equiv 1 \bmod 4$.

(d) In this case, δ can easily be calculated: if $p \equiv 1 \bmod 4$, then $\delta(p^\kappa) = \kappa + 1$ is just the number of all divisors of p^κ. Moreover, $\gcd(m, n) = 1$ implies $\delta(mn) = \delta(m)\delta(n)$. It remains to show that ρ has analogous properties.

(e) Let $\tilde{\rho}(n) = \text{card}\{\,(a,b) \in \mathbb{N}^2 \mid a^2 + b^2 = n \wedge (a,b) = 1\}$ be the number of *primitive* solutions, and put $\tilde{\rho}(1) = 1$. By the very definition, $\rho(n) = \sum_{d^2 \mid n} \tilde{\rho}(d^{-2}n)$.

(f) For a prime $p \equiv 1 \bmod 4$, the equation $a^2 + b^2 = p$ has only primitive solutions, and $\rho(p) = \tilde{\rho}(p) = 2$ by theorem 34.7. In step (g) we will show that $\tilde{\rho}(p^\kappa) = 2$ for each $\kappa \in \mathbb{N}$. With (e) this implies for even as well as for odd κ that $\rho(p^\kappa) = \kappa + 1$.

(g) Suppose that $p^\kappa = u^2 + v^2$ with $\gcd(u,v) = 1$, and let $p = c\bar{c}$, where $c \in \mathbb{J}$. By 34.21, the elements $c, \bar{c}$ are primes in $\mathbb{J}$ and $\mathbb{J}$ is a unique factorization domain. Hence $w = u + vi$ is of the form $c^{\kappa-\lambda}\bar{c}^\lambda$ (up to a unit i^μ). If $\lambda \neq 0, \kappa$, then $c\bar{c} = p$ divides w. This means that $w = pz$ for some $z \in \mathbb{J}$, and $p \mid u, v$, which is a contradiction. Therefore, up to a unit, $w = c^\kappa$ or $w = \bar{c}^\kappa$ and the pair $\{u, v\}$ is essentially unique. Consequently, $\tilde{\rho}(p^\kappa) = 2$.

(h) Let $n = \prod_{\nu=1}^{r} p_\nu^{\kappa_\nu} = u^2 + v^2$ with $p_\nu \equiv 1 \bmod 4$ and $\gcd(u,v) = 1$. We have to show that $\tilde{\rho}(n) = 2^r$. Consider a prime divisor c_ν of p_ν in $\mathbb{J}$ and put $w = u + vi$. By the argument of step (g) it follows that $w = i^\mu \prod_{\nu=1}^{r} \hat{c}_\nu^{\kappa_\nu}$, where $\hat{c}_\nu \in \{c_\nu, \bar{c}_\nu\}$. As only positive solutions are counted, this yields exactly 2^r pairs (u, v).

(i) Assume, finally, that $\gcd(m, n) = 1$ and that $m \cdot n$ has only prime factors $p \equiv 1 \bmod 4$. Then $\tilde{\rho}(m \cdot n) = \tilde{\rho}(m) \cdot \tilde{\rho}(n)$ by (h). The formula $\rho(m) = \sum_{d^2 | m} \tilde{\rho}(d^{-2}m)$ from (e) implies immediately that $\rho(m \cdot n) = \rho(m) \cdot \rho(n)$. $\quad\square$

34.23 Corollary (Leibniz 1673) $\frac{\pi}{4} = 1 - \frac{1}{3} + \frac{1}{5} - \frac{1}{7} + \frac{1}{9} - + \cdots$.

Proof The number of lattice points in a circle of radius r is

$$\gamma(r) = \operatorname{card}\{ (x, y) \in \mathbb{Z}^2 \mid x^2 + y^2 \le r^2 \} = 1 + 4 \sum_{n \le r^2} \rho(n).$$

The unit squares centred at these lattice points approximate the area of the circle. More precisely, $(r - 1)^2 \pi \le \gamma(r) \le (r + 1)^2 \pi$, and $\lim_{r \to \infty} r^{-2}(\gamma(r) - 1) = \pi$. By the last theorem, this is equivalent to $\pi/4 = \lim_{r \to \infty} r^{-2} \sum_{n \le r^2} \delta(n)$.

This sum can be determined by counting how often each odd number d divides some natural number $n \le r^2$. We obtain

$$
\begin{aligned}
\sum_{n \le r^2} \delta(n) &= \left[r^2\right] - \left[3^{-1}r^2\right] + \left[5^{-1}r^2\right] - + \cdots \\
&= r^2 - 3^{-1}r^2 + - \cdots - (4k - 1)^{-1}r^2 + \psi(r) .
\end{aligned}
$$

(As customary, $[x]$ denotes the greatest integer $z \le x$.) It is easy to find a bound for the error $|\psi(r)|$: if only the first $2k$ terms of the upper sum are considered, the error is at most $\left[(4k + 1)^{-1}r^2\right]$ because the terms of the sum decrease and have alternating signs. For the same reason, the two sums differ by less than k. Choose k such that $r \le 2k < r + 2$. Then $|\psi(r)| < (2r)^{-1}r^2 + r/2 + 1 = r + 1$. Hence $r^{-2}|\psi(r)|$ converges to 0 and $\pi/4 = \sum_{\nu=0}^{\infty}(-1)^\nu(2\nu + 1)^{-1}$.

See also HILBERT–COHN-VOSSEN 1932 §6.1. For the problem of finding the precise order of magnitude of $\gamma(r) - r^2\pi$, see GROSSWALD 1985 Chapter 2 §7. $\quad\square$

34.24 Multiplicative quadratic forms Diophant's identity 34.4 or the equation $N(xy) = (Nx)(Ny)$ for quaternions are special cases of more general results mainly due to Pfister.

Let $V \cong F^n$ be a vector space over a field F of characteristic $\ne 2$. Then a quadratic form φ on V given by $\varphi(x) = xAx^t$ (with a matrix $A \in F^{n \times n}$) is said to be *multiplicative* if there exists a vector $z = (z_1, \ldots, z_n)$ with rational functions $z_\nu \in F(x, y)$ such that the equation $\varphi(x) \cdot \varphi(y) = \varphi(z)$ holds in $F(x, y)$; if z depends linearly on y for each vector x, then φ is called *strictly multiplicative*. It is a rare phenomenon that z is even a bilinear function of (x, y); in that case we say that φ is a *composition form*.

Let $\langle a_1, \ldots, a_n \rangle$ denote the form defined by the diagonal matrix $A = \mathrm{diag}(a_1, \ldots, a_n)$. By definition, the tensor product of two forms is given by the tensor product of the corresponding matrices.

Proofs of the following theorem can be found in PFISTER 1995; see also SCHARLAU 1985 Chapter 2 §10.

Theorem (1) *For each choice of $a_1, \ldots, a_k \in F^\times$, the 2^k-dimensional form $\langle 1, a_1 \rangle \otimes \cdots \otimes \langle 1, a_k \rangle$ is strictly multiplicative.*

(2) *Every anisotropic multiplicative form is equivalent to a form $\langle 1, a_1 \rangle \otimes \cdots \otimes \langle 1, a_k \rangle$.*

(3) *Any isotropic form is multiplicative.*

(4) *Non-degenerate composition forms exist only in the dimensions $n \in \{1, 2, 4, 8\}$.*

The last statement (4) is essentially due to HURWITZ 1898; see LAM 2005 V Theorem 5.10 for a *proof*. For early developments compare EBBINGHAUS *et al.* 1991 Chapter 10 §1.

Exercises

(1) Show that $\mathbb{Q}$ has non-isomorphic quadratic field extensions with isomorphic additive and isomorphic multiplicative groups.

(2) In the rational plane $\mathbb{Q}^2$, the circle with the equation $x^2 + y^2 = 3$ is empty.

(3) Let $C = \{\, (x, y) \in \mathbb{R}^2 \mid x^2 + y^2 = 1 \,\}$ denote the unit circle in $\mathbb{R}^2$. Show that $C \cap \mathbb{Q}^2$ is dense in C.

(4) Determine the rotation group Δ of the rational unit circle $C \cap \mathbb{Q}^2$.

(5) Let $n = \prod_p p^{v_p} \in \mathbb{N}$ be a sum of two squares of integers. Show that n is a sum of two *positive* squares of integers if, and only if, v_2 is odd or $v_p > 0$ for some prime $p \equiv 1 \bmod 4$.

(6) Verify Jacobi's theorem 34.19 in the case $n = 30$.

(7) If $a, b \in \mathbb{J} := \mathbb{Z} + \mathbb{Z}i$ and $b \neq 0$, there exists $z \in \mathbb{J}$ such that $N(a - bz) < Nb$ (Euclidean algorithm in $\mathbb{J}$).

35   Ordered groups of rational numbers

The additive group $\mathbb{Q}^+$ and the multiplicative group $\mathbb{Q}^\times_{\mathrm{pos}}$ of positive rational numbers are ordered groups in the sense of 7.1 with the usual ordering; these ordered groups will sometimes be denoted by $\mathbb{Q}^+_<$ and $(\mathbb{Q}^\times_{\mathrm{pos}})_<$ for brevity. It is easy to see that they are both Archimedean (compare 7.4 for this notion).

Note that in the multiplicative case it makes no sense to include the negative rationals since an ordered group is always torsion free (7.3).

In the ordered group $\mathbb{Q}_<^+$ the group structure determines the ordering (see 35.1). For $\mathbb{Q}_{\mathrm{pos}}^\times$ this is not true (see 35.2).

In the case of real numbers, one knows for the isomorphic ordered groups $\mathbb{R}_<^+$ and $(\mathbb{R}_{\mathrm{pos}}^\times)_<$ that, conversely, the ordering determines the group operation: it follows from completeness that an ordered group whose underlying ordered set is $\mathbb{R}$ (with the usual ordering) is isomorphic to the ordered group $\mathbb{R}$ with the usual addition and the usual ordering; see 7.10. For rational numbers, an analogous result cannot be expected; we shall present several counter-examples of increasingly bad behaviour.

Concerning automorphisms of these structures, we shall prove that the ordered group $(\mathbb{Q}_{\mathrm{pos}}^\times)_<$ is rigid, that is, it has no automorphism except the identity. The proof uses quite deep results from the theory of transcendental numbers.

35.1 Proposition *The only ordering relations on $\mathbb{Q}$ making $\mathbb{Q}^+$ an ordered group are the usual ordering and its converse.*

Proof Let $\prec$ be such an ordering relation. We may assume that $0 \prec 1$; if not, we replace $\prec$ by the converse ordering. By monotonicity of addition and transitivity of an ordering relation, it follows that $0 \prec n$ for all $n \in \mathbb{N}$. For $0 < x \in \mathbb{Q}$, there is $m \in \mathbb{N}$ such that $mx \in \mathbb{N}$, so that $0 \prec mx$ and hence $0 \prec x$ (since $x \prec 0$ would imply $mx \prec 0$). By monotonicity, it follows that $-x \prec 0$. Thus, the positive elements with respect to $\prec$ are precisely the positive rational numbers in the usual sense, but the ordering of an ordered group is determined by the set of positive elements. $\square$

35.2 Example *There are ordering relations making $\mathbb{Q}_{\mathrm{pos}}^\times$ a non-Archimedean ordered group.*

Since the usual ordering on $\mathbb{Q}_{\mathrm{pos}}$ is Archimedean, this shows in particular that there are essentially different ways of turning $\mathbb{Q}_{\mathrm{pos}}^\times$ into an ordered group, in contrast with 35.1.

Instead of constructing non-Archimedean ordering relations on $\mathbb{Q}_{\mathrm{pos}}^\times$, we may do so on the isomorphic group $\mathbb{Z}^{(\mathbb{N})}$, the direct sum of infinitely many copies of the group $(\mathbb{Z}, +)$ indexed by $\mathbb{N}$; see 32.1. The elements of $\mathbb{Z}^{(\mathbb{N})}$ are the maps $\alpha : \mathbb{N} \to \mathbb{Z}$ such that there exist only finitely many $n \in \mathbb{N}$ with $\alpha(n) \neq 0$. We define the lexicographic ordering $\prec$ on $\mathbb{Z}^{(\mathbb{N})}$: For different elements $\alpha, \beta : \mathbb{N} \to \mathbb{Z}$ of $\mathbb{Z}^{(\mathbb{N})}$ and

$$m := \min\{ n \in \mathbb{N} \mid \alpha(n) \neq \beta(n) \}, \text{ let } \alpha \prec \beta \iff \alpha(m) < \beta(m),$$

and $\beta \prec \alpha$ otherwise. It is straightforward to verify that this is an ordering relation and that $\mathbb{Z}^{(\mathbb{N})}$ with this ordering relation is an ordered group. However, it is not Archimedean; indeed, if $\varepsilon_k : \mathbb{N} \to \mathbb{Z}$ is the characteristic map of $\{k\}$ mapping k to 1 and the other integers to 0, then $0 \prec \varepsilon_2$ and $n \cdot \varepsilon_2 \prec \varepsilon_1$ for all $n \in \mathbb{N}$.

We now present several counter-examples showing that in the ordered groups $\mathbb{Q}_{<}^{+}$ and $(\mathbb{Q}_{\text{pos}}^{\times})_{<}$, the ordering does not determine the group structure. We do so by constructing ordered groups such that the underlying ordered sets are order isomorphic to $\mathbb{Q}$ and hence to $\mathbb{Q}_{\text{pos}}$ (with the usual ordering), but they are not isomorphic to $\mathbb{Q}_{<}^{+}$ or to $(\mathbb{Q}_{\text{pos}}^{\times})_{<}$. Note that $\mathbb{Q}$ and $\mathbb{Q}_{\text{pos}}$ are order isomorphic according to 3.4 since $\mathbb{Q}_{\text{pos}}$ is countable and strongly dense in itself.

35.3 Example The set

$$\mathbb{Q}(\sqrt{2}) = \{ a + b\sqrt{2} \mid a, b \in \mathbb{Q} \}$$

is a subgroup of the additive group of real numbers, hence it is an ordered group with the addition and the usual ordering of real numbers. It is countable and, as an ordered set, it is strongly dense in itself, hence it is order isomorphic to $\mathbb{Q}$ by 3.4.

The group $(\mathbb{Q}(\sqrt{2}), +)$ is, however, not isomorphic to $\mathbb{Q}^{+}$, since the latter group is locally cyclic (1.6c, 31.3), but the former is not: 1 and $\sqrt{2}$ generate a subgroup of $(\mathbb{Q}(\sqrt{2}), +)$ isomorphic to the group $\mathbb{Z} \times \mathbb{Z}$, which is not cyclic.

The group $(\mathbb{Q}(\sqrt{2}), +)$ is not isomorphic to $\mathbb{Q}_{\text{pos}}^{\times}$ either, since it is clearly divisible (in the sense of 1.7), but $\mathbb{Q}_{\text{pos}}^{\times}$ is not, being isomorphic to a direct sum of copies of $\mathbb{Z}$ by 32.1.

35.4 Example The set $\mathbb{Q} \times \mathbb{Q}$ with componentwise addition and the lexicographic ordering $\prec$ defined by

$$(r_1, r_2) \prec (s_1, s_2) \iff r_1 < s_1 \text{ or } (r_1 = s_1 \text{ and } r_2 < s_2)$$

is an ordered group. As an ordered set, it is order isomorphic to $\mathbb{Q}$ by 3.4 since it is countable and strongly dense in itself.

As a group, $\mathbb{Q} \times \mathbb{Q}$ is isomorphic to the group $\mathbb{Q}(\sqrt{2})$ considered in the previous example via the isomorphism $(a, b) \mapsto a + b\sqrt{2}$. As explained there, these groups are not isomorphic either to $\mathbb{Q}^{+}$ or to $\mathbb{Q}_{\text{pos}}^{\times}$. So, $\mathbb{Q} \times \mathbb{Q}$ is another example of an ordered group not isomorphic to $\mathbb{Q}_{<}^{+}$ nor to $(\mathbb{Q}_{\text{pos}}^{\times})_{<}$ whose underlying ordered set is isomorphic to $\mathbb{Q}$. This example is essentially different from 35.3, although the underlying groups are isomorphic. Indeed, it is straightforward that $\mathbb{Q}(\sqrt{2})$ is an Archimedean ordered group, whereas $\mathbb{Q} \times \mathbb{Q}$ is not Archimedean: $0 = (0,0) \prec (0,1)$, but $n \cdot (0,1) = (0,n) \prec (1,0)$ for all $n \in \mathbb{N}$.

35.5 Example We finally present an example of a non-commutative ordered group such that the underlying ordered set is order isomorphic to $\mathbb{Q}$. Being non-commutative, this group cannot be isomorphic to $\mathbb{Q}^+$ nor to $\mathbb{Q}_{\mathrm{pos}}^{\times}$.

The underlying group will be the subgroup $\mathrm{A}_1^+(\mathbb{Q})$ of $\mathrm{GL}_2\mathbb{R}$ consisting of all matrices $\begin{pmatrix} 1 & 0 \\ a & \alpha \end{pmatrix}$ with $a \in \mathbb{Q}$ and $\alpha \in \mathbb{Q}_{\mathrm{pos}}$. It is isomorphic to the group of order-preserving affine transformations $\mathbb{Q} \to \mathbb{Q} : x \mapsto a + \alpha x$ of $\mathbb{Q}$; compare 9.3 and 9.4, where such groups are considered for the field $\mathbb{R}$ instead of $\mathbb{Q}$ (and without restrictions concerning the ordering).

We endow the group $\mathrm{A}_1^+(\mathbb{Q})$ with an ordering relation by carrying the lexicographic ordering on $\mathbb{Q}_{\mathrm{pos}} \times \mathbb{Q}$ to $\mathrm{A}_1^+(\mathbb{Q})$ via the bijection

$$\mathbb{Q}_{\mathrm{pos}} \times \mathbb{Q} \to \mathrm{A}_1^+(\mathbb{Q}) : (\alpha, a) \mapsto \begin{pmatrix} 1 & 0 \\ a & \alpha \end{pmatrix} .$$

One may easily verify that $\mathrm{A}_1^+(\mathbb{Q})$ with this ordering relation is an ordered group. The underlying ordered set is order isomorphic to $\mathbb{Q}_{\mathrm{pos}} \times \mathbb{Q}$ and hence to $\mathbb{Q}$ by 3.4, as in the previous examples. Clearly, $\mathrm{A}_1^+(\mathbb{Q})$ is not commutative; e.g., for $0 \neq a \in \mathbb{Q}$ and $1 \neq \alpha \in \mathbb{Q}_{\mathrm{pos}}$, the matrices $\begin{pmatrix} 1 & 0 \\ 0 & \alpha \end{pmatrix}$ and $\begin{pmatrix} 1 & 0 \\ a & \alpha \end{pmatrix}$ do not commute.

Next, we consider the automorphisms of the ordered groups $\mathbb{Q}_<^+$ and $(\mathbb{Q}_{\mathrm{pos}}^{\times})_<$, that is, group automorphisms which preserve the ordering. For $\mathbb{Q}_<^+$ it is easy to single them out from the endomorphisms of $\mathbb{Q}^+$, which have been determined in 31.10.

35.6 $\mathrm{Aut}\,\mathbb{Q}_<^+ = \{\, x \mapsto rx \mid r \in \mathbb{Q}_{\mathrm{pos}} \,\} \cong \mathbb{Q}_{\mathrm{pos}}^{\times}$.

For $(\mathbb{Q}_{\mathrm{pos}}^{\times})_<$, the analogous question is much more difficult to solve. The result is the following.

35.7 Theorem *The only order-preserving monomorphisms of $\mathbb{Q}_{\mathrm{pos}}^{\times}$ into itself are the mappings $x \mapsto x^n$ for $n \in \mathbb{N}$.*

Corollary *The ordered group $(\mathbb{Q}_{\mathrm{pos}}^{\times})_<$ has no automorphism except the identity.*

By 7.13, an order-preserving monomorphism of $\mathbb{Q}_{\mathrm{pos}}^{\times}$ into itself is of the form $x \mapsto x^t$ for some real number $t > 0$. Therefore, Theorem 35.7 is equivalent to the following.

35.8 Theorem *Let t be a positive real number such that $r^t \in \mathbb{Q}_{\mathrm{pos}}$ for all $r \in \mathbb{Q}_{\mathrm{pos}}$. Then t is a natural number.*

Assume that, for some natural number n, the map $x \mapsto x^n$ is even an automorphism of $\mathbb{Q}_{\mathrm{pos}}^{\times}$. Then the inverse mapping $x \mapsto x^{1/n}$ is also an automorphism. By 35.8, we conclude that $1/n$ is a natural number, as well, hence $n = 1$, which establishes the corollary of Theorem 35.7.

Theorem 35.8 shall be derived from the following powerful result in transcendental number theory, which we use without proof; see 35.11 for references.

35.9 Theorem *Let c_1, c_2, d_1, d_2, d_3 be complex numbers such that c_1, c_2 are linearly independent over $\mathbb{Q}$ as well as d_1, d_2, d_3. Then at least one of the six numbers $e^{c_\mu d_\nu}$ for $\mu \in \{1, 2\}$, $\nu \in \{1, 2, 3\}$ is transcendental, that is, not a solution of a polynomial equation with rational coefficients.* □

In our application of this theorem, linear independence will be granted by the following fact:

35.10 Lemma *The set $\{\ln p \mid p \in \mathbb{P}\}$ of logarithms of prime numbers is linearly independent over $\mathbb{Q}$.*

Proof Consider n distinct primes p_ν, $1 \le \nu \le n$, and rational numbers a_ν with $\sum_{\nu=1}^{n} a_\nu \cdot \ln p_\nu = 0$. Write $a_\nu = m_\nu/d$ where $m_\nu \in \mathbb{Z}$ and $d \in \mathbb{N}$ (a common denominator of the a_ν). Then $0 = \sum_{\nu=1}^{n} m_\nu \cdot \ln p_\nu = \ln(\prod_{\nu=1}^{n} p_\nu^{m_\nu})$ and $\prod_{\nu=1}^{n} p_\nu^{m_\nu} = 1$. By uniqueness of prime factor decomposition, it follows that $m_\nu = 0$ and hence $a_\nu = 0$ for all ν. □

Proof of Theorem 35.8. Let $t > 0$ be a real number such that $r^t \in \mathbb{Q}_{\mathrm{pos}}$ for all $r \in \mathbb{Q}_{\mathrm{pos}}$. It suffices to show that t must be rational, for then, we may argue as follows: Write $t = m/n$ with natural numbers m, n, then $(2^t)^n = 2^m$. This shows that in the decomposition of the rational number 2^t as a product of positive and negative powers of finitely many primes, the exponent k of the prime 2 satisfies $kn = m$, so that $t = m/n = k$ is a natural number, as asserted.

In order to prove that t is rational, we distinguish two cases.

Assume first that t and t^2 are linearly independent over $\mathbb{Q}$. This is where transcendental number theory comes in. Take any three different primes p_1, p_2, p_3. By 35.10, the numbers $\ln p_1, \ln p_2, \ln p_3$ are linearly independent over $\mathbb{Q}$. By Theorem 35.9, therefore, at least one of the numbers $e^{(\ln p_\nu) \cdot t^\mu} = p_\nu^{t^\mu}$ for $\mu \in \{1, 2\}$, $\nu \in \{1, 2, 3\}$ is transcendental and *a fortiori* not rational. But this is a contradiction to the hypothesis on t.

In the remaining case, t and t^2 are linearly dependent over $\mathbb{Q}$, so that $at + bt^2 = 0$ for some rational coefficients a, b not both equal to 0. If $b = 0$, then $a \neq 0$ and $t = 0$; if $b \neq 0$, then $t = 0$ or $t = -a/b$. Thus t is rational, as asserted.. $\square$

35.11 Remarks   (1) Theorem 35.9 was discovered independently by Siegel, Lang and Ramachandra. Proofs may be found in LANG 1966 Chapter II §1 Theorem 1 p. 8 and BAKER 1975 Theorem 12.3 p. 119. A survey on the subject with a sketch of proof is LANG 1971, see in particular 1.6 p. 638 and pp. 640ff. Theorem 35.8 is a special case of a corollary derived from Theorem 35.9 in LANG 1966 Chapter II.

(2) The arguments presented here leading to Theorem 35.7 are taken from the lecture notes SALZMANN 1973. Practically the same approach can be found in GLASS–RIBENBOIM 1994, together with various generalizations. They point out that instead of 35.9 the following easier statement suffices for the proof of 35.8 and, hence, of 35.7:

If $0 < t \in \mathbb{R} \smallsetminus \mathbb{Q}$, then at least one of the three numbers $2^t, 3^t, 5^t$ is not rational (in fact, transcendental).

As to a proof of the latter statement, GLASS–RIBENBOIM 1994 cite HALBERSTAM 1974, who shows that $2^t, 3^t, 5^t \in \mathbb{N}$ implies $t \in \mathbb{N}$ and mentions that his proof together with some results on algebraic numbers yields the transcendency statement above.

36   Addition and topologies of the rational numbers

The additive group $\mathbb{Q}^+$ endowed with the usual topology is a topological group (see Definition 8.1), since it is a subgroup of the topological group $\mathbb{R}^+$; see 8.2. However, there are quite different group topologies on $\mathbb{Q}^+$, that is, topologies making $\mathbb{Q}^+$ into a topological group. An important class of such topologies is obtained from the p-adic metrics on $\mathbb{Q}$ which we shall present. Furthermore, using continuous characters of $\mathbb{Q}^+$, we shall construct a plethora of group topologies on $\mathbb{Q}^+$ that cannot be obtained from a metric. Before going into these general constructions of group topologies, we shall ascertain that there are non-discrete group topologies on $\mathbb{Q}^+$ that are strictly coarser than the usual topology.

36.1 Example   Consider the factor group $\mathbb{R}/\mathbb{Z}$ of the additive group $\mathbb{R}$ of real numbers; it is a topological group when endowed with the quotient topology, which makes the canonical projection homomorphism

$$q : \mathbb{R} \to \mathbb{R}/\mathbb{Z} : x \mapsto x + \mathbb{Z}$$

continuous and open (see 62.6 and 62.11). The continuous homomorphism

$$w : \mathbb{Q}^+ \to \mathbb{R}/\mathbb{Z} : r \mapsto q(r\sqrt{2})$$

is injective since for $0 \neq r \in \mathbb{Q}$ the irrational number $\sqrt{2}r$ is not an element of the kernel $\mathbb{Z}$ of q. The image group $Q = w(\mathbb{Q}) \leq \mathbb{R}/\mathbb{Z}$ is a topological group with the subspace topology inherited from $\mathbb{R}/\mathbb{Z}$. The topology τ on $\mathbb{Q}$ induced by w, whose open sets are the preimages of open sets of $\mathbb{R}/\mathbb{Z}$, makes w a homeomorphism of $\mathbb{Q}$ onto Q and makes $\mathbb{Q}^+$ a topological group.

The topology τ is coarser than the usual topology, since w is continuous. In fact, it is strictly coarser; more precisely, we shall see that $(\mathbb{Q}^+, \tau)$ is not isomorphic as a topological group to $\mathbb{Q}^+$ with the usual topology. For this, we consider cyclic subgroups of $\mathbb{Q}^+$. In the usual topology the cyclic subgroups are not dense, whereas we shall find a cyclic subgroup which is dense in the topology τ. The subgroup $\langle 1, \sqrt{2} \rangle$ of $\mathbb{R}$ generated by 1 and $\sqrt{2}$ is dense by 1.6(b). The image subgroup $q(\langle 1, \sqrt{2} \rangle)$ therefore is dense in $\mathbb{R}/\mathbb{Z}$; since 1 belongs to the kernel of q, this subgroup is just the cyclic group generated by $q(\sqrt{2}) = w(1)$, in other words, the image of the subgroup $\mathbb{Z}$ of $\mathbb{Q}^+$ under the isomorphism $(\mathbb{Q}^+, \tau) \to Q$ of topological groups induced by w. Via this isomorphism, it follows that $\mathbb{Z}$ is dense in the topology τ.

The usual topology on $\mathbb{Q}$ is induced by the metric $d(a, b) = |b - a|$ (33.1(c)). With respect to this metric, $\mathbb{Q}^+$ is even a metric group in the following sense:

36.2 Definition A *metric group* is a group G together with a metric d on G which is invariant under right and left translations in the group G, that is (in additive notation) for all $a, x, y \in G$ we have $d(a+x, a+y) = d(x, y) = d(x + a, y + a)$.

It is immediate that every metric group is a topological group with respect to the topology induced by the metric.

Next, we shall construct other invariant metrics on $\mathbb{Q}^+$.

36.3 The p-adic metric Fix a prime number p. Every non-zero rational number x can be written in the form

$$x = p^{e(x)} \frac{a}{b}$$

with integers $e(x) \in \mathbb{Z}$ and $a, b \in \mathbb{Z} \setminus \{0\}$ such that a and b are not divisible by p. Here $e(x)$ is uniquely determined by x, as a consequence of uniqueness of prime factorization. Note that $e(-x) = e(x)$ and $e(x) \geq 0$ if $x \in \mathbb{Z} \smallsetminus \{0\}$. We assert for $x, y \in \mathbb{Q} \smallsetminus \{0\}$ that

$$e(xy) = e(x) + e(y) \tag{1}$$
$$e(x + y) \geq \min\{e(x), e(y)\} \quad \text{if} \quad x + y \neq 0 \tag{2}$$
$$e(x + y) = \min\{e(x), e(y)\} \quad \text{if} \quad e(x) \neq e(y) . \tag{3}$$

Assertion (1) is immediate. For (2) and (3), in case $x + y \neq 0$, write $y = p^{e(y)} c/d$ with $c, d \in \mathbb{Z} \smallsetminus p\mathbb{Z}$, and assume $e(x) \leq e(y)$. Then

$$x + y = p^{e(x)} \left(\frac{a}{b} + \frac{p^{e(y)-e(x)} c}{d} \right) = p^{e(x)} \frac{g}{bd}$$

where $0 \neq g = ad + p^{e(y)-e(x)} bc \in \mathbb{Z}$, so that $e(x+y) = e(x) + e(g) \geq e(x)$, which proves (2). If $e(x) < e(y)$, then g is not divisible by p, so that $e(g) = 0$, and we have obtained (3).

We now define the *p-adic absolute value* $|\ |_p$ on $\mathbb{Q}$ by $|0|_p = 0$ and

$$|x|_p = p^{-e(x)}$$

for $0 \neq x \in \mathbb{Q}$. According to (1), (2) and (3), this absolute value satisfies

$$|xy|_p = |x|_p |y|_p \tag{1'}$$
$$|x + y|_p \leq \max\{|x|_p, |y|_p\} \tag{2'}$$
$$|x + y|_p = \max\{|x|_p, |y|_p\} \quad \text{if} \quad |x|_p \neq |y|_p \tag{3'}$$

for $x, y \in \mathbb{Q}$. We note that (3') could also be deduced from (2') by a general argument; see 55.3.

The inequality (2') is called the *ultrametric* property. It implies the weaker *triangle inequality* $|x + y|_p \leq |x|_p + |y|_p$. One verifies easily that therefore the map $d_p : \mathbb{Q} \times \mathbb{Q} \to [0, \infty[$ defined by

$$d_p(x, y) = |x - y|_p$$

is a metric on $\mathbb{Q}$, the *p-adic metric*, which is translation invariant. It induces a topology on $\mathbb{Q}$, the *p-adic topology*. With this topology, $\mathbb{Q}$ is a topological field; this can be shown by exactly the same arguments as for $\mathbb{R}$ with the standard absolute value; see 8.2 and 9.1.

In the *p*-adic topology, a sequence $(x_n)_{n \in \mathbb{N}}$ converges to 0 if, and only if, $|x_n|_p$ tends to 0 in the usual topology, in other words, if the *p*-exponent $e(x_n)$ of x_n tends to ∞. A typical example of such a sequence is $(p^n)_{n \in \mathbb{N}}$ (which in the usual topology tends to ∞ itself). In particular, the subgroup $\mathbb{Z}$ of $\mathbb{Q}^+$ is not discrete in the *p*-adic topology, in contrast with the usual topology.

36.4 The endomorphisms of $\mathbb{Q}^+$ are the maps $x \mapsto rx$ for $r \in \mathbb{Q}$; see 31.10. Except for the zero endomorphism, they are automorphisms. They are continuous not only in the usual topology, but also in the p-adic topology, since by the multiplicative property $(1')$ of the p-adic absolute value one has $d_p(rx, ry) = |r|_p d_p(x, y)$.

But the automorphisms are not continuous when considered as maps of $\mathbb{Q}$ with the p-adic topology into $\mathbb{Q}$ with the usual topology or the q-adic topology for a different prime $q \neq p$. Indeed, p^n tends to 0 in the p-adic topology, but for $r \neq 0$ the sequence (rp^n) does not converge to 0 in the ordinary topology or in the q-adic topology, since $|rp^n|_q = |r|_q \neq 0$ is constant.

36.5 Corollary *For different primes p and q, the p-adic topology, the q-adic topology and the usual topology on $\mathbb{Q}^+$ give rise to non-isomorphic topological groups. The same is true for the subgroup $\mathbb{Z}^+$.* □

Thus, we have found countably many essentially different ways of making $\mathbb{Q}^+$ into a metric group. But we may go much further.

36.6 Construction From the p-adic metrics, we shall construct continuously many translation invariant metrics on $\mathbb{Q}^+$ such that the topological groups obtained from the corresponding topologies on $\mathbb{Q}^+$ are pairwise non-isomorphic.

Let X be an infinite set of primes. The p-adic distances $d_p(x, y)$ of two fixed rational numbers x, y assume only finitely many values, since $|x - y|$ has only finitely many prime factors, and it is easily seen that the convergent sum

$$d_X(x, y) = \sum_{p \in X} 2^{-p} \cdot d_p(x, y)$$

defines a metric d_X on $\mathbb{Q}$. This metric is obviously translation invariant, since the metrics d_p are, so that the corresponding topology makes $\mathbb{Q}^+$ into a topological group. Let $X = \{ p_n \mid n \in \mathbb{N} \}$; then the sequence $(p_1 p_2 \ldots p_n)^n$ converges to 0 with respect to the metric d_X.

Now we consider another infinite set Y of primes, and we assume that $q \in Y \setminus X$. The image $r(p_1 p_2 \ldots p_n)^n$ of the sequence $(p_1 p_2 \ldots p_n)^n$ under an automorphism $x \mapsto rx$, $r \neq 0$, of $\mathbb{Q}^+$ does not converge to 0 with respect to d_Y, since

$$d_Y(r(p_1 p_2 \ldots p_n)^n, 0) \geq 2^{-q} \cdot |r(p_1 p_2 \ldots p_n)^n|_q = 2^{-q} \cdot |r|_q > 0$$

is bounded below by a non-zero constant. Thus, the group $\mathbb{Q}^+$ with the topology obtained from the metric d_X is not isomorphic as a topological group to $\mathbb{Q}^+$ with the topology obtained from d_Y.

Calling two invariant metrics on $\mathbb{Q}^+$ essentially different if the result-ing topological groups are non-isomorphic, we thus have obtained that there are at least as many essentially different invariant metrics on $\mathbb{Q}^+$ as there are infinite subsets of the set of primes, that is, continuously many, since the set of primes is countably infinite (32.7).

In the sequel we shall define further group topologies on $\mathbb{Q}^+$ using the following general procedure.

36.7 Construction *Let G be a group (written additively), T a topo-logical group (written multiplicatively), and X a non-empty set of ho-momorphisms of G to T. Let τ_{X} be the topology on G induced by X, that is, the topology having the set of preimages of open sets of T under elements of X as subbasis. Then the following statements hold:*
  (a) *(Universal property) Let Y be any topological space. Then a map $f : Y \to G$ is continuous with respect to the topology τ_{X} if, and only if, the composition $\chi \circ f : Y \to T$ is continuous for all $\chi \in \mathsf{X}$.*
  (b) *With the topology τ_{X}, the group G is a topological group.*
  (c) *A neighbourhood base of 0 in G in the topology τ_{X} is given by the intersections $\bigcap_{\chi \in \Phi} \chi^{-1}(V)$ where $\Phi \subseteq \mathsf{X}$ varies over the finite subsets of X and V over a neighbourhood base of 1 in T.*

Proof (a) Clearly the maps $\chi \in \mathsf{X}$ are continuous with respect to τ_{X}. Thus, continuity of f implies continuity of $\chi \circ f$ for all $\chi \in \mathsf{X}$.

Assume, conversely, the continuity of the maps $\chi \circ f$ with $\chi \in \mathsf{X}$. Then for an open subset U of T the preimage $f^{-1}(\chi^{-1}(U)) = (\chi \circ f)^{-1}(U)$ is open in Y. Thus, the elements of the subbasis of the topology τ_{X} have open preimages under f, so that f is continuous.

(b) One has to show that the maps

$$\alpha : G \times G \to G : (x, y) \mapsto x + y \quad \text{and} \quad \iota : G \to G : x \to -x$$

are continuous with respect to the topology τ_{X} and the product topology on $G \times G$. For the topological group T, the corresponding maps

$$\alpha' : T \times T \to T : (w, z) \mapsto wz \quad \text{and} \quad \iota' : T \to T : z \to z^{-1}$$

are known to be continuous. For $\chi \in \mathsf{X}$, the product map $\chi \times \chi : G \times G \to T \times T : (x, y) \mapsto (\chi(x), \chi(y))$ is continuous. The fact that $\chi : G \to T$ is a homomorphism means that $\chi \circ \alpha = \alpha' \circ (\chi \times \chi)$ and $\chi \circ \iota = \iota' \circ \chi$. Hence, all these maps are continuous. By the universal property (a), it follows that α and ι are continuous.

(c) is clear. $\square$

We now specialize the preceding construction to $G = \mathbb{Q}^+$.

36.8 Construction   Let $\mathbb{T} = \mathbb{S}_1 \leq \mathbb{C}^\times$ be the 1-dimensional torus group. For $s \in \mathbb{R}$, the map

$$\chi_s : \mathbb{Q} \to \mathbb{T} : x \mapsto e^{2\pi i s x}$$

is a homomorphism of $\mathbb{Q}^+$ to $\mathbb{T}$ (a character of $\mathbb{Q}^+$). For a non-empty subset $H \subseteq \mathbb{R}$, let

$$\mathsf{X}_H = \{\chi_s \mid s \in H\}.$$

Consider the topology $\tau_H = \tau_{\mathsf{X}_H}$ on $\mathbb{Q}$ induced by X_H as described in 36.7. With this topology, $\mathbb{Q}^+$ is a topological group.

For a finite subset $F \subseteq H$ and $\varepsilon > 0$, let

$$U_{\varepsilon,F}(0) = \left\{ x \in \mathbb{Q} \mid \forall_{s \in F} \ |e^{2\pi i s x} - 1| < \varepsilon \right\}.$$

According to 36.7(c), these subsets form a neighbourhood base of 0 in the topology τ_H if F varies over the finite subsets of H and ε assumes arbitrarily small positive values.

If H is countable, then we have a countable neighbourhood base with $\varepsilon = 1/n$ for $n \in \mathbb{N}$, and the topology τ_H might be (and in fact is) induced by a metric. If H is uncountable, then τ_H is a non-metric group topology on $\mathbb{Q}^+$ by 36.11(b) below.

36.9   *If $H \not\subseteq \mathbb{Q}$, then τ_H is a Hausdorff topology on $\mathbb{Q}$.*

Proof   Since $\mathbb{Q}^+$ is a topological group in this topology, it suffices to show that the intersection of all neighbourhoods of 0 is $\{0\}$ (see 62.4). In the notation of 36.8, this intersection is

$$\bigcap_{\varepsilon > 0} U_{\varepsilon,H}(0) = \left\{ x \in \mathbb{Q} \mid \forall_{s \in H} \ e^{2\pi i s x} = 1 \right\}.$$

If $s \in H \smallsetminus \mathbb{Q}$, then $sx \notin \mathbb{Q}$ for $0 \neq x \in \mathbb{Q}$, so that $e^{2\pi i s x} \neq 1$, and the intersection above consists of 0 only.   □

36.10 Lemma   *Let B be a basis of $\mathbb{R}$ as a vector space over $\mathbb{Q}$ (a Hamel basis, see 1.13) such that $1 \in B$, and F a finite subset of B. Then for $u \in B \smallsetminus (F \cup \{1\})$ and $\delta, \varepsilon \in \ ]0, 1[$, the basic 0-neighbourhoods specified in 36.8 satisfy*

$$U_{\varepsilon,F}(0) \not\subseteq U_{\delta,\{u\}}(0) \quad \text{and} \quad U_{\varepsilon,F}(0) \cap \mathbb{Z} \neq \{0\}.$$

Proof   By the second version 5.70 of Kronecker's Theorem, one may find integers z_s for $s \in F \cup \{u\}$ and an integer k such that $ks - z_s$ is arbitrarily close to 0 for $s \in F$ and $ku - z_u$ is arbitrarily close to $1/2$. Now $e^{2\pi i(ks-z_s)} = e^{2\pi iks}$, $e^0 = 1$ and $e^{2\pi i \cdot 1/2} = e^{\pi i} = -1$. Thus, by continuity of the exponential function, the integers z_s and k may be chosen such that $|e^{2\pi iks} - 1| < \varepsilon$ for $s \in F$ and $|e^{2\pi iku} - (-1)| < \varepsilon$. The first inequality means that $k \in U_{\varepsilon,F}(0)$, whereas the second inequality together with $\delta, \varepsilon < 1$ implies that $0 \neq k \notin U_{\delta,\{u\}}(0)$.   □

36.11 Theorem   *Let B be a basis of $\mathbb{R}$ as a vector space over $\mathbb{Q}$ such that $1 \in B$, and let $H \subseteq B \smallsetminus \{1\}$. Then the following assertions about the topology τ_H on $\mathbb{Q}$ introduced in 36.8 hold.*

(a) *The topology τ_H is not discrete.*

(b) *If H is uncountable, then 0 does not have a countable neighbourhood base in the topology τ_H. In particular, this topology is not induced by any metric.*

(c) *For distinct subsets H and H' of $B \smallsetminus \{1\}$, the topological groups $(\mathbb{Q}^+, \tau_H)$ and $(\mathbb{Q}^+, \tau_{H'})$ are not isomorphic.*

Remark   The basis B has the cardinality of the continuum; see 1.12.

Proof (a) Every neighbourhood of 0 contains a basic neighbourhood $U_{\varepsilon,F}(0)$ for a finite subset $F \subseteq B$ and $0 < \varepsilon < 1$ and hence an element different from 0 by Lemma 36.10.

(b) Assume that 0 has a countable neighbourhood base. Then there is a countable neighbourhood base consisting of basic neighbourhoods of the form $U_{\varepsilon_n,F_n}(0)$ with finite subsets $F_n \subseteq H$ and $\varepsilon_n \in {]0,1[}$, $n \in \mathbb{N}$. As H is uncountable, there exists $u \in H$ not contained in any of the F_n. By 36.10 the neighbourhood $U_{\delta,\{u\}}(0)$ for $0 < \delta < 1$ does not contain any of the neighbourhoods $U_{\varepsilon_n,F_n}(0)$, contradicting our assumption that these form a neighbourhood base.

(c) A group automorphism of $\mathbb{Q}^+$ is of the form $x \mapsto rx$ for $0 \neq r \in \mathbb{Q}$; see 31.10. Assume that there is $u \in H' \smallsetminus H$. Then we show that the automorphism $x \mapsto rx$ is not continuous as a map $(\mathbb{Q}, \tau_H) \to (\mathbb{Q}, \tau_{H'})$. This then proves that the topological groups $(\mathbb{Q}^+, \tau_H)$ and $(\mathbb{Q}^+, \tau_{H'})$ are not isomorphic.

More specifically, we show for $0 < \delta < 1$ that $U_{\delta,\{u\}}$, which is a neighbourhood of 0 in the topology $\tau_{H'}$, does not contain the image of any basic neighbourhood $U_{\varepsilon,F}(0)$ in the topology τ_H, for $\varepsilon > 0$ and a finite subset $F \subseteq H$. The image of the latter neighbourhood is easily seen to be $r \cdot U_{\varepsilon,F}(0) = U_{\varepsilon,r^{-1}F}(0)$. That this set is not contained in $U_{\delta,\{u\}}$ is obtained by applying 36.10 to the basis $\{1,u\} \cup r^{-1} \cdot (B \smallsetminus \{1,u\})$ instead of B.   □

36.12 Corollary *The cardinality of the set of isomorphism classes of Hausdorff topological groups with $\mathbb{Q}^+$ as underlying group is $2^{2^{\aleph_0}}$.*

Proof Let c be the cardinality in question. Any topology on $\mathbb{Q}$ is given by specifying the open sets, that is, by a subset of the power set $2^{\mathbb{Q}}$. Hence, $c \leq 2^{2^{\aleph_0}}$, the cardinality of the power set of $2^{\mathbb{Q}}$.

Let B be a basis of $\mathbb{R}$ as a vector space over $\mathbb{Q}$ such that $1 \in B$. Then $(B \smallsetminus \{1\}) \cap \mathbb{Q} = \emptyset$, so that the topology τ_H constructed from a subset $H \subseteq B \smallsetminus \{1\}$ according to 36.8 makes $\mathbb{Q}^+$ a Hausdorff topological group by 36.9 and 36.7. Now B has the cardinality $2^{\aleph_0}$ of the continuum (see 1.12 and 1.10), and the same holds for $B \smallsetminus \{1\}$. Therefore assertion (c) of Theorem 36.11 shows that $c \geq 2^{2^{\aleph_0}}$. The converse estimate has been established initially. $\qquad\square$

Exercises

(1) Let $S \neq \{0\}$ be a subspace of the rational vector space $\mathbb{R}$, and consider $S \leq \mathbb{R}^+$ as a topological group with the topology inherited from $\mathbb{R}$. Then $S \cong \mathbb{Q}$ if, and only if, each automorphism of S is continuous.

(2) Determine the open and the closed subgroups of $\mathbb{Q}^+$ in the usual topology.

37 Multiplication and topologies of the rational numbers

The multiplicative group $\mathbb{Q}^\times$ of non-zero rational numbers endowed with the usual topology is a topological group, a subgroup of the topological group $\mathbb{R}^\times$ studied in Section 9.

The positive rational numbers form an open subgroup $\mathbb{Q}^\times_{\text{pos}}$ of $\mathbb{Q}^\times$. It will be shown (37.3) that in fact this is the only proper open subgroup. We then establish that the only automorphisms of the topological group $\mathbb{Q}^\times$ are the identity and inversion (37.4), by reducing the question to the analogous question about the ordered group $(\mathbb{Q}^\times_{\text{pos}})_<$, which has been discussed in Section 35. The proof uses approximation of irrational numbers by continued fractions.

$\mathbb{Q}^\times$ also becomes a topological group when endowed with the p-adic topology introduced in 36.3. The topology induced by the p-adic topology on certain cyclic subgroups of $\mathbb{Q}^\times$ will be studied using some elementary number theory. The result will be employed to show that for different primes p and q, the topological groups obtained by the p-adic and the q-adic topology on $\mathbb{Q}^\times$ are not isomorphic.

37.1 Proposition *A subgroup of $\mathbb{Q}_{pos}^{\times}$ is either cyclic or dense (with respect to the ordinary topology).*

Proof Let $A \leq \mathbb{Q}_{pos}^{\times}$ be a subgroup which is not cyclic. Then A is dense in $\mathbb{R}_{pos}^{\times}$ (and then *a fortiori* in $\mathbb{Q}_{pos}^{\times}$) by 1.4, since $\mathbb{R}_{pos}^{\times}$ is isomorphic to $\mathbb{R}^{+}$ as a topological group (see 2.1).   □

37.2 Example *For two different primes p and q, the subgroup $p^{\mathbb{Z}}q^{\mathbb{Z}}$ of $\mathbb{Q}_{pos}^{\times}$ generated by p and q is not cyclic and hence dense.*

Proof The subgroup in question is the image group of the homomorphism $\mathbb{Z}^2 \to \mathbb{Q}^{\times} : (m,n) \mapsto p^m q^n$. This homomorphism is injective due to uniqueness of prime factorization. Thus, $p^{\mathbb{Z}}q^{\mathbb{Z}}$ is isomorphic to the group $\mathbb{Z}^2$, which is not cyclic.   □

37.3 Theorem *The only proper open subgroup of $\mathbb{Q}^{\times}$ is $\mathbb{Q}_{pos}^{\times}$.*

Proof An open subgroup A is also closed; see 62.7. The open subgroup $A \cap \mathbb{Q}_{pos}^{\times}$ cannot be cyclic, so that $A \cap \mathbb{Q}_{pos}^{\times}$ is dense in $\mathbb{Q}_{pos}^{\times}$ by 37.1. Hence $A \cap \mathbb{Q}_{pos}^{\times} = \mathbb{Q}_{pos}^{\times}$ since it is closed, in other words $\mathbb{Q}_{pos}^{\times} \subseteq A$. If moreover A contains any negative number, then $A = \mathbb{Q}^{\times}$.   □

37.4 Theorem *The only automorphisms of the topological groups $\mathbb{Q}^{\times}$ and $\mathbb{Q}_{pos}^{\times}$ are the identity and the inversion map $x \mapsto x^{-1}$.*

Proof An automorphism α of the topological group $\mathbb{Q}^{\times}$ maps the only proper open subgroup $\mathbb{Q}_{pos}$ onto itself. Moreover, α is determined by its restriction on $\mathbb{Q}_{pos}$ since $\mathbb{Q}^{\times} = \mathbb{Q}_{pos} \cdot \{1, -1\}$. Thus it suffices to prove the theorem for an automorphism τ of the topological group $\mathbb{Q}_{pos}^{\times}$.

We shall reduce the problem to the analogous problem for the ordered group $\mathbb{Q}_{pos}^{\times}$, which has been studied in Section 35. To this end, we shall show that τ preserves or reverses the ordering. In the first case τ is the identity by Theorem 35.7. In the second case we compose τ with the inversion map to obtain an automorphism which preserves the ordering and hence, again by 35.7, is the identity, so that τ is the inversion map. (We point out that the proof of 35.7 depends on the difficult theorem 35.9 from transcendental number theory.)

In order to prove that τ preserves or reverses the ordering, we compare the images of two arbitrary distinct prime numbers p and q. For this, we use the techniques of continued fractions introduced in Section 4, in particular the following facts: every irrational number ζ may be expanded into an infinite continued fraction $[c_0; c_1, c_2, \dots]$ with $c_0 \in \mathbb{Z}$ and $c_\nu \in \mathbb{N}$

for $\nu \geq 1$; the approximating finite continued fractions

$$[c_0; c_1, c_2, \ldots, c_n] = p_n/q_n$$

(where $p_n, q_n \in \mathbb{N}$ are relatively prime) converge to ζ, and the numbers p_n, q_n have the following properties:

(i) the sequence $(q_n)_{n\in\mathbb{N}}$ is strictly increasing

(ii) $p_n q_{n+1} - p_{n+1} q_n \in \{1, -1\}$

(iii) $|\zeta - p_n/q_n| < 1/(q_n q_{n+1}) < 1/q_n^2$

(for (i) and (ii), see 4.2(†), and 4.3 for (iii)).

Now let p and q be different primes. The ratio $(\ln p)/(\ln q)$ is irrational, for if it could be expressed as a fraction m/n of natural numbers m, n, then $\ln p^n = n \ln p = m \ln q = \ln q^m$, so that $p^n = q^m$, contradicting unique prime factorization. Let p_n/q_n be the fractions approximating $\zeta = (\ln p)/(\ln q)$ as described above. From (iii) we infer that $|(\ln p)/(\ln q) - p_n/q_n| < 1/q_n^2$, and we obtain $|\ln (p^{q_n}/q^{p_n})| = |\ln p^{q_n} - \ln q^{p_n}| < (\ln q)/q_n$. By (i), $1/q_n$ tends to 0 for $n \to \infty$, so that $\lim_{n\to\infty} p^{q_n}/q^{p_n} = 1$. This property is preserved under a continuous multiplicative homomorphism, hence $\lim_{n\to\infty} \tau(p)^{q_n}/\tau(q)^{p_n} = 1$. By taking logarithms we infer that

$$\lim_{n\to\infty} (q_n \ln \tau(p) - p_n \ln \tau(q)) = 0 .$$

Since the sequence (q_n) is strictly increasing by property (i), one has $\lim_{n\to\infty} 1/(q_n \ln \tau(q)) = 0$ as well. By multiplying, we obtain

$$\lim_{n\to\infty} \left(\frac{\ln \tau(p)}{\ln \tau(q)} - \frac{p_n}{q_n} \right) = 0 , \quad \text{hence} \quad \frac{\ln \tau(p)}{\ln \tau(q)} = \lim_{n\to\infty} \frac{p_n}{q_n} = \frac{\ln p}{\ln q} .$$

Thus $t := (\ln \tau(p))/(\ln p) = (\ln \tau(q))/(\ln q)$ is a real constant independent of the primes p, q. Exponentiation gives $\tau(p) = p^t$ for all primes p. Since every positive rational number is a finite product of (positive and negative) powers of primes and since τ is multiplicative, it follows that $\tau(x) = x^t$ for all $x \in \mathbb{Q}_{\text{pos}}$. Now it is clear that τ preserves or reverses the ordering. $\qquad\square$

Remark The fact that every automorphism of the topological group $\mathbb{Q}_{\text{pos}}^\times$ preserves or reverses the ordering, which in the proof above has been obtained by using continued fractions, can be established in a more systematic way by invoking the concept of completion of topological groups. Here, we anticipate results from Section 43. The completion of the topological group $\mathbb{Q}_{\text{pos}}^\times$ is $\mathbb{R}_{\text{pos}}^\times$ (see 43.11), and every automorphism can be uniquely extended to the completion (43.24). Now $\mathbb{R}^{\times}_{\mathrm{pos}}$ is isomorphic to the additive group $\mathbb{R}^{+}$, both as a topological group and as an ordered group; indeed, the exponential function is an isomorphism $\mathbb{R}^{+} \to \mathbb{R}^{\times}_{\mathrm{pos}}$ and a homeomorphism and it preserves the ordering. Thus it suffices to see that every automorphism of the topological group $\mathbb{R}^{+}$ preserves or reverses the ordering; but it is well known that this is true more generally for every continuous bijection (essentially due to the intermediate value theorem).

Multiplication of rational numbers and p-adic topologies

We consider, for a prime p, the p-adic topology on $\mathbb{Q}$ which was introduced in 36.3 via the p-adic absolute value $|\ |_p$. As we remarked there, $\mathbb{Q}$ is a topological field with this topology. In particular, the multiplicative group $\mathbb{Q}^{\times}$ with the topology induced by the p-adic topology is a topological group.

The following number-theoretic lemma, which we shall use as a technical tool in the sequel, improves upon Theorems II and III in BIRKHOFF–VANDIVER 1904.

37.5 Lemma *Let p be a prime and $n, u, v \in \mathbb{Z}$ such that $u \neq v$ and $u \equiv v \not\equiv 0 \bmod p$.*

(i) *If $p \neq 2$ or if $u \equiv v \bmod p^2$ or if n is odd, then*

$$\left|u^n - v^n\right|_p = |u - v|_p \cdot |n|_p .$$

(ii) *In the remaining case, with $p = 2$, $u \not\equiv v \bmod 4$ and n even, one has*

$$\left|u^n - v^n\right|_2 = |u + v|_2 \cdot |n|_2 .$$

Proof We may assume that $n \in \mathbb{N}$, as $|-n|_p = |n|_p$ and $\left|u^{-n} - v^{-n}\right|_p = \left|u^n - v^n\right|_p$, in view of $|u|_p = 1 = |v|_p$.

(i) For $2 \leq m \in \mathbb{N}$, we obtain by binomial expansion

$$
\begin{aligned}
\frac{u^m - v^m}{u - v} &= \frac{(v + u - v)^m - v^m}{u - v} \\
&= mv^{m-1} + \binom{m}{2}v^{m-2}(u - v) + \sum_{j=3}^{m} \binom{m}{j} v^{m-j}(u - v)^{j-1} \\
&\equiv mv^{m-1} + \binom{m}{2}v^{m-2}(u - v) \bmod p^2 \\
&\equiv mv^{m-1} \bmod p .
\end{aligned}
$$

Assume that $m \not\equiv 0 \bmod p$. Then $\left|(u^m - v^m)/(u - v)\right|_p = 1 = |m|_p$, and this is trivially true for $m = 1$, as well. For $n = mp^k$ one then has

$$\left| \frac{u^n - v^n}{u - v} \right|_p = \left| \frac{u^n - v^n}{u^m - v^m} \right|_p = \prod_{i=0}^{k-1} \left| \frac{u^{mp^{i+1}} - v^{mp^{i+1}}}{u^{mp^i} - v^{mp^i}} \right|_p$$

and $|n|_p = |p|_p^k$. Thus it suffices to prove (i) in the special case $n = p$. In this case, $p \neq 2$ or $u \equiv v \bmod p^2$, hence $\binom{p}{2} v^{p-2}(u - v) \equiv 0 \bmod p^2$. The above considerations (for $m = p$) show that $(u^p - v^p)/(u - v) \equiv p v^{p-1} \bmod p^2$, so that $|(u^p - v^p)/(u - v)|_p = p^{-1} = |p|_p$, which proves (i) for $n = p$.

Assertion (ii) follows from (i): the assumptions of (ii) imply that $u^2 - v^2 = (u + v)(u - v) \equiv 0 \bmod 4$ and $|u^2 - v^2|_2 = |2(u + v)|_2$. Hence by (i) we have $|u^n - v^n|_2 = |(u^2)^{n/2} - (v^2)^{n/2}|_2 = |u^2 - v^2|_2 \cdot |n/2|_2 = |u + v|_2 \cdot |n|_2$. $\qquad \square$

Let $1 \neq a \in \mathbb{Q}_{pos}$ and endow the subgroup $\langle a \rangle = a^{\mathbb{Z}} \cong \mathbb{Z}^+$ of $\mathbb{Q}^\times$ with the p-adic topology. One may ask when $a^{\mathbb{Z}}$ (with the p-adic topology) is isomorphic as a topological group to the group $\mathbb{Z}^+$, also endowed with the p-adic topology, or when $a^{\mathbb{Z}}$ is discrete in the p-adic topology. Note that the p-adic topology on $\mathbb{Z}^+$ is not discrete; see the end of 36.3.

37.6 Theorem   *Let p be a prime, and $a = u/v$ where $u \neq v$ are relatively prime positive integers. Then the following assertions hold.*

(i) *The group $a^{\mathbb{Z}}$ is discrete in the p-adic topology if, and only if, the prime p divides u or v.*

(ii) *The groups $a^{\mathbb{Z}}$ and $\mathbb{Z}^+$, both endowed with the p-adic topology, are isomorphic as topological groups if, and only if, $u \equiv v \not\equiv 0 \bmod p$.*

(ii') *In particular, if $u \not\equiv 0 \not\equiv v \bmod p$, then the subgroup $(a^{p-1})^{\mathbb{Z}}$ generated by a^{p-1} and the group $\mathbb{Z}^+$, both endowed with the p-adic topology, are isomorphic topological groups.*

(iii) *Assume that $u \not\equiv 0 \not\equiv v \bmod p$, and consider another prime q and another positive rational number b. If the group $a^{\mathbb{Z}}$ with the p-adic topology and the group $b^{\mathbb{Z}}$ with the q-adic topology are isomorphic as topological groups, then necessarily $q = p$.*

(iv) *If $p \neq q$, then the group $a^{\mathbb{Z}}$ with the p-adic topology is not isomorphic to the group $\mathbb{Z}^+$ with the q-adic topology.*

Proof (i) First we make the following *Observation*: if p divides u, then it does not divide v, so that $a^n = u^n/v^n$ tends to 0 for $n \to \infty$ in the p-adic topology.

Assume now that $a^{\mathbb{Z}}$ is not discrete in the p-adic topology. Then there is an unbounded sequence $(m_\nu)_{\nu \in \mathbb{N}}$ of integers such that a^{m_ν} converges

p-adically to a^m for some $m \in \mathbb{Z}$, and $a^{m_\nu - m} = a^{m_\nu} a^{-m}$ converges to 1. By a mere change of notation we may assume that a^{m_ν} converges to 1. Moreover we may assume that $m_\nu > 0$ for all ν (by passing to inverses, if necessary) and that m_ν tends to ∞. Then our initial observation shows that p cannot divide u. Since a and a^{-1} generate the same subgroup of $\mathbb{Q}^\times$, we also obtain that p cannot divide v if $a^\mathbb{Z}$ is not discrete. Thus, we have proved the 'if' part of (i) by contraposition.

The 'only if' part will be postponed and proved after the proof of (ii).

(ii) If the groups $\mathbb{Z}^+$ and $a^\mathbb{Z}$ are isomorphic as topological groups with the p-adic topology, then in the first place $a^\mathbb{Z}$ is not discrete, so that by the part of (i) that is already proved we know that $u \not\equiv 0 \not\equiv v \bmod p$.

Specifically, an isomorphism maps the generator 1 of $\mathbb{Z}$ to one of the generators a or a^{-1} of $a^\mathbb{Z}$, and since the inversion map is an automorphism of the topological group $a^\mathbb{Z}$ with the p-adic topology, there is an isomorphism $\mathbb{Z}^+ \to a^\mathbb{Z}$ of topological groups mapping 1 to a and hence $m \in \mathbb{Z}$ to a^m. In other words, the group isomorphism

$$\mathbb{Z}^+ \to a^\mathbb{Z} : m \mapsto a^m \tag{1}$$

is a homeomorphism with respect to the p-adic topology. Now the sequence $(p^\mu)_{\mu \in \mathbb{N}}$ converges to 0 in the p-adic topology; see the end of 36.3. Hence, a^{p^μ} converges to 1. In other words, $a^{p^\mu} - 1 = (u^{p^\mu} - v^{p^\mu})/v^{p^\mu}$ converges p-adically to 0. Hence, for μ sufficiently large, p divides $u^{p^\mu} - v^{p^\mu}$. Since by Fermat's little theorem $u^p \equiv u \bmod p$, it follows that $0 \equiv u^{p^\mu} - v^{p^\mu} \equiv u - v \bmod p$, that is, $u \equiv v \bmod p$. Thus the 'only if' part of (ii) is proved.

Assume now conversely that $u \equiv v \not\equiv 0 \bmod p$. We show that the group isomorphism (1) is a homeomorphism with respect to the p-adic topology on both groups. Since $|a|_p = 1 = |v|_p$ we obtain from 37.5 for $m, n \in \mathbb{Z}$ and $k := |m - n|$ that

$$|a^m - a^n|_p = |a^k - 1|_p = |(u^k - v^k)v^{-k}|_p = |u^k - v^k|_p = |m-n|_p \cdot |u \pm v|_p .$$

This shows that the bijection (1) is a homeomorphism, as $|u \pm v|_p \neq 0$. Thus, the 'if' part of (ii) is also established.

By Fermat's little theorem again, $u^{p-1} \equiv 1 \bmod p$ if $u \not\equiv 0$. Hence, (ii') immediately follows from (ii).

Thus if $u \not\equiv 0 \not\equiv v \bmod p$, then $a^{(p-1)\mathbb{Z}}$ is not discrete in the p-adic topology, hence the subgroup $a^\mathbb{Z}$ containing this non-discrete subgroup is not discrete. This is the contraposition of the 'only if' part of (i).

(iii) As in the proof of (ii), one sees that if $a^\mathbb{Z}$ with the p-adic topology and $b^\mathbb{Z}$ with the q-adic topology are isomorphic topological groups, then the map $a^m \mapsto b^m$ for $m \in \mathbb{Z}$ is an isomorphism. This map then induces an isomorphism of the subgroup $(a^{(p-1)(q-1)})^{\mathbb{Z}}$ with the p-adic topology onto the subgroup $(b^{(p-1)(q-1)})^{\mathbb{Z}}$ with the q-adic topology. By (ii$'$), the first group is isomorphic to $\mathbb{Z}^+$ with the p-adic topology. In particular, both subgroups are not discrete. By (i) and (ii$'$) again, the second subgroup is isomorphic to $\mathbb{Z}^+$ with the q-adic topology. Thus, the p-adic and the q-adic topology on $\mathbb{Z}$ give rise to isomorphic topological groups. By 36.5, it follows that $p = q$.

(iv) If p divides u or v, then $a^{\mathbb{Z}}$ is discrete in the p-adic topology by (i); in particular, it cannot be isomorphic to $\mathbb{Z}^+$ with the q-adic topology.

If $u \not\equiv 0 \not\equiv v \bmod p$, and if $a^{\mathbb{Z}}$ with the p-adic topology is isomorphic to $\mathbb{Z}^+$ with the q-adic topology, then by (ii$'$) these groups contain cyclic subgroups isomorphic to $\mathbb{Z}^+$ with the p-adic topology. Thus, there exists $m \in \mathbb{Z} \smallsetminus \{0\}$ such that the sequence $(p^\mu m)_{\mu \in \mathbb{N}}$ converges to 0 in the q-adic topology, which means that, for μ sufficiently large, $p^\mu m$ is divisible by arbitrarily large powers of q. This is possible only if $q = p$. □

37.7 Corollary *For distinct primes p and q, the p-adic topology, the q-adic topology and the usual topology on $\mathbb{Q}^\times$ give rise to non-isomorphic topological groups. The same is true for $\mathbb{Q}^\times_{\mathrm{pos}}$.*

Proof First, $\mathbb{Q}^\times_{\mathrm{pos}}$ contains cyclic subgroups which are not discrete in the p-adic topology according to 37.6. In contrast, every cyclic subgroup of $\mathbb{Q}^\times$ is discrete in the ordinary topology, so that $\mathbb{Q}^\times_{\mathrm{pos}}$ and $\mathbb{Q}^\times$ with the ordinary topology cannot be isomorphic as topological groups to their counterparts with the p-adic topology.

Now assume that for primes p and q there is an isomorphism φ of $\mathbb{Q}^\times_{\mathrm{pos}}$ (or $\mathbb{Q}^\times$) with the p-adic topology onto $\mathbb{Q}^\times_{\mathrm{pos}}$ (or $\mathbb{Q}^\times$, respectively) with the q-adic topology. We show that then $p = q$, so that our corollary is proved. Let $m \in \mathbb{N} \smallsetminus p\mathbb{Z}$. Then $\varphi(m^2) = \varphi(m)^2 \in \mathbb{Q}^\times_{\mathrm{pos}}$. Via φ, the subgroup $(m^2)^{\mathbb{Z}}$ of $\mathbb{Q}^\times_{\mathrm{pos}}$ with the p-adic topology is isomorphic to the subgroup $\varphi(m^2)^{\mathbb{Z}}$ of $\mathbb{Q}^\times_{\mathrm{pos}}$ with the q-adic topology. Since $m^2 \in \mathbb{N} \smallsetminus p\mathbb{Z}$, as well, 37.6(iii) says that indeed $q = p$. □

Exercises

(1) Multiplication defines a map $\mu : \bigoplus_{p \in \mathbb{P}} p^{\mathbb{Z}} \to \mathbb{Q}^\times_{\mathrm{pos}}$ from a subspace of a product of discrete groups into $\mathbb{R}$. Is μ or μ^{-1} continuous?

(2) The topological groups $\mathbb{Q}^+$ and $\mathbb{Q}^\times_{\mathrm{pos}}$ have isomorphic character groups.

4

Completion

In Chapter 1, the basic properties of the real numbers were taken for granted. In Chapter 3, devoted to the rational numbers, we sometimes used the fact that they are embedded in the real numbers. In the present chapter, we shall discuss standard procedures which allow us to construct the domain of real numbers from the domain of rational numbers by so-called completion. (A non-standard procedure was already presented in Section 23.)

The rational numbers are not complete, either with respect to their ordering (there are non-empty bounded sets which have no supremum within the rational numbers), or as a topological group (there are Cauchy sequences of rational numbers which do not converge to a rational number). Completion processes remedy these defects by a cautious enlargement which supplies the missing suprema or limits, without introducing new incompleteness problems. Corresponding to the two facets of incompleteness of the rational numbers, there are two types of completion, one for ordered structures and, more specifically, for ordered groups, and another one for certain topological structures, in particular for topological groups. These completion principles will be presented here. (We do not, however, discuss completion of metric spaces or, more generally, of uniform spaces without algebraic structure.)

For the rational numbers, both kinds of completion, the completion of $\mathbb{Q}$ as an ordered group and its completion as a topological group, will lead to the same mathematical object (up to isomorphism), the additive group $\mathbb{R}$ of real numbers, with its ordering and its topology. In the case of an ordered field and of a topological field, one may also extend the multiplication to the completion of the additive group. In particular, this allows us to endow $\mathbb{R}$ with the structure of an ordered field and of a topological field in a canonical way.

41   Completion of chains

In this section we shall discuss the completion of chains, as a basis for the next section on the completion of ordered groups.

41.1 Basic notions We recall from 3.1 that a chain is called complete if every non-empty subset A that is bounded above has a least upper bound, the supremum $\sup A$. In a complete chain C, every non-empty subset B that is bounded below has a greatest lower bound, the infimum $\inf B$. Indeed, the set $A := \{\, x \in C \mid x \le B \,\}$ of all lower bounds of B is non-empty and bounded above (by any element of B), and it is an immediate consequence of the definitions that $\sup A$ is the greatest lower bound of B.

A subset D of a chain C is called *coterminal* if for every $c \in C$ there are $x, y \in D$ such that $x \le c \le y$.

A subset D of C is called *weakly dense* if it is coterminal and for every $c \in C$ we have $\sup\{\, d \in D \mid d \le c \,\} = c = \inf\{\, d \in D \mid c \le d \,\}$. Clearly, this is equivalent to saying that every element of C is the supremum of *some* non-empty subset of D and the infimum of *some* non-empty subset of D. Yet another way of putting this is that for all $c_1 < c_2$ in C there are $d_1, d_2, d_3, d_4 \in D$ such that $d_1 \le c_1 \le d_2 < d_3 \le c_2 \le d_4$. In 3.1 and 3.2, this was compared with other density notions. In particular, weak density is stronger than topological density of D in C, which means that every non-empty open interval of C contains an element of D.

An *ideal* of a chain C is a non-empty subset I with the property that for every element $a \in I$, every element smaller than a is also contained in I, that is, I contains the interval $\,]\,, a] = \{\, x \in C \mid x \le a \,\}$.

As in Section 3, a map $\varphi : C \to C'$ between two chains is said to *preserve the ordering* (or to be *order-preserving*) if for all $x, y \in C$ such that $x \le y$ one has $\varphi(x) \le \varphi(y)$.

41.2 Definition A *completion* of a chain C is a complete chain $\widehat{C}$ together with an order-preserving injective map $\iota : C \to \widehat{C}$ whose image $\iota(C)$ is weakly dense in $\widehat{C}$.

41.3 Construction *Every chain C has a completion.*

Proof (1) We call an ideal of C *admissible*, if it is bounded above and contains its supremum if this supremum exists, and we define $\widehat{C}$ to be the set of all admissible ideals of C.

The ordering relation on $\widehat{C}$ will be inclusion of sets. The simplest admissible ideals are the intervals $\,]\,, c]$ for $c \in C$; we may therefore embed C into $\widehat{C}$ by the map

$$\iota : C \to \widehat{C} : c \mapsto \,] \,,c]\,,$$

which clearly is injective and preserves the ordering.

The admissible ideals can be obtained in the following way. For an arbitrary ideal I of C, we define

$$\bar{I} = \begin{cases} I \cup \{\sup I\} & \text{if } I \text{ has a supremum} \\ I & \text{otherwise.} \end{cases}$$

One verifies that this is an ideal again. Obviously, if I has a supremum, then

$$\sup I = \sup \bar{I}, \quad \text{so that } \bar{\bar{I}} = \bar{I}\,. \tag{1}$$

Moreover, one verifies for ideals I, J that

$$I \subseteq J \Longrightarrow \bar{I} \subseteq \bar{J}\,. \tag{2}$$

If the ideal I is bounded above, then $\bar{I}$ is admissible in view of equation (1), and an admissible ideal A satisfies $\bar{A} = A$. Thus, the admissible ideals are precisely the ideals $\bar{I}$ obtained from an ideal I which is bounded above:

$$\widehat{C} = \{\,\bar{I} \mid I \text{ ideal of } C \text{ and bounded above}\,\}\,.$$

(2) We remark that for an admissible ideal A the pair $(A, C \smallsetminus A)$ constitutes what is called a Dedekind cut of the chain C. We shall not make further use of this notion.

(3) It is easy to see that $\widehat{C}$ is a chain. Indeed, if $I \neq J$ are any ideals of C and if $J \smallsetminus I \neq \emptyset$, say, we obtain for $a \in J \smallsetminus I$ that $I \leq a$ since I is an ideal and then $I \subseteq J$ since J is an ideal.

(4) Now it will be shown that the chain $\widehat{C}$ is complete. Let $\mathcal{A} \subseteq \widehat{C}$ be bounded above, that is, $\mathcal{A}$ consists of admissible ideals all contained in some admissible ideal W. It is immediate from the definition that the union $U := \bigcup \mathcal{A} = \bigcup \{A \mid A \in \mathcal{A}\}$ is an ideal again; it is bounded above since it is contained in W which is bounded above. Clearly, $\overline{U}$ is an element of $\widehat{C}$ which is an upper bound of $\mathcal{A}$. We show that it is the supremum of $\mathcal{A}$, which means that it is contained in every admissible ideal W containing all the elements of $\mathcal{A}$. Clearly $U \subseteq W$; using equations (2) and (1), we conclude that $\overline{U} \subseteq \overline{W} = W$.

(5) Finally we verify that $\iota(C)$ is weakly dense in $\widehat{C}$. Let $A, B \in \widehat{C}$ such that $A \subseteq B$, $A \neq B$. An element $b \in B \smallsetminus A$ is an upper bound of A, but there is still another bound $u \in C$ of A such that $u < b$

(else $b = \sup A \in A$). Finally, let $v \in C$ be an upper bound of B and $a \in A$. Then $]\,,a] \subseteq A \subseteq]\,,u] \subset]\,,b] \subseteq B \subseteq]\,,v]$, in other words $\iota(a) \subseteq A \subseteq \iota(u) \subset \iota(b) \subseteq B \subseteq \iota(v)$. This shows that $\iota(C)$ is weakly dense in $\widehat{C}$, according to one of the alternative descriptions of this notion in 41.1.   □

In order to show that essentially this is the only way to construct a completion of a chain, we study some properties of completions with respect to extensions of order-preserving maps. It is instructive to distinguish the roles played by the two constituents of the notion of a completion, completeness and density.

41.4 Lemma   *Let $\iota : C \to L$ be an order-preserving injective map between chains such that $\iota(C)$ is coterminal in L, and let R be a complete chain. Then every order-preserving map $\varphi : C \to R$ admits an extension over ι, in other words there exists an order-preserving map $\widetilde{\varphi} : L \to R$ such that $\widetilde{\varphi} \circ \iota = \varphi$.*

Proof For $x \in L$, the preimage $\iota^{-1}(]\,,x])$ of the interval $]\,,x] \subseteq L$ is non-empty and bounded above in C since $\iota(C)$ is coterminal in L. Hence, $\varphi(\iota^{-1}(]\,,x]))$ is bounded above in R, so that the map

$$\widetilde{\varphi} : L \to R : x \mapsto \sup \varphi(\iota^{-1}(]\,,x]))$$

is well-defined. It is immediate that $\widetilde{\varphi}$ has the stated properties.   □

41.5 Lemma   *Let $\iota : C \to L$ be an order-preserving injective map between chains such that $\iota(C)$ is weakly dense in L. Then the identity map $\psi = id$ is the only map $\psi : L \to L$ that preserves the ordering and satisfies $\psi \circ \iota = \iota$.*

Proof Let ψ be such a map. The condition $\psi \circ \iota = \iota$ says that ψ induces the identity map on $\iota(C)$. Now suppose that $x \neq \psi(x)$ for some $x \in L$, say $x < \psi(x)$ (the other case is treated analogously). Then $x \notin \iota(C)$. By weak density, there is $c \in C$ such that $x < \iota(c) < \psi(x)$, and then ψ reverses the order of the elements $\iota(c) = \psi(\iota(c))$ and x, which is a contradiction.   □

41.6 Theorem: Universal property of completions   Let C_i for $i = 1, 2$ be a chain, and $\iota_i : C_i \to \widehat{C}_i$ a completion of C_i.

Then for every order-preserving bijection $\varphi : C_1 \to C_2$ there is a unique extension $\widehat{\varphi} : \widehat{C}_1 \to \widehat{C}_2$ of φ to the completions, that is an order-preserving map such that $\widehat{\varphi} \circ \iota_1 = \iota_2 \circ \varphi$. Moreover, the extension $\widehat{\varphi}$ is bijective.

Proof Such a map $\widehat{\varphi}$ is obtained by applying 41.4 to $\iota_2 \circ \varphi : C_1 \to \widehat{C}_2$. We replace φ by $\psi := \varphi^{-1}$ and permute the roles of ι_1 and ι_2; this gives an order-preserving map in the opposite direction $\widehat{\psi} : \widehat{C}_2 \to \widehat{C}_1$ such that $\widehat{\psi} \circ \iota_2 = \iota_1 \circ \psi$. The composition $\widehat{\psi} \circ \widehat{\varphi} : \widehat{C}_1 \to \widehat{C}_1$ satisfies $\widehat{\psi} \circ \widehat{\varphi} \circ \iota_1 = \widehat{\psi} \circ \iota_2 \circ \varphi = \iota_1 \circ \psi \circ \varphi = \iota_1$, so that $\widehat{\psi} \circ \widehat{\varphi} = \mathrm{id}$ by 41.5. Again by permuting roles one obtains analogously that $\widehat{\varphi} \circ \widehat{\psi} = \mathrm{id}$. Thus, $\widehat{\varphi}$ is bijective with inverse map $\widehat{\psi}$. This holds for every pair of extensions $\widehat{\varphi}$ and $\widehat{\psi}$; hence these extensions are unique.   □

41.7 Corollary: Uniqueness of completion   *Let $\iota : C \to \widehat{C}$ and $\widetilde{\iota} : C \to \widetilde{C}$ be completions of a chain C. Then there is a unique order-preserving bijection $\widehat{\varphi} : \widehat{C} \to \widetilde{C}$ such that $\widehat{\varphi} \circ \iota = \widetilde{\iota}$.*

Proof This is the special case of 41.6 with $C_1 = C = C_2$ and $\varphi = \mathrm{id}$.   □

The following will provide a useful tool for handling elements of the completion without knowing the completion explicitly.

41.8 Lemma   *Let $\iota : C \to \widehat{C}$ be a completion of a chain C, and X, Y non-empty subsets of C that are bounded above. Then $\sup \iota(X) \le \sup \iota(Y)$ if and only if every upper bound $u \in C$ of Y is an upper bound of X as well.*

Remark   Note that the upper bounds u are restricted to C. The suprema of $\iota(X)$ and $\iota(Y)$ exist in the completion.

Proof We remark that for $u \in C$ the condition $\sup \iota(X) \le \iota(u)$ is equivalent to $X \le u$. The 'only if' part is then clear. For the 'if' part, we consider the set $U(X) = \{ u \in C \mid X \le u \}$ of all upper bounds of X in C. Since $\iota(C)$ is weakly dense in $\widehat{C}$, the supremum of $\iota(X)$ is the infimum of the set $\{ \iota(u) \mid u \in C, \sup \iota(X) \le \iota(u) \}$, which is just $\iota(U(X))$ by the initial remark. Hence, if by assumption $U(Y) \subseteq U(X)$, then $\sup \iota(X) = \inf \iota(U(X)) \le \inf \iota(U(Y)) = \sup \iota(Y)$.   □

42   Completion of ordered groups and fields

We now turn to the completion problem for ordered groups. (The definition and basic properties of an ordered group may be found in Section 7.) When constructing the completion of the underlying chain of an ordered group as in 41.3, one may try to extend the group operation to the completion and hope to obtain an ordered group again. Now according to 7.5, a completely ordered group is Archimedean, so this programme can

only be successful for Archimedean groups. Furthermore, recall that
by 7.7 an Archimedean ordered group is commutative. This is why all
groups considered here will be written additively.

42.1 Definitions An ordered group is called *completely ordered* if the
underlying chain is complete. A *completion* of an ordered group G is
a completely ordered group $\widehat{G}$ together with an order-preserving group
monomorphism $\iota : G \to \widehat{G}$ whose image $\iota(G)$ is weakly dense in $\widehat{G}$ (so
that $\iota : G \to \widehat{G}$ is a completion of the chain underlying G).

The following lemma gives a hint as to how the group operation of an
ordered group should be extended to the chain completion.

42.2 Lemma *Let X, Y be non-empty subsets of an ordered group G.*
(a) *If Y has a supremum in G, and if $z \in G$ satisfies $X + Y \leq z$, then*
    $X + \sup Y \leq z$.
(b) *If the suprema of X and Y exist in G, then the same holds for*
    $X + Y$, *and* $\sup(X + Y) = \sup X + \sup Y$.

Proof (a) The assumption $X + Y \leq z$ implies that $Y \leq -x + z$ for all
$x \in X$ so that $\sup Y \leq -x + z$ and hence $X + \sup Y \leq z$.

(b) It follows that $X \leq z - \sup Y$ and hence, if $\sup X$ exists, then
$\sup X \leq z - \sup Y$, $\sup X + \sup Y \leq z$ for all upper bounds z of $X + Y$.
On the other hand, it is clear that $\sup X + \sup Y$ is itself an upper bound
of $X + Y$, hence it is the least upper bound.   □

42.3 Construction *Every Archimedean ordered group G has a completion.*

Proof (1) Let $\iota : G \to \widehat{G}$ be any completion of G as a chain. Using the
addition of G, we define an addition on $\widehat{G}$ as follows.

For $a \in \widehat{G}$, let $A(a) := \{\, x \in G \mid \iota(x) \leq a \,\}$. Since $\iota(G)$ is weakly dense
in $\widehat{G}$, we have

$$a = \sup \iota(A(a)) .$$

Since $\iota(G)$ is coterminal in $\widehat{G}$, the subset $A(a)$ is bounded above in G.
Now consider a second element $b \in \widehat{G}$. Since G is an ordered group,
$A(a) + A(b)$ is bounded above. With view to 42.2, we define

$$a + b := \sup \iota(A(a) + A(b)) .$$

It is clear that this addition satisfies the monotonicity law required for
an ordered group since for $c \in \widehat{G}$ the relation $a \leq c$ translates into
$A(a) \subseteq A(c)$.

(2) We now prove more generally that this addition satisfies the following equation for arbitrary non-empty subsets $X, Y \subseteq G$ that are bounded above:

$$\sup \iota(X) + \sup \iota(Y) = \sup \iota(X + Y) \qquad (1)$$

Let $a = \sup \iota(X)$, $b = \sup \iota(Y)$. Then $X \subseteq A(a), Y \subseteq A(b)$, so that $\sup \iota(X + Y) \leq \sup \iota(A(a) + A(b)) = a + b$.

As to the converse inequality $\sup \iota(X + Y) \geq \sup \iota(A(a) + A(b))$, according to 41.8, it suffices to show for $z \in G$ that

$$X + Y \leq z \Longrightarrow A(a) + A(b) \leq z . \qquad (2)$$

Let $x' \in A(a)$. If $\iota(x') < a = \sup \iota(X)$, there is $x \in X$ with $x' < x$. If $\iota(x') = a = \sup \iota(X)$, then $x' = \sup X$. Hence there is $\tilde{x} \in G$ such that $x' \leq \tilde{x}$ and $\tilde{x} \in X$ or $\tilde{x} = \sup X$. Likewise, for $y' \in A(b)$, there is $\tilde{y} \in G$ such that $y' \leq \tilde{y}$ and $\tilde{y} \in Y$ or $\tilde{y} = \sup Y$. By 42.2, it follows that $x' + y' \leq \tilde{x} + \tilde{y} \leq z$, which proves the implication (2) and equation (1).

(3) If in equation (1) we take X and Y to be singletons, we see at once that ι is a homomorphism with respect to the addition defined on $\widehat{G}$. If one of X, Y is the singleton $\{0\}$, it follows that $\iota(0)$ is a neutral element for the addition in $\widehat{G}$. Moreover, this addition is associative, because the group G is associative: For $a, b, c \in \widehat{G}$, repeated use of equation (1) gives $a + (b + c) = \sup \iota(A(a)) + \sup \iota(A(b) + A(c)) = \sup \iota(A(a) + (A(b) + A(c))) = \sup \iota((A(a) + A(b)) + A(c)) = \sup \iota(A(a) + A(b)) + \sup \iota(A(c)) = (a + b) + c$.

(4) Up to now, we have not used the hypothesis that G is Archimedean. This will be crucial for the existence of inverses in $\widehat{G}$. For $a \in \widehat{G}$, consider the set $A(a)$ defined above and

$$A'(a) = \{ t \in G \mid t \leq -A(a) \} .$$

We show that $a' := \sup A'(a)$ satisfies $a + a' = \iota(0)$, which by equation (1) is equivalent to $\sup \iota(A(a) + A'(a)) = \iota(0)$.

It is clear from the definitions that $A(a) + A'(a) \leq 0$, which implies that $\sup \iota(A(a) + A'(a)) \leq \iota(0)$. In order to prove the converse inequality, we need to show, according to 41.8, that $0 \leq e$ for every upper bound $e \in G$ of $A(a) + A'(a)$. Now for $x \in A(a), t \in A'(a)$ we have $x + t \leq e$ by the choice of e. Thus, $t - e \leq -x$, and since this holds for all $x \in A(a), t \in A'(a)$, we have shown that $A'(a) - e \subseteq A'(a)$. Induction yields $A'(a) - ne \subseteq A'(a)$ for all $n \in \mathbb{N}$, so that for $x \in A(a), t \in A'(a)$ we obtain $n(-e) \leq -t - x$. Since G is Archimedean, it follows that $-e \leq 0$, so that $e \geq 0$, which was to be shown.   □

42.4 Theorem: Universal property of group completions *Let* $\iota : G \to \widehat{G}$ *and* $\iota' : H \to \widehat{H}$ *be completions of ordered groups* G, H. *Then for every order-preserving group isomorphism* $\varphi : G \to H$ *the unique extension* $\widehat{\varphi} : \widehat{G} \to \widehat{H}$ *of* φ *to the completions according to 41.6, that is the order-preserving map satisfying* $\widehat{\varphi} \circ \iota = \iota' \circ \varphi$, *is a group isomorphism.*

Proof For elements $a, b \in \widehat{G}$ there are non-empty subsets $X, Y \subseteq G$ bounded above in G such that $a = \sup \iota(X)$ and $b = \sup \iota(Y)$. The order-preserving extension $\widehat{\varphi}$ is a bijection by 41.6; therefore $\widehat{\varphi}$ respects suprema, e.g., $\widehat{\varphi}(a) = \sup \widehat{\varphi}(\iota(X))$. Now we can use 42.2 to obtain that $\widehat{\varphi}(a+b) = \widehat{\varphi}(\sup(\iota(X)+\iota(Y))) = \sup \widehat{\varphi}(\iota(X+Y)) = \sup \iota'(\varphi(X+Y)) = \sup(\iota'(\varphi(X))+\iota'(\varphi(Y))) = \sup \iota'(\varphi(X)) + \sup \iota'(\varphi(Y)) = \sup \widehat{\varphi}(\iota(X)) + \sup \widehat{\varphi}(\iota(Y)) = \widehat{\varphi}(\sup \iota(X)) + \widehat{\varphi}(\sup \iota(Y)) = \widehat{\varphi}(a) + \widehat{\varphi}(b)$. $\square$

42.5 Corollary: Uniqueness of group completions *Let* $\iota : G \to \widehat{G}$ *and* $\widetilde{\iota} : G \to \widetilde{G}$ *be completions of an ordered group* G. *Then there is a unique order-preserving group isomorphism* $\widehat{\varphi} : \widehat{G} \to \widetilde{G}$ *such that* $\widehat{\varphi} \circ \iota = \widetilde{\iota}$.

Proof This is the special case of 42.4 with $H = G$ and $\varphi = \mathrm{id}$. $\square$

Finally, we study the completion of Archimedean ordered fields. For the definition of an ordered skew field, see 11.1; the additive group of an ordered skew field F is an ordered group, and the set P of positive elements is a subgroup of the multiplicative group and an ordered group, as well (11.5). The characteristic of F is zero (see 11.2), thus the prime field of F is isomorphic to $\mathbb{Q}$ and consists of the elements $m \cdot (n \cdot 1)^{-1}$ for $m, n \in \mathbb{Z}$, $n \neq 0$, which we write as m/n for short.

42.6 Lemma *If the additive group of an ordered skew field F is Archimedean, then the same holds for the group P of positive elements under multiplication, and F is commutative.*

In this case, we call F an *Archimedean ordered field*.

Proof For $1 < a \in F$, we have $-1 + a > 0$. Since the additive group is Archimedean, there is $n \in \mathbb{N}$ such that $n(-1 + a) > 1$, that is $-1 + a > 1/n$, $a > 1 + 1/n$. It follows for $m \in \mathbb{N}$ that $a^m > (1 + 1/n)^m = 1 + m/n + \cdots > m/n$. Now for $0 < b \in K$, by the Archimedean property again, we may choose m in such a way that $m/n = m \cdot 1/n > b$, so that finally $a^m > b$. This proves that P is Archimedean (and is part of a solution of Exercise 4 in Section 11).

By 7.7, it follows that P is commutative. Hence $F^\times = P \times \{\pm 1\}$ is commutative, as well.   □

42.7 Definitions An ordered field is called *completely ordered* if the underlying chain is complete. A *completion* of an ordered field F is a completely ordered field $\widehat{F}$ together with an order-preserving field monomorphism $\iota : F \to \widehat{F}$ whose image $\iota(F)$ is weakly dense in $\widehat{F}$ (so that $\iota : F \to \widehat{F}$ is a chain completion of F).

42.8 Construction *Every Archimedean ordered field F has a completion.*

Proof (1) Let $\iota : F \to \widehat{F}$ be a group completion of the additive group of F, which is an ordered group; such a group completion exists by 42.3. By 7.5 and 7.7, the ordered group $\widehat{F}$ is Archimedean and therefore commutative; its group operation will be denoted and referred to as addition.

By identifying the elements of F with their images under ι, we may assume that $F \subseteq \widehat{F}$ and that ι is the set theoretic inclusion map. This will simplify notation.

We shall extend the multiplication of F to $\widehat{F}$ in order to obtain a multiplication of $\widehat{F}$ which makes $\widehat{F}$ an ordered field. One way would be to write down the multiplication explicitly for suprema of subsets of $\iota(F)$, similarly as for the addition of an Archimedean ordered group in 42.3. Here, we choose an alternative way. We use the completion of the ordered group P of positive elements of F, which is a subgroup of the multiplicative group; the completion of P as a chain is just the set of positive elements of $\widehat{F}$. After extending the multiplication to the whole of $\widehat{F}$ we use the universal property 42.2 of group completions as a conceptual method for the necessary verifications in order to show that the completion is again an ordered field. Now for the details.

(2) The set $\widehat{P} = \{ a \in \widehat{F} \mid a > 0 \}$ is a complete chain, since $\widehat{F}$ is, and $P \subseteq \widehat{P}$ is weakly dense in $\widehat{P}$. Hence, $\widehat{P}$ is the completion of the chain P. By 42.6, the group P is Archimedean, so that is has a completion as an ordered group according to 42.3. In other words, there is a multiplication on $\widehat{P}$ extending the multiplication of P. (Of course, this multiplication can be established explicitly as in step (2) of the proof of 42.3 by declaring, for non-empty subsets X and Y of P which are bounded above, $\sup X \cdot \sup Y$ to be $\sup(X \cdot Y)$. However, one can just as well rely on the existence and uniqueness of the chain completion and the group completion; see 41.7 and 42.5.)

We now extend the multiplication of $\widehat{P}$ to the whole of $\widehat{F}$ in several steps, respecting the multiplication on F. For $0 < a \in \widehat{F}$, that is, $a \in \widehat{P}$, we consider the map

$$\lambda_a : \widehat{F} \to \widehat{F} : b \mapsto \begin{cases} ab & \text{if } b > 0 \\ 0 & \text{if } b = 0 \\ -a(-b) & \text{if } b < 0 \, . \end{cases}$$

The restriction of λ_a to $\widehat{P}$ is an order-preserving bijection of $\widehat{P}$ into itself, since $\widehat{P}$ is an ordered group. For the same reason, and since the bijection

$$\mu : \widehat{F} \to \widehat{F} : b \mapsto -b$$

reverses the ordering (being the inversion map of the additive group of $\widehat{F}$, which is an ordered group), λ_a induces an order-preserving bijection of $-\widehat{P}$ onto itself. Thus we obtain that λ_a is an order-preserving bijection of the whole of $\widehat{F}$ onto itself.

For $a < 0$, we define

$$\lambda_a := \mu \circ \lambda_{-a} \, ,$$

obviously an order-reversing bijection of $\widehat{F}$ onto itself. Using the maps λ_a, we construct the multiplication of $\widehat{F}$ as follows:

$$ab := \begin{cases} \lambda_a(b) & \text{if } a \neq 0 \\ 0 & \text{if } a = 0 \, . \end{cases}$$

It is clear from the definition of λ_a for $a > 0$ that this multiplication extends the multiplication of $\widehat{P}$ so that no confusion will arise from using the same notation for both multiplications.

Next, it will be shown that

$$(-a)b = -ab = a(-b) \tag{1}$$

for all $a, b \in \widehat{F}$. For $a = 0$, this is trivial. For $a \neq 0$, equation (1) can be translated into

$$\mu \circ \lambda_a = \lambda_{-a} = \lambda_a \circ \mu \tag{2}$$

and will be proved in this form. For $a > 0$, the first equality is just the definition of λ_{-a}; for $a < 0$, by definition, $\lambda_a = \mu \circ \lambda_{-a}$, from which the first equality follows by composition with the involutory map μ. The second equality for $a > 0$ is obtained directly from the definitions of λ_a and of λ_{-a}; for $a < 0$, we may then conclude that $\lambda_a \circ \mu = \mu \circ \lambda_{-a} \circ \mu = \mu \circ \lambda_a = \lambda_{-a}$.

(3) We now show that $\widehat{F} \setminus \{0\}$ with the multiplication defined above is a commutative group. Recall first that $\widehat{P}$ is a group under multiplication and is commutative, being a completely ordered group and hence Archimedean; see 7.5 and 7.7. By equation (1), the group properties and commutativity carry over directly from $\widehat{P}$ to $\widehat{F} \setminus \{0\}$; indeed (1) says that $\widehat{F} \setminus \{0\}$ is the direct product of $\widehat{P}$ and of the group $\{1, -1\}$.

In order to establish that $\widehat{F}$ is a field it remains to prove distributivity. Then $\widehat{F}$ is an ordered field as is immediate from the stated monotonicity properties of the maps λ_a.

(4) For $u \in F$, the restriction of λ_u to F is the map $F \mapsto F : x \mapsto ux$ obtained from the original multiplication in F. This is so because the multiplication of $\widehat{P}$ extends that of $P \subseteq F$ and since equation (1) holds for the multiplication of F. Thus, the multiplication of $\widehat{F}$ extends that of F, and the inclusion map $F \to \widehat{F}$ is a homomorphism with respect to both multiplication and addition.

(5) For two positive elements $u, v \in F$ both the maps λ_{u+v} and $\lambda_u + \lambda_v$ are order-preserving extensions of the order-preserving bijection $F \to F : x \mapsto (u + v)x = ux + vx$. By the universal property 41.6 of completions, these two extensions coincide, which means that

$$(u + v)a = ua + va \tag{3}$$

for $a \in \widehat{F}$. For $u > v > 0$ it follows that $(u - v)a + va = ua$, so that $(u - v)a = ua - va$. From this and equation (3) and using equation (1) one obtains that equation (3) holds for arbitrary $u, v \in F$, which means that for fixed $a \in \widehat{F} \setminus \{0\}$, the set Fa is a subgroup of the additive group of $\widehat{F}$ and that the bijection

$$\lambda_a : \widehat{F} \to \widehat{F} : b \mapsto ab = ba$$

induces an isomorphism of the additive group $(F, +)$ onto $(Fa, +)$.

For $a > 0$, the map λ_a is order-preserving. Since P is weakly dense in $\widehat{P}$, we infer that Fa is weakly dense in $\widehat{F}$, so that $\widehat{F}$ is a completion of both the ordered groups F and Fa. By the universal property 42.4 of group completions, λ_a is a group automorphism of $(\widehat{F}, +)$, being an extension of an order-preserving group isomorphism $F \to Fa$.

For $a < 0$, we thus know that λ_{-a} is an automorphism of the additive group. The same is true for the inversion map μ, since addition is commutative; hence $\lambda_a = \mu \circ \lambda_{-a}$ is an automorphism.

Thus the distributive law $a(b + c) = ab + ac$ holds for all $a, b, c \in \widehat{F}$. (Again, the case $a = 0$ needs a separate, but trivial verification.) All the properties of a field completion are now established.   □

42.9 Uniqueness theorem *Up to isomorphism, there is exactly one completely ordered field. In other words, for any two completely ordered fields F_1 and F_2, there is an order-preserving field isomorphism of F_1 onto F_2.*

Remark An ordered field F is a topological field with the topology induced by the ordering. If F is completely ordered, F is also complete as a topological field (43.10). The converse does not hold in general, but is true for Archimedean ordered fields; see 43.29. Thus, Theorem 42.9 may be rephrased as follows. Up to isomorphism, there is exactly one Archimedean ordered field which is complete as a topological field. An Archimedean ordered field is complete as a topological field if and only if every Cauchy sequence converges. Recall that order completeness implies the Archimedean property. For a comprehensive discussion of these matters, see PRIESS-CRAMPE 1983 Chapter III Section 1, BLYTH 2005 Section 10.2 or DALES–WOODIN 1996 Section 3.

The proof of Theorem 42.9 will make use of the following lemma.

42.10 Lemma *In an Archimedean ordered field, the prime field is weakly dense.*

Remark The assertion of Lemma 42.10 is contained in Theorem 11.14. But there, the field of real numbers is involved; also, the assertion of 42.10 is obtained rather indirectly there, for the sake of further related results. Here, we give a direct proof which avoids the use of real numbers, as one of the aims of the present section is the *construction* of the real numbers; see 42.11.

Proof Let F be an Archimedean ordered field and $a \in F$. Since F is Archimedean, every element lies between two elements of the prime field. Hence it suffices to show that for all $a, b \in F$ such that $a < b$ there is an element q of the prime field such that $a < q < b$. We may assume that b is positive; if not, we apply the order-reversing bijection $x \mapsto -x$. Furthermore, we may assume that $a \geq 0$; else we replace a by 0.

Then, since F is Archimedean, there is $n \in \mathbb{N}$ such that $n \cdot 1 > (b-a)^{-1}$, so that $1/n < b-a$, and likewise there is $m \in \mathbb{N}$ such that $(m+1) \cdot 1/n \geq b$. If m is the smallest natural number with this property, then $m \cdot 1/n < b$ and $m \cdot 1/n \geq b - 1/n > a$. Thus, $m \cdot 1/n$ is an element of the prime field lying between a and b. $\qquad\square$

Proof of Theorem 42.9. There is a completely ordered field, for instance the field completion according to 42.8 of the ordered field $\mathbb{Q}$.

Now let F_1 and F_2 be completely ordered fields; we show that they are isomorphic. Let Q_1, Q_2 be the prime fields of F_1 and F_2. Since a completely ordered field is Archimedean, these prime fields are weakly dense by 42.10, so that F_1 and F_2 are field completions of Q_1 and Q_2. The prime fields are isomorphic to $\mathbb{Q}$, and the natural bijection $\varphi : Q_1 \to Q_2$ mapping the element $m/n = m \cdot (n \cdot 1)^{-1}$ of Q_1 to the corresponding element of Q_2 is order-preserving and a field isomorphism. By the universal properties 41.6 and 42.4 of completions, φ extends to an order-preserving bijection $\widehat{\varphi} : F_1 \to F_2$ which is a group isomorphism of the additive groups of F_1 and F_2.

The set P_1 of positive elements of F_1 is a group under multiplication. It is also completely ordered, and the set $Q_1 \cap P_1$ of positive elements of Q_1 is weakly dense in P_1. The same holds for the group P_2 of positive elements of F_2, so that P_1 and P_2 are group completions of $Q_1 \cap P_1$ and $Q_2 \cap P_2$, respectively. The natural map $\varphi : Q_1 \to Q_2$ induces an order-preserving group isomorphism between the groups $Q_1 \cap P_1$ and $Q_2 \cap P_2$. The extension $P_1 \to P_2$ of this map obtained by restricting $\widehat{\varphi}$ therefore is an isomorphism (with respect to multiplication) by the universal property 42.4 of group completions. Since every product of non-zero elements of F_1 can be reduced to a product of positive elements up to sign, it follows that $\widehat{\varphi}$ respects multiplication not only on P_1, but throughout F_1. We conclude that $\widehat{\varphi}$ is an order-preserving field automorphism.   □

42.11 Construction of the real numbers On the basis of the uniqueness theorem 42.9, we may now define the field $\mathbb{R}$ of real numbers to be any completely ordered field. Since this determines $\mathbb{R}$ only up to isomorphism, one may wish to specify a concrete construction of such a completely ordered field. For instance, one may define the underlying chain to be the completion of $\mathbb{Q}$ obtained by the explicit construction described in 41.3. This chain is the underlying chain for a group completion of $(\mathbb{Q}, +)$ obtained by 42.3, which will be the additive group of $\mathbb{R}$. On this completely ordered group, a multiplication is established according to the construction in 42.8, which finally produces the completely ordered field $\mathbb{R}$.

In Chapter 1 of this book, the known properties of the real numbers were used without further justification, but these properties just express the fact that $\mathbb{R}$ is a completely ordered field, or can be derived easily from this. Sometimes it is even helpful to recall that this comprises all we know about $\mathbb{R}$. For instance, 11.8 can be seen as an immediate corollary of the uniqueness theorem 42.9 for completely ordered fields; in Section 11, it was proved differently. Another example is Corollary 42.12 below.

In the present section, we have sometimes used arguments from Chapter 1, in particular the simple facts about ordered groups and ordered fields from Sections 7 and 11. We have been careful, however, to use only such arguments which in Chapter 1 are derived directly from the definitions of these structures, without making use of the real numbers. So, our conclusive construction of the real numbers based on these arguments does not suffer from a vicious circle.

For a different approach to the completion of ordered fields and the construction of the ordered field $\mathbb{R}$ avoiding many of the verifications in 42.8 see BANASCHEWSKI 1998.

In the next two sections, topological completion methods are discussed which offer an alternative construction of $\mathbb{R}$ from $\mathbb{Q}$. Another possibility for the construction of $\mathbb{R}$ using non-standard methods has already been described in Section 23; see also the introduction to Chapter 1.

42.12 Corollary *Every Archimedean ordered field admits an order-preserving isomorphism onto a subfield of $\mathbb{R}$ endowed with the induced ordering.*

Proof The completion of an Archimedean ordered field is (isomorphic to) the field of real numbers by 42.9. □

Remark In Theorem 11.14, this is the implication (d) $\Rightarrow$ (a). The proof there uses the same ideas as here in an ad hoc manner.

43 Completion of topological abelian groups

A metric space is said to be complete if every Cauchy sequence in this space has a limit point. Recall that a Cauchy sequence is a sequence $(x_\nu)_\nu$ with the property that the diameters of the sequence 'tails' $\{\, x_\nu \mid \nu \geq n \,\}$ become arbitrarily small for large $n \in \mathbb{N}$. Such a notion can only be expressed if one has the possibility of comparing the size of subsets at different places of the space. A metric offers such a possibility, whereas the notion of a topological space does not.

In a topological group, however, the concept of a Cauchy sequence can be formulated adequately, since the neighbourhoods of an arbitrary element are obtained by translation from the neighbourhoods of the neutral element. For topological groups which do not have countable neighbourhood bases, however, Cauchy sequences are not sufficient to express the property of completeness; instead of sequences, one has to use nets or, as we do here, filterbases.

In this section, we show that every Hausdorff abelian topological group G can be embedded as a dense subgroup into a Hausdorff abelian topological group $\widehat{G}$ which is complete in the sense that every Cauchy filterbase has a limit point; $\widehat{G}$ will be called the completion of G.

Our interest is not so much the completion of topological groups but rather the completion of topological fields, the completion of their additive groups being the first step; see Section 44. For this reason, we deal with abelian topological groups only. It is not difficult, however, to treat completion of topological groups in general on the same lines; see WARNER 1989, STROPPEL 2006 and the hints in 43.27.

43.1 Filterbases and filters A *filterbase* on a set X is a non-empty set $\mathfrak{B}$ of non-empty subsets of X such that the intersection of any two members of $\mathfrak{B}$ contains a member of $\mathfrak{B}$. As we know from 21.1, a *filter* is a filterbase $\mathfrak{F}$ such that every subset of X containing a member of $\mathfrak{F}$ belongs to $\mathfrak{F}$, as well. In particular, the intersection of any two members of $\mathfrak{F}$ then belongs to $\mathfrak{F}$. For a filterbase $\mathfrak{B}$, the smallest filter containing $\mathfrak{B}$ as a subset is called the filter *generated by* $\mathfrak{B}$; it consists of the subsets of X containing a member of $\mathfrak{B}$.

For filterbases $\mathfrak{A}$, $\mathfrak{B}$ on X, we say that $\mathfrak{A}$ is *finer* than $\mathfrak{B}$ and write $\mathfrak{A} < \mathfrak{B}$ if every member of $\mathfrak{B}$ contains a member of $\mathfrak{A}$. Equivalently, this means that the filter generated by $\mathfrak{B}$ is a subset of the filter generated by $\mathfrak{A}$. We say that $\mathfrak{A}$ and $\mathfrak{B}$ are *equivalent* if $\mathfrak{A} < \mathfrak{B}$ and $\mathfrak{B} < \mathfrak{A}$; this is true if, and only if, the filters generated by $\mathfrak{A}$ and $\mathfrak{B}$ coincide.

A sequence $(x_\nu)_\nu$ in X determines a filterbase consisting of all 'tails' $\{x_\nu \mid \nu \geq n\}$ for $n \in \mathbb{N}$. In this way, filterbases generalize the notion of a sequence.

43.2 Convergence of filterbases (a) If X is a topological space, we say that a filterbase $\mathfrak{B}$ on X *converges* to $x \in X$, and write $\mathfrak{B} \to x$, if every neighbourhood of x contains a member of $\mathfrak{B}$. With the terms introduced above, this may be expressed by saying that $\mathfrak{B}$ is finer than the filter $\mathfrak{V}_x$ of neighbourhoods of x. The same then holds for the filter generated by $\mathfrak{B}$. The point x is also called a *limit point* of $\mathfrak{B}$. A filterbase $\mathfrak{A}$ which is finer than $\mathfrak{B}$ then converges to x as well.

(b) A sequence $(x_\nu)_\nu$ in X converges to x in the usual sense if, and only if, the filterbase consisting of all 'tails' $\{x_\nu \mid \nu \geq n\}$ for $n \in \mathbb{N}$ converges to x.

(c) If X is a Hausdorff space, then a filterbase $\mathfrak{B}$ on X converging to $x \in X$ has no other limit point except x. We may therefore write $x = \lim \mathfrak{B}$ to characterize this situation.

Indeed, assume that there were a second limit point $y \in X$. Then, since $\mathfrak{B}$ would be finer than $\mathfrak{V}_x$ and finer than $\mathfrak{V}_y$ at the same time, every neighbourhood of x would intersect every neighbourhood of y, which contradicts the Hausdorff separation property.

(d) If $\varphi : X \to Y$ is a continuous map between two topological spaces, and if $\mathfrak{B}$ is a filterbase on X which converges to $x \in X$, then it is straightforward that $\varphi(\mathfrak{B}) = \{\, \varphi(B) \mid B \in \mathfrak{B} \,\}$ is a filterbase on Y which converges to $\varphi(x)$.

In what follows, G will be an abelian topological group, written additively, with neutral element 0.

43.3 Concentrated filterbases (a) A filterbase $\mathfrak{C}$ on G is said to be *concentrated* if for every neighbourhood U of 0 there is $C \in \mathfrak{C}$ such that $C - C \subseteq U$. (Commonly, such filterbases are also called Cauchy filterbases.) It is clear that the filter generated by a concentrated filterbase $\mathfrak{C}$ is concentrated, as well, and so is, more generally, any filterbase which is finer than $\mathfrak{C}$.

(b) A sequence $(x_\nu)_\nu$ in G is called a *Cauchy sequence* if for every neighbourhood U of 0 there is $n \in \mathbb{N}$ such that for all $\mu, \nu \in \mathbb{N}$ satisfying $\mu \geq n$ and $\nu \geq n$ one has $x_\mu - x_\nu \in U$. This just means that the filterbase consisting of all 'tails' $\{\, x_\nu \mid \nu \geq n \,\}$ for $n \in \mathbb{N}$ is concentrated.

43.4 Neighbourhood filters Let $a \in G$. Continuity of the difference map $G \times G \to G : (x, y) \mapsto x - y$ at (a, a) says that for every neighbourhood U of 0 there is a neighbourhood N of a such that $N - N \subseteq U$, in other words, the neighbourhood filter $\mathfrak{V}_a$ is concentrated.

It follows that a filterbase which converges to some element is concentrated, since it is finer than the neighbourhood filter of this element. The converse is not true, in general. This motivates the following notion.

43.5 Definition: Completeness The topological group G is called *complete* if each concentrated filterbase converges. Then, in particular, every Cauchy sequence converges; see 43.3(b) and 43.2(b).

It is clear from the definition that a closed subgroup of a complete group G is complete, as well.

Trivially, a group G with the discrete topology is a complete topological group. Indeed, a filterbase $\mathfrak{C}$ on G is concentrated if and only if some element of $\mathfrak{C}$ is a singleton $\{x\}$, and then $\mathfrak{C}$ converges to x.

43.6 Lemma   *A filterbase* $\mathfrak{C}$ *on* G *converges to* $c \in G$ *if, and only if, it is concentrated and* $c \in \bigcap \{\overline{C} \mid C \in \mathfrak{C}\}$ (*the intersection of the closures of the members of* $\mathfrak{C}$).

In particular, a Cauchy sequence in G converges to c if, and only if, c is an accumulation point of the sequence.

Proof Assume that $\mathfrak{C}$ converges to c. We have already noted that a convergent filterbase is concentrated. Every neighbourhood of c contains a member of $\mathfrak{C}$, which in turn has non-empty intersection with every member $C \in \mathfrak{C}$, so that $c \in \overline{C}$.

Conversely, assume that $\mathfrak{C}$ is concentrated and let c be an element of the intersection above. Every neighbourhood N of c contains a neighbourhood of the form $U + c$ where U is a neighbourhood of 0. Let V be a neighbourhood of 0 such that $V + V \subseteq U$. Since $\mathfrak{C}$ is concentrated, there is $C \in \mathfrak{C}$ such that $C - C \subseteq V$, and $(V + c) \cap C \neq \emptyset$ since $c \in \overline{C}$. For $x \in (V + c) \cap C$, we infer that $C - x \subseteq V$ and $C \subseteq V + x \subseteq V + V + c \subseteq U + c \subseteq N$. Thus, $\mathfrak{C} \to c$. $\qquad\square$

43.7 Corollary   *Let* $\mathfrak{C}, \mathfrak{D}$ *be two filterbases on* G *such that* $\mathfrak{C} < \mathfrak{D}$. *If* $\mathfrak{C}$ *converges to* $c \in G$ *and if* $\mathfrak{D}$ *is concentrated, then* $\mathfrak{D}$ *converges to* c, *as well.*

Proof By 43.6, we have $c \in \bigcap\{\overline{C} \mid C \in \mathfrak{C}\}$. Since $\mathfrak{C} < \mathfrak{D}$, clearly $\bigcap\{\overline{C} \mid C \in \mathfrak{C}\} \subseteq \bigcap\{\overline{D} \mid D \in \mathfrak{D}\}$. Hence c belongs to the latter intersection, as well, and the assertion follows from 43.6. $\qquad\square$

43.8 Lemma   *Let* $\mathfrak{C}$ *be a concentrated filterbase on* G. *If there is a member* $C_0 \in \mathfrak{C}$ *with compact closure* $\overline{C_0}$, *then* $\mathfrak{C}$ *converges.*

Proof By the properties of a filterbase, $\mathfrak{C}$ has the finite intersection property in the sense that any collection of finitely many members of $\mathfrak{C}$ has non-empty intersection. The same then is true for the set $\{\overline{C} \cap \overline{C_0} \mid C \in \mathfrak{C}\}$ of closed subsets of $\overline{C_0}$. By compactness, therefore, the whole set has non-empty intersection, which clearly equals $\bigcap\{\overline{C} \mid C \in \mathfrak{C}\}$. If the filterbase $\mathfrak{C}$ is concentrated, then it converges to every point of this intersection by 43.6. $\qquad\square$

43.9 Corollary   *A locally compact topological group is complete.*

Proof Let U be a compact neighbourhood of 0, and V a neighbourhood of 0 such that $-V + V \subseteq U$. A concentrated filterbase $\mathfrak{C}$ contains a member C_0 such that $C_0 - C_0 \subseteq V$ and hence $C_0 \subseteq V + c$ for $c \in C_0$.

For $x \in \overline{C_0}$, the intersection $(V+x) \cap C_0$ is not empty, and for an element $d \in (V+x) \cap C_0$ we find that $x \in -V + d \subseteq -V + V + c \subseteq U + c$. Hence $\overline{C_0}$ is contained in the compact subset $U + c$ and therefore is compact itself. By 43.8, we conclude that $\mathfrak{C}$ converges.   □

43.10 Completeness of ordered groups Let G be an ordered group (see 7.1) that is complete as such. It is a topological group with the topology induced by the ordering; see 8.4. Note that G is commutative by 7.7.

Proposition *If an ordered group G is complete as an ordered group, it is also complete as a topological group.*

Remarks (1) The converse need not be true; see 43.29.

(2) The result applies in particular to the additive group of $\mathbb{R}$, if we think of $\mathbb{R}$ as the completion of the ordered field $\mathbb{Q}$; see 42.11.

Proof The assertion is an immediate consequence of 43.9 since G is locally compact according to 5.2.

Because of the importance of our result for the structure of $\mathbb{R}$, we give another, more straightforward proof. Let $\mathfrak{C}$ be a concentrated filterbase on G. For $0 < a \in G$, there is $C \in \mathfrak{C}$ such that $C - C \subseteq\ ]-a, a[$. For $c \in C$, then, $C \subseteq\ ]-a + c, a + c[$. Thus, $\mathfrak{C}$ has members which are bounded (in the chain G) and hence have a least upper bound and a greatest lower bound, since G is order complete. For bounded members $C, C' \in \mathfrak{C}$, we have $\inf C' \leq \sup C$, for else $C \cap C' = \emptyset$, contradicting the properties of a filterbase. Thus the set $\{\sup C \mid C \in \mathfrak{C}, C \text{ bounded}\}$ is bounded below and has a greatest lower bound

$$a := \inf\{\sup C \mid C \in \mathfrak{C}, C \text{ bounded}\} .$$

We show that $\mathfrak{C}$ converges to a. Let $C \in \mathfrak{C}$; by 43.6 if suffices to show that $a \in \overline{C}$. For $0 < \varepsilon \in G$ there is a bounded member $B \in \mathfrak{C}$ such that $a \leq \sup B < a + \varepsilon$. We may assume that $B \subseteq C$ by making B smaller if necessary. There is $b \in B$ such that $\sup B - \varepsilon < b \leq \sup B$, and then $a - \varepsilon < b < a + \varepsilon$. Since $b \in C$, this shows that C intersects every neighbourhood of a, so that $a \in \overline{C}$.   □

As we have remarked at the beginning of the preceding proof, the complete ordered group $\mathbb{R}$ is locally compact; see also Theorem 5.3. Since the multiplicative group $\mathbb{R} \setminus \{0\}$ and the subgroup $\mathbb{R}_{\mathrm{pos}}$ of positive real numbers are open in $\mathbb{R}$, they are locally compact as well. Thus, Corollary 43.9 implies the following.

43.11 Examples   *The two groups* $\mathbb{R} \smallsetminus \{0\}$ *and* $\mathbb{R}_{\mathrm{pos}}$ *are complete topological groups.* For $\mathbb{R} \smallsetminus \{0\}$, this might seem strange on first sight since there is a 'hole'. Completeness of $\mathbb{R} \smallsetminus \{0\}$ as a topological group does not mean completeness as a metric space with the usual metric inherited from $\mathbb{R}$.

Concentrated filterbases are preserved under continuous homomorphisms and, more generally, under maps of the following type.

43.12 Uniform continuity   A map $\varphi : G \to H$ between topological groups (not necessarily a homomorphism) is said to be *uniformly continuous* if for every neighbourhood V of 0 in H there is a neighbourhood U of 0 in G such that for all $x \in G$ one has $\varphi(U + x) \subseteq V + \varphi(x)$.

This is obviously satisfied if φ is a continuous homomorphism: choose U in such a way that $\varphi(U) \subseteq V$.

If φ is uniformly continuous and $\mathfrak{C}$ a concentrated filterbase on G, then the filterbase $\varphi(\mathfrak{C})$ on H is concentrated, as well. Indeed, for a neighbourhood V of 0 in H choose a neighbourhood U of 0 in G as above and a member $C \in \mathfrak{C}$ such that $C - C \subseteq U$. Then for all $c \in C$ one has $C \subseteq U + c$ and consequently $\varphi(C) \subseteq \varphi(U + c) \subseteq V + \varphi(c)$, so that $\varphi(C) - \varphi(C) \subseteq V$.

In particular, the image sequence of a Cauchy sequence in G under the uniformly continuous map φ is a Cauchy sequence again.

43.13 Definition   An *embedding* of a topological group G into a topological group H is a monomorphism $\varphi : G \to H$ which induces a homeomorphism of G onto the image group $\varphi(G)$ endowed with the topology induced from H.

43.14 Definition   A *completion* of a topological group G is a complete topological group $\widehat{G}$ together with an embedding $\iota : G \to \widehat{G}$ such that $\iota(G)$ is dense in $\widehat{G}$. A *Hausdorff completion* is a completion that is a Hausdorff group. Of course, a Hausdorff completion will only exist for Hausdorff topological groups.

For every abelian topological group G, a complete abelian topological group will be constructed below, which will give a completion if G is a Hausdorff group. The elements of the complete group constructed from G will be certain concentrated filters on G; after all, if such a filter does not converge in G, why not add it to the elements of G and let it converge to itself? We do not use all concentrated filters, however, as different filters may converge to the same element. We focus on concentrated filters of a special type which we discuss now.

43.15 Minimal concentrated filters *Let G be a topological group, $\mathfrak{V}_0$ the filter of neighbourhoods of 0, and $\mathfrak{C}$ a concentrated filterbase on G. Then $\mathfrak{V}_0 + \mathfrak{C} := \{U + C \mid U \in \mathfrak{V}_0, C \in \mathfrak{C}\}$ is a concentrated filterbase, as well.*

Let $\widehat{\mathfrak{C}}$ be the filter generated by $\mathfrak{V}_0 + \mathfrak{C}$ (which is concentrated, too). Then $\mathfrak{C} < \widehat{\mathfrak{C}}$; moreover, if $\mathfrak{D}$ is any concentrated filterbase such that $\mathfrak{C} < \mathfrak{D}$, then $\mathfrak{D} < \widehat{\mathfrak{C}}$.

Remarks The last statement, when expressed for a concentrated filter $\mathfrak{F}$ instead of the filterbase $\mathfrak{D}$, says that $\mathfrak{C} < \mathfrak{F}$ implies $\widehat{\mathfrak{C}} \subseteq \mathfrak{F}$. In other words, $\widehat{\mathfrak{C}}$ is the smallest filter among the concentrated filters $\mathfrak{F}$ such that $\mathfrak{C} < \mathfrak{F}$ holds.

Note that the existence of filters having this minimality property is guaranteed by the explicit construction of $\widehat{\mathfrak{C}}$, not just by a transfinite principle like Zorn's lemma.

When working with sequences or nets instead of filterbases, an analogous construction is not possible; it would amount to producing a smallest subnet from a Cauchy net.

Proof For $C_1, C_2 \in \mathfrak{C}$ there is $C \in \mathfrak{C}$ such that $C \subseteq C_1 \cap C_2$. For $U_1, U_2 \in \mathfrak{V}_0$ it is then clear that $(U_1 \cap U_2) + C \subseteq (U_1 + C_1) \cap (U_2 + C_2)$. This shows that $\mathfrak{V}_0 + \mathfrak{C}$ is a filterbase.

In order to verify that $\mathfrak{V}_0 + \mathfrak{C}$ is concentrated, let U be a neighbourhood of 0. There is a neighbourhood V of 0 and a member $C \in \mathfrak{C}$ such that $V + V - V \subseteq U$ and $C - C \subseteq V$. Then $(V + C) - (V + C) = V + C - C - V \subseteq V + V - V \subseteq U$. Thus the filterbase $\mathfrak{V}_0 + \mathfrak{C}$ is concentrated, and so is the filter $\widehat{\mathfrak{C}}$ generated by it.

For $U \in \mathfrak{V}_0$ and $C \in \mathfrak{C}$ clearly $C \subseteq U + C$; hence $\mathfrak{C} < \mathfrak{V}_0 + \mathfrak{C} < \widehat{\mathfrak{C}}$.

For a concentrated filterbase $\mathfrak{D}$ such that $\mathfrak{C} < \mathfrak{D}$ we have to show that every element $U + C \in \mathfrak{V}_0 + \mathfrak{C} \subseteq \widehat{\mathfrak{C}}$ contains a member of $\mathfrak{D}$. Find $D \in \mathfrak{D}$ such that $D - D \subseteq U$. Since D contains a member of $\mathfrak{C}$, it intersects C; let $d \in C \cap D$. Then $D \subseteq U + d \subseteq U + C$.   □

43.16 Lemma *Let $\mathfrak{C}$ and $\mathfrak{D}$ be concentrated filterbases on a topological group. If $\mathfrak{C} < \mathfrak{D}$, then $\widehat{\mathfrak{C}} = \widehat{\mathfrak{D}}$. In particular, $\widehat{\widehat{\mathfrak{C}}} = \widehat{\mathfrak{C}}$.*

Proof If $\mathfrak{C} < \mathfrak{D}$, the minimality principle of 43.15 says that $\mathfrak{D} < \widehat{\mathfrak{C}}$, which implies that $\widehat{\mathfrak{C}} < \widehat{\mathfrak{D}}$. Since $\mathfrak{C} < \widehat{\mathfrak{C}}$, it follows that $\mathfrak{C} < \widehat{\mathfrak{D}}$. The same minimality principle yields $\widehat{\mathfrak{D}} < \widehat{\mathfrak{C}}$; thus the two filters $\widehat{\mathfrak{D}}$ and $\widehat{\mathfrak{C}}$ are equivalent as filterbases and hence coincide, which proves the first statement. The second statement follows since $\mathfrak{C} < \widehat{\mathfrak{C}}$.   □

43.17 Neighbourhood filters again   *For an element x of a topological group, the set consisting of just the singleton $\{x\}$ obviously is a concentrated filterbase, and $\widehat{\{\{x\}\}} = \mathfrak{V}_x$, the neighbourhood filter of the element x. Moreover, a concentrated filterbase $\mathfrak{C}$ converges to x if, and only if, $\widehat{\mathfrak{C}} = \mathfrak{V}_x$.*

Proof By definition, $\widehat{\{\{x\}\}}$ is the filter which is generated by the filterbase $\mathfrak{V}_0 + \{x\}$, which is just the neighbourhood filter $\mathfrak{V}_x$. A concentrated filterbase $\mathfrak{C}$ converges to x precisely if it is finer than $\mathfrak{V}_x$; the second statement now follows from 43.16.   □

More generally, the filters constructed in 43.15 can be perceived as distinguished representatives of equivalence classes of filterbases for a certain equivalence relation which we now elucidate.

43.18 Lemma   *For two concentrated filterbases $\mathfrak{C}_1, \mathfrak{C}_2$ on a topological group the following statements are equivalent.*
 (i) $\widehat{\mathfrak{C}_1} = \widehat{\mathfrak{C}_2}$
 (ii) $C_1' \cap C_2' \neq \emptyset$ *for all* $C_1' \in \widehat{\mathfrak{C}_1}, C_2' \in \widehat{\mathfrak{C}_2}$
 (iii) *The set* $\mathfrak{C} := \{ C_1 \cup C_2 \mid C_1 \in \mathfrak{C}_1, C_2 \in \mathfrak{C}_2 \}$ *(which clearly is a filterbase again) is concentrated.*

Proof It is clear that $\mathfrak{C}$ is a filterbase since $\mathfrak{C}_1$ and $\mathfrak{C}_2$ are; and obviously, (i) implies (ii).

(ii) implies (iii): For a neighbourhood U of 0 let V be a neighbourhood of 0 such that $V - V \subseteq U$. Since $\widehat{\mathfrak{C}_1}$ and $\widehat{\mathfrak{C}_2}$ are concentrated, there are $C_1' \in \widehat{\mathfrak{C}_1}, C_2' \in \widehat{\mathfrak{C}_2}$ such that $C_1' - C_1' \subseteq V$ and $C_2' - C_2' \subseteq V$. By (ii), there is an element $c \in C_1' \cap C_2'$. Then $C_1' \cup C_2' \subseteq V + c$ and $(C_1' \cup C_2') - (C_1' \cup C_2') \subseteq V + c - (V + c) = V - V \subseteq U$. There are $C_1 \in \mathfrak{C}_1$ and $C_2 \in \mathfrak{C}_2$ such that $C_1 \subseteq C_1', C_2 \subseteq C_2'$, and $(C_1 \cup C_2) - (C_1 \cup C_2) \subseteq U$, as well. This proves (iii).

(iii) implies (i): Clearly $\mathfrak{C}_1 < \mathfrak{C}$, and $\mathfrak{C} < \widehat{\mathfrak{C}}$ by 43.15. By 43.16, hence, $\widehat{\mathfrak{C}_1} = \widehat{\mathfrak{C}}$. In the same way one obtains $\widehat{\mathfrak{C}_2} = \widehat{\mathfrak{C}}$, and (i) is proved.   □

Now we provide two technical tools.

43.19 Lemma   *Let $\iota : G \to \widetilde{G}$ be an embedding of a topological group G into a topological group $\widetilde{G}$ such that $\iota(G)$ is dense in $\widetilde{G}$. Then $\widetilde{G}$ is complete if only for every concentrated filterbase $\mathfrak{C}$ on G, the filterbase $\iota(\mathfrak{C})$ on $\widetilde{G}$ converges.*

Remark   This criterion makes it easier to prove completeness: one does not have to consider all concentrated filterbases on $\widetilde{G}$, but only those which come from filterbases on G via ι.

Proof By $\widetilde{\mathfrak{V}}_0$ we denote the filter of neighbourhoods of 0 in $\widetilde{G}$. Let $\mathfrak{D}$ be a concentrated filterbase on $\widetilde{G}$; we have to show that it converges.

It is finer than the filterbase $\widetilde{\mathfrak{V}}_0 + \mathfrak{D}$ formed according to 43.15, which is concentrated, as well. A member $V + D$ of this filterbase for $V \in \widetilde{\mathfrak{V}}_0$, $D \in \mathfrak{D}$ is a neighbourhood of every element of D. Since $\iota(G)$ is dense in $\widetilde{G}$, the intersection $(V+D) \cap \iota(G)$ is non-empty, and $(\widetilde{\mathfrak{V}}_0 + \mathfrak{D}) \cap \iota(G) :=$ $\{ (V+D) \cap \iota(G) \mid V \in \widetilde{\mathfrak{V}}_0, D \in \mathfrak{D} \}$ is a filterbase which is finer than $\widetilde{\mathfrak{V}}_0 + \mathfrak{D}$. Since $\iota^{-1} : \iota(G) \to G$ is a continuous homomorphism by assumption, the filterbase

$$\iota^{-1}((\widetilde{\mathfrak{V}}_0 + \mathfrak{D}) \cap \iota(G)) = \{ \iota^{-1}((V + D) \cap \iota(G)) \mid V \in \widetilde{\mathfrak{V}}_0, D \in \mathfrak{D} \}$$

on G is concentrated. By assumption, its image filterbase, which is just $(\widetilde{\mathfrak{V}}_0 + \mathfrak{D}) \cap \iota(G)$, converges in $\widetilde{G}$. Since it is finer than $\widetilde{\mathfrak{V}}_0 + \mathfrak{D}$, the latter converges as well by 43.7.

Thus the filterbase $\mathfrak{D}$, which is finer than $\widetilde{\mathfrak{V}}_0 + \mathfrak{D}$, converges too. □

The next result will help to extend algebraic properties to a completion. The formulation covers a sufficiently general situation for later use in the completion of topological rings.

43.20 Lemma *Let $(G, *)$ and $(G', *')$ be topological spaces with continuous binary operations, and $\iota : G \to G'$ a continuous homomorphism such that $\iota(G)$ is dense in G'. Moreover, G' is assumed to be a Hausdorff space.*

(i) *If the operation $*$ on G is associative or commutative, then so is the operation $*'$ on G'.*

(ii) *If e is a neutral element of $(G, *)$, then $\iota(e)$ is a neutral element of $(G', *')$.*

(ii') *Assume in situation (ii) that there is a continuous map $\nu : G \to G$ such that for $x \in G$ the image $\nu(x)$ is an inverse of x in the sense that $x * \nu(x) = e$. Assume furthermore that there is a continuous map $\nu' : G' \to G'$ such that $\nu' \circ \iota = \iota \circ \nu$.*
Then for every $x' \in G'$ the image $\nu'(x')$ is an inverse of x' as well, that is $x' *' \nu'(x') = \iota(e)$.

Proof (i) Let $*$ be associative. We want to prove that $*'$ is associative as well, which means that the two continuous maps

$$G' \times G' \times G' \to G' \quad : \quad (x', y', z') \mapsto (x' *' y') *' z'$$
$$G' \times G' \times G' \to G' \quad : \quad (x', y', z') \mapsto x' *' (y' *' z')$$

coincide. Since G' is Hausdorff, the set of coincidence is closed. Therefore it suffices to show that the two maps coincide on the dense subset

$\iota(G) \times \iota(G) \times \iota(G)$. But this is associativity on $\iota(G)$, which holds since associativity of G is transported by the homomorphism ι. In exactly the same way one proves that $*'$ is commutative if $*$ is.

(ii) Assume that e is a neutral element of $(G, *)$. To prove that $\iota(e)$ is a neutral element of $(G', *')$ means to show that the continuous map $G' \to G' : x' \mapsto x' *' \iota(e)$ coincides with the identity map. As in (i), it suffices to have coincidence on the dense subset $\iota(G)$; but this is clear since ι is a homomorphism.

(ii') Here, again, we deduce the coincidence of the two continuous maps $G' \to G' : x' \mapsto x' *' \nu(x')$ and $G' \to G' : x' \mapsto \iota(e)$ from the fact that they coincide on the dense subset $\iota(G)$, as can be easily seen: indeed, for $x \in G$ one has $\iota(x) *' \nu'(\iota(x)) = \iota(x) *' \iota(\nu(x)) = \iota(x * \nu(x)) = \iota(e)$.   □

43.21 Construction *Every abelian topological Hausdorff group has an abelian Hausdorff completion.*

Proof (1) Let G be an abelian topological group (it need not be a Hausdorff group for the moment). We shall construct a completion $\widehat{G}$ explicitly. The underlying set will consist of the minimal concentrated filters on G constructed in 43.15:

$$\widehat{G} = \{\widehat{\mathfrak{C}} \mid \mathfrak{C} \text{ concentrated filterbase on } G\} .$$

By 43.17 there is a natural map

$$\iota : G \to \widehat{G} : x \mapsto \widehat{\{\{x\}\}} = \mathfrak{V}_x .$$

(2) *A topology* on $\widehat{G}$ will be defined using the following sets as a basis. For $A \subseteq G$, let

$$\widehat{A} = \{\widehat{\mathfrak{C}} \in \widehat{G} \mid A \in \widehat{\mathfrak{C}}\} .$$

We remark that by construction every member of a minimal concentrated filter $\widehat{\mathfrak{C}}$ contains an open subset. Hence $\widehat{A}$ is non-empty if, and only if, A contains an open subset; for instance, x is contained in the interior of A if, and only if, $\mathfrak{V}_x \in \widehat{A}$.

One verifies directly for a subset $B \subseteq G$ that $\widehat{A \cap B} = \widehat{A} \cap \widehat{B}$ since the elements of $\widehat{G}$ are filters and not just filterbases. Thus, there is a topology on $\widehat{G}$ having $\{\widehat{A} \mid A \subseteq G\}$ as a basis. A neighbourhood base of $\widehat{\mathfrak{C}} \in \widehat{G}$ is given by $\{\widehat{B} \mid B \in \widehat{\mathfrak{C}}\}$.

The space $\widehat{G}$ is a Hausdorff space whether G is a Hausdorff space or not. Indeed, for $\widehat{\mathfrak{C}}, \widehat{\mathfrak{D}} \in \widehat{G}$, if $\widehat{\mathfrak{C}} \neq \widehat{\mathfrak{D}}$ then by 43.18(ii) there are $C \in \widehat{\mathfrak{C}}$ and $D \in \widehat{\mathfrak{D}}$ such that $C \cap D = \emptyset$, and $\widehat{C}, \widehat{D}$ are disjoint neighbourhoods of $\widehat{\mathfrak{C}}$ and $\widehat{\mathfrak{D}}$, respectively.

(3) *Properties of* $\iota : G \to \widehat{G}$. The map ι is continuous. For this, it suffices to show that the preimage of every basic open set $\widehat{A}$ is open. Now we have seen above that $\iota^{-1}(\widehat{A}) = \{ x \in G \mid \mathfrak{V}_x \in \widehat{A} \}$ is the interior of A, which is open.

In order to see that $\iota(G)$ is dense in $\widehat{G}$, we verify that every non-empty basic open set $\widehat{A}$ intersects $\iota(G)$. As remarked above, the interior of A is not empty, and for an element x of the interior one has $\iota(x) = \mathfrak{V}_x \in \widehat{A}$.

The map ι is injective if G is a Hausdorff group. We show that then the bijection $G \to \iota(G)$ induced by ι is a homeomorphism. As ι is continuous, it remains to show that this map is open. For an open subset $U \subseteq G$ the image set $\iota(U) = \{ \mathfrak{V}_x \mid x \in U \} = \{ \mathfrak{V}_x \mid U \in \mathfrak{V}_x \} = \widehat{U} \cap \iota(G)$ is an open subset of $\iota(G)$ in the topology induced from $\widehat{G}$.

(4) *Group structure on* $\widehat{G}$. We now define an addition which makes $\widehat{G}$ an abelian topological group. For $\widehat{\mathfrak{C}}, \widehat{\mathfrak{D}} \in \widehat{G}$, it is clear that

$$\widehat{\mathfrak{C}} \oplus \widehat{\mathfrak{D}} := \{ C + D \mid C \in \widehat{\mathfrak{C}}, D \in \widehat{\mathfrak{D}} \}$$

is a filterbase. We verify that this filterbase is concentrated since $\widehat{\mathfrak{C}}, \widehat{\mathfrak{D}}$ are. Indeed, for a neighbourhood U of 0 in G let V be a neighbourhood of 0 such that $V + V \subseteq U$. There are $C \in \widehat{\mathfrak{C}}, D \in \widehat{\mathfrak{D}}$ such that $C - C \subseteq V$ and $D - D \subseteq V$. Then $(C+D) - (C+D) = C - C + D - D \subseteq V + V \subseteq U$; here we have used that G is abelian.

Now we may define addition on $\widehat{G}$ by

$$\widehat{\mathfrak{C}} + \widehat{\mathfrak{D}} := \widehat{\widehat{\mathfrak{C}} \oplus \widehat{\mathfrak{D}}} \ ,$$

and we show that the map $\iota : G \to \widehat{G}$ is a homomorphism. For $x, y \in G$ we have to verify that $\widehat{\mathfrak{V}_x \oplus \mathfrak{V}_y} = \mathfrak{V}_{x+y}$. By 43.16, it suffices to show that $\mathfrak{V}_x \oplus \mathfrak{V}_y < \mathfrak{V}_{x+y}$. By continuity of addition in G, for $W \in \mathfrak{V}_{x+y}$ there are neighbourhoods $U \in \mathfrak{V}_x, V \in \mathfrak{V}_y$ such that $U + V \subseteq W$; this is our claim.

Next we prove that addition in $\widehat{G}$ is continuous. Let $\widehat{A}$ be a basic neighbourhood of $\widehat{\mathfrak{C}} + \widehat{\mathfrak{D}}$, which means that $A \in \widehat{\mathfrak{C}} + \widehat{\mathfrak{D}}$. By definition of the addition of $\widehat{G}$, there are $C \in \widehat{\mathfrak{C}}, D \in \widehat{\mathfrak{D}}$ and a neighbourhood U of 0 in G such that $U + C + D \subseteq A$. The sets $\widehat{C}, \widehat{D}$ are neighbourhoods of $\widehat{\mathfrak{C}}$ and $\widehat{\mathfrak{D}}$, respectively. Continuity of addition is proved if for all $\widehat{\mathfrak{X}} \in \widehat{C}, \widehat{\mathfrak{Y}} \in \widehat{D}$ we show that $\widehat{\mathfrak{X}} + \widehat{\mathfrak{Y}}$ belongs to the given neighbourhood $\widehat{A}$ of $\widehat{\mathfrak{C}} + \widehat{\mathfrak{D}}$. Now $C \in \widehat{\mathfrak{X}}, D \in \widehat{\mathfrak{Y}}$, so that $U + C + D \in \widehat{\mathfrak{X}} + \widehat{\mathfrak{Y}}$. Since $U + C + D \subseteq A$, it follows that $A \in \widehat{\mathfrak{X}} + \widehat{\mathfrak{Y}}$, in other words, that $\widehat{\mathfrak{X}} + \widehat{\mathfrak{Y}} \in \widehat{A}$, indeed.

By the general lemma 43.20(i, ii), addition on $\widehat{G}$ inherits associativity and commutativity from the addition on G, and $\iota(0)$ is a neutral element of $\widehat{G}$.

Thus, in order to prove that $\widehat{G}$, like G, is a topological abelian group, we finally have to show that every element $\widehat{\mathfrak{C}} \in \widehat{G}$ (for a concentrated filter $\mathfrak{C}$ on G) has an inverse which depends continuously on $\widehat{\mathfrak{C}}$. Since G is abelian, the map

$$\nu : G \to G : x \to -x$$

is a homomorphism, and therefore $-\mathfrak{C} = \{ -C \mid C \in \mathfrak{C} \}$ is concentrated (43.12). Since ν permutes the set of neighbourhoods of 0 in G, it is clear from the construction in 43.15 that $-\widehat{\mathfrak{C}} = \widehat{-\mathfrak{C}}$; in particular, this is an element of $\widehat{G}$ again. We verify that the map

$$\widehat{\nu} : \widehat{G} \to \widehat{G} : \widehat{\mathfrak{C}} \mapsto -\widehat{\mathfrak{C}}$$

is continuous. It suffices to show that the preimage $\widehat{\nu}^{-1}(\widehat{A})$ of a basic open set $\widehat{A}$ for $A \subseteq G$ is open. Now

$$\widehat{\mathfrak{C}} \in \widehat{\nu}^{-1}(\widehat{A}) \iff -\widehat{\mathfrak{C}} \in \widehat{A} \iff A \in -\widehat{\mathfrak{C}} \iff -A \in \widehat{\mathfrak{C}} \iff \widehat{\mathfrak{C}} \in \widehat{-A} \,,$$

so that $\widehat{\nu}^{-1}(\widehat{A}) = \widehat{-A}$ is a basic open set, again.

The two maps ν and $\widehat{\nu}$ are linked via ι in the sense that $\iota \circ \nu = \widehat{\nu} \circ \iota$. For $x \in G$, indeed, $\iota(\nu(x)) = \mathfrak{V}_{-x}$ consists of the neighbourhoods of $-x$, and the involutory homeomorphism ν exchanges the neighbourhoods of x and of $-x$, so that $\mathfrak{V}_{-x} = -\mathfrak{V}_x = \widehat{\nu}(\iota(x))$.

From these properties of $\widehat{\nu}$, we infer by the general lemma 43.20(ii') that $\widehat{\nu}(\widehat{\mathfrak{C}}) = -\widehat{\mathfrak{C}}$ is an inverse of $\widehat{\mathfrak{C}}$ in $\widehat{G}$; and we have shown that it depends continuously on $\widehat{\mathfrak{C}}$. Thus, $\widehat{G}$ is an abelian topological group.

(5) *Completeness.* In order to prove that the topological group $\widehat{G}$ is complete, we use the criterion 43.19. Let $\mathfrak{C}$ be a concentrated filterbase on G; we show that $\iota(\mathfrak{C})$ converges in $\widehat{G}$ to $\widehat{\mathfrak{C}}$, or equivalently, that for every $A \in \widehat{\mathfrak{C}}$ the basic neighbourhood $\widehat{A}$ of $\widehat{\mathfrak{C}}$ contains a member of $\iota(\mathfrak{C})$. According to the construction of $\widehat{\mathfrak{C}}$, there are $C \in \mathfrak{C}$ and a neighbourhood U of 0 in G such that $U + C \subseteq A$. In particular, A is a neighbourhood of every element $c \in C$, that is $A \in \mathfrak{V}_c$. Hence, $\iota(c) = \mathfrak{V}_c \in \widehat{A}$ for all $c \in C$, so that $\iota(C) \subseteq \widehat{A}$.

Thus $\widehat{G}$ has every property of a completion of G. $\qquad\qquad\square$

The following is a technical preparation for our further study of completions.

43.22 Special filterbases for dense embeddings   Let $\iota : G \to \widetilde{G}$ be an embedding of a topological group G into a topological group $\widetilde{G}$ such that $\iota(G)$ is dense in $\widetilde{G}$. For $\tilde{x} \in \widetilde{G}$, we consider the neighbourhood filter $\widetilde{\mathfrak{V}}_{\tilde{x}}$ of $\tilde{x}$. For $\widetilde{N} \in \widetilde{\mathfrak{V}}_{\tilde{x}}$, the intersection $\widetilde{N} \cap \iota(G)$ is non-empty since $\iota(G)$

is dense, and the same is true for the preimage $\iota^{-1}(\widetilde{N})$. Hence we have a filterbase

$$\widetilde{\mathfrak{V}}_{\tilde{x}} \cap \iota(G) := \{\,\widetilde{N} \cap \iota(G) \mid \widetilde{N} \in \widetilde{\mathfrak{V}}_{\tilde{x}}\,\}$$

on $\widetilde{G}$, which converges to $\tilde{x}$, since it is finer than $\widetilde{\mathfrak{V}}_{\tilde{x}}$; in particular, it is concentrated. Hence, when $\widetilde{\mathfrak{V}}_{\tilde{x}} \cap \iota(G)$ is considered as a filterbase on $\iota(G)$, it is concentrated, as well. Furthermore, we have a filterbase

$$\mathfrak{W}_{\tilde{x}} := \iota^{-1}(\widetilde{\mathfrak{V}}_{\tilde{x}}) = \{\,\iota^{-1}(\widetilde{N}) \mid \widetilde{N} \in \widetilde{\mathfrak{V}}_{\tilde{x}}\,\}$$

on G. Now $\iota^{-1}(\widetilde{N}) = \iota^{-1}(\widetilde{N} \cap \iota(G))$ for $\widetilde{N} \subseteq \widetilde{G}$, so that

$$\mathfrak{W}_{\tilde{x}} = \iota^{-1}(\widetilde{\mathfrak{V}}_{\tilde{x}}) = \iota^{-1}(\widetilde{\mathfrak{V}}_{\tilde{x}} \cap \iota(G))\,.$$

This filterbase is also concentrated, since by assumption $\iota^{-1} : \iota(G) \to G$ is continuous and hence uniformly continuous; see 43.12.

Conversely, $\iota(\iota^{-1}(\widetilde{N})) = \widetilde{N} \cap \iota(G))$ for $\widetilde{N} \subseteq \widetilde{G}$; hence

$$\iota(\mathfrak{W}_{\tilde{x}}) = \iota(\iota^{-1}(\widetilde{\mathfrak{V}}_{\tilde{x}})) = \widetilde{\mathfrak{V}}_{\tilde{x}} \cap \iota(G) \to \tilde{x}\,.$$

43.23 Theorem: Extensions of uniformly continuous maps *Let G and $\widetilde{G}$ be topological groups together with an embedding $\iota : G \to \widetilde{G}$ such that $\iota(G)$ is dense in $\widetilde{G}$, let H be a complete Hausdorff topological group, and let $\varphi : G \to H$ be a uniformly continuous map.*

Then φ has a unique continuous extension $\widetilde{\varphi} : \widetilde{G} \to H$ over ι, that is, a continuous map such that $\widetilde{\varphi} \circ \iota = \varphi$, and $\widetilde{\varphi}$ is uniformly continuous, as well. This holds in particular if φ is a continuous homomorphism; in this case, the extension $\widetilde{\varphi}$ is also a homomorphism.

Proof (1) We first prove uniqueness of the extension, using the filterbases obtained in 43.22 from the neighbourhood filters $\widetilde{\mathfrak{V}}_{\tilde{x}}$ of elements $\tilde{x} \in \widetilde{G}$. Since $\widetilde{\mathfrak{V}}_{\tilde{x}} \cap \iota(G) \to \tilde{x}$, continuity implies that $\widetilde{\varphi}(\widetilde{\mathfrak{V}}_{\tilde{x}} \cap \iota(G)) \to \widetilde{\varphi}(\tilde{x})$. Now $\widetilde{\varphi}(\widetilde{\mathfrak{V}}_{\tilde{x}} \cap \iota(G)) = \widetilde{\varphi}(\iota(\iota^{-1}(\widetilde{\mathfrak{V}}_{\tilde{x}}))) = \varphi(\iota^{-1}(\widetilde{\mathfrak{V}}_{\tilde{x}}))$, so that

$$\varphi(\iota^{-1}(\widetilde{\mathfrak{V}}_{\tilde{x}})) \to \widetilde{\varphi}(\tilde{x})\,. \tag{1}$$

Thus, since in Hausdorff spaces filterbases have at most one limit point, $\widetilde{\varphi}$ is uniquely determined.

(2) We prove that a map $\widetilde{\varphi} : \widetilde{G} \to H$ having property (1) for every $\tilde{x} \in \widetilde{G}$ is uniformly continuous, without using any other property of $\widetilde{\varphi}$.

Let V_1 be a neighbourhood of 0 in H. We claim that there is a neighbourhood $\widetilde{U}$ of 0 in $\widetilde{G}$ such that $\widetilde{\varphi}(\widetilde{U} + \tilde{x}) \subseteq V_1 + \widetilde{\varphi}(\tilde{x})$ for all $\tilde{x} \in \widetilde{G}$. Choose a neighbourhood V_2 of 0 in H such that $-V_2 + V_2 + V_2 \subseteq V_1$. Since φ is uniformly continuous, there is a neighbourhood V of 0 in G such that $\varphi(V + x) \subseteq V_2 + \varphi(x)$ for all $x \in G$. As ι induces a homeomorphism $G \to \iota(G)$, there is a neighbourhood $\widetilde{U}_1$ of 0 in $\widetilde{G}$ whose preimage under ι satisfies $\iota^{-1}(\widetilde{U}_1) \subseteq V$. Let $\widetilde{U}_2$ be a neighbourhood of 0 in $\widetilde{G}$ such that $\widetilde{U}_2 - \widetilde{U}_2 \subseteq \widetilde{U}_1$; in addition, we may assume $\widetilde{U}_2$ to be open in $\widetilde{G}$.

Now let $\tilde{x} \in \widetilde{G}$ and $\tilde{y} \in \widetilde{U}_2 + \tilde{x}$. By (1), there is a neighbourhood $\widetilde{N}$ of $\tilde{y}$ in $\widetilde{G}$ such that $\varphi(\iota^{-1}(\widetilde{N})) \subseteq V_2 + \widetilde{\varphi}(\tilde{y})$ and $\widetilde{N} \subseteq \widetilde{U}_2 + \tilde{x}$. For $y \in \iota^{-1}(\widetilde{N}) \subseteq \iota^{-1}(\widetilde{U}_2 + \tilde{x})$ then $\varphi(y) \in V_2 + \widetilde{\varphi}(\tilde{y})$. In the same way, for $\tilde{x}$ instead of $\tilde{y}$, we find $x \in \iota^{-1}(\widetilde{U}_2 + \tilde{x})$ such that $\varphi(x) \in V_2 + \widetilde{\varphi}(\tilde{x})$. Then $\iota(y) - \iota(x) \in \widetilde{U}_2 - \widetilde{U}_2 \subseteq \widetilde{U}_1$, so that $y \in \iota^{-1}(\widetilde{U}_1) + x \subseteq V + x$ and $\varphi(y) \subseteq V_2 + \varphi(x)$. Hence $\widetilde{\varphi}(\tilde{y}) \in -V_2 + \varphi(y) \in -V_2 + V_2 + \varphi(x) \subseteq -V_2 + V_2 + V_2 + \widetilde{\varphi}(\tilde{x}) \subseteq V_1 + \widetilde{\varphi}(\tilde{x})$. Since $\tilde{y} \in \widetilde{U}_2 + \tilde{x}$ was arbitrary, we have obtained that $\widetilde{\varphi}(\widetilde{U}_2 + \tilde{x}) \subseteq V_1 + \widetilde{\varphi}(\tilde{x})$, which is our claim with $\widetilde{U} = \widetilde{U}_2$.

(3) In order to prove that an extension $\widetilde{\varphi}$ with the required properties exists, we use (1) to *define* a map $\widetilde{\varphi}$. Indeed the filterbase in (1) is concentrated, since $\iota^{-1}(\mathfrak{V}_{\tilde{x}})$ is concentrated (see 43.22) and since φ is uniformly continuous. In the complete Hausdorff topological group H, this filterbase converges to a unique element, which is defined to be the image point $\widetilde{\varphi}(\tilde{x})$ of $\tilde{x}$, for every element $\tilde{x} \in \widetilde{G}$. According to step (2), the map $\widetilde{\varphi}$ is uniformly continuous.

We now verify that $\widetilde{\varphi} \circ \iota = \varphi$. For $x \in G$, by continuity of ι, the neighbourhood filter $\mathfrak{V}_x$ of x is finer than $\iota^{-1}(\mathfrak{V}_{\iota(x)})$, so that $\varphi(\mathfrak{V}_x)$ is finer than $\varphi(\iota^{-1}(\mathfrak{V}_{\iota(x)}))$. By (1), consequently, $\varphi(\mathfrak{V}_x) \to \widetilde{\varphi}(\iota(x))$. On the other hand, clearly $\mathfrak{V}_x \to x$ and hence $\varphi(\mathfrak{V}_x) \to \varphi(x)$. Thus $\widetilde{\varphi}(\iota(x)) = \varphi(x)$, again since H is a Hausdorff group.

(4) Now let φ be a continuous homomorphism. Then φ is uniformly continuous, with a continuous extension $\widetilde{\varphi}$ as above. We show that in this case $\widetilde{\varphi}$ is also a homomorphism, by the density argument employed in 43.20. The continuous maps $\widetilde{G} \times \widetilde{G} \to \widetilde{G} : (x, y) \mapsto \widetilde{\varphi}(x + y)$ and $\widetilde{G} \times \widetilde{G} \to \widetilde{G} : (x, y) \mapsto \widetilde{\varphi}(x) + \widetilde{\varphi}(y)$ coincide on the dense subgroup $\iota(G)$, since ι and φ are homomorphisms; hence they coincide everywhere, and $\widetilde{\varphi}$ is a homomorphism. $\square$

43.24 Corollary: Universal property of completions *Let G_i for $i = 1, 2$ be a Hausdorff topological group, and $\iota_i : G_i \to \widehat{G}_i$ a Hausdorff completion of G_i.*

Then every continuous homomorphism $\varphi : G_1 \to G_2$ extends to a unique continuous map $\widehat{\varphi} : \widehat{G}_1 \to \widehat{G}_2$ such that $\widehat{\varphi} \circ \iota_1 = \iota_2 \circ \varphi$, and $\widehat{\varphi}$ is a homomorphism.

If φ is an isomorphism of topological groups, then so is $\widehat{\varphi}$.

Proof Existence and uniqueness of $\widehat{\varphi}$ is an immediate consequence of
43.23 with $G = G_1$, $\widehat{G} = \widehat{G}_1$, $H = \widehat{G}_2$, and the map $\iota_2 \circ \varphi$ instead of φ.

Now assume that φ is an isomorphism, and let $\widehat{\psi} : \widehat{G}_2 \to \widehat{G}_1$ be the
unique extension of the continuous homomorphism $\psi := \varphi^{-1} : G_2 \to G_1$
satisfying $\widehat{\psi} \circ \iota_2 = \iota_1 \circ \psi$. The homomorphism $\widehat{\vartheta} := \widehat{\psi} \circ \widehat{\varphi} : \widehat{G}_1 \to \widehat{G}_1$
is the extension of the identity map $\mathrm{id}_{G_1} : G_1 \to G_1$ in the sense that
$\widehat{\vartheta} \circ \iota_1 = \iota_1 \circ \mathrm{id}_{G_1}$, but the same holds also for the identity map of $\widehat{G}_1$
instead of $\widehat{\vartheta}$. Hence, again by uniqueness of extensions, we obtain that
$\widehat{\psi} \circ \widehat{\varphi} = \mathrm{id}_{\widehat{G}_1}$. Likewise $\widehat{\varphi} \circ \widehat{\psi} : \widehat{G}_2 \to \widehat{G}_2$ is the extension of id_{G_2}, and hence
equals the identity map of $\widehat{G}_2$. Thus the continuous homomorphism $\widehat{\psi}$ is
inverse to the homomorphism $\widehat{\varphi}$, which therefore is an isomorphism. $\square$

The universal property above, applied in the special case $G_1 = G_2$ and
$\varphi = \mathrm{id}$, immediately shows that Hausdorff completions are essentially
unique.

43.25 Corollary: Uniqueness of completion *Let $\iota : G \to \widehat{G}$ and
$\widetilde{\iota} : G \to \widetilde{G}$ be Hausdorff completions of a Hausdorff topological group G.
Then there is a unique isomorphism $\widehat{\varphi} : \widehat{G} \to \widetilde{G}$ of topological groups
such that $\widehat{\varphi} \circ \iota = \widetilde{\iota}$.* $\square$

The following is a slight refinement.

43.26 Corollary *Let $\iota : G \to \widehat{G}$ be a Hausdorff completion of a
Hausdorff topological group G. Then the topology of $\widehat{G}$ is minimal
among the topologies which make $\widehat{G}$ a Hausdorff topological group and
ι an embedding.*

Proof Let $\widetilde{G}$ be obtained from $\widehat{G}$ by endowing it with such a topology
that is coarser. We have to show that this topology is in fact the original
topology of $\widehat{G}$. Since the topology of $\widetilde{G}$ is coarser, $\iota(G)$ is still dense
in $\widetilde{G}$. We show that $\widetilde{G}$ is complete. It suffices to restrict attention to a
concentrated filterbase $\mathfrak{C}$ on G and to show that $\iota(\mathfrak{C})$ converges in $\widetilde{G}$; see
43.19. Now $\iota(\mathfrak{C})$ converges in $\widehat{G}$, hence it converges in $\widetilde{G}$ since the latter
has a coarser topology. Thus, $\iota : G \to \widetilde{G}$ is a completion of G, as well.
We apply 43.24 in the special case $G_1 = G_2 = G$ and $\varphi = \mathrm{id} : G \to G$
obtaining that there is a unique continuous homomorphism $\widehat{\varphi} : \widehat{G} \to \widetilde{G}$
such that $\widehat{\varphi} \circ \iota = \iota$. But of course the identity map id of $\widehat{G}$ is such a
homomorphism, so that $\widehat{\varphi} = \mathrm{id}$. The last assertion of 43.24 finally says
that this is an isomorphism of topological groups, so that the topologies
of $\widetilde{G}$ and $\widehat{G}$ coincide. $\square$

The following will be used in Section 44 in order to identify completions.

43.27 Lemma *Let $(G_i)_{i\in I}$ be a family of complete topological groups. Then the direct product group $\bigtimes_{i\in I} G_i$ with the product topology is also a complete topological group.*

Proof For $k \in I$, let $\pi_k : \bigtimes_{i\in I} G_i \to G_k : (x_i)_{i\in I} \mapsto x_k$ be the canonical projection. For $x, y \in \bigtimes_{i\in I} G_i$, the projection $\pi_k(x + y) = \pi_k(x) + \pi_k(y)$ depends continuously on x and y, and $\pi_k(-x) = -\pi_k(x)$ depends continuously on x, so that the maps $(x, y) \mapsto x + y$ and $x \mapsto -x$ are continuous. Thus $\bigtimes_{i\in I} G_i$ is a topological group.

Now let $\mathfrak{C}$ be a concentrated filterbase on $\bigtimes_{i\in I} G_i$. Then for $i \in I$, the filterbase $\pi_i(\mathfrak{C})$ is concentrated as well by 43.12 since π_i is a continuous homomorphism. By the completeness assumption, $\pi_i(\mathfrak{C})$ converges to an element $c_i \in G_i$. We show that $\mathfrak{C}$ converges to $c = (c_i)_{i\in I}$; then we have proved that $\bigtimes_{i\in I} G_i$ is complete. A neighbourhood U of c contains a neighbourhood of the form $\bigtimes_{i\in I} U_i$ where U_i is a neighbourhood of c_i and $I' = \{\, i \in I \mid U_i \neq G_i \,\}$ is finite. For $i \in I'$, there is $C_i \in \mathfrak{C}$ such that $\pi_i(C_i) \subseteq U_i$. Let C be a member of $\mathfrak{C}$ such that $C \subseteq \bigcap_{i\in I'} C_i$. Then for all $i \in I$ we have $\pi_i(C) \subseteq U_i$, and $C \subseteq \bigtimes_{i\in I} U_i \subseteq U$. $\square$

43.28 Completion of non-commutative topological groups If the topological group G is not commutative, then one has to distinguish between left concentrated filterbases and right concentrated filterbases.

A filterbase $\mathfrak{C}$ on G is said to be *left concentrated* (or *right concentrated*) if for every neighbourhood U of 0 there exists $C \in \mathfrak{C}$ such that $-C + C \subseteq U$ (or $C - C \subseteq U$), respectively. (In Definition 43.3, the right concentrated filters were simply called concentrated.) A filterbase is called *bilaterally concentrated* if it is both left and right concentrated. The topological group is called *left, right* or *bilaterally complete* if every left, right or bilaterally concentrated filterbase, respectively, converges. Note that under inversion left concentrated filterbases are mapped to right concentrated filterbases and vice versa, so that a topological group is left complete if, and only if, it is right complete.

DIEUDONNÉ 1944 gives an example of a Hausdorff topological group admitting a right concentrated filterbase $\mathfrak{C}$ whose image under inversion is not right concentrated, so that $\mathfrak{C}$ itself is not left concentrated.

Every Hausdorff topological group has a bilateral completion which is essentially unique; see WARNER 1989 Theorems 5.9 p. 35 and 5.2 p. 32. With arguments as in the remark on p. 33 following the latter theorem and using loc. cit. Theorem 4.4 p. 26 one can conclude that a Hausdorff topological group with a right concentrated filterbase which is not left concentrated (as in the example above) has no right completion.

43.29 Completion of ordered groups as topological groups Let G be an ordered group. Considering it as a topological group (see 8.4), we may ask about its completion.

If the ordered group G is Archimedean, then it has a completion $\iota : G \to \widehat{G}_{\mathrm{ord}}$ as an ordered group; see 42.3. The complete ordered group $\widehat{G}_{\mathrm{ord}}$ with the topology induced by the ordering is also complete as a topological group (43.10). As $\iota(G)$ is weakly dense in $\widehat{G}_{\mathrm{ord}}$, it is topologically dense. Thus, in the Archimedean case, $\iota : G \to \widehat{G}_{\mathrm{ord}}$ is a completion of G also as a topological group.

Now what if G is not Archimedean? In BANASCHEWSKI 1957 Section 4 pp. 56ff it is proved that G has a completion $G \to \widehat{G}_{\mathrm{top}}$ as a topological group, which can be retrieved within the completion $\iota : G \to \widehat{G}_{\mathrm{ord}}$ of G as an ordered set. Note that $\widehat{G}_{\mathrm{ord}}$ is not a group if G is not Archimedean, since complete ordered groups are Archimedean; see 7.5. But $\widehat{G}_{\mathrm{ord}}$ is a semigroup, as can be seen from the proof of 42.3, and $\widehat{G}_{\mathrm{top}}$ can be obtained as the largest subgroup of this semigroup containing $\iota(G)$. Thus, $\widehat{G}_{\mathrm{top}}$ carries an ordering, and it turns out that its topology is just the topology induced by the ordering and that $\widehat{G}_{\mathrm{top}}$ is an ordered group. If G is not Archimedean, $\widehat{G}_{\mathrm{top}}$ cannot be complete as an ordered group. In particular, this shows that there are ordered groups that are complete as topological groups but not as ordered groups.

44   Completion of topological rings and fields

Now, the completion of topological abelian groups will be used for the additive group of a topological ring. Its multiplication can be extended to the completion in such a way that a topological ring results.

When this is applied to a topological (skew) field F, the resulting complete topological ring $\widehat{F}$ will not necessarily be a (skew) field again. However, we shall see that $\widehat{F}$ is a (skew) field if for instance the topology of F comes from an absolute value, like the p-adic absolute value on $\mathbb{Q}$; see 36.3, and 55.1 for the general notion.

The completion of the field $\mathbb{Q}$ with the usual topology (coming from the usual absolute value) is the topological field $\mathbb{R}$. The completion of the field $\mathbb{Q}$ with the p-adic topology is the field $\mathbb{Q}_p$ of p-adic numbers, which will be discussed in detail in Sections 51–54.

44.1 Definition   A topological ring R is said to be *complete* if its additive group $(R, +)$ is a complete topological group.

44.2 Definition   Let R be a topological ring. A *ring completion* (or *completion* for short) of R is a complete topological ring $\widehat{R}$ together with an embedding $\iota : R \to \widehat{R}$ of topological rings (that is, a monomorphism of rings which induces a homeomorphism $R \to \iota(R)$) such that $\iota(R)$ is dense in $\widehat{R}$.

We shall presently see that for a Hausdorff topological ring such a ring completion always exists. First, we need some some technical preparations.

44.3 Lemma   *Let $\mathfrak{C}$ be a concentrated filterbase on a topological ring. Then for every neighbourhood U of 0 there is $C \in \mathfrak{C}$ and another neighbourhood V of 0 such that $VC \subseteq U$ and $CV \subseteq U$.*

Proof By continuity of the operations, there are neighbourhoods U_1, U_2 of 0 such that $U_1 + U_1 \subseteq U$ and $U_2 U_2 \subseteq U_1$. Since $\mathfrak{C}$ is concentrated, there is $C \in \mathfrak{C}$ such that $C - C \subseteq U_2$. Choose $c \in C$; then $C \subseteq U_2 + c$. By continuity again, there is a neighbourhood V of 0 such that $Vc \subseteq U_1$ and $cV \subseteq U_1$, and we may assume that $V \subseteq U_2$. Then $VC \subseteq V(U_2 + c) \subseteq U_2 U_2 + U_1 \subseteq U_1 + U_1 \subseteq U$, and similarly $CV \subseteq U$. $\qquad\square$

44.4 Lemma   *Let $\mathfrak{C}$ and $\mathfrak{D}$ be concentrated filterbases on a topological ring. Then $\mathfrak{C}\mathfrak{D} := \{ CD \mid C \in \mathfrak{C}, D \in \mathfrak{D} \}$ is a concentrated filterbase, as well (where of course $CD = \{ cd \mid c \in C, d \in D \}$).*

Proof It is clear that $\mathfrak{C}\mathfrak{D}$ is a filterbase. We have to show that for a neighbourhood U of 0 there are $C \in \mathfrak{C}$, $D \in \mathfrak{D}$ such that $CD - CD \subseteq U$. Choose a neighbourhood V of 0 such that $V + V \subseteq U$. By 44.3, there are $C \in \mathfrak{C}$ and $D \in \mathfrak{D}$ and a neighbourhood W of 0 such that $CW \subseteq V$ and $WD \subseteq V$. By making C and D still smaller, if necessary, we may obtain that also $C - C \subseteq W$ and $D - D \subseteq W$. Then $CD - CD \subseteq (C - C)D + C(D - D) \subseteq WD + CW \subseteq V + V \subseteq U$. $\qquad\square$

44.5 Construction   *Every Hausdorff topological ring R has a Hausdorff ring completion $\iota : R \to \widehat{R}$. If R is commutative, then so is $\widehat{R}$.*

Proof (1) The additive group of R is a Hausdorff abelian topological group and therefore has a group completion $\iota : R \to \widehat{R}$ such that $\widehat{R}$ is an abelian Hausdorff group; see 43.21. We shall define a multiplication on $\widehat{R}$ which makes $\widehat{R}$ a topological ring and such that ι is a ring monomorphism (and hence an embedding of topological rings).

The image $\iota(R)$ is dense in $\widehat{R}$. As in 43.22, we may therefore represent the elements of $\widehat{R}$ by special filterbases in R. For $x' \in \widehat{R}$, let $\widehat{\mathfrak{V}}_{x'}$ be the neighbourhood filter of x' in $\widehat{R}$. Then

$$\widehat{\mathfrak{V}}_{x'} \cap \iota(R) = \{\, N' \cap \iota(R) \mid N' \in \widehat{\mathfrak{V}}_{x'} \,\}$$

is a filterbase on $\iota(R) \subseteq \widehat{R}$ which converges to x', and

$$\mathfrak{W}_{x'} = \iota^{-1}(\widehat{\mathfrak{V}}_{x'}) = \iota^{-1}(\widehat{\mathfrak{V}}_{x'} \cap \iota(R))$$

is a concentrated filterbase on R.

(2) For a further element $y' \in \widehat{R}$, we consider the product filterbase $\mathfrak{W}_{x'}\mathfrak{W}_{y'}$ on R, which according to Lemma 44.4 is concentrated. Since ι is uniformly continuous, the image filterbase $\iota(\mathfrak{W}_{x'}\mathfrak{W}_{y'})$ is concentrated as well (43.12) and hence converges in the complete group $\widehat{R}$. We define a multiplication on $\widehat{R}$ by

$$x'y' := \lim \iota(\mathfrak{W}_{x'}\mathfrak{W}_{y'}) \ .$$

(Recall that in the Hausdorff space $\widehat{R}$ the limit is unique.)

We first prove that $\iota : R \to \widehat{R}$ is a homomorphism with respect to multiplication. Since ι induces a homeomorphism $R \to \iota(R)$, the neighbourhoods of an element $x \in R$ are the inverse images of neighbourhoods of $\iota(x)$ in $\widehat{R}$. In our notation this says that for $x, y \in R$ the filterbases $\mathfrak{W}_{\iota(x)}$ and $\mathfrak{W}_{\iota(y)}$ are just the neighbourhood filters of x and y in R. The continuity of multiplication in R implies that the filterbase $\mathfrak{W}_{\iota(x)}\mathfrak{W}_{\iota(y)}$ converges to xy. Hence, by continuity of ι, it follows that

$$\iota(xy) = \iota(\lim(\mathfrak{W}_{\iota(x)}\mathfrak{W}_{\iota(y)})) = \lim \iota(\mathfrak{W}_{\iota(x)}\mathfrak{W}_{\iota(y)}) = \iota(x) \cdot \iota(y) \ . \qquad (1)$$

(3) We now show that the multiplication defined in (2) is continuous. For fixed $x'_0, y'_0 \in \widehat{R}$, let N' be a neighbourhood of $x'_0 y'_0$. Since $\widehat{R}$ is regular (62.4), there is a neighbourhood N'_1 of $x'_0 y'_0$ such that $\overline{N'_1} \subseteq N'$. By definition of $x'_0 y'_0$ there are $X_0 \in \mathfrak{W}_{x'_0}$ and $Y_0 \in \mathfrak{W}_{y'_0}$ such that $\iota(X_0 Y_0) \subseteq N'_1$. Let $X'_0 \in \widehat{\mathfrak{V}}_{x'_0}$ and $Y'_0 \in \widehat{\mathfrak{V}}_{y'_0}$ be neighbourhoods such that $X_0 = \iota^{-1}(X'_0)$, $Y_0 = \iota^{-1}(Y'_0)$, and choose smaller neighbourhoods $X' \in \widehat{\mathfrak{V}}_{x'_0}$ and $Y' \in \widehat{\mathfrak{V}}_{y'_0}$ such that X'_0 is a neighbourhood of all points of X' and that Y'_0 is a neighbourhood of all points of Y'. Then for $x' \in X'$ and $y' \in Y'$ we have $X_0 \in \mathfrak{W}_{x'}$ and $Y_0 \in \mathfrak{W}_{y'}$. Consequently, the product $x'y' = \lim \iota(\mathfrak{W}_{x'}\mathfrak{W}_{y'})$ is contained in the closure $\overline{\iota(X_0 Y_0)}$ by 43.6 and hence in $\overline{N'_1} \subseteq N'$. Thus, the neighbourhoods X' of x'_0 and Y' of y'_0 satisfy $X'Y' \subseteq N'$, and continuity of the multiplication is proved.

(4) It remains to show that the multiplication of $\widehat{R}$ has the properties of a ring multiplication, associativity and distributivity over the addition

of $\widehat{R}$. This is obtained from the corresponding properties of the multiplication of the ring R by continuity via the usual density argument,
which we have already used in 43.20. Associativity is proved there, and
also that the multiplication of $\widehat{R}$ is commutative if the multiplication of
R is, and that the image $\iota(1)$ of the unit element 1 of R is a unit element
of the multiplication of $\widehat{R}$. Distributivity is proved in the same way.

Thus, $\widehat{R}$ is a ring, in fact topological ring, since the additive group is
a topological group, and multiplication is continuous by step (3). Moreover, ι is a ring homomorphism: it is an embedding of topological groups
of the additive group of R into that of $\widehat{R}$, and it respects multiplication;
see equation (1). The other properties of a ring completion are clear.   □

44.6 Theorem: Universal property of ring completions   *For*
$i = 1, 2$, let R_i be a Hausdorff topological ring and $\iota_i : R_i \to \widehat{R}_i$ a
Hausdorff ring completion.

Then every continuous ring homomorphism $\varphi : R_1 \to R_2$ extends to
a unique continuous map $\widehat{\varphi} : \widehat{R}_1 \to \widehat{R}_2$ such that $\widehat{\varphi} \circ \iota_1 = \iota_2 \circ \varphi$, and $\widehat{\varphi}$ is
a ring homomorphism. If φ is an isomorphism of topological rings, that
is, a ring homomorphism and a homeomorphism, then so is $\widehat{\varphi}$.

Proof A ring completion is in particular a group completion of the additive group. Thus, existence of the continuous extension $\widehat{\varphi}$ follows from
the analogous universal property 43.24 of group completions, and we also
obtain that $\widehat{\varphi}$ is a homomorphism of the additive groups, and a homeomorphism if φ is. That $\widehat{\varphi}$ respects multiplication, as well, is proved by a
density argument again, analogously to step (4) of the proof of 43.23.   □

The universal property above, applied in the special case $R_1 = R_2$ and
$\varphi = \mathrm{id}$, immediately gives that Hausdorff ring completions are essentially
unique:

44.7 Corollary: Uniqueness of ring completion   Let $\iota : R \to \widehat{R}$
and $\widetilde{\iota} : R \to \widetilde{R}$ be Hausdorff ring completions of a Hausdorff topological
ring R. Then there is an isomorphism $\widehat{\varphi} : \widehat{R} \to \widetilde{R}$ of topological rings
such that $\widehat{\varphi} \circ \iota = \widetilde{\iota}$.   □

The process of ring completion may of course be applied to a topological skew field F, and one might hope that the completion $\widehat{F}$, which
is a topological ring, is even a topological skew field, but this is not true
in general. The problem is not the continuity of the multiplicative inversion map, but more fundamentally the fact that the completion may
contain elements which are not invertible. In 44.12 we shall see that the

completion may even contain zero divisors. For the more subtle case of a topological field whose ring completion is an integral domain yet still not a field, see HECKMANNS 1991 and WARNER 1989 Exercises 13.2–13.14 pp. 101ff.

Here is a criterion for the completion of a topological skew field to be a skew field again. Recall from 13.4 that a topological skew field is a Hausdorff space if it is not indiscrete.

44.8 Theorem *Let F be a topological skew field whose topology is not the indiscrete topology, and $\iota : F \to \widehat{F}$ its Hausdorff ring completion. Then $\widehat{F}$ is a skew field if, and only if, the following condition holds in F:*

For every concentrated filterbase $\mathfrak{C}$ on F which does not converge to 0, the filterbase $\mathfrak{C}^{-1} := \{ (C \smallsetminus \{0\})^{-1} \mid C \in \mathfrak{C} \}$ is concentrated as well.

If this holds, then the inversion map $\widehat{F} \smallsetminus \{0\} \to \widehat{F} \smallsetminus \{0\} : x' \mapsto x'^{-1}$ is continuous, so that $\widehat{F}$ is a topological skew field.

Remarks (1) If the filterbase $\mathfrak{C}$ does not converge to 0, then by 43.6 there is $C \in \mathfrak{C}$ such that $0 \notin \overline{C}$. In particular we have for every $C \in \mathfrak{C}$ that $C \smallsetminus \{0\} \neq \emptyset$, so that in the statement of the theorem $\mathfrak{C}^{-1}$ can be formed without problem and is a filterbase.

(2) The condition of Theorem 44.8 is satisfied for topological skew fields of type V, as defined in 57.5 (see Exercise 6 of Section 57). These include the skew fields whose topology is described by an absolute value (see Exercise 1 of Section 57); this case will be treated explicitly in 44.9.

Proof (1) For simplicity, we denote the unit element $\iota(1)$ of $\widehat{F}$ by 1. As before, we shall use filterbases within F to represent elements of $\widehat{F}$ in the following way. For $x' \in \widehat{F}$, let $\widehat{\mathfrak{V}}_{x'}$ be the neighbourhood filter of x' in $\widehat{F}$. As explained in 43.22,

$$\mathfrak{W}_{x'} = \iota^{-1}(\widehat{\mathfrak{V}}_{x'}) = \{ \iota^{-1}(N') \mid N' \in \widehat{\mathfrak{V}}_{x'} \}$$

is a concentrated filterbase on F whose image filterbase $\iota(\mathfrak{W}_{x'})$ converges to x'.

(2) We first prove that the stated condition is sufficient for $\widehat{F}$ to be a skew field. We have to show that every element $x' \in \widehat{F} \smallsetminus \{0\}$ has a multiplicative inverse. Let $\mathfrak{C}$ be any concentrated filterbase on F such that the image filterbase $\iota(\mathfrak{C})$ converges to x' (one could use the filterbase $\mathfrak{W}_{x'}$ above, but it does not matter which). $\mathfrak{C}$ does not converge to 0, or else $\iota(\mathfrak{C})$ would have to converge to $\iota(0) = 0$. By assumption, the filterbase $\mathfrak{C}^{-1}$ on F is concentrated again, and so is the image filterbase $\iota(\mathfrak{C}^{-1})$, since ι is uniformly continuous. Hence $\iota(\mathfrak{C}^{-1})$ converges to an

element y' of the completion $\widehat{F}$. We shall show that $x'y' = 1$. By 44.4, the product filterbase $\mathfrak{C}\mathfrak{C}^{-1}$ is concentrated, as well. We verify that each of its members contains 1. Indeed, for $C_1, C_2 \in \mathfrak{C}$ the intersection $C_1 \cap C_2$ contains a further member $C \in \mathfrak{C}$, and $1 \in C \cdot (C \smallsetminus \{0\})^{-1} \subseteq C_1 \cdot (C_2 \smallsetminus \{0\})^{-1}$. According to 43.6, therefore, $\mathfrak{C}\mathfrak{C}^{-1}$ converges to 1. By continuity of ι and of multiplication it follows that $1 = \lim \iota(\mathfrak{C}\mathfrak{C}^{-1}) = \lim \iota(\mathfrak{C}) \cdot \iota(\mathfrak{C}^{-1}) = \lim \iota(\mathfrak{C}) \cdot \lim \iota(\mathfrak{C}^{-1}) = x'y'$, so that y' is a multiplicative inverse of x'.

(3) For the second part of the proof, let $\widehat{F}$ be a skew field. We want to show that the condition stated in the theorem holds. Let $\mathfrak{C}$ be a concentrated filterbase on F which does not converge to 0. The image filterbase $\iota(\mathfrak{C})$ is concentrated again and thus converges to an element x' of the completion $\widehat{F}$. Since the embedding ι induces a homeomorphism of F onto $\iota(F)$, we can be sure that $x' \neq 0 = \iota(0)$, or else $\mathfrak{C} = \iota^{-1}(\iota(\mathfrak{C}))$ would have to converge to $\iota^{-1}(\iota(0)) = 0$, contrary to our assumption. As above, there is a concentrated filterbase $\mathfrak{D}$ on F whose image filterbase $\iota(\mathfrak{D})$ converges to the inverse x'^{-1}. We consider the product filterbase $\mathfrak{C}\mathfrak{D}$; for continuity reasons, its image $\iota(\mathfrak{C}\mathfrak{D}) = \iota(\mathfrak{C}) \cdot \iota(\mathfrak{D})$ converges to $x'x'^{-1} = 1$. Again it follows that $\mathfrak{C}\mathfrak{D}$ converges to 1 in F, since ι is an embedding. In order to prove that $\mathfrak{C}^{-1}$ is concentrated, let U be a neighbourhood of 0 in F. By continuity of addition, there is a neighbourhood V of 0 such that $V + V \subseteq U$. By 44.3, there is a neighbourhood N of 0 and a member $D \in \mathfrak{D}$ such that $DN \subseteq V$. By continuity of the operations in F, there are neighbourhoods W_0 of 0 and W_1 of 1 such that $0 \notin W_1$, $W_0 W_1^{-1} \subseteq V$ and $W_1^{-1} - W_1^{-1} \subseteq N$. Furthermore, we may choose D so small that $D - D \subseteq W_0$. Since $\mathfrak{C}\mathfrak{D}$ converges to 1, there are $C \in \mathfrak{C}$ and $D' \in \mathfrak{D}$ such that $CD' \subseteq W_1$. Replacing D and D' by a member of $\mathfrak{D}$ contained in their intersection, we may assume that $D' = D$. Then $(C \smallsetminus \{0\})^{-1} - (C \smallsetminus \{0\})^{-1} \subseteq DW_1^{-1} - DW_1^{-1} \subseteq (D - D)W_1^{-1} + D\left(W_1^{-1} - W_1^{-1}\right) \subseteq W_0 W_1^{-1} + DN \subseteq V + V \subseteq U$.

(4) It remains to prove continuity of the inversion map (still under the assumption that $\widehat{F}$ is a skew field). By 8.3 it suffices to show continuity at 1. Let N_1' be a neighbourhood of 1 in $\widehat{F}^\times$. By continuity of multiplication, there exists a neighbourhood N_2' of 1 such that $N_2' N_2' \subseteq N_1'$. Since inversion is continuous on $F^\times$, there exists a neighbourhood U of 1 in $F^\times$ such that $U^{-1} \subseteq \iota^{-1}(N_2')$. As ι is an embedding, we find a neighbourhood Q_1' of 1 in $\widehat{F}^\times$ such that $Q_1' \cap \iota(F) \subseteq \iota(U)$. Then $(Q_1' \cap \iota(F))^{-1} \subseteq N_2'$. Again, there is a neighbourhood Q_2' of 1 such that $Q_2' Q_2' \subseteq Q_1'$ and $Q_2' \subseteq N_2'$. We now show that $Q_2'^{-1} \subseteq N_1'$, which finishes the proof.

Let $x' \in Q_2'$. Since $\iota(F)$ is dense, the neighbourhood $Q_2'x'$ of x' meets $\iota(F)$, hence there exists $y' \in Q_2'$ such that $y'x' \in \iota(F)$. Then we have $y'x' \in Q_1' \cap \iota(F)$ and $y' \in N_2'$, which gives $(x')^{-1} = (y'x')^{-1}y' \in (Q_1' \cap \iota(F))^{-1}y' \subseteq N_2'N_2' \subseteq N_1'$. □

44.9 Completion of fields with absolute value Let F be a field with an absolute value, that is, a map $\varphi : F \to [0, \infty[\subseteq \mathbb{R}$ which satisfies the following properties for $x, y \in F$: $\varphi(x) = 0$ if, and only if, $x = 0$ (*definiteness*), $\varphi(x + y) \leq \varphi(x) + \varphi(y)$ (*triangle inequality*), and $\varphi(xy) = \varphi(x)\varphi(y)$ (*multiplicativity*); see 55.1. Here, $\mathbb{R}$ is meant to be the completion of $\mathbb{Q}$ as an ordered field; see 42.11.

A metric d on F is defined by $d(x, y) = \varphi(x - y)$ for $x, y \in F$. With the topology induced by this metric F is a topological field; this can be shown exactly as in the special case of the standard absolute value on $\mathbb{R}$; see 8.2 and 9.1.

Theorem *Let $\iota : F \to \widehat{F}$ be the Hausdorff ring completion of a field F with an absolute value φ. Then $\widehat{F}$ is a topological field. In fact, its topology comes from an absolute value on $\widehat{F}$ which is an extension of φ.*

Proof (1) That $\widehat{F}$ is a topological field can be verified via the criterion in Theorem 44.8. Let $\mathfrak{C}$ be a concentrated filterbase in F which does not converge to 0; one has to show that $\mathfrak{C}^{-1}$ is concentrated, as well. Let U be a neighbourhood of 0 in F and $\varepsilon > 0$ such that $B_\varepsilon(0) = \{x \in F \mid \varphi(x) < \varepsilon\} \subseteq U$. According to 43.6, there is $C \in \mathfrak{C}$ such that $0 \notin \overline{C}$, in other words there is $\delta > 0$ such that for all $c \in C$ one has $\varphi(c) \geq \delta$. Since $\mathfrak{C}$ is concentrated, we may assume that $C - C \subseteq B_{\delta^2\varepsilon}(0)$. For $c, d \in C$ then $\varphi(c^{-1} - d^{-1}) = \varphi(d^{-1}(d - c)c^{-1}) = \varphi(d^{-1}) \cdot \varphi(d - c) \cdot \varphi(c^{-1}) = \varphi(d)^{-1} \cdot \varphi(d - c) \cdot \varphi(c)^{-1} < \delta^{-1}\delta^2\varepsilon\delta^{-1} = \varepsilon$, so that $C^{-1} - C^{-1} \subseteq U$.

(2) We now extend the absolute value of F to $\widehat{F}$. For simplicity, we shall write $\iota(0) = 0$ and $\iota(1) = 1$. We observe that the absolute value φ is uniformly continuous, as a direct consequence of the triangle inequality. By 43.10, the additive group $\mathbb{R}$ is complete. According to 43.23, we may extend φ to a uniformly continuous map $\widehat{\varphi} : \widehat{F} \to \mathbb{R}$ such that

$$\widehat{\varphi}(\iota(x)) = \varphi(x) \tag{1}$$

for all $x \in F$. Since $\iota(F)$ is dense in $\widehat{F}$, the image $\widehat{\varphi}(\widehat{F})$ is contained in the closure of $\varphi(\iota(F))$, and since $[0, \infty[$ is closed in $\mathbb{R}$, it follows that $\widehat{\varphi}(x') \geq 0$ for all $x' \in \widehat{F}$, as well.

(3) We prove that $\widehat{\varphi}$ satisfies the triangle inequality. This amounts to verifying that the set $\widehat{T} = \{(x', y') \in \widehat{F} \times \widehat{F} \mid \widehat{\varphi}(x' + y') \leq \widehat{\varphi}(x') + \widehat{\varphi}(y')\}$

equals $\widehat{F} \times \widehat{F}$. Since ι is a ring homomorphism and since φ satisfies the triangle inequality, equation (1) shows that $\widehat{T}$ contains $\iota(F) \times \iota(F)$, which is a dense subset of $\widehat{F} \times \widehat{F}$. By continuity of addition and of $\widehat{\varphi}$ it is clear that $\widehat{T}$ is closed in $\widehat{F} \times \widehat{F}$, hence $\widehat{T} = \widehat{F} \times \widehat{F}$, indeed. In the same way, one obtains that $\widehat{\varphi}$ is multiplicative since φ is: one verifies that $\{\, (x', y') \in \widehat{F} \times \widehat{F} \mid \widehat{\varphi}(x'y') = \widehat{\varphi}(x')\widehat{\varphi}(y') \,\} = \widehat{F} \times \widehat{F}$.

(4) Next we show that $\widehat{\varphi}$ also is definite. It is clear that $\widehat{\varphi}(0) = 0$. Assume that $\widehat{\varphi}(x') = 0$. Since $\iota(F)$ is dense in $\widehat{F}$, there is a concentrated filterbase $\mathfrak{W}_{x'}$ on F such that $\iota(\mathfrak{W}_{x'})$ converges to x'; see 43.22. We show that $\mathfrak{W}_{x'}$ is finer than the neighbourhood filter $\mathfrak{V}_0$ of 0 in F. Then $\iota(\mathfrak{W}_{x'})$ is finer than the filterbase $\iota(\mathfrak{V}_0)$, which converges to $\iota(0) = 0$, so that $\iota(\mathfrak{W}_{x'})$ converges to both x' and 0, and $x' = 0$ as $\widehat{F}$ is a Hausdorff space. In order to show that $\mathfrak{W}_{x'}$ is finer than $\mathfrak{V}_0$, let $U \in \mathfrak{V}_0$. There is $\varepsilon > 0$ such that $B_\varepsilon(0) \subseteq U$. Since $\iota(\mathfrak{W}_{x'})$ converges to x' and since $\widehat{\varphi}(x') = 0$, there is $W \in \mathfrak{W}_{x'}$ such that all $w \in W$ satisfy $\widehat{\varphi}(\iota(w)) < \varepsilon$, by continuity of $\widehat{\varphi}$. Because of equation (1) this means $W \subseteq B_\varepsilon(0) \subseteq U$.

(5) Due to steps (3) and (4), $\widehat{\varphi}$ is an absolute value on $\widehat{F}$. The topology defined by the corresponding metric $\widehat{d}$ metric makes $\widehat{F}$ a topological field, as well. We compare this metric topology with the topology of $\widehat{F}$ as completion of F, the 'completion' topology. Since $\widehat{\varphi}$ is continuous in the completion topology, the metric topology is coarser. Equation (1) says that ι is an isometric embedding of (F, d) into the metric space $(\widehat{F}, \widehat{d})$. It now follows from 43.26 that the metric topology coincides with the completion topology. $\square$

44.10 Completion of fields with absolute values in terms of Cauchy sequences
In the situation discussed above in 44.9, the completion $\widehat{F}$ may be described using Cauchy sequences in F. Since the topologies of the topological fields F and $\widehat{F}$ are defined by the translation invariant metrics d and $\widehat{d}$, the notion of a Cauchy sequence can be expressed in the usual way by these metrics. It is an easy special case of 44.4 that the set C of all Cauchy sequences of F is closed under termwise addition and multiplication and hence is a ring. The isometry ι maps a Cauchy sequence of F to a Cauchy sequence of $\widehat{F}$. By completeness, the image sequence has a (unique) limit in $\widehat{F}$. Thus we have a map

$$C \to \widehat{F} : (x_\nu)_\nu \mapsto \lim \iota(x_\nu) \,. \tag{1}$$

Since ι is a continuous ring homomorphism and since the ring operations in F and $\widehat{F}$ are continuous, the map (1) is a ring homomorphism, as well. It is surjective; indeed, since $\iota(F)$ is dense in $\widehat{F}$, every element of $\widehat{F}$ is the limit of a sequence in $\iota(F)$, which is a Cauchy sequence with respect to the metric $\widehat{d}$, and hence is the image sequence of a Cauchy sequence of F, as ι is an isometry. For the same reason, a Cauchy sequence of F converges to 0 if, and only if, its image sequence converges to 0. Hence, the kernel of the map (1) is the ideal Z consisting of the sequences of F converging to 0. The map (1) factors through the factor ring C/Z and yields a ring isomorphism

$$C/Z \to \widehat{F} : (x_\nu)_\nu + Z \mapsto \lim \iota(x_\nu) . \tag{2}$$

In order to have a self-contained description of the completion of F in terms of Cauchy sequences, we prove once more that C/Z is a field.

The sequence all of whose terms are 1 is the unit element of C. A Cauchy sequence $(x_\nu)_\nu$ which does not converge to 0 cannot even accumulate at 0 (see 43.6). Hence there are only finitely many ν such that $x_\nu = 0$, and the sequence $(y_\nu)_\nu$ defined by $y_\nu = x_\nu$ if $x_\nu \neq 0$ and $y_\nu = 1$ if $x_\nu = 0$ is still a Cauchy sequence which does not accumulate at 0. It has no zero terms any more, and clearly $(x_\nu)_\nu + Z = (y_\nu)_\nu + Z$. If we ascertain that the inverse elements y_ν^{-1} form a Cauchy sequence again, then it is clear that $(y_\nu^{-1})_\nu$ is an inverse of $(y_\nu)_\nu$ in C, and $(x_\nu)_\nu + Z = (y_\nu)_\nu + Z$ has the inverse $(y_\nu^{-1})_\nu + Z$.

The following argument is a Cauchy sequence version of step (1) in the proof of Theorem 44.9. The multiplicativity of absolute values implies that $\varphi(y_\mu^{-1} - y_\nu^{-1}) = \varphi(y_\nu^{-1}(y_\nu - y_\mu)y_\mu^{-1}) = \varphi(y_\nu)^{-1} \cdot \varphi(y_\nu - y_\mu) \cdot \varphi(y_\mu)^{-1}$. The real numbers $\varphi(y_\nu)$ are bounded away from 0, since $(y_\nu)_\nu$ does not accumulate at 0, hence their inverses $\varphi(y_\nu)^{-1}$ are bounded. Thus the above equation shows that $\varphi(y_\mu^{-1} - y_\nu^{-1})$ is arbitrarily small if μ, ν are sufficiently large, since $(y_\nu)_\nu$ is a Cauchy sequence. Thus, indeed, $(y_\nu^{-1})_\nu$ is a Cauchy sequence, as well.

Furthermore, we pull back the absolute value $\widehat{\varphi}$ from $\widehat{F}$ to C/Z using the ring isomorphism (2). For a Cauchy sequence $(x_\nu)_\nu$ continuity of $\widehat{\varphi}$ implies that $\widehat{\varphi}(\lim \iota(x_\nu)) = \lim \widehat{\varphi}(\iota(x_\nu)) = \lim \varphi(x_\nu)$. Thus, the absolute value Φ on C/Z corresponding to the absolute value $\widehat{\varphi}$ on $\widehat{F}$ under the ring isomorphism (2) is given by

$$\Phi((x_\nu)_\nu + Z) = \lim \varphi(x_\nu) .$$

If one prefers, one may verify directly that this defines an absolute value on C/Z.

With the topology defined by this absolute value, C/Z is a topological field which is isomorphic to $\widehat{F}$ by the map (2). In particular, C/Z is complete. Under the isomorphism (2), the embedding $\iota : F \to \widehat{F}$ corresponds to the map $F \to C/Z : x \to (x)_\nu + Z$, where $(x)_\nu$ is the constant sequence all of whose terms are equal to x. With this embedding, C/Z can be viewed as the completion of the topological field F.

We remark that above (a countable form of) the axiom of choice has been used without mention to show that the map (2) is surjective. This could be avoided by using concentrated filters instead of Cauchy sequences in an analogous way.

If one employs C/Z to *construct* the completion of F, as is often done, then the axiom of choice is needed for proving that C/Z is complete. In contrast, our proof of 44.9 does not use the axiom of choice.

44.11 Examples (1) The real numbers The ordered field $\mathbb{R}$ is a topological field with the ordering topology (see 8.2 and 9.1), and it is a complete topological field (by 43.10). In 42.11, we have obtained $\mathbb{R}$ as the completion of the Archimedean ordered field $\mathbb{Q}$ of rational numbers. In particular, $\mathbb{Q}$ may be considered as a subfield of $\mathbb{R}$ which is weakly dense (see 41.1 about this density notion for chains). This implies that $\mathbb{Q}$ is topologically dense in $\mathbb{R}$. Thus, the topological field $\mathbb{R}$ is the completion of $\mathbb{Q}$ as a topological field with the usual topology.

(2) The p-adic numbers Let p be a prime number. The completion of the topological field $\mathbb{Q}$ endowed with the p-adic topology (compare 36.3) is the field $\mathbb{Q}_p$ of *p-adic numbers*. This field will be discussed in Sections 51–54.

(3) Laurent series fields (1) Let F be a field. The elements of the Laurent series field $F((t))$ are those formal series $\sum_{\nu \in \mathbb{Z}} \xi_\nu t^\nu$ with $\xi_\nu \in F$ whose *support* $\{\nu \in \mathbb{Z} \mid \xi_\nu \neq 0\}$ is bounded below; see 64.23. If n is a lower bound of the support, we also write the above Laurent series as $\sum_{\nu \geq n} \xi_\nu t^\nu$. Addition of Laurent series is defined termwise, and multiplication is the formal Cauchy product.

The so-called *power series* are the elements of $F((t))$ of the form $\sum_{\nu \geq 0} \xi_\nu t^\nu$ (the Laurent series whose support is contained in $\{0\} \cup \mathbb{N}$). These power series form a subring $F[[t]]$ of $F((t))$.

Let $A := \{ \sum_{\nu=1}^{n} \xi_\nu t^{-\nu} \mid n \in \mathbb{N}, \xi_\nu \in F \}$ be the F-linear span of $\{ t^{-n} \mid n \in \mathbb{N} \}$ in $F((t))$. The map

$$A \times F[[t]] \to F((t)) : (a, b) \mapsto a + b \tag{1}$$

is a group isomorphism with respect to addition.

We endow $F((t))$ with a topology. $F[[t]]$ can be considered as the Cartesian product $F^{\{0\} \cup \mathbb{N}}$ since the power series $\sum_{\nu \geq 0} \xi_\nu t^\nu$ is just a suggestive way of writing the sequence $(\xi_\nu)_{\nu\in\{0\}\cup\mathbb{N}}$. We consider the discrete topology on F, the product topology on $F[[t]]$, the discrete topology on A and finally the product topology on $A \times F[[t]]$. The topology on $F((t))$ will be the topology which makes the map (1) a homeomorphism.

Since discrete groups are complete (43.5), the additive group of $F((t))$ is a complete topological group by 43.27.

If F is finite, then the Cartesian product $F[[t]] = F^{\{0\}\cup\mathbb{N}}$ is compact by Tychonoff's theorem; hence $F((t))$ is locally compact.

(2) Next, we give another description of the topology of $F((t))$ using the valuation

$$v : F((t)) \smallsetminus \{0\} \to \mathbb{Z} : v\big(\textstyle\sum_{\nu\in\mathbb{Z}} \xi_\nu t^\nu\big) = \min\{\nu \in \mathbb{Z} \mid \xi_\nu \neq 0\} \ .$$

The term valuation means that $v(x + y) \geq \min\{v(x), v(y)\}$ for $x, y \in F((t))^\times$ such that $y \neq -x$ and $v(xy) = v(x) + v(y)$ for $x, y \in F((t))^\times$; compare 56.1. These properties imply that the map $\varphi : F((t)) \to \mathbb{R}$ with $\varphi(x) = 2^{-v(x)}$ for $x \neq 0$ and $\varphi(0) = 0$ is an absolute value on $F((t))$ satisfying the ultrametric property $\varphi(x + y) \leq \max\{\varphi(x), \varphi(y)\}$, which is stronger than the ordinary triangle inequality.

Let d be the metric on $F((t))$ defined by $d(x, y) = \varphi(x - y)$. With the topology induced by this metric (the so-called valuation topology, compare Section 56), the field $F((t))$ is a topological field as in 44.9. For $m \in \mathbb{N}$, the Laurent series whose values under v are at least m constitute a neighbourhood

$$\{\textstyle\sum_{\nu\geq m} \xi_\nu t^\nu \mid \xi_\nu \in F\} = t^m F[[t]]$$

of 0 which is an ideal of $F[[t]]$. Moreover, the ideals $t^m F[[t]]$ with $m \in \mathbb{N}$ form a neighbourhood base of 0 in $F[[t]]$ and in $F((t))$ for the valuation topology; see also 13.2(b). This neighbourhood base is also a neighbourhood base of 0 for the topology defined initially in step (1) using product topologies. Since the additive group $F((t))$ is a topological group for both topologies, it follows that the two topologies coincide.

(3) The polynomial ring $F[t]$ is the subring of $F[[t]]$ consisting of the power series with finite support. From the definition of the product topology, it is clear that $F[t]$ is dense in $F[[t]]$. By continuity of multiplication in the topological field $F((t))$, it follows that the ring $F[t, t^{-1}] = \bigcup_{n\geq 0} t^{-n} F[t]$ is dense in $\bigcup_{n\geq 0} t^{-n} F[[t]] = F((t))$. The field $F((t))$ contains (an isomorphic copy of) the field of fractions $F(t)$ of $F[t]$ as a subfield which in turn contains $F[t, t^{-1}]$ and hence is dense, as well. Thus, $F((t))$ is the field completion of $F(t)$.

Here, we must of course consider $F(t)$ as a topological ring with the topology induced by the topology of $F((t))$. This topology is defined by the restriction to $F(t)$ of the absolute value φ, which comes from the valuation v. For a polynomial $f \in F[t]$, the value $v(f) \in \mathbb{N} \cup \{0\}$ is the lowest exponent of t appearing in f. For a non-zero polynomial g, we infer by the properties of a valuation that $v(f/g) = v(f) - v(g)$. This is the valuation v_t of $F(t)$ constructed as in 56.3(c) using the irreducible polynomial $p = t \in F[t]$. Thus, $F((t))$ is the completion of $F(t)$ for the topology defined by this valuation.

(4) Another prominent valuation on $F(t)$ is the *degree valuation* v_∞ given by $v_\infty(f/g) = -\deg f + \deg g$ for polynomials $f, g \in F[t]$; compare 56.3(c) again. We shall see that the unique field automorphism ϑ : $F(t) \to F(t)$ mapping t to $1/t$ and fixing every element of F satisfies $v_t \circ \vartheta = v_\infty$. Thus, ϑ is an isomorphism of topological fields of $F(t)$ with the topology defined by v_t onto $F(t)$ with the topology defined by v_∞.

Indeed, for $f \in F[t]$, we have $\vartheta(f) = f(1/t)$. The polynomial $t^d f(1/t)$ with $d = \deg f$ has non-zero absolute term, hence $v_t(t^d f(1/t)) = 0$ and consequently $v_t(\vartheta(f)) = v_t(f(1/t)) = -v_t(t^d) = -d = v_\infty(f)$. This shows that the valuations $v_t \circ \vartheta$ and v_∞ coincide on $F[t]$, hence they coincide on the field of fractions $F(t)$.

The isomorphism ϑ of topological fields which we have just established extends to an isomorphism of the completions of these topological fields according to 44.7. Thus, $F((t))$ can be considered as completion of $F(t)$ not only for the topology defined by v_t, but also for the topology defined by v_∞. Note that these topologies do not coincide: the sequence $(t^\nu)_{\nu \in \mathbb{N}}$ converges to 0 in the topology defined by v_t, but not in the topology defined by v_∞.

44.12 Example: A topological field whose completion has zero divisors We consider the 2-adic and the 3-adic topology on $\mathbb{Q}$ defined by the 2-adic and the 3-adic absolute value $| \ |_2$ and $| \ |_3$; see 36.3. With each of these topologies, $\mathbb{Q}$ is a topological field. We endow the Cartesian product $\mathbb{Q} \times \mathbb{Q}$ with the product topology of the 2-adic topology on the first factor and the 3-adic topology on the second factor. This topology on $\mathbb{Q} \times \mathbb{Q}$ will be called the hexadic topology. With componentwise addition and multiplication $\mathbb{Q} \times \mathbb{Q}$ is a topological ring.

The diagonal $D = \{(x, x) \mid x \in \mathbb{Q}\} \subseteq \mathbb{Q} \times \mathbb{Q}$ with the topology induced by the hexadic topology is a topological field (as a field, it is isomorphic to $\mathbb{Q}$). We shall see that the ring completion $\widehat{D}$ of D has zero divisors.

For $n \in \mathbb{N}$, $n \to \infty$, the rational numbers

$$x_n = \frac{(2/3)^n}{1 + (2/3)^n} = \frac{1}{1 + (3/2)^n}$$

converge to 0 in the 2-adic topology and to 1 in the 3-adic topology. Indeed, the definition in 36.3 yields $|(2/3)^n|_2 = 2^{-n} \to 0$, so that $|x_n|_2$ converges to 0 in the 2-adic topology. In the same way, $|(3/2)^n|_3 = 3^{-n} \to 0$, hence x_n converges to 1 in the 3-adic topology. The sequence $(x_n, x_n)_{n \in \mathbb{N}}$ in D converges in $\mathbb{Q} \times \mathbb{Q}$ to $(0,1) \notin D$. Exchanging the primes 2 and 3, we obtain a sequence $(y_n)_{n \in \mathbb{N}}$ in $\mathbb{Q}$ converging to 1 in the 2-adic topology and to 0 in the 3-adic topology, and the sequence $(y_n, y_n)_{n \in \mathbb{N}}$ in D converges in $\mathbb{Q} \times \mathbb{Q}$ to $(1,0) \notin D$. Thus, the closure of D in the hexadic topology of $\mathbb{Q} \times \mathbb{Q}$ contains $(1,0)$ and $(0,1)$.

In order to determine the completion $\widehat{D}$, we embed $\mathbb{Q} \times \mathbb{Q}$ with the hexadic topology into the Cartesian product $\mathbb{Q}_2 \times \mathbb{Q}_3$ of the completions of $\mathbb{Q}$ for the 2-adic and the 3-adic topology; this Cartesian product is a topological ring with componentwise addition and multiplication, and it is complete by 43.27, since the factors are complete. The closure $\overline{D}$ of D in $\mathbb{Q}_2 \times \mathbb{Q}_3$ is a subgroup of the additive group of $\mathbb{Q}_2 \times \mathbb{Q}_3$; see 62.5. The latter is a topological vector space over $\mathbb{Q}$ with componentwise scalar multiplication, and D is a linear subspace. For $r \in \mathbb{Q}$, we obtain that $r \cdot \overline{D} \subseteq \overline{rD} \subseteq \overline{D}$, so that $\overline{D}$ is a $\mathbb{Q}$-linear subspace, as well, but as we have seen above, $\overline{D}$ contains $(0,1)$ and $(1,0)$ and hence $\mathbb{Q} \times \mathbb{Q}$. Since $\mathbb{Q} \times \mathbb{Q}$ is dense in $\mathbb{Q}_2 \times \mathbb{Q}_3$, we infer that the diagonal D of $\mathbb{Q} \times \mathbb{Q}$ embeds as a dense subset into $\mathbb{Q}_2 \times \mathbb{Q}_3$ (this fact is generalized in Exercise 2 of Section 55). Thus, the ring completion of the field D is $\widehat{D} = \mathbb{Q}_2 \times \mathbb{Q}_3$. Since $\mathbb{Q}_2$ and $\mathbb{Q}_3$ are fields (see 44.11 Example (2) and 44.9), the zero divisors of this ring are the elements of $(\mathbb{Q}_2 \times \{0\}) \cup (\{0\} \times \mathbb{Q}_3)$.

44.13 Additive versus multiplicative completeness of topological skew fields One may ask if the multiplicative group $F^\times$ of a complete topological skew field F is right complete as a topological group. (See 43.28 for the notions of right, left and bilateral completeness of noncommutative topological groups.) Completeness of F is expressed using the additive group, whereas the question about completeness of $F^\times$ is concerned with right concentrated filterbases on the group $F^\times$, with multiplication as group operation. We do not know if such filterbases are necessarily concentrated in the additive group.

A positive answer can be given if one asks for bilateral completeness of $F^\times$ instead of right completeness; see WARNER 1989 Theorem 14.11 p. 111. He proves more generally that if in a complete topological ring R

the group $R^\times$ of invertible elements is open and if inversion is continuous on it, then $R^\times$ is bilaterally complete. In the special case of topological fields (with commutative multiplication), there is no distinction between left, right and bilateral completeness. The argument of WARNER 1989 loc. cit. can be adapted to show that the multiplicative group of a locally bounded complete topological skew field (see 57.1) is even right complete.

Conversely, it is rather trivial that a filterbase on the multiplicative group of a topological field which is concentrated in the additive group need not be concentrated in the multiplicative group. To see this, we take a topological field F whose topology is neither discrete nor the indiscrete topology and whose multiplicative group is complete, and we consider the neighbourhood filter $\mathfrak{V}_0$ of 0 in F. It converges to 0. Since the topology of F is not discrete, $\{0\}$ is not a neighbourhood of 0, so that every neighbourhood V of 0 intersects $F^\times$. Hence $\mathfrak{V}_0 \cap F^\times = \{V \cap F^\times \mid V \in \mathfrak{V}_0\}$ is a filterbase which still converges to 0 and so is concentrated in the additive group. However, as a filterbase of $F^\times$, it cannot be concentrated; else, by completeness, it would have to converge to an element of $F^\times$, but instead it converges to 0.

5

The p-adic numbers

The idea of p-adic numbers is due to Hensel, who was inspired by local power series expansions of meromorphic functions (see WARNER 1989 p. 469f, EBBINGHAUS *et al.* 1991 Chapter 6 and ULLRICH 1998). We treat the p-adic numbers as relatives of the real numbers. In fact, completion of the rational field $\mathbb{Q}$ with respect to an absolute value leads either to the reals $\mathbb{R}$ or to a field $\mathbb{Q}_p$ of p-adic numbers, where p is a prime number (see 44.9, 44.10, 51.4, 55.4), and these fields are locally compact. We consider the additive and the multiplicative group of $\mathbb{Q}_p$ in Sections 52 and 53, and we study squares and quadratic forms over $\mathbb{Q}_p$ in Section 54. It turns out (see 53.2) that the additive and the multiplicative group of $\mathbb{Q}_p$ are locally isomorphic, in the sense that some open (and compact) subgroup of $\mathbb{Q}_p^\times$ is isomorphic to an open subgroup of $\mathbb{Q}_p^+$; this is similar to the situation for $\mathbb{R}$.

Comparing $\mathbb{R}$ and $\mathbb{Q}_p$, it appears that the structure of $\mathbb{Q}_p$ is dominated much more by algebraic and number theoretic features. A major topological difference between the locally compact fields $\mathbb{R}$ and $\mathbb{Q}_p$ is the fact that $\mathbb{R}$ is connected and $\mathbb{Q}_p$ is totally disconnected (51.10). Moreover, $\mathbb{Q}_p$ cannot be made into an ordered field (54.2).

In Sections 55–58 we put the fields $\mathbb{Q}_p$ in the context of general topological fields: we study absolute values, valuations and the corresponding topologies. Section 58 deals with the classification of all locally compact fields and skew fields. If $\mathbb{Q}$ is a dense subfield of a non-discrete locally compact field F, then $F \cong \mathbb{R}$ or $F \cong \mathbb{Q}_p$ for some prime p; see 58.7.

A notable omission in this chapter is Hensel's Lemma on fields F which are complete with respect to a valuation v; this lemma allows to lift roots of suitable polynomials in the residue field F_v to roots in F, by a method related to Newton approximation. We could have used Hensel's Lemma on several occasions (compare ROBERT 2000 1.6, GOUVÊA 1997

278

3.4, LANG 1970 II §2, GREENBERG 1969 Chapter 5), but we chose to deviate from well-trodden paths and to find out how far we can get without it. See also RIBENBOIM 1985, PRIESS-CRAMPE–RIBENBOIM 2000 and ENGLER–PRESTEL 2005 p. 20f.

51   The field of *p*-adic numbers

Here we construct the field $\mathbb{Q}_p$ of p-adic numbers (51.4) and the ring $\mathbb{Z}_p$ of p-adic integers (51.6), and we derive some basic properties of $\mathbb{Q}_p$ and of $\mathbb{Z}_p$. Everything hinges on the p-adic absolute value $|\ |_p : \mathbb{Q} \to \mathbb{Q}$ and its extension $|\ |_p : \mathbb{Q}_p \to \mathbb{Q}$.

51.1 Definition   Fix a prime number p. We recall some facts from 36.3. Every non-zero rational number x can be written in the form

$$x = p^n \cdot \frac{a}{b}$$

with integers $n, a, b \in \mathbb{Z}$ such that a and b are not divisible by p. The *p-adic absolute value* $|\ |_p$ on $\mathbb{Q}$ is defined by

$$|x|_p = p^{-n}$$

for $0 \neq x \in \mathbb{Q}$, and $|0|_p = 0$. The mapping $|\ |_p$ is multiplicative and ultrametric, i.e. $|xy|_p = |x|_p|y|_p$ and $|x + y|_p \leq \max\{|x|_p, |y|_p\}$ for all $x, y \in \mathbb{Q}$; see 36.3. (Hence $|\ |_p$ is a non-Archimedean absolute value on $\mathbb{Q}$ as defined in 55.1.) The multiplicative group $\mathbb{Q}^\times$ is the direct product of $\{\pm 1\}$ and of the free abelian group generated by all prime numbers (see 32.1), and the p-adic absolute value is just the projection $\mathbb{Q}^\times \to p^{\mathbb{Z}}$ onto the factor $p^{\mathbb{Z}}$ of $\mathbb{Q}^\times$, followed by inversion $p^n \mapsto p^{-n}$.

The ordinary absolute value $|x| = \max\{x, -x\}$ on $\mathbb{Q}$ is often associated with the 'prime $p = \infty$' (but CONWAY 1997 argues that it is associated rather to the 'prime $p = -1$'). These definitions give the product formula

$$|x| \cdot \prod_{p \in \mathbb{P}} |x|_p = 1$$

for all $x \in \mathbb{Q}^\times$. One could also define $|x|_p$ to be 2^{-n} instead of p^{-n}, but then the product formula would become slightly more complicated.

51.2 Metric and topology   As in 36.3, we define on $\mathbb{Q}$ the p-adic metric d_p by $d_p(x, y) = |x - y|_p$ for $x, y \in \mathbb{Q}$. This is in fact an ultrametric, i.e., we have $d_p(x, z) \leq \max\{d_p(x, y), d_p(y, z)\}$ for $x, y, z \in \mathbb{Q}$, as a consequence of the ultrametric inequality for $|\ |_p$. Ultrametric spaces have some unexpected geometric properties: all triangles are isosceles, and spheres have many centres. The metric topology defined by d_p is the *p-adic topology* of $\mathbb{Q}$. One defines *p*-adic convergence and *p*-adic Cauchy sequences in the usual manner. For example, $\lim_{n\to\infty} p^n = 0$ in the *p*-adic topology.

The metric space $(\mathbb{Q}, d_p)$ is not complete; indeed, the integers $a_n = \sum_{k=0}^{n} p^{k^2}$ or $a_n = \sum_{k=0}^{n} p^{k!}$ form *p*-adic Cauchy sequences (since in both cases $|a_n - a_m|_p \leq p^{-\min\{m,n\}}$) which have no limit in $\mathbb{Q}$; see Proposition 51.11.

The following observation is useful when forming the completion of $(\mathbb{Q}, d_p)$.

51.3 Lemma *Let* $(a_n)_n \in \mathbb{Q}^{\mathbb{N}}$ *be a p-adic Cauchy sequence.*

(i) *Then the sequence* $(|a_n|_p)_n$ *is a Cauchy sequence in* $\mathbb{R}$.

(ii) *If the sequence* $(a_n)_n$ *does not converge to 0 in the p-adic topology, then* $|a_n|_p$ *is finally constant (and non-zero).*

Proof (i) follows from the inequality $\big||x|_p - |y|_p\big| \leq |x - y|_p$, which is a consequence of the triangle inequality $|x|_p = |x - y + y|_p \leq |x - y|_p + |y|_p$.

(ii) The values of $|\ |_p$ form the set $p^{\mathbb{Z}} \cup \{0\}$, and this closed subset of $\mathbb{R}$ has zero as its only accumulation point.   □

51.4 Definition We recall the construction of the *p*-adic completion of $\mathbb{Q}$ from 44.9, 44.10. Let $C(p)$ be the set of all *p*-adic Cauchy sequences in $\mathbb{Q}$, and let $Z(p)$ be the set of all sequences in $\mathbb{Q}$ with *p*-adic limit 0. Then $C(p)$ is a ring (with component-wise addition and multiplication), and $Z(p)$ is an ideal in $C(p)$ (since Cauchy sequences are bounded; see 51.3). As shown in 44.10, the quotient ring

$$\mathbb{Q}_p := C(p)/Z(p)$$

is a field, the *field* $\mathbb{Q}_p$ *of p-adic numbers*; the field property depends on the observation that a sequence in $C(p) \smallsetminus Z(p)$ cannot accumulate at 0; see 43.6 or 51.3(ii).

Mapping $q \in \mathbb{Q}$ to the constant sequence $(q, q, \dots)$ gives a natural embedding $\mathbb{Q} \to \mathbb{Q}_p$, as in 44.10. We identify $\mathbb{Q}$ with its image in $\mathbb{Q}_p$, i.e. with the prime field of $\mathbb{Q}_p$.

By 44.10 or 51.3(i), the ring $C(p)$ admits the map $\varphi : C(p) \to \mathbb{R} :$ $\varphi(a) = \lim_n |a_n|_p$, which is multiplicative and ultrametric. Furthermore, φ is constant on each additive coset of $Z(p)$, since $a \in C(p)$, $b \in Z(p)$ imply $\varphi(b) = 0$ and $\varphi(a) = \varphi(a + b - b) \leq \varphi(a + b) \leq \varphi(a)$. Thus we obtain a well-defined extension of $|\ |_p$ to $\mathbb{Q}_p = C(p)/Z(p)$ by defining

$$|a + Z(p)|_p = \lim_{n\to\infty} |a_n|_p \quad \text{for } a \in C(p) .$$

This extension is again multiplicative and ultrametric; it is the *p-adic absolute value* of $\mathbb{Q}_p$; compare 44.9, 44.10. The corresponding metric $d_p\big(a + Z(p), b + Z(p)\big) = |a - b|_p = \lim_{n\to\infty} |a_n - b_n|_p$ on $\mathbb{Q}_p$ extends the *p*-adic metric on $\mathbb{Q}$ and defines a metric topology on $\mathbb{Q}_p$, the *p-adic topology* of $\mathbb{Q}_p$.

From 44.9, 44.10 we know that $\mathbb{Q}$ is dense in $\mathbb{Q}_p$, and that $\mathbb{Q}_p$ is complete. So far, we have proved the following result.

51.5 Theorem *For each prime p there exists a field extension $\mathbb{Q}_p$ of $\mathbb{Q}$ and a map $| \ |_p : \mathbb{Q}_p \to p^{\mathbb{Z}} \cup \{0\}$ with the following properties:*

(i) *The map $| \ |_p$ is multiplicative and ultrametric and extends the p-adic absolute value of $\mathbb{Q}$.*

(ii) *The field $\mathbb{Q}$ is dense in $\mathbb{Q}_p$ and $\mathbb{Q}_p$ is complete (with respect to the p-adic metric $d_p(x, y) = |x - y|_p$ on $\mathbb{Q}_p$).*   □

We point out that $\mathbb{Q}_p$ is determined uniquely by the properties stated in 51.5, in a strong sense: every field with the same properties is isomorphic to $\mathbb{Q}_p$ by a unique isomorphism, and this isomorphism preserves the absolute values. Indeed, the inclusion of $\mathbb{Q}$ in any field F with these properties preserves the *p*-adic absolute value and has therefore a unique extension to an isomorphism $\mathbb{Q}_p \to F$; see 44.6. This means that one can derive all results about $\mathbb{Q}_p$ from 51.5, without recourse to the actual construction of $\mathbb{Q}_p$. The rest of this section relies only on 51.5.

We remark that $\mathbb{Q}_p$ endowed with the *p*-adic topology is a topological field (as defined in 13.1); this is a special case of 13.2(b) and of 44.9. We use this remark often without mentioning.

51.6 Definition Multiplicativity together with the ultrametric inequality implies that the set

$$\mathbb{Z}_p := \big\{ x \in \mathbb{Q}_p \mid |x|_p \leq 1 \big\}$$

is a subring of $\mathbb{Q}_p$, the ring of *p-adic integers*. (It is the valuation ring of the valuation belonging to $| \ |_p$; see Section 56.) Since $|p|_p = p^{-1}$, we have

$$p^n \mathbb{Z}_p = \big\{ x \in \mathbb{Q}_p \mid |x|_p \leq p^{-n} \big\} = \big\{ x \in \mathbb{Q}_p \mid |x|_p < p^{-n+1} \big\}$$

for all $n \in \mathbb{Z}$. This shows that each set $p^n \mathbb{Z}_p$ is simultaneously open and closed in $\mathbb{Q}_p$.

51.7 Lemma *The ring $\mathbb{Z}_p$ has a unique maximal ideal, namely $p\mathbb{Z}_p = \{ x \in \mathbb{Z}_p \mid |x|_p < 1 \}$. Furthermore $\mathbb{Z}_p = \mathbb{Z} + p^n \mathbb{Z}_p$ for each $n \in \mathbb{N}$. The integers $0, 1, \ldots, p^n - 1$ are representatives for the additive cosets of*

$p^n\mathbb{Z}_p$ in $\mathbb{Z}_p$, and $\mathbb{Z}_p/p^n\mathbb{Z}_p \cong \mathbb{Z}/p^n\mathbb{Z}$ as rings. In particular, the quotient ring

$$\mathbb{Z}_p/p\mathbb{Z}_p \cong \mathbb{Z}/p\mathbb{Z} = \mathbb{F}_p$$

is isomorphic to the finite field $\mathbb{F}_p$ with p elements.

Proof The set $\mathbb{Z}_p \smallsetminus p\mathbb{Z}_p = \{\, x \in \mathbb{Z}_p \mid |x|_p = 1 \,\}$ is the group of all units of $\mathbb{Z}_p$, hence the ideal $p\mathbb{Z}_p$ is the largest proper ideal of $\mathbb{Z}_p$.

Since $\mathbb{Q}$ is dense in $\mathbb{Q}_p$, the intersection $\mathbb{Q} \cap \mathbb{Z}_p$ is dense in the open set $\mathbb{Z}_p$. Thus, for every $x \in \mathbb{Z}_p$, we find $a, b \in \mathbb{Z}$ with b not divisible by p such that $|x - ab^{-1}|_p < 1$. Because $\mathbb{F}_p = \mathbb{Z}/p\mathbb{Z}$ is a field, we find $b' \in \mathbb{Z}$ with $1 - bb' \in p\mathbb{Z}$. Hence $|ab^{-1} - ab'|_p = |ab^{-1}(1 - bb')|_p < 1$ and

$$|x - ab'|_p \le \max\{|x - ab^{-1}|_p, |ab^{-1} - ab'|_p\} < 1 \,,$$

which shows that $x - ab' \in p\mathbb{Z}_p$, hence $x \in \mathbb{Z} + p\mathbb{Z}_p$. Thus $\mathbb{Z}_p = \mathbb{Z} + p\mathbb{Z}_p$, and an easy induction gives $\mathbb{Z}_p = \mathbb{Z} + p^n\mathbb{Z}_p$ for $n \in \mathbb{N}$. The description of $p^n\mathbb{Z}_p$ in 51.6 shows that $\mathbb{Z} \cap p^n\mathbb{Z}_p = p^n\mathbb{Z}$. We infer that the quotient $\mathbb{Z}_p/p^n\mathbb{Z}_p = (\mathbb{Z} + p^n\mathbb{Z}_p)/p^n\mathbb{Z}_p \cong \mathbb{Z}/(\mathbb{Z} \cap p^n\mathbb{Z}_p) = \mathbb{Z}/p^n\mathbb{Z}$ is represented by the integers $0, 1, \ldots, p^n - 1$. $\qquad\square$

51.8 Proposition *We have $\mathbb{Z}_p = \left\{\, \sum_{n=0}^\infty c_n p^n \mid c_n \in \{0, 1, \ldots, p-1\} \,\right\}$ and*

$$\mathbb{Q}_p = \left\{\, \sum_{n=k}^\infty c_n p^n \mid k \in \mathbb{Z},\ c_n \in \{0, 1, \ldots, p-1\} \,\right\} \,.$$

Each of these series converges, and the coefficients c_n are determined uniquely up to omitting some leading zeros.

Proof We have $\mathbb{Q}_p = \bigcup_{n \ge 0} p^{-n}\mathbb{Z}_p$, hence it suffices to prove the statement about $\mathbb{Z}_p$. For each $x \in \mathbb{Z}_p$ there exists by 51.7 a sequence of integers $x_k \in \{0, 1, \ldots, p^k - 1\}$ such that $x \in x_k + p^k\mathbb{Z}_p$. Since $x_k \equiv x_{k+1} \bmod p^k$ for all $k \in \mathbb{N}$, we can write $x_k = \sum_{n=0}^{k-1} c_n p^n$ with integers $c_n \in \{0, 1, \ldots, p-1\}$, where $n \in \mathbb{N}_0$. By construction we have $|x - x_k|_p \le p^{-k}$ for all $k \in \mathbb{N}$, hence $x = \lim_k x_k = \sum_{n=0}^\infty c_n p^n$. The coefficients c_n are uniquely determined by $x \in \mathbb{Z}_p$, since c_0 is the representative of x modulo $p\mathbb{Z}_p$, $c_1 p$ represents $x - c_0$ modulo $p^2\mathbb{Z}_p$, etc; in general, $c_k p^k$ represents $x - \sum_{n=0}^{k-1} c_n p^n$ modulo $p^{k+1}\mathbb{Z}_p$.

Finally, each of these series converges in $\mathbb{Q}_p$, because the partial sums $\sum_{n=0}^k c_n p^n$ form a p-adic Cauchy sequence in $\mathbb{Z}$ for arbitrary $c_n \in \{0, 1, \ldots, p-1\}$, as a consequence of the estimate $\left| \sum_{n=a}^b c_n p^n \right|_p \le \max_{a \le n \le b} |c_n p^n|_p \le p^{-a}$. $\qquad\square$

51.9 Corollary *The ring* $\mathbb{Z}_p$ *is the closure of* $\mathbb{N}$ *and of* $\mathbb{Z}$ *in the p-adic topology of* $\mathbb{Q}_p$.

Proof Clearly $\mathbb{N} \subseteq \mathbb{Z}_p$, and $\mathbb{N}$ is dense in the closed set $\mathbb{Z}_p$ by 51.8. □

51.10 Proposition *The sets* $p^n \mathbb{Z}_p$ *with* $n \in \mathbb{N}$ *form a neighbourhood base at 0 for the p-adic topology of* $\mathbb{Q}_p$. *Each of these sets, in particular* $\mathbb{Z}_p$, *is compact and totally disconnected.* $\mathbb{Q}_p$ *is locally compact, totally disconnected, and not discrete.*

Proof The first statement holds since $p^n \mathbb{Z}_p = \{\, x \in \mathbb{Q}_p \mid |x|_p \leq p^{-n}\,\}$. The relation $\lim_{n \to \infty} p^n = 0$ shows that $\mathbb{Q}_p$ and $\mathbb{Z}_p$ are not discrete. Thus it suffices to prove that the ring $\mathbb{Z}_p$ is compact and totally disconnected.

The series representation in 51.8 gives a natural bijection of $\mathbb{Z}_p$ onto the Cartesian power $\{0, 1, \ldots, p-1\}^{\aleph_0}$. Transferring the p-adic metric d_p via this bijection to $\{0, 1, \ldots, p-1\}^{\aleph_0}$, we obtain the metric d with

$$d(x, y) = p^{-\min\{n \mid x_n \neq y_n\}}$$

for distinct sequences $x, y \in \{0, 1, \ldots, p-1\}^{\aleph_0}$. We endow the finite set $\{0, 1, \ldots, p-1\}$ with the discrete topology; then d describes the product topology of $\{0, 1, \ldots, p-1\}^{\aleph_0}$, which is compact (by Tychonoff's theorem) and totally disconnected (since any projection of a non-empty connected set is a singleton). Hence also $\mathbb{Z}_p$ is compact and totally disconnected.

Alternatively, one can derive the compactness of $\mathbb{Z}_p$ from that fact that it is complete and totally bounded; see GOUVÊA 1997 3.3.8. □

We mention that each set $p^n \mathbb{Z}_p$ is in fact homeomorphic to the product space $\{0, 1\}^{\mathbb{N}}$, hence homeomorphic to Cantor's triadic set; see 5.48 (for $p = 2$ we have just shown this; compare also Exercise 8). Furthermore $\mathbb{Q}_p$ is homeomorphic to Cantor's triadic set minus a point, for every prime p; see CHRISTENSON–VOXMAN 1977 6.C.11, HEWITT–ROSS 1963 9.15 or WITT 1975.

Finally, we characterize the p-adic series which represent rational numbers; the result is similar to the corresponding result for decimal expansions.

51.11 Proposition *A p-adic number* $x = \sum_{n=k}^{\infty} c_n p^n$, *with coefficients* $c_n \in \{0, 1, \ldots, p-1\}$, *is a rational number if, and only if, the sequence* $(c_n)_{n \geq k}$ *is finally periodic:*

$$x \in \mathbb{Q} \iff \exists_{s, t \in \mathbb{N}} \, \forall_{m \geq t} \; c_m = c_{m+s} \,.$$

Proof We may assume that $k = 0$. If the sequence $(c_n)_{n \geq 0}$ is finally periodic, then using s and t as above we can write $x = a + \sum_{n=0}^{\infty} bp^{sn}$ with $a \in \mathbb{Z}$ and $b = c_t p^t + c_{t+1} p^{t+1} + \cdots + c_{t+s-1} p^{t+s-1} \in \mathbb{Z}$. Since $\sum_{n=0}^{m-1} p^{sn} = (1 - p^{sm})/(1 - p^s)$ converges p-adically to $1/(1 - p^s)$ for $m \to \infty$, we infer that $x = a + b/(1 - p^s) \in \mathbb{Q}$.

Conversely, let $x = \sum_n c_n p^n$ with $c_n \in \{0, 1, \ldots, p - 1\}$ be rational, say $x = a/b$ with $a, b \in \mathbb{Z}$ and $b \neq 0$. For $m \geq 0$ we define

$$z_m := b \sum_{n \geq m} c_n p^n = a - b \sum_{n=0}^{m-1} c_n p^n \in \mathbb{Z} .$$

Since $|z_m|_p \leq p^{-m}$, the integer z_m is divisible by p^m, and for $m \geq 0$ we have the following estimate of ordinary absolute values

$$|p^{-m} z_m| = |p^{-m} a - b \sum_{n=0}^{m-1} c_n p^{n-m}| \leq |a| + |b| \sum_{n \geq 1} (p-1) p^{-n} = |a| + |b| ,$$

which does not depend on m. Thus the set $\{\, p^{-m} z_m \mid m \geq 0 \,\}$ of integers is finite, hence $p^{-t} z_t = p^{-t-s} z_{t+s}$ for suitable numbers $s, t \in \mathbb{N}$. We conclude that

$$b \sum_{n \geq t} c_n p^n = z_t = p^{-s} z_{t+s} = b \sum_{n \geq t+s} c_n p^{n-s} = b \sum_{n \geq t} c_{n+s} p^n,$$

and the uniqueness of the coefficients (51.8) implies that $c_n = c_{n+s}$ for all $n \geq t$. □

51.12 Other constructions of $\mathbb{Q}_p$ The ring $\mathbb{Z}_p$ may be constructed also as the inverse limit of the finite rings $\mathbb{Z}/p^n \mathbb{Z}$ with $n \in \mathbb{N}$; see SERRE 1973, BOREVICH–SHAFAREVICH 1966, NEUKIRCH 1992 II.2.5, HEWITT–ROSS 1963 §10 or ROBERT 2000 1.4.7. This construction is a modern version of Hensel's approach.

By 52.9 the endomorphism ring of the Prüfer group C_{p^∞} is isomorphic to $\mathbb{Z}_p$, which gives another possibility to define $\mathbb{Z}_p$ (and then $\mathbb{Q}_p$ as well, as the field of fractions); compare LÜNEBURG 1973 p. 113.

Exercise 3 below gives direct algebraic definitions of $\mathbb{Z}_p$ and $\mathbb{Q}_p$.

Each field $\mathbb{Q}_p$ is isomorphic to a subfield of $\mathbb{C}$, since the algebraic closure $\mathbb{Q}_p^\natural$ of $\mathbb{Q}_p$ is isomorphic to $\mathbb{C}$ by 64.21. One can show that $\mathbb{Q}_p^\natural$ has dimension $\aleph_0$ over $\mathbb{Q}_p$; see 58.2.

Analysis in the locally compact field $\mathbb{Q}_p$ has many aspects that are quite different from the usual analysis in $\mathbb{R}$. For details we refer the reader to the books by ROBERT 2000, KOBLITZ 1977, GOUVÊA 1997 and to BURGER–STRUPPECK 1996, and we mention only the following result of Mahler. A map $f : \mathbb{N}_0 \to \mathbb{Z}_p$ is continuous with respect to the p-adic topology on $\mathbb{N}_0$ if, and only if, the values $a_n := \sum_{k=0}^n (-1)^k \binom{n}{k} f(n-k)$ tend to 0 as $n \to \infty$. Moreover, if this holds, then f has a unique continuous extension $g : \mathbb{Z}_p \to \mathbb{Z}_p$, given by the series $g(x) = \sum_{n \geq 0} a_n \binom{x}{n}$ in the binomial polynomials $\binom{x}{n} := x(x-1) \cdots (x-n+1)/n!$ and $\binom{x}{0} := 1$. For a proof see COHN 2003a 9.3.8, ROBERT 2000 4.2 or CASSELS 1986 12.7.

Exercises

(1) Show that the series $\sum_{n=0}^{\infty} p^n$ converges in $\mathbb{Q}_p$ to $(1-p)^{-1}$.

(2) Express as a p-adic series: $-1 \in \mathbb{Q}_p$, $1/10 \in \mathbb{Q}_3$.

(3) Show that the ring $\mathbb{Z}_p$ is isomorphic to the quotient $\mathbb{Z}[[x]]/(x-p)$ of the formal power series ring $\mathbb{Z}[[x]]$ modulo the ideal generated by $x-p$. Show also that the field $\mathbb{Q}_p$ is isomorphic to $\mathbb{Z}((x))/(x-p)$, where $\mathbb{Z}((x))$ denotes the ring of integral Laurent series (see 64.23).
(For the real analogue, see Exercise 2 of Section 6.)

(4) Let $a \in \mathbb{Z} \smallsetminus p\mathbb{Z}$. Show that the sequence of powers a^n contains a subsequence which converges in $\mathbb{Q}_p$ to 1.

(5) A series $\sum_n a_n$ converges in $\mathbb{Q}_p$ if, and only if, $\lim_{n \to \infty} a_n = 0$. Deduce that a series in $\mathbb{Q}_p$ may be rearranged arbitrarily, without changing the convergence or the value of the series.

(6) A sequence $(a_n)_n$ in $\mathbb{Q}_p$ converges if, and only if, $\lim_n |a_{n+1} - a_n|_p = 0$.

(7) The topological closure in $\mathbb{Q}_p$ of the set $\mathbb{P}$ of primes is $(\mathbb{Z}_p \smallsetminus p\mathbb{Z}_p) \cup \{p\}$ (use Dirichlet's theorem on primes, which says that an arithmetic progression contains infinitely many primes if it contains two coprime numbers).

(8) Let $k \in \mathbb{N}$. Define a map $f : \{0, 1, \ldots, k\}^{\mathbb{N}} \to \{0, 1\}^{\mathbb{N}}$ by replacing each entry $i < k$ in a sequence from $\{0, 1, \ldots, k\}^{\mathbb{N}}$ by i ones followed by a single zero, and by replacing each occurrence of k by k ones. Show that f is a homeomorphism of these two product spaces.

(9) Find all elements $a \in \mathbb{Q}_p^{\times}$ such that the function $f : \mathbb{N} \to \mathbb{Q}_p$ given by $f(n) = a^n$ is continuous with respect to the p-adic topology on $\mathbb{N}$.

(10) Let $f \in \mathbb{Q}[x]$ be a polynomial with rational coefficients which assumes on $\mathbb{N}$ only integer values, i.e. $f(\mathbb{N}) \subseteq \mathbb{Z}$. Show that $f(\mathbb{Z}) \subseteq \mathbb{Z}$.

(11) For every $k \in \mathbb{N}$, the p-adic number $\sum_{n \geq 0} n^k p^n$ belongs to $\mathbb{Q}$.

52   The additive group of p-adic numbers

In this section, we consider $\mathbb{Q}_p^+$ and $\mathbb{Z}_p^+$ as abstract groups and as topological groups. As usual, p denotes a prime number. The Prüfer groups $C_{p^{\infty}}$ introduced in 1.26 will play an important role.

52.1 Proposition *For each prime p the additive group $\mathbb{Q}_p^+$ is isomorphic to $\mathbb{R}^+$ (but compare 52.4 below).*

Proof $\mathbb{Q}_p$ has characteristic zero, hence $\mathbb{Q}_p^+$ is divisible and torsion free. Furthermore card $\mathbb{Q}_p =$ card $\mathbb{R}$ by 51.8, hence $\mathbb{Q}_p^+ \cong \mathbb{R}^+$ by 1.14. □

The topological groups $\mathbb{R}^+$ and $\mathbb{Q}_p^+$ are both locally compact, but $\mathbb{Q}_p^+$ is totally disconnected, since the subgroups $p^n\mathbb{Z}_p$ with $n \in \mathbb{N}$ form a neighbourhood base at 0; see 51.10.

52.2 Lemma *For each $n \in \mathbb{Z}$, the quotient group $\mathbb{Q}_p^+/p^n\mathbb{Z}_p$ is isomorphic to the Prüfer group C_{p^∞}. Moreover, one has the following isomorphisms of abstract groups:*

$$\mathbb{Q}_p^+/\mathbb{Z} \cong C_{p^\infty} \oplus (\mathbb{Z}_p/\mathbb{Z}) \cong \mathbb{R}/\mathbb{Z} \quad \text{and} \quad \mathbb{Z}_p^+/\mathbb{Z} \cong \underset{q \in \mathbb{P} \smallsetminus \{p\}}{\mathsf{X}} C_{q^\infty} \ .$$

Proof For the first assertion it suffices to consider the case $n = 0$, since $x \mapsto p^n x$ is an automorphism of $\mathbb{Q}_p^+$. Denote by $R := \mathbb{Z}[p^{-1}] = \{\, ap^{-m} \mid a \in \mathbb{Z}, m \in \mathbb{N}\}$ the subring of $\mathbb{Q}$ generated by p^{-1}. Then $R \cap \mathbb{Z}_p = \mathbb{Z}$, and the series representations in 51.8 show that $\mathbb{Q}_p = R + \mathbb{Z}_p$. Hence $\mathbb{Q}_p/\mathbb{Z}_p = (R + \mathbb{Z}_p)/\mathbb{Z}_p \cong R/(R \cap \mathbb{Z}_p) = R/\mathbb{Z}$, and $R/\mathbb{Z}$ is the p-primary torsion subgroup C_{p^∞} of $\mathbb{Q}/\mathbb{Z}$ as defined in 1.26.

For the second assertion we derive from $\mathbb{Q}_p = R + \mathbb{Z}_p$ and $R \cap \mathbb{Z}_p = \mathbb{Z}$ the equation

$$\mathbb{Q}_p/\mathbb{Z} = (R/\mathbb{Z}) \oplus (\mathbb{Z}_p/\mathbb{Z}) = C_{p^\infty} \oplus (\mathbb{Z}_p/\mathbb{Z}) \ ,$$

hence $\mathbb{Z}_p/\mathbb{Z} \cong (\mathbb{Q}_p/\mathbb{Z})/C_{p^\infty}$. From 52.1 and 1.31 we infer that $\mathbb{Q}_p/\mathbb{Z} \cong \mathbb{R}/\mathbb{Z} \cong \mathsf{X}_q C_{q^\infty}$. This product of all Prüfer groups C_{q^∞} has only one subgroup isomorphic to C_{p^∞}, namely its p-primary torsion subgroup, hence $\mathbb{Z}_p/\mathbb{Z} \cong \underset{q \in \mathbb{P} \smallsetminus \{p\}}{\mathsf{X}} C_{q^\infty}$. □

52.3 Proposition (i) *The proper open subgroups of $\mathbb{Q}_p^+$ are precisely the compact groups $p^n\mathbb{Z}_p$ with $n \in \mathbb{Z}$.*

(ii) *The closed subgroups of $\mathbb{Q}_p^+$ are precisely the $\mathbb{Z}_p$-submodules of $\mathbb{Q}_p^+$, i.e. the subgroups $\{0\}$, $\mathbb{Q}_p$ and $p^n\mathbb{Z}_p$ with $n \in \mathbb{Z}$.*

(iii) *Each non-zero ideal of the ring $\mathbb{Z}_p$ is of the form $p^n\mathbb{Z}_p$ with $n \in \mathbb{N}_0$.*

Proof (i) By 51.10 each proper open subgroup U of $\mathbb{Q}_p^+$ contains $p^m\mathbb{Z}_p$ for some $m \in \mathbb{N}$; and $U/p^m\mathbb{Z}_p$ is a proper subgroup of $\mathbb{Q}_p/p^m\mathbb{Z}_p$, which is isomorphic to C_{p^∞} by 52.2. Each proper subgroup of C_{p^∞} is finite and uniquely determined by its order p^k; see 1.26. Hence $U = p^{m-k}\mathbb{Z}_p$.

(ii) Each closed subgroup U is a $\mathbb{Z}_p$-submodule by the density (51.9) of $\mathbb{Z}$ in $\mathbb{Z}_p$. If $U \neq \{0\}$, then the $\mathbb{Z}_p$-submodule U is open, and the assertion follows from (i).

(iii) is a consequence of (ii). □

We remark that the groups $\mathbb{Z}_p^+$ and $\mathbb{Q}_p^+$ are the only non-discrete locally compact abelian Hausdorff groups with the property that every non-trivial closed subgroup is open; see ROBERTSON–SCHREIBER 1968. Moreover, the groups $\mathbb{Z}_p^+$ are the only infinite compact Hausdorff groups such that all non-trivial closed subgroups are topologically isomorphic (see MORRIS–OATES-WILLIAMS 1987), and these groups are also the only non-discrete locally compact Hausdorff groups such that all non-trivial closed subgroups have finite index (see DIKRANJAN 1979, MORRIS *et al.* 1990).

52.4 Corollary *Let p and q be distinct primes. Then the topological groups $\mathbb{Q}_p^+$ and $\mathbb{Q}_q^+$ are not isomorphic.*

Proof This is a consequence of 52.3(i): the quotient $p^n\mathbb{Z}_p/p^{n+1}\mathbb{Z}_p \cong \mathbb{Z}_p/p\mathbb{Z}_p$ of two neighbours in the chain of all proper open subgroups of $\mathbb{Q}_p^+$ is cyclic of order p (see 51.7). One could also use 52.3(i) and 52.2, or 52.3(i) and 52.6 below.   □

Incidentally, 52.4 provides another proof for the fact (36.5) that the group $\mathbb{Q}^+$ with the p-adic topology is not isomorphic to $\mathbb{Q}^+$ with the q-adic topology (using the existence and uniqueness of completions; see 43.21 and 43.25).

Now we focus on properties of the group $\mathbb{Z}_p^+$.

52.5 Definition Modifying the concept of divisibility (1.7), we say that an abelian group A^+ is *p-divisible*, if $pA = A$; this means that each element of A is of the form pa for some $a \in A$.

52.6 Lemma *The group $\mathbb{Z}_p^+$ has no p-divisible subgroup except $\{0\}$, and $\mathbb{Z}_p^+$ is q-divisible for every prime $q \neq p$.*

Proof If $U \leq \mathbb{Z}_p^+$ is p-divisible, then $U = p^n U \subseteq p^n\mathbb{Z}_p$ for each $n \in \mathbb{N}$, hence $U \subseteq \bigcap_n p^n\mathbb{Z}_p = \{0\}$; see 51.6. Furthermore, $|q|_p = 1$, so q is a unit of the ring $\mathbb{Z}_p$, and $q\mathbb{Z}_p = \mathbb{Z}_p$.   □

In particular, the group $\mathbb{Z}_p^+$ is reduced, i.e., $\{0\}$ is the only divisible subgroup of $\mathbb{Z}_p^+$. Lemma 52.6 shows that the abstract groups $\mathbb{Z}_p^+$ with $p \in \mathbb{P}$ are mutually not isomorphic; in fact, more is true.

52.7 Corollary *Let p and q be different primes. Then every group homomorphism $\varphi : \mathbb{Z}_p \to \mathbb{Z}_q$ is trivial.*

Proof The image $\varphi(\mathbb{Z}_p) = \varphi(q\mathbb{Z}_p) = q\varphi(\mathbb{Z}_p)$ is a q-divisible subgroup of $\mathbb{Z}_q^+$, hence trivial by 52.6.   □

52.8 Proposition *Each endomorphism of the group $\mathbb{Z}_p^+$ is continuous and of the form $z \mapsto az$ for some $a \in \mathbb{Z}_p$. The endomorphism ring $\operatorname{End}\mathbb{Z}_p^+$ is isomorphic to the ring $\mathbb{Z}_p$ via the isomorphism*

$$\operatorname{End}\mathbb{Z}_p^+ \to \mathbb{Z}_p : \alpha \mapsto \alpha(1).$$

Proof By 51.10 the sets $p^n\mathbb{Z}_p$ with $n \in \mathbb{N}$ form a neighbourhood base at 0. If $\alpha \in \operatorname{End}\mathbb{Z}_p^+$, then $\alpha(x + p^n\mathbb{Z}_p) = \alpha(x) + p^n\alpha(\mathbb{Z}_p) \subseteq \alpha(x) + p^n\mathbb{Z}_p$ for each $x \in \mathbb{Z}_p$, hence α is continuous (in fact, non-expanding with respect to the p-adic absolute value).

One has $\alpha(z) = \alpha(z \cdot 1) = z\alpha(1)$ for each integer $z \in \mathbb{Z}$. This equation holds also for all $z \in \mathbb{Z}_p$, because $\mathbb{Z}$ is dense in $\mathbb{Z}_p$ by 51.9 and because α and the multiplication of $\mathbb{Z}_p$ are continuous. Hence the evaluation map $\alpha \mapsto \alpha(1)$ is injective. It is also surjective, since $z \mapsto za$ is an endomorphism of $\mathbb{Z}_p^+$ for every $a \in \mathbb{Z}_p$.   □

Now we determine the endomorphisms, the automorphisms and the characters of the Prüfer group C_{p^∞}; see Section 63 for general information on characters. Recall that the endomorphism ring of the finite cyclic group C_{p^n} is isomorphic to the ring $\mathbb{Z}/p^n\mathbb{Z}$ (see LANG 1993 II Theorem 2.3).

52.9 Theorem *The endomorphism ring of the group C_{p^∞} is isomorphic to the ring $\mathbb{Z}_p$, even as topological rings, if we endow $\operatorname{End}C_{p^\infty}$ with the compact-open topology derived from the discrete topology on C_{p^∞}.*

The automorphism group of C_{p^∞} is isomorphic to the group $\mathbb{Z}_p^\times = \left\{ x \in \mathbb{Q}_p \mid |x|_p = 1 \right\}$ of units of $\mathbb{Z}_p$ (see 53.3).

The character group $C_{p^\infty}^* = \operatorname{Hom}(C_{p^\infty}, \mathbb{T})$, as usual endowed with the compact-open topology, is isomorphic to the topological group $\mathbb{Z}_p^+$.

Proof Each $a \in \mathbb{Z}_p$ yields an endomorphism α_a of $\mathbb{Q}_p/\mathbb{Z}_p = C_{p^\infty}$ simply by multiplication: $\alpha_a(x + \mathbb{Z}_p) := ax + \mathbb{Z}_p$. The ring homomorphism $\alpha : \mathbb{Z}_p \to \operatorname{End}C_{p^\infty} : a \mapsto \alpha_a$ is injective, since $\alpha_a = 0$ implies $a\mathbb{Q}_p \subseteq \mathbb{Z}_p$, hence $a = 0$.

Now we show that α is also surjective. Let φ be an endomorphism of $C_{p^\infty} = \mathbb{Q}_p/\mathbb{Z}_p$. This group can be written as the union of the groups $\mathbb{Z}_p p^{-n}/\mathbb{Z}_p$ with $n \geq 0$. The unique (cyclic) subgroup $\mathbb{Z}_p p^{-n}/\mathbb{Z}_p$ of order p^n in $\mathbb{Q}_p/\mathbb{Z}_p$ is generated by $p^{-n} + \mathbb{Z}_p$ and invariant under φ, hence we find integers $z_n \in \{0, 1, \ldots, p^n - 1\}$ with $\varphi(p^{-n} + \mathbb{Z}_p) = z_n p^{-n} + \mathbb{Z}_p$ for each $n \geq 0$. Applying φ to the relation $p(p^{-n-1} + \mathbb{Z}_p) = p^{-n} + \mathbb{Z}_p$ in $\mathbb{Q}_p/\mathbb{Z}_p$ we obtain $z_{n+1} \equiv z_n \bmod p^n$ for $n \geq 0$, hence $z_{n+1} = z_n + c_n p^n$ with suitable coefficients $c_n \in \{0, 1, \ldots, p - 1\}$. Now the p-adic integer

$a := \sum_{n=0}^{\infty} c_n p^n$ satisfies $\alpha_a(p^{-m} + \mathbb{Z}_p) = (\sum_{j=0}^{m-1} c_j p^j) p^{-m} + \mathbb{Z}_p = z_m p^{-m} + \mathbb{Z}_p = \varphi(p^{-m} + \mathbb{Z}_p)$ for each $m \geq 0$, hence $\alpha_a = \varphi$.

For a discrete topological space X, the compact-open topology of X^X coincides with the 'point-open' topology, which is just the product topology. Since we endow C_{p^∞} with the discrete topology, the compact-open topology of $\operatorname{End} C_{p^\infty}$ has a subbasis consisting of the sets $\{ \varphi \mid \varphi(x + \mathbb{Z}_p) = y + \mathbb{Z}_p \}$ with $x, y \in \mathbb{Q}_p$. Under α^{-1} such a set corresponds to $\{ a \in \mathbb{Z}_p \mid ax \in y + \mathbb{Z}_p \}$, which is open in $\mathbb{Z}_p$. Hence the bijection $\alpha : \mathbb{Z}_p \to \operatorname{End} C_{p^\infty}$ is continuous, in fact a homeomorphism, as $\mathbb{Z}_p$ is compact.

The image of any group homomorphism $C_{p^\infty} \to \mathbb{T} = \mathbb{R}/\mathbb{Z}$ consists of elements whose order is a power of p, hence that image is contained in the p-primary torsion subgroup C_{p^∞} of $\mathbb{T}$. Thus $\operatorname{Hom}(C_{p^\infty}, \mathbb{T}) = \operatorname{Hom}(C_{p^\infty}, C_{p^\infty})$ is just the additive group of $\operatorname{End} C_{p^\infty}$.   □

52.10 Some infinite Galois groups For a prime p and $n \in \mathbb{N}$ we denote by $W_{p,n} = \{ \exp(2\pi i a/p^n) \mid a \in \{0, 1, \ldots, p^n - 1\} \}$ the group of all complex roots of unity of order dividing p^n. The automorphism group $\operatorname{Aut} \mathbb{Q}(W_{p,n})$ of the cyclotomic field $\mathbb{Q}(W_{p,n})$ is isomorphic to the group of units $(\mathbb{Z}/p^n\mathbb{Z})^\times$, via restriction to the cyclic group $W_{p,n}$ of order p^n; compare COHN 2003a 7.7.5 or LANG 1993 VI Theorem 3.1. The multiplicative group $W_p := \bigcup_n W_{p,n}$ is isomorphic to C_{p^∞}, and the field $\mathbb{Q}(W_p) = \bigcup_n \mathbb{Q}(W_{p,n})$ is an infinite Galois extension of $\mathbb{Q}$. Each field automorphism of $\mathbb{Q}(W_p)$ restricts to a group automorphism of W_p. The restriction $\operatorname{Aut} \mathbb{Q}(W_p) \to \operatorname{Aut}(W_p)$ is injective, and also surjective by the above remark on $\operatorname{Aut} \mathbb{Q}(W_{p,n})$. Thus we infer from 52.9 that

$$\operatorname{Aut} \mathbb{Q}(W_p) \cong \operatorname{Aut} C_{p^\infty} \cong \mathbb{Z}_p^\times .$$

For the group-theoretic structure of $\mathbb{Z}_p^\times$ see 53.3.

Denote by $\mathbb{A} := \mathbb{Q}(\exp(2\pi i \mathbb{Q})) = \mathbb{Q}(\bigcup_p W_p)$ the field generated by all complex roots of unity. The Kronecker–Weber Theorem (compare CASSELS 1986 10.12, LANG 1970 X §3 or NEUKIRCH 1992 V.1.10) implies that $\mathbb{A}|\mathbb{Q}$ is the largest Galois extension of $\mathbb{Q}$ with abelian Galois group. The Galois group $\operatorname{Aut} \mathbb{A}$ is isomorphic to the Cartesian product $\bigtimes_p \operatorname{Aut} \mathbb{Q}(W_p)$; see RIBENBOIM 1999 11.2. From the previous paragraph we obtain that

$$\operatorname{Aut} \mathbb{A} \cong \bigtimes_{p \in \mathbb{P}} \mathbb{Z}_p^\times = \widehat{\mathbb{Z}}^\times$$

is isomorphic to the group of units of the product ring $\widehat{\mathbb{Z}} := \bigtimes_p \mathbb{Z}_p$ (see also WEIL 1967 XIII §4 Corollary 2).

This product ring $\widehat{\mathbb{Z}}$ is isomorphic to the ring $\mathrm{End}(\mathbb{Q}^+/\mathbb{Z})$ considered in 31.11 and 31.12: by 1.28, we have the direct decomposition $\mathbb{Q}/\mathbb{Z} \cong \bigoplus_p C_{p^\infty}$, and each primary component C_{p^∞} is invariant under every endomorphism, hence $\mathrm{End}(\mathbb{Q}^+/\mathbb{Z}) \cong \mathsf{X}_{p\in\mathbb{P}}\, \mathrm{End}\, C_{p^\infty} \cong \mathsf{X}_{p\in\mathbb{P}}\, \mathbb{Z}_p$ by 52.9. The ring $\widehat{\mathbb{Z}}$ may also be described as the completion (44.5) of $\mathbb{Z}$ with respect to the topology generated by all non-zero ideals.

In fact, infinite Galois groups carry a natural compact, totally disconnected (or profinite) group topology, and the isomorphisms obtained here can be read as isomorphisms of topological groups; see RIBENBOIM 1999 11.1, 11.2.

For prime numbers p, q let $F_{p,q} := \bigcup_{n\in\mathbb{N}} \mathbb{F}_{p^{q^n}}$ (in a fixed algebraic closure $\mathbb{F}_p^\natural$ of $\mathbb{F}_p$). The automorphism group of the field $F_{p,q}$ is isomorphic to $\mathbb{Z}_q^+$ (Exercise 7). The algebraic closure $\mathbb{F}_p^\natural$ is generated by its subfields $F_{p,q}$ with $q \in \mathbb{P}$, and one can show that

$$\mathrm{Aut}\,\mathbb{F}_p^\natural \cong \mathsf{X}_{q\in\mathbb{P}}\, \mathbb{Z}_q^+ \ ;$$

see RIBENBOIM 1999 11.2 pp. 316–320. The product group appearing here is the additive group of the ring $\widehat{\mathbb{Z}}$ as above, and the Frobenius automorphism $x \mapsto x^p$ generates a dense cyclic subgroup of $\mathrm{Aut}\,\mathbb{F}_p^\natural$; compare BOURBAKI 1990 V.12.3, NEUKIRCH 1992 IV.2.

Finally we show that the compact group topology of $\mathbb{Z}_p$ is uniquely determined by the abstract group $\mathbb{Z}_p^+$; the following result is due to SOUNDARARAJAN 1969; see also CORWIN 1976, KALLMAN 1976.

52.11 Theorem *The only locally compact Hausdorff group topologies of $\mathbb{Z}_p^+$ are the natural p-adic topology, which is compact, and the discrete topology.*

Proof For every $n \in \mathbb{N}$ the map $x \mapsto p^n x : \mathbb{Z}_p \to \mathbb{Z}_p$ is continuous with respect to any group topology τ on $\mathbb{Z}_p$. If τ is a compact Hausdorff topology, then the image $p^n\mathbb{Z}_p$ is compact and hence closed in $\mathbb{Z}_p$ with respect to τ. By 51.7 the index $|\mathbb{Z}_p/p^n\mathbb{Z}_p| = |\mathbb{Z}/p^n\mathbb{Z}| = p^n$ is finite, and $p^n\mathbb{Z}_p$ is also open with respect to τ. The p-adic topology τ_p of $\mathbb{Z}_p$ has $\{p^n\mathbb{Z}_p \mid n \in \mathbb{N}\}$ as a neighbourhood base at 0. Therefore the identity id : $(\mathbb{Z}_p, \tau) \to (\mathbb{Z}_p, \tau_p)$ is continuous at 0, hence everywhere. This bijective map between two compact Hausdorff spaces is also closed, hence a homeomorphism, which means that $\tau = \tau_p$.

Now let τ be a non-discrete locally compact Hausdorff group topology of $\mathbb{Z}_p^+$. We shall show that τ is compact, which completes the proof. Lemma 52.6 implies that the divisible groups $\mathbb{R}^n$ with $n > 0$ do not occur as subgroups of $\mathbb{Z}_p^+$, hence the Splitting Theorem 63.14 entails that $(\mathbb{Z}_p, \tau)$ has a compact open subgroup $C \neq \{0\}$. We are going to use the character group C^* as defined in Section 63 and the Pontryagin duality $C^{**} \cong C$; see 63.20.

Since $\mathbb{Z}_p$ and C are torsion free, the character group C^* is divisible by 63.32, and C^* is discrete by 63.5. If C^* were torsion free, then $C^{**} \cong C$ would be divisible, again by 63.32; but this would contradict 52.6. Thus C^* contains a torsion element of prime order q, hence by divisibility also a Prüfer subgroup $\Pi \cong C_{q^\infty}$ (compare 31.7). By 1.23 and 1.22 this group Π is a direct summand of C^*, hence Π^* is a subgroup (even a direct summand) of $C^{**} \cong C$ (see 63.8). From 52.9 we know that $\Pi^* \cong \mathbb{Z}_q^+$, and 52.7 implies that $q = p$.

We conclude that C contains a subgroup $P \cong \Pi^* \cong \mathbb{Z}_p^+$. Each (abstract) isomorphism $\mathbb{Z}_p \to P$ is a non-zero endomorphism of $\mathbb{Z}_p$, hence the image P has finite index in $\mathbb{Z}_p$ by 52.8. Thus $C \supseteq P$ has finite index in $\mathbb{Z}_p$, whence τ is compact. $\square$

Exercises

(1) Show that $\mathbb{Q}_p^+/\mathbb{Q} \cong \mathbb{R}^+$ as abstract groups.

(2) Each subgroup of finite index in $\mathbb{Z}_p^+$ is of the form $p^n \mathbb{Z}_p$ with $n \in \mathbb{N}_0$, and $\mathbb{Z}_p^+/p^n \mathbb{Z}_p$ is cyclic of order p^n.

(3) The elements of finite order in $\mathbb{Z}_p/\mathbb{Z}$ form a subgroup isomorphic to the direct sum of all Prüfer groups C_{q^∞} with $q \in \mathbb{P} \setminus \{p\}$.

(4) The composition of the natural maps $\mathbb{Q}_p \to \mathbb{Q}_p/\mathbb{Z}_p \cong C_{p^\infty} \to \mathbb{C}^\times$ is the function $\chi : \mathbb{Q}_p \to \mathbb{C}^\times$ with $\chi(\sum_k c_k p^k) = \exp(2\pi i \sum_{k<0} c_k p^k)$ for $c_k \in \{0, 1, \ldots, p-1\}$. Show that χ is a continuous homomorphism of $\mathbb{Q}_p^+$ into the multiplicative group $\mathbb{C}^\times$, and determine the kernel and the image of χ. Conclude that χ is a non-trivial character of $\mathbb{Q}_p^+$ (the Tate character).

(5) Let F be a locally compact field which is neither discrete nor indiscrete, let $\chi_1 : F^+ \to \mathbb{R}/\mathbb{Z}$ be a non-trivial character, and define χ_a by $\chi_a(x) = \chi(ax)$ for $a, x \in F$. Show that the map $a \mapsto \chi_a$ is an isomorphism of the topological group F^+ onto the character group of F^+. In particular, the character group of $\mathbb{Q}_p^+$ is isomorphic to $\mathbb{Q}_p^+$.

(6) The p-adic solenoid S_p can be defined as the topological quotient group $S_p := (\mathbb{R} \times \mathbb{Z}_p)/\{(z, z) \mid z \in \mathbb{Z}\}$. Show that S_p is a compact connected Hausdorff group, and determine the torsion subgroup of S_p.

(7) Let p, q be prime numbers. Show that the automorphism group of the field $\bigcup_n \mathbb{F}_{p^{q^n}}$ is isomorphic to $\mathbb{Z}_q^+$.

53   The multiplicative group of *p*-adic numbers

Here we consider the structure of the multiplicative group $\mathbb{Q}_p^\times$ of non-zero
p-adic numbers. It turns out (in Theorem 53.2) that $\mathbb{Q}_p^+$ and $\mathbb{Q}_p^\times$ have
open subgroups which are isomorphic (as topological groups), similarly
as for $\mathbb{R}$; compare 9.2. Furthermore we show in 53.5 that $\mathbb{Q}_p$ has no field
endomorphism apart from the identity (like $\mathbb{R}$; see 6.4).

Let *p* always denote a prime number.

53.1 Proposition (i) *For each* $n \in \mathbb{N}$, *the set* $U_n := 1 + p^n\mathbb{Z}_p$ *is a*
multiplicative subgroup of $\mathbb{Q}_p^\times$, *and* U_n *is open and closed in* $\mathbb{Q}_p$.
Furthermore $U_n/U_{n+1} \cong \mathbb{F}_p^+ \cong C_p$.

(ii) *The field* $\mathbb{Q}_p$ *contains a root of unity* ζ *of order* $p - 1$, *and the set*
$\{0, 1, \zeta, \zeta^2, \ldots, \zeta^{p-2}\}$ *is a system of representatives for* $\mathbb{Z}_p/p\mathbb{Z}_p$.

(iii) *The group* $\mathbb{Q}_p^\times$ *is the direct product* $\mathbb{Q}_p^\times = \langle p \rangle \times \langle \zeta \rangle \times U_1$ *of the*
three subgroups $\langle p \rangle$, $\langle \zeta \rangle$ *and* U_1, *and* $\mathbb{Z}_p^\times = \{x \in \mathbb{Q}_p \mid |x|_p = 1\} =$
$\langle \zeta \rangle \times U_1$.

Proof (i) Clearly the natural ring epimorphism $\mathbb{Z}_p \to \mathbb{Z}_p/p^n\mathbb{Z}_p \neq \{0\}$
restricted to $\mathbb{Z}_p^\times = \mathbb{Z}_p \smallsetminus p\mathbb{Z}_p$ is multiplicative, hence the kernel $U_n =$
$1 + p^n\mathbb{Z}_p$ of that restriction is a subgroup of $\mathbb{Q}_p^\times$. By 51.6, the subgroup
U_n is open and closed in $\mathbb{Q}_p$.

The mapping $f : U_n \to \mathbb{Z}_p^+/p\mathbb{Z}_p$ defined by $f(1 + p^n x) = x + p\mathbb{Z}_p$ for
$x \in \mathbb{Z}_p$ is an epimorphism of groups, since f maps $(1 + p^n x)(1 + p^n y) =$
$1 + p^n(x + y + p^n xy)$ onto $x + y + p^n xy + p\mathbb{Z}_p = x + y + p\mathbb{Z}_p$. The kernel
of f is $1 + p^n p\mathbb{Z}_p = U_{n+1}$, hence $U_n/U_{n+1} \cong \mathbb{Z}_p/p\mathbb{Z}_p \cong \mathbb{F}_p^+$ by 51.7.

(ii) Since $\mathbb{F}_p^\times$ is cyclic (Section 64, Exercise 1), we find an integer
$i \in \{1, 2, \ldots, p - 1\}$ which represents a generator of $\mathbb{F}_p^\times = (\mathbb{Z}/p\mathbb{Z})^\times$.
For each $n \in \mathbb{N}$ the group of units of the finite ring $\mathbb{Z}/p^n\mathbb{Z}$ has order
$p^n - p^{n-1}$ (see LANG 1993 II.2), and *i* represents such a unit, hence
$i^{p^n - p^{n-1}} \equiv 1 \bmod p^n$ and

$$i^{p^n} \equiv i^{p^{n-1}} \bmod p^n .$$

This means that the two *p*-adic series representing i^{p^n} and $i^{p^{n-1}}$, respectively, have the same initial coefficients $c_0, c_1, \ldots, c_{n-1}$. Thus there exists
a sequence of integer coefficients $c_k \in \{0, 1, \ldots, p - 1\}$ with

$$i^{p^n} \equiv \sum_{k=0}^{n-1} c_k p^k \bmod p^n$$

for each $n \geq 0$.

We define $\zeta \in \mathbb{Z}_p$ by $\zeta = \sum_{k \geq 0} c_k p^k$ (thus $\zeta = \lim_{n \to \infty} i^{p^n}$); note
that $\zeta \neq 0$, as $c_0 = i$. Modulo $p^{n+1}\mathbb{Z}_p$ we have $\zeta \equiv i^{p^n}$ and therefore
$\zeta^p - \zeta \equiv i^{p^{n+1}} - i^{p^n} \equiv 0$ for each $n \in \mathbb{N}$, whence $\zeta^p - \zeta = 0$ and $\zeta^{p-1} = 1$.

As $\zeta + p\mathbb{Z}_p = i + p\mathbb{Z}_p$ has multiplicative order $p-1$ in $(\mathbb{Z}_p/p\mathbb{Z}_p)^\times \cong \mathbb{F}_p^\times$ (see 51.7), we infer that ζ is a root of unity of order exactly $p-1$.

Other methods to find ζ can be based on Hensel's Lemma or Newton's rule; see COHN 2003a p. 326, compare also SERRE 1973 p. 16 or WEIL 1967 I.4 Theorem 7 p. 16.

(iii) The group epimorphism $| \ |_p : \mathbb{Q}_p^\times \to p^{\mathbb{Z}}$ has the kernel $\{ x \in \mathbb{Q}_p \mid |x|_p = 1 \} = \mathbb{Z}_p^\times$, hence $\mathbb{Q}_p^\times = \langle p \rangle \times \mathbb{Z}_p^\times$. We infer from (ii) that $\mathbb{Z}_p^\times$ is the disjoint union of the cosets $\zeta^j + p\mathbb{Z}_p = \zeta^j(1 + p\mathbb{Z}_p) = \zeta^j U_1$ with $0 \le j \le p - 2$ (note that $\zeta\mathbb{Z}_p = \mathbb{Z}_p$). Hence $\mathbb{Z}_p^\times = \langle \zeta \rangle \times U_1$.   □

The following result is crucial for relating the multiplicative group $\mathbb{Q}_p^\times$ to the additive group of p-adic numbers. Often this is achieved using the series for the exponential function exp (see Exercise 2). The proof below employs a variant of the usual exponential function: we use the basis $1+p$ instead of e (see WARNER 1989 22.2f for exponential functions with arbitrary basis).

53.2 Theorem *If $p \ne 2$, then $U_1 = 1 + p\mathbb{Z}_p \le \mathbb{Q}_p^\times$ is isomorphic to $\mathbb{Z}_p^+$ as a topological group. For $p = 2$, we have $U_1 = 1 + 2\mathbb{Z}_2 = \langle -1 \rangle \times U_2$, and $U_2 = 1 + 4\mathbb{Z}_2$ is isomorphic to $\mathbb{Z}_2^+$ as a topological group.*

Proof The mapping $\beta_0 : \mathbb{Z} \to U_1 : z \mapsto (1+p)^z$ is a group monomorphism of $\mathbb{Z}^+$ into the multiplicative group U_1. Moreover β_0 is continuous when $\mathbb{Z}$ is provided with the p-adic topology, since $|(1 + p)^z - 1|_p \le p^{-1}|z|_p$ by 37.5. As $U_1 = 1 + p\mathbb{Z}_p$ is compact, β_0 has a unique extension to a continuous group homomorphism $\beta : \mathbb{Z}_p \to U_1$; see 43.9, 43.23. (By Exercise 3, this extension is given by $\beta(x) = (1+p)^x := \sum_{n \ge 0} \binom{x}{n} p^n$, but we do not need this series representation.)

The kernel of β is a closed subgroup of $\mathbb{Z}_p^+$. Furthermore β is injective on the dense subset $\mathbb{N}$, hence the kernel cannot be open in $\mathbb{Z}_p$. Thus by 52.3(i, ii) the kernel is trivial and β is injective on $\mathbb{Z}_p$.

The image $\beta(\mathbb{Z}_p)$ is the topological closure of the cyclic group $\beta(\mathbb{Z}) = (1 + p)^{\mathbb{Z}}$. We have $|U_1/U_{n+1}| = p^n$ by 53.1(i). If $p \ne 2$ and $n \ge 1$, then 37.5 shows that

$$|(1 + p)^{p^{n-1}} - 1|_p = |p|_p|p^{n-1}|_p = p^{-n} ,$$

hence $(1 + p)^{p^{n-1}} \notin 1 + p^{n+1}\mathbb{Z}_p = U_{n+1}$ (see also Exercise 4). Thus $(1 + p)U_{n+1}$ has order p^n in U_1/U_{n+1} and is therefore a generator of U_1/U_{n+1}. Since the subgroups U_n with $n \ge 1$ form a neighbourhood base at 1 in U_1, this implies that $1 + p$ generates a dense subgroup of U_1, whence $\beta(\mathbb{Z}_p) = U_1$. By compactness β is a homeomorphism.

It remains to deal with the case $p = 2$. Here $(1 + p)^{\mathbb{Z}} = 3^{\mathbb{Z}}$ is not dense in U_1 (in fact, $3U_3$ is an involution in the elementary abelian group $U_1/U_3 \cong (\mathbb{Z}/8\mathbb{Z})^{\times}$ of order 4). We have $U_1 = \langle -1 \rangle \times U_2$, because $\mathbb{Z}_2 = 2\mathbb{Z}_2 \cup (1+2\mathbb{Z}_2)$ by 51.7, hence $U_1 = 1 + 2\mathbb{Z}_2 = (1+4\mathbb{Z}_2) \cup (3+4\mathbb{Z}_2) = U_2 \cup -U_2$. We replace β_0 by the mapping $\gamma_0 : \mathbb{Z} \to U_2 : z \mapsto (1+4)^z$ and obtain by extension as above a continuous monomorphism $\gamma : \mathbb{Z}_2 \to U_2$ of groups. The image $\gamma(\mathbb{Z}_2)$ is the topological closure of $5^{\mathbb{Z}}$ in U_2. From 37.5 we infer that

$$|(1+4)^{2^{n-1}} - 1|_2 = 2^{-(n+1)}$$

for $n \geq 1$, hence $(1+4)^{2^{n-1}} \notin 1 + 2^{n+2}\mathbb{Z}_2 = U_{n+2}$ (see also Exercise 4). Thus 5 represents an element of order 2^n in the quotient group U_2/U_{n+2}, which has order 2^n by 53.1(i). Therefore 5 generates a dense subgroup of U_2, and $\gamma : \mathbb{Z}_2 \to U_2$ is surjective.   □

53.3 Corollary *Let ζ be as in 53.1, and write C_n for a cyclic group of order n. If $p \neq 2$, then*

$$\mathbb{Q}_p^{\times} = \langle p \rangle \times \langle \zeta \rangle \times (1 + p\mathbb{Z}_p) \cong \mathbb{Z} \times C_{p-1} \times \mathbb{Z}_p^{+} \quad \text{and} \quad \mathbb{Z}_p^{\times} \cong C_{p-1} \times \mathbb{Z}_p^{+}.$$

For $p = 2$, we have

$$\mathbb{Q}_2^{\times} = \langle 2 \rangle \times \langle -1 \rangle \times (1 + 4\mathbb{Z}_2) \cong \mathbb{Z} \times C_2 \times \mathbb{Z}_2^{+} \quad \text{and} \quad \mathbb{Z}_2^{\times} \cong C_2 \times \mathbb{Z}_2^{+}.$$

The isomorphisms can be read as isomorphisms of topological groups, where the cyclic factors are discrete.

Proof This is a direct consequence of 53.1 and 53.2.   □

53.4 Corollary *Let p and q be distinct primes. Then the abstract groups $\mathbb{Q}_p^{\times}$ and $\mathbb{Q}_q^{\times}$ are not isomorphic. In particular, the fields $\mathbb{Q}_p$ and $\mathbb{Q}_q$ are not isomorphic.*

Proof By 53.3 and 52.6, $\mathbb{Q}_p^{\times}$ contains an infinite (in fact, uncountable) r-divisible subgroup only for the prime $r = p$.

Alternatively, 53.3 shows that the torsion subgroup of $\mathbb{Q}_p^{\times}$ (i.e. the group of all roots of unity of $\mathbb{Q}_p$) has order $p - 1$ or 2; then it remains to consider the case $p = 2$, $q = 3$. By 54.1, the group $\mathbb{Q}_3^{\times}$ consists of four square classes, but $\mathbb{Q}_2^{\times}$ has eight square classes.   □

53.5 Theorem *Let $\alpha : \mathbb{Q}_p \to \mathbb{Q}_q$ be a ring homomorphism, with $\alpha(1) = 1$. Then $p = q$ and α is the identity. In particular, the identity is the only non-zero field endomorphism of $\mathbb{Q}_p$.*

Proof The kernel of α is a proper ideal, hence α is injective. The image $\alpha(\mathbb{Q}_p^\times) \cong \mathbb{Q}_p^\times$ contains a subgroup isomorphic to $\mathbb{Z}_p^+$ by 53.2. But $\mathbb{Q}_q^\times$ has such a subgroup only if $q = p$; see 53.3 and 52.6.

Results 53.2, 53.3 and 52.6 imply that $U_1 = 1 + p\mathbb{Z}_p$ is the set of all non-zero p-adic numbers that admit r-th roots for each $r \in \mathbb{N}$ not divisible by p:

$$1 + p\mathbb{Z}_p = \left\{ x \in \mathbb{Q}_p^\times \mid \forall_{r\in\mathbb{N}\setminus p\mathbb{N}} \exists_y \ x = y^r \right\} = \bigcap_{r\in\mathbb{N}\setminus p\mathbb{N}} (\mathbb{Q}_p^\times)^r .$$

As α is multiplicative, we infer that $\alpha(1+p\mathbb{Z}_p) \subseteq 1+p\mathbb{Z}_p$. Since α is also additive, we obtain $\alpha(p\mathbb{Z}_p) \subseteq p\mathbb{Z}_p$ and then $\alpha(p^n\mathbb{Z}_p) \subseteq p^n\mathbb{Z}_p$ for $n \geq 1$. (See Exercise 5 for an alternative proof of the inclusion $\alpha(\mathbb{Z}_p) \subseteq \mathbb{Z}_p$.) This shows that α is continuous at 0, hence everywhere (by additivity).

Each rational number is fixed by α, and $\mathbb{Q}$ is dense in $\mathbb{Q}_p$ by 51.5, whence α is the identity. $\square$

Exercises

(1) Show that $\mathbb{Z}_p^\times/U_1 \cong \mathbb{F}_p^\times$ and $U_n \cong \mathbb{Z}_p^+$ for $n \geq 2$.

(2) Determine the domain of p-adic convergence of the exponential series $\exp x = \sum_{n\geq 0} x^n/n!$.

(3) Consider the binomial coefficient $\binom{x}{n} = x(x-1)\cdots(x-n+1)/n!$ with $n \in \mathbb{N}_0$ as a polynomial in x, where $\binom{x}{0} := 1$. Show that the binomial series $\beta(x) = (1+p)^x := \sum_{n\geq 0} \binom{x}{n} p^n$ converges for $x \in \mathbb{Z}_p$ and defines a continuous homomorphism $\beta : \mathbb{Z}_p \to U_1$ of groups.

(4) Let $n \in \mathbb{N}$. Show that $(1+p)^{p^{n-1}} \equiv 1 + p^n \bmod p^{n+1}$ for $p \neq 2$, and $(1+4)^{2^{n-1}} \equiv 1 + 2^{n+1} \bmod 2^{n+2}$.

(5) Show that $\mathbb{Z}_p = \{ x \in \mathbb{Q}_p \mid 1 + px^2 \text{ is a square in } \mathbb{Q}_p \}$ for $p \neq 2$, and that $\mathbb{Z}_2 = \{ x \in \mathbb{Q}_2 \mid 1 + 2x^3 \text{ is a cube in } \mathbb{Q}_2 \}$. Use these facts to give another proof of 53.5.

(6) Let $p \neq 2$ and $n \in \mathbb{N} \setminus p\mathbb{N}$. Show that a non-zero p-adic number $p^k \sum_{m\geq 0} c_m p^m$ with $c_0 \neq 0$ is an nth power in $\mathbb{Q}_p^\times$ if, and only if, n divides k and c_0 is an nth power modulo p.

54   Squares of p-adic numbers and quadratic forms

For any field F we denote by $F^\square := \{ x^2 \mid x \in F^\times \}$ the group of all non-zero squares of F, and we call each coset $aF^\square \in F^\times/F^\square$ a *square class* of F. The field $\mathbb{R}$ of real numbers has only two square classes, the positive and the negative real numbers. For p-adic numbers the situation is slightly more complicated. As usual, p is a prime number.

54.1 Theorem *For $p \neq 2$ we have $\mathbb{Q}_p^\square = \langle p^2 \rangle \times \langle \zeta^2 \rangle \times (1 + p\mathbb{Z}_p)$, and $\mathbb{Q}_p^\times / \mathbb{Q}_p^\square$ consists of four square classes, with representatives 1, ζ, p, ζp, where ζ is a root of unity of order $p - 1$ as in 53.1. Another system of representatives is given by 1, r, p, rp, where $r \in \mathbb{Z}$ is not a square modulo p.*

For $p = 2$ we have $\mathbb{Q}_2^\square = \langle 4 \rangle \times (1 + 8\mathbb{Z}_2)$, and $\mathbb{Q}_2^\times / \mathbb{Q}_2^\square$ consists of eight square classes, with representatives ± 1, ± 2, ± 3, ± 6; also ± 1, ± 2, ± 5, ± 10 are representatives.

Proof For $p \neq 2$, this product decomposition of $\mathbb{Q}_p^\square$ follows directly from the decomposition of $\mathbb{Q}_p^\times$ in 53.3; note that $1 + p\mathbb{Z}_p \cong \mathbb{Z}_p^+$ is 2-divisible by 52.6. For $p = 2$, result 53.3 implies that $\mathbb{Q}_2^\square = \langle 4 \rangle \times U_2^2$, where $U_n = 1 + 2^n \mathbb{Z}_2$, and 53.1(i) yields $|U_2 / U_3| = 2$. Hence $U_2^2 \subseteq U_3$, and we have in fact equality, as $U_2 \cong \mathbb{Z}_2^+$ by 53.2 and $|\mathbb{Z}_2 / 2\mathbb{Z}_2| = 2$ by 51.7.

The systems of representatives are obtained by combining coset representatives of factors of $\mathbb{Q}_2^\square$ in the corresponding factors of $\mathbb{Q}_p^\times$. $\qquad\square$

54.2 Corollary *The field $\mathbb{Q}_p$ is not formally real, hence it cannot be made into an ordered field.*

Proof An ordered field has only ± 1 as roots of unity. By 53.1(ii), it remains to consider the two cases $p = 2, 3$. Using 54.1 (or 53.2 and 52.6) we see that $-2 = 1 + 3(-1)$ is a square in $\mathbb{Q}_3$ and that $-7 = 1 + 8(-1)$ is a square in $\mathbb{Q}_2$ (more generally, $1 - p^3 \in \mathbb{Q}_p^\square$ for any prime p). Hence none of these fields is formally real; compare 12.3.

We give a second, topological proof. Assume that P is a domain of positivity (11.3) of $\mathbb{Q}_p$. Then both P and $-P$ are unions of square classes of $\mathbb{Q}_p$, hence open in $\mathbb{Q}_p$ by 54.1. Thus $P \cup \{0\}$ is topologically closed in $\mathbb{Q}_p$ and therefore contains the infinite sum $\sum_{n \geq 0} (p - 1)p^n = -1$, which is a contradiction (alternatively, $P \cup \{0\}$ contains $\mathbb{N}$ and its topological closure $\mathbb{Z}_p \supseteq \mathbb{Z}$; see 51.9). $\qquad\square$

54.3 Corollary *If $p \neq 2$, then $\mathbb{Q}_p$ has precisely three quadratic field extensions, namely $\mathbb{Q}_p(\sqrt{a})$ with $a \in \{p, r, pr\}$, where $r \in \mathbb{Z}$ is not a square modulo p. The field $\mathbb{Q}_2$ has precisely seven quadratic extensions, namely $\mathbb{Q}_2(\sqrt{a})$ with $a \in \{-1, \pm 2, \pm 3, \pm 6\}$.*

Proof Every quadratic field extension of $\mathbb{Q}_p$ is of the form $\mathbb{Q}_p(\sqrt{a})$ with $a \in \mathbb{Q}_p \setminus \mathbb{Q}_p^\square$. Hence 54.1 shows that we have listed all quadratic extensions of $\mathbb{Q}_p$. Furthermore, an element $b \in \mathbb{Q}_p^\times$ is a square in $\mathbb{Q}_p(\sqrt{a})$ if, and only if, b belongs to the square class of 1 or of a in $\mathbb{Q}_p$. This shows that the listed extensions are distinct. $\qquad\square$

Corollary 54.3 is a special case of the following general fact: the field $\mathbb{Q}_p$ has only finitely many field extensions of given degree n, for each $n \in \mathbb{N}$; see 58.2.

Now we consider sums of squares in $\mathbb{Q}_p$ and, more generally, quadratic forms over $\mathbb{Q}_p$.

54.4 Lemma Let $a \in \mathbb{Q}_p^\times$ and $S_a := \{\, x^2 - ay^2 \mid x, y \in \mathbb{Q}_p \,\}$. If a is a square in $\mathbb{Q}_p$, then $S_a = \mathbb{Q}_p$. If a is not a square in $\mathbb{Q}_p$, then $S_a \setminus \{0\}$ is a subgroup of index 1 or 2 in $\mathbb{Q}_p^\times$.

Proof If a is a square, then $S_a = \{\, x^2 - y^2 \mid x, y \in \mathbb{Q}_p \,\}$ contains all elements of the form $(y+1)^2 - y^2 = 2y + 1$ with $y \in \mathbb{Q}_p$, hence $S_a = \mathbb{Q}_p$ (this argument is a special case of 54.8 below).

The equation $(x^2 - ay^2)(z^2 - aw^2) = (xz + ayw)^2 - a(xw + yz)^2$, which is related to Diophant's identity (34.4), shows that S_a is closed under multiplication. S_a contains all squares of $\mathbb{Q}_p^\times$, hence the formula $x^{-1} = x(x^{-1})^2$ implies that $S_a \setminus \{0\}$ is a subgroup of $\mathbb{Q}_p^\times$.

Clearly $-a \in S_a$. If $-a$ is not a square, then S_a contains at least two square classes, hence by 54.1 the index of $S_a \setminus \{0\}$ in $\mathbb{Q}_p^\times$ is at most 2, provided that $p \neq 2$. If $-a$ is a square, then $S_a = \{\, x^2 + y^2 \mid x, y \in \mathbb{Q}_p \,\}$ does contain a non-square (and therefore at least two square classes), as we show by an indirect argument: otherwise $S_a = \{\, x^2 \mid x \in \mathbb{Q}_p \,\}$ would be closed under addition, which is a contradiction, as p is not a square by 54.1.

It remains to deal with the case $p = 2$. By 54.1 it suffices to consider the seven cases $a = -1, \pm 2, \pm 3, \pm 6$. Now $1, -a, 1 - a, 4 - a \in S_a$, and one easily verifies in each of these seven cases that $1, -a, 1 - a, 4 - a$ represent at least three distinct square classes (SCHARLAU 1985 p. 188 and CONWAY 1997 p. 120 enumerate the square classes in $S_a \subseteq \mathbb{Q}_2$ for each value a). Since $|\mathbb{Q}_2^\times/\mathbb{Q}_2^\square| = 8$, the subgroup $S_a \setminus \{0\}$ has index at most 2 in $\mathbb{Q}_2^\times$. $\qquad\square$

Let $a \in \mathbb{Q}_p$ be a non-square. Then S_a is the set of all norms of the quadratic Galois extension $\mathbb{Q}_p(\sqrt{a})|\mathbb{Q}_p$. Improving on 54.4, one can show that $S_a \setminus \{0\}$ is always a subgroup of index 2 in $\mathbb{Q}_p^\times$, and one can describe this subgroup in terms of the Hilbert symbol (which is a certain bilinear form on the vector space $\mathbb{Q}_p^\times/\mathbb{Q}_p^\square$ over $\mathbb{F}_2$); see SERRE 1973 p. 19/20, BOREVICH–SHAFAREVICH 1966 Chapter 1 §6, CASSELS 1978 p. 56 or LAM 2005 VI.2.

54.5 Corollary If $p \equiv 1 \bmod 4$, then every p-adic number is a sum of two (non-zero) squares of $\mathbb{Q}_p$.

Proof With ζ as in 54.1, we have $-1 = \zeta^{(p-1)/2}$. Hence -1 is a square in $\mathbb{Q}_p$, and $S_{-1} = \mathbb{Q}_p$ by 54.4. The formulae $a^2 = (3a/5)^2 + (4a/5)^2$ and $0 = 1 + (-1)$ show that every square in $\mathbb{Q}_p$ is a sum of two non-zero squares. $\qquad\square$

In the following we consider quadratic forms $f(x) = \sum_{i=1}^{n} a_i x_i^2$ in n variables over a field F with coefficients $a_i \in F$. We say that f is *isotropic* over F, if $f(x) = 0$ has a solution $x \in F^n$ with $x \neq 0$.

54.6 Lemma *Let $p \neq 2$. Then each quadratic form $ax^2 + by^2 + cz^2$ with $a, b, c \in \mathbb{Q}_p$ and $|a|_p = |b|_p = |c|_p$ is isotropic over $\mathbb{Q}_p$.*

Proof Apart from the trivial case $a = b = c = 0$, we may assume that $|a|_p = |b|_p = |c|_p = 1$, that is, $a, b, c \in \mathbb{Z}_p \smallsetminus p\mathbb{Z}_p$. Then $a + p\mathbb{Z}_p$ and $b + p\mathbb{Z}_p$ are non-zero elements of $\mathbb{Z}_p/p\mathbb{Z}_p \cong \mathbb{F}_p$; see 51.7. By 34.15 we find integers $x, y \in \mathbb{Z}$ such that

$$ax^2 + by^2 \in -c + p\mathbb{Z}_p = -c(1 + p\mathbb{Z}_p) \ .$$

As $p \neq 2$, the set $1 + p\mathbb{Z}_p$ consists of squares of $\mathbb{Q}_p^{\times}$ by 54.1, hence $ax^2 + by^2 = -cz^2$ for some $z \in \mathbb{Q}_p^{\times}$. $\qquad\square$

The following result shows a remarkable contrast between $\mathbb{Q}_p$ and $\mathbb{R}$. To mention just one of the consequences of 54.7, there is no octonion (Cayley–Dickson) division algebra over $\mathbb{Q}_p$ (see LAM 2005 X.2 p. 327).

54.7 Theorem *Each quadratic form $\sum_{i=1}^{n} a_i x_i^2$ with $a_i \in \mathbb{Q}_p$ and $n \geq 5$ is isotropic over $\mathbb{Q}_p$.*

Proof We may assume that $n = 5$ and $a_i \in \mathbb{Q}_p^{\times}$. By a substitution $x_i \mapsto p^{n(i)} x_i$ we can achieve that $|a_i|_p \in \{1, p\}$ for $1 \leq i \leq 5$. Then at least three of the absolute values $|a_i|_p$ coincide, and 54.6 gives the assertion for $p \neq 2$.

For the rest of the proof let $p = 2$. We show by ad hoc arguments that $f(x) = \sum_{i=1}^{5} a_i x_i^2$ is isotropic over $\mathbb{Q}_2$. If the five coefficients a_i belong to five different square classes of $\mathbb{Q}_2^{\times}$, then by 54.1 the quotient of two of them, say a_i and a_j, belongs to the square class of -1, i.e., $-a_i a_j^{-1}$ is a square in $\mathbb{Q}_2$, and then the forms $a_i x_i^2 + a_j x_j^2$ and f are isotropic. Hence we may assume that $a_1 = a_2 = 1$ after changing f by a factor.

The quadratic form $x_1^2 + x_2^2$ takes the values 1 (hence also the squares -7 and -15; see 54.1), $1+1 = 2, 1+4 = 5, 1+9 = 10$. If $a_3 x_3^2 + a_4 x_4^2$ takes one of the values $-1, -2, -5, -10$, then the forms $x_1^2 + x_2^2 + a_3 x_3^2 + a_4 x_4^2$ and f are isotropic. Hence we may assume that $a_3 x_3^2 + a_4 x_4^2$ avoids the square classes of $-1, -2, -5, -10$. Lemma 54.4 implies that $a_3 x_3^2 + a_4 x_4^2$

takes as values at least four square classes of $\mathbb{Q}_2$, hence it takes the
values $1, 2, 5, 10$; compare 54.1.

Thus the sum $x_1^2 + x_2^2 + a_3 x_3^2 + a_4 x_4^2$ takes the values $2 + 5 = 7, -7 + 2 =$
$-5, -7 + 5 = -2$ and $-15 + 5 = -10$, in addition to the values $1, 2, 5, 10$
taken by $x_1^2 + x_2^2$. As -7 is a square, 7 belongs to the square class of -1.
In view of 54.1 this shows that $x_1^2 + x_2^2 + a_3 x_3^2 + a_4 x_4^2$ takes values in any
square class of $\mathbb{Q}_2^\times$, in particular it takes the value $-a_5$. This implies
that f is isotropic.   □

Quadratic forms $\sum_{i=1}^{n} a_i x_i^2$ over $\mathbb{Q}_p$ are well understood; the non-
degenerate ones (all $a_i \neq 0$) are classified by the number n of variables,
the square class of $\prod_i a_i$ and the Hasse (–Minkowski) invariant; see
CASSELS 1978, SCHARLAU 1985 and LAM 2005 for more information.

The following general result says that non-degenerate isotropic quadratic forms are surjective:

54.8 Lemma *Let F be a field with characteristic distinct from 2, and
let $f(x) = \sum_{i=1}^{n} a_i x_i^2$ be a quadratic form with $a_i \in F^\times$. If f is isotropic
over F, then $f : F^n \to F$ is surjective.*

Proof By assumption, $f(x) = 0$ for some non-zero vector $x \in F^n$, say
$x_1 \neq 0$. We have $f(tx_1 + 1, tx_2, \ldots, tx_n) = t^2 f(x) + a_1(2tx_1 + 1) =$
$a_1(2tx_1 + 1)$ for every $t \in F$, hence f is surjective.   □

54.9 Corollary *Let $a_1, a_2, a_3, a_4 \in \mathbb{Q}_p^\times$. Then the quadratic form
$f : \mathbb{Q}_p^4 \to \mathbb{Q}_p$ with $f(x) = \sum_{i=1}^{4} a_i x_i^2$ is surjective.*

Proof Let $c \in \mathbb{Q}_p$. By 54.7 the quadratic form $f(x) - cx_5^2$ is isotropic,
hence $f(x) - cx_5^2 = 0$ for some non-zero vector $(x, x_5) \in \mathbb{Q}_p^5$. If $x_5 \neq 0$,
then $f(x_5^{-1} x) = c$. If $x_5 = 0$, then $x \neq 0$ and f is isotropic, hence
surjective by 54.8.   □

54.10 Corollary *For $p \neq 2$, every element of $\mathbb{Q}_p$ is a sum of three
squares of $\mathbb{Q}_p$. Every element of $\mathbb{Q}_2$ is a sum of four squares of $\mathbb{Q}_2$.*

Proof By 54.6 the form $x^2 + y^2 + z^2$ is isotropic over $\mathbb{Q}_p$ for $p \neq 2$,
hence surjective by 54.8. The assertion about $\mathbb{Q}_2$ is a consequence of
54.9 (with $a_i = 1$).   □

The last corollary shows again that the fields $\mathbb{Q}_p$ are not formally real
(54.2). The fact that every p-adic number is a sum of four squares of
p-adic numbers may also be inferred from Lagrange's theorem (34.18) by
a topological argument: the set $S := \{ a^2 + b^2 + c^2 + d^2 \mid a, b, c, d \in \mathbb{Q}_p \}$

is the union of $\{0\}$ and of finitely many square classes of $\mathbb{Q}_p$, hence S is topologically closed in $\mathbb{Q}_p$ by 54.1; furthermore $\mathbb{N} \subseteq S$ by 34.18, hence $\mathbb{Z}_p \subseteq S$ by 51.9, and $\mathbb{Q}_p = \bigcup_{n \geq 0} p^{-2n}\mathbb{Z}_p \subseteq S$.

Exercises

(1) Find all primes p such that -1 is a square in $\mathbb{Q}_p$.

(2) For $p \neq 2$, a p-adic number $p^r(c_0 + c_1 p + c_2 p^2 + \cdots)$ with $r \in \mathbb{Z}$, $c_i \in \{0, 1, \ldots, p-1\}$ is a square in $\mathbb{Q}_p$ if and only if r is even and c_0 is a square modulo p. For $p = 2$ the squares are characterized by the condition that r is even and $c_0 + 2c_1 + 4c_2 \equiv 1 \bmod 8$.

(3) Characterize the square-free integers $d \in \mathbb{Z}$ such that $\mathbb{Q}(\sqrt{d})$ is isomorphic to a subfield of $\mathbb{Q}_p$.

(4) Show that -1 is not a sum of three squares in $\mathbb{Q}_2$; in fact, $x^2 + y^2 + z^2$ represents all square classes of $\mathbb{Q}_2$ except the square class of -1.

(5) Let $f(x) = \sum_{i=1}^{n} a_i x_i^2$ be a quadratic form with integer coefficients $a_i \in \mathbb{Z}$ and $n \geq 5$. Then f is isotropic over the finite ring $\mathbb{Z}/m\mathbb{Z}$ for every integer $m \geq 2$.

55   Absolute values

In this section, we define the general concept of absolute values, and we prove two results of Ostrowski which determine all absolute values of the field $\mathbb{Q}$ of rational numbers, and all fields with an Archimedean absolute value.

55.1 Definition   An *absolute value* φ of a field (or skew field) F is a non-trivial homomorphism $\varphi : F^\times \to \mathbb{R}_{\mathrm{pos}}$ of multiplicative groups which is *subadditive*, which means that

$$\varphi(x + y) \leq \varphi(x) + \varphi(y)$$

for all $x, y \in F^\times$ with $x + y \in F^\times$. It is convenient to define $\varphi(0) = 0$; then $\varphi(xy) = \varphi(x)\varphi(y)$ and $\varphi(x + y) \leq \varphi(x) + \varphi(y)$ for all $x, y \in F$.

An absolute value φ is said to be *Archimedean*, if $\varphi(n \cdot 1) > 1$ for some $n \in \mathbb{N}$. Otherwise, we have $\varphi(n \cdot 1) \leq 1$ for all $n \in \mathbb{N}$, and then we say that φ is *non-Archimedean*.

The multiplicative group $\mathbb{R}_{\mathrm{pos}}$ is torsion free, hence every absolute value φ satisfies $\varphi(r) = 1$ whenever $r \in F$ is a root of unity. In particular $\varphi(-1) = 1$, and this gives $\varphi(-x) = \varphi(x)$ for every $x \in F$. If the characteristic of F is not zero, then the non-zero elements of the prime field of F are roots of unity, hence every absolute value of F is non-Archimedean.

55.2 Examples The ordinary absolute values on $\mathbb{Q}$, $\mathbb{R}$ or $\mathbb{C}$ are defined by $|x| = \max\{x, -x\}$ for $x \in \mathbb{R}$ and by $|z| = (z\bar{z})^{1/2} = (a^2 + b^2)^{1/2}$ for $z = a + ib \in \mathbb{C}$, $a, b \in \mathbb{R}$. These are examples of Archimedean absolute values. The p-adic absolute value $|\ |_p$ on $\mathbb{Q}$ and its extension to $\mathbb{Q}_p$ are non-Archimedean absolute values; compare 36.3, 44.9 and 51.4. These are the main examples in the context of this book.

Further examples of non-Archimedean absolute values can be obtained from valuations; see 56.2 and 56.3. See also 58.5, Section 23 and 44.11.

Non-Archimedean absolute values satisfy a stronger version of subadditivity:

55.3 Lemma *Let F be a skew field. A non-trivial homomorphism $\varphi : F^{\times} \to \mathbb{R}_{\mathrm{pos}}$ is a non-Archimedean absolute value if, and only if, φ is ultrametric, that is,*

$$\varphi(x + y) \leq \max\{\varphi(x), \varphi(y)\} \ \text{for all } x, y \in F .$$

In this case, $\varphi(x) \neq \varphi(y)$ implies that $\varphi(x \pm y) = \max\{\varphi(x), \varphi(y)\}$.

Proof The ultrametric inequality implies by induction that $\varphi(n \cdot 1) = \varphi(1 + 1 + \cdots + 1) \leq \varphi(1) = 1$ for every $n \in \mathbb{N}$, whence φ is non-Archimedean.

Conversely, suppose that φ is a non-Archimedean absolute value. Let $\varphi(x) \leq 1$. Then $\varphi(x + 1) \leq 2$, hence $m := \sup\{\varphi(x + 1) \mid \varphi(x) \leq 1\}$ is finite, and $\varphi(x + y) \leq m \cdot \max\{\varphi(x), \varphi(y)\}$ for all $x, y \in F$ by the multiplicativity of φ. Furthermore $\varphi(x) \leq 1$ implies

$$\varphi(x+1)^3 = \varphi\big(x^3 + 1 + 3x(x+1)\big) \leq m \cdot \max\{\varphi(x^3 + 1), \varphi(x+1)\} \leq m^2 ,$$

as $\varphi(3) \leq 1$ and $\varphi(x^3) \leq 1$. Thus $m^3 = \sup\{\varphi(x+1)^3 \mid \varphi(x) \leq 1\} \leq m^2$, hence $m \leq 1$, and φ is ultrametric.

As an alternative, we infer from $\varphi(x) \leq 1$ that

$$\varphi(x + 1)^n = \varphi\big(\textstyle\sum_{k=0}^{n} \binom{n}{k} x^k\big) \leq \sum_{k=0}^{n} \varphi(x)^k \leq n + 1$$

for each $n \in \mathbb{N}$. Taking n-th roots and using $\lim_n \sqrt[n]{n + 1} = 1$ we see that $m = \sup\{\varphi(x + 1) \mid \varphi(x) \leq 1\} \leq 1$, hence φ is ultrametric.

Finally, let φ be ultrametric and $\varphi(x) > \varphi(y)$. In view of $\varphi(y) = \varphi(-y)$ it suffices to compute $\varphi(x + y)$. We have $\varphi(x) = \varphi(x + y - y) \leq \max\{\varphi(x + y), \varphi(y)\}$, so this maximum is $\varphi(x + y)$, and we infer that $\varphi(x + y) = \varphi(x)$. □

By 55.3 the ultrametric character of an absolute value is revealed already in the prime field.

If φ is an absolute value, then the function φ^s defined by $\varphi^s(x) = \varphi(x)^s$ is also an absolute value for $0 < s \le 1$ (see Exercise 4), and if φ is non-Archimedean, then one can even admit all $s > 0$.

The following theorem determines all absolute values of the field $\mathbb{Q}$ of rational numbers.

55.4 Theorem (Ostrowski) *Let φ be an absolute value of $\mathbb{Q}$. If φ is Archimedean, then $\varphi = | \; |^s$ for some real number s with $0 < s \le 1$. If φ is non-Archimedean, then $\varphi = | \; |_p^s$ for some prime number p and some $s > 0$.*

Proof First, let φ be non-Archimedean. Then $S := \{ n \in \mathbb{Z} \mid \varphi(n) < 1 \}$ is an additive subgroup of $\mathbb{Z}$ (see 55.3), and S is not trivial, because $\varphi(\mathbb{Q}^\times) \ne \{1\}$ by definition. Thus, $S = p\mathbb{Z}$ for some $p \in \mathbb{N}$, and we claim that p is a prime number. Otherwise $p = ab$ with natural numbers $a, b \notin p\mathbb{Z} = S$. But then $\varphi(a) = 1 = \varphi(b)$, a contradiction to $\varphi(a)\varphi(b) = \varphi(p) < 1$.

We can now write $\varphi(p) = p^{-s}$ with a real number $s > 0$, and we show that $\varphi = | \; |_p^s$. Every prime number $q \ne p$ satisfies $q \notin p\mathbb{Z} = S$, hence $\varphi(q) = 1 = |q|_p^s$. This shows that φ and $| \; |_p^s$ coincide on all primes. The primes together with -1 generate the multiplicative group $\mathbb{Q}^\times$, as we have seen in 32.1. Therefore the two multiplicative functions φ and $| \; |_p^s$ are equal.

Now let φ be Archimedean. We consider integers $a, b \in \mathbb{N}$ with $b \ge 2$, and we define $C := \max\{ \varphi(c) \mid c = 0, 1, \ldots, b - 1 \}$. For every natural number n there is a unique $m \in \mathbb{N}_0$ with $b^m \le a^n < b^{m+1}$, hence $m \le n \cdot \log a / \log b$. From the representation

$$a^n = \sum_{k=0}^m c_k b^k$$

of a^n with basis b, with digits $c_k \in \{0, 1, \ldots, b - 1\}$, we infer that

$$\varphi(a)^n \le \sum_{k=0}^m \varphi(c_k)\varphi(b)^k \le (m + 1)C \max\{1, \varphi(b)\}^m$$

$$\le (n \cdot \log a / \log b + 1)C \max\{1, \varphi(b)\}^{n \log a / \log b} .$$

Taking the n-th root and observing that $\lim_n \sqrt[n]{n\alpha + \beta} = 1$ for $\alpha \ge 0$, we obtain

$$\varphi(a) \le \max\{1, \varphi(b)\}^{\log a / \log b}$$

for all $a, b \in \mathbb{N}$ with $b \ge 2$. Since φ is Archimedean, we have $\varphi(a) > 1$ for some a, hence $\varphi(b) > 1$ for each $b \ge 2$, and our inequality gives

$\varphi(a)^{1/\log a} \leq \varphi(b)^{1/\log b}$ for all $a, b \in \mathbb{N}$ with $a, b \geq 2$. By symmetry, we conclude that $\varphi(a)^{1/\log a}$ is a constant $e^s > 1$, for some $s > 0$. Thus,

$$\varphi(a) = e^{s \log a} = a^s = |a|^s$$

for every $a \in \mathbb{N}$, which implies that $\varphi = |\ |^s$.   □

Theorem 55.6 below describes all Archimedean absolute values of $\mathbb{R}$. The non-Archimedean absolute values of $\mathbb{R}$ defy classification; indeed, for every transcendency basis T of $\mathbb{R}$ over $\mathbb{Q}$, there are many possibilities to extend a p-adic absolute value of $\mathbb{Q}$ to $\mathbb{Q}(T)$ (compare Exercise 5 of Section 56), and then further extensions to $\mathbb{R}$ are possible (see JACOBSON 1989 9.9, LANG 1993 XII §3 or WARNER 1989 Theorem 26.6). We may also infer from 56.15 that the p-adic absolute value of $\mathbb{Q}_p$ extends to the algebraic closure $\mathbb{Q}_p^\natural$; now $\mathbb{Q}_p^\natural \cong \mathbb{C}$ by 64.21, and each of the many embeddings of $\mathbb{R}$ into $\mathbb{C}$ (see 14.15, 14.9) gives a non-Archimedean absolute value on $\mathbb{R}$.

55.5 Remark: Metric and topology Every absolute value φ of a field F yields a metric d which is defined by $d(x, y) = \varphi(x - y)$ for $x, y \in F$. The corresponding topology is a field topology of F; see 13.2(b).

The following theorem describes all fields with an Archimedean absolute value: they are just the subfields of the complex field $\mathbb{C}$.

55.6 Theorem (Ostrowski) *Let F be a field with an Archimedean absolute value φ. Then F is isomorphic to a dense subfield of $\mathbb{R}$ or $\mathbb{C}$, and there exists a monomorphism ι of F into $\mathbb{C}$ and a real number s with $0 < s \leq 1$ such that*

$$\varphi(x) = |\iota(x)|^s \text{ for every } x \in F ,$$

where $|\ |$ is the ordinary absolute value of $\mathbb{C}$.

Proof Since φ is Archimedean, F has $\mathbb{Q}$ as its prime field, and the restriction $\varphi|_{\mathbb{Q}^\times}$ is not identically 1, hence an Archimedean absolute value of $\mathbb{Q}$. Theorem 55.4 implies that $\varphi|_\mathbb{Q} = |\ |^s$ with $0 < s \leq 1$. Therefore the completion of $\mathbb{Q}$ with respect to $\varphi|_\mathbb{Q}$ is isomorphic to the field of real numbers; see 44.11 (1). Thus the topological closure of $\mathbb{Q}$ in the completion $\widehat{F}$ of F with respect to φ is a copy $\mathbb{R}$ of the real numbers. By 44.9, the topology of $\widehat{F}$ is described by an absolute value $\widehat{\varphi} : \widehat{F} \to \mathbb{R}$ that is a continuous extension of φ, and $\widehat{\varphi}$ coincides on $\mathbb{R}$ with $|\ |^s$, as $\mathbb{Q}$ is dense in $\mathbb{R}$.

We show below that every element $a \in \widehat{F}$ satisfies a quadratic equation over $\mathbb{R}$. Since $\mathbb{C}$ is the only proper algebraic extension of $\mathbb{R}$, this implies

$\widehat{F} = \mathbb{R}$ or $\widehat{F} = \mathbb{C}$, and then the embedding of F into its completion $\widehat{F}$ gives the assertion of 55.6 (note that $\varphi(\zeta) = 1$ for every root of unity ζ, hence φ is uniquely determined by $\varphi|_{\mathbb{Q}}$).

Let $a \in \widehat{F}$. We essentially follow NEUKIRCH 1992 II 4.2 and define a continuous map $f : \mathbb{C} \to \mathbb{R}$ by $f(z) = \widehat{\varphi}(a^2 - (z + \bar{z})a + z\bar{z})$ for $z \in \mathbb{C}$; note that $z + \bar{z}$, $z\bar{z} \in \mathbb{R} \subseteq \widehat{F}$. The triangle inequality yields $f(z) \geq \widehat{\varphi}(z\bar{z}) - \widehat{\varphi}((z + \bar{z})a) - \widehat{\varphi}(a^2) = |z\bar{z}|^s - |z + \bar{z}|^s \widehat{\varphi}(a) - \widehat{\varphi}(a^2)$, hence $\lim_{z \to \infty} f(z) = \infty$. Therefore f attains its lower bound m, and the preimage $f^{-1}(m) \subseteq \mathbb{C}$ is non-empty and compact. Choose $z_0 \in f^{-1}(m)$ with $|z_0| = \max |f^{-1}(m)|$. The real polynomial

$$q(x) := x^2 - (z_0 + \overline{z_0})\, x + z_0 \overline{z_0}$$

satisfies $q(x) \geq 0$ for every $x \in \mathbb{R}$, since $(z_0 - \overline{z_0})^2 \leq 0$. If $m = 0$, then $q(a) = 0$.

Now we assume that $m > 0$ and aim for a contradiction. Let $\mu = (m/2)^{1/s}$. The real polynomial $q(x) + \mu$ is strictly positive on $\mathbb{R}$, hence it has roots $w, \overline{w} \in \mathbb{C}$. These roots satisfy $|w|^2 = w\overline{w} = z_0\overline{z_0} + \mu > |z_0|^2$, hence $f(w) > m$ by our choice of z_0.

For any odd number $n \in \mathbb{N}$, the real polynomial $g(x) = q(x)^n + \mu^n$ of degree $2n$ is strictly positive on $\mathbb{R}$, and $g(w) = 0$. Thus we have the real factorization $g(x) = \prod_{j=1}^{n}(x^2 - (w_j + \overline{w_j})x + w_j\overline{w_j})$, where $w_1 = w$, $\overline{w_1}$, w_2, $\overline{w_2}$, $\ldots$, w_n, $\overline{w_n}$ are the complex roots of $g(x)$. Substituting a for x, we obtain $g(a) = \prod_{j=1}^{n}(a^2 - (w_j + \overline{w_j})a + w_j\overline{w_j}) \in \widehat{F}$ and therefore

$$\widehat{\varphi}(g(a)) = \prod_{j=1}^{n} f(w_j) \geq f(w) \cdot m^{n-1} .$$

On the other hand,

$$\widehat{\varphi}(g(a)) \leq \widehat{\varphi}(q(a))^n + \widehat{\varphi}(\mu)^n = f(z_0)^n + \mu^{sn} = m^n + (m/2)^n .$$

We infer that $f(w) \leq m(1 + 2^{-n})$. This holds for all odd numbers $n \in \mathbb{N}$, hence $f(w) \leq m$, a contradiction to $f(w) > m$. □

For other proofs of 55.6 see JACOBSON 1989 9.5, EBBINGHAUS et al. 1991 Chapter 8 §4, BOURBAKI 1972 VI.6.6 p. 410 and CASSELS 1986 Chapter 3. The related Theorem of Gelfand and Mazur says that every extension field of $\mathbb{R}$ which is complete with respect to a norm is isomorphic to $\mathbb{R}$ or $\mathbb{C}$ (for the definition of a norm φ on a field, replace the multiplicativity in 55.1 by submultiplicativity: $\varphi(xy) \leq \varphi(x)\varphi(y)$). For proofs see EBBINGHAUS et al. 1991 Chapter 8 §4, RIBENBOIM 1999 Theorem 3 p. 37, LANG 1993 XII.2, ENGLER–PRESTEL 2005 1.2.4, BOURBAKI 1972 VI.6.4 p. 407 or WARNER 1989 26.10 p. 262 and the remarks on pp. 499–501. The last two references allow also skew fields, and then Hamilton's quaternions $\mathbb{H}$ appear as the only further possibility.

The following characterization of absolute values will be used in 57.4 and 58.5.

55.7 Lemma (Artin's trick)   *Let F be a skew field and $\varphi : F \to \mathbb{R}$ be a multiplicative map such that $\varphi(0) = 0$, $\varphi(1) = 1$ and $\varphi(x + y) \le 2 \max\{\varphi(x), \varphi(y)\}$ for all $x, y \in F$. Then φ is an absolute value of F.*

Proof We have $\varphi(x_1 + \cdots + x_{2^k}) \le 2^k \max\{\varphi(x_1), \ldots, \varphi(x_{2^k})\}$ for $x_i \in F$, as an easy induction on k shows; in particular $\varphi(2^k \cdot 1_F) \le 2^k$. This implies that $\varphi(m \cdot 1_F) \le 2m$ for all $m \in \mathbb{N}$ (one needs to add at most m zeros in order to write $m \cdot 1_F$ as a sum of 2^k terms $x_i \in \{0, 1_F\}$). For $x \in F$ and $n = 2^k - 1$ we compute by binomial expansion that

$$\varphi(1 + x)^n \le 2^k \max_{0 \le i \le n} \varphi(\tbinom{n}{i})\varphi(x^i)$$
$$\le 2^{k+1} \textstyle\sum_{i=0}^{n} \tbinom{n}{i}\varphi(x)^i = 2(n+1)(1 + \varphi(x))^n .$$

Taking the root of order $n = 2^k - 1$ we obtain for $k \to \infty$ that $\varphi(1 + x) \le 1 + \varphi(x)$ for all $x \in F$. If $x, y \in F$ with $0 \ne \varphi(x) \ge \varphi(y)$, then

$$\varphi(x + y) \le (1 + \varphi(yx^{-1}))\varphi(x) = \varphi(x) + \varphi(y) .$$

As φ is not constant, multiplicativity implies that $\varphi(x) = 0 \Leftrightarrow x = 0$. Hence φ is an absolute value of F. $\square$

Exercises

(1) Two absolute values φ_1 and φ_2 of a field F define the same topology, if, and only if, $\varphi_1(a) < 1 \Leftrightarrow \varphi_2(a) < 1$ for $a \in F$, and this occurs precisely if $\varphi_1 = \varphi_2^s$ for some positive real number s.

(2) Let φ_1 and φ_2 be absolute values of a field F that induce different topologies on F. Show that the diagonal $\{(a, a) \mid a \in F\}$ is dense in the product space $F_1 \times F_2$, where F_i denotes the set F endowed with the topology induced by φ_i. (This is a special case of the Approximation Theorem of Artin–Whaples.)

(3) Assume that φ is a non-Archimedean absolute value on a field F, and let $\widehat{\varphi}$ be the continuous extension of φ to the completion $\widehat{F}$ (compare 44.9). Show that $\widehat{\varphi}(\widehat{F}) = \varphi(F)$.

(4) If φ is an absolute value on a field F and $0 < s \le 1$, then also the map φ^s defined by $\varphi^s(a) = \varphi(a)^s$ is an absolute value on F.

56   Valuations

We define general valuations (which are sometimes called Krull valuations), and we obtain some basic results of valuation theory. For more comprehensive accounts of valuation theory see COHN 2003a Chapter 9, JACOBSON 1989 Chapter 9, RIBENBOIM 1999, WARNER 1989 Chapter V, BOURBAKI 1972 VI and ENGLER–PRESTEL 2005.

56.1 Definition   Let Γ be a non-trivial ordered abelian group. A *valuation* v of a field (or skew field) F with *value group* Γ is an epimorphism $v : F^\times \to \Gamma$ of groups which satisfies

$$v(x + y) \geq \min\{v(x), v(y)\}$$

for all $x, y \in F^\times$ with $x + y \in F^\times$. If $\Gamma \cong \mathbb{Z}$, one speaks of a *principal valuation* (or of a discrete rank 1 valuation). In that case, one identifies Γ with the ordered group $(\mathbb{Z}, <)$, and one calls each $a \in F$ with $v(a) = 1$ a *prime element* (or a uniformizer) of v.

By convention, Γ is written additively. Often one defines $v(0) = \infty$, where ∞ is a new element with $\gamma < \infty$ and $\infty + \gamma = \gamma + \infty = \infty + \infty = \infty$ for each $\gamma \in \Gamma$. Then the map $v : F \to \Gamma \cup \{\infty\}$ satisfies $v(xy) = v(x) + v(y)$ and $v(x + y) \geq \min\{v(x), v(y)\}$ for all $x, y \in F$.

Since Γ is torsion free (7.3), we have $v(x) = 0$ whenever $x \in F$ is a root of unity. In particular, $v(-1) = 0$, hence $v(-x) = v(x)$ for all $x \in F$. Furthermore it is easy to show that $v(x) \neq v(y)$ implies $v(x + y) = \min\{v(x), v(y)\}$; compare the proof of 55.3.

56.2 Absolute values and valuations   Let φ be a non-Archimedean absolute value of a field F, as defined in 55.1. Then $v(x) = -\log_a \varphi(x)$ gives a valuation v of F, for every basis $a > 1$, and the value group $v(F^\times) \subseteq \mathbb{R}$ is an Archimedean ordered group.

Conversely, let v be a valuation of a field F such that the value group Γ is Archimedean. By 7.8 we may consider Γ as a subgroup of the ordered group $\mathbb{R}^+$, and then $\varphi(x) = a^{-v(x)}$ defines a non-Archimedean absolute value φ of F, for every real number $a > 1$.

Thus a valuation with Archimedean value group is essentially the same thing as a non-Archimedean absolute value (an unfortunate clash of terminology, which is sometimes avoided by saying 'real valuation' instead of 'valuation with Archimedean value group'). In this sense, valuations generalize non-Archimedean absolute values.

The negative sign (and the corresponding switch from max to min; compare 55.3 and 56.1) is motivated by the idea that $v(x)$ should describe a degree of divisibility of x, as in the following examples.

56.3 Examples (a) We fix a prime number p. The p-adic valuation v_p of $\mathbb{Q}$ is defined by

$$v_p\left(p^n \cdot \frac{a}{b}\right) = n$$

for $n \in \mathbb{Z}$ and integers $a, b \in \mathbb{Z}$ that are not divisible by p (see 36.3). An extension to $\mathbb{Q}_p$ may be defined by the same formula, for $a, b \in \mathbb{Z}_p \setminus p\mathbb{Z}_p$. These principal valuations correspond to the p-adic absolute values of $\mathbb{Q}$ and $\mathbb{Q}_p$, via $v_p = -\log_p |\ |_p$; compare also 51.1, 51.4, 56.2 and 56.7.

(b) Let R be a unique factorization domain, F its field of fractions, and fix an irreducible element $p \in R$. Every element $x \in F^\times$ has the form

$$x = p^n \cdot \frac{a}{b}$$

with $n \in \mathbb{Z}$ and $a, b \in R \setminus pR$. The integer n is uniquely determined by x, and it is easy to verify that $v : F^\times \to \mathbb{Z} : x \mapsto n$ is a principal valuation of F with prime element p. Example (a) is the special case where $R = \mathbb{Z}$.

(c) We describe some valuations which are relevant in algebraic geometry. Let $F = k(x)$ be a simple transcendental extension of k (compare 64.19). Then F is the field of fractions of the polynomial ring $k[x]$, which is a unique factorization domain. Hence for each irreducible polynomial $p \in k[x]$, example (b) yields a principal valuation v_p of F defined by

$$v_p\left(p^n \cdot \frac{a}{b}\right) = n$$

for $n \in \mathbb{Z}$ and polynomials $a, b \in k[x]$ which are not divisible by p.

The degree of polynomials leads to another valuation v_∞ of F, the *degree valuation*, which is is given by

$$v_\infty\left(\frac{a}{b}\right) = -\deg a + \deg b$$

for non-zero polynomials $a, b \in k[x]$. The field $F = k(x)$ is also the field of fractions of its subring $k[x^{-1}] \cong k[x]$, and the irreducible element $x^{-1} \in k[x^{-1}]$ gives a valuation $v_{x^{-1}}$ on F with prime element x^{-1}. The equation

$$a = a' \cdot (x^{-1})^{-\deg a}$$

with $a' = ax^{-\deg a} \in k[x^{-1}]$ shows that v_∞ coincides with $v_{x^{-1}}$.

In fact, we have just described all valuations of $k(x)$ which are trivial on k (see Exercise 2). If k is algebraically closed, then all the irreducible polynomials $p \in k[x]$ have degree 1, hence in this case the valuations of

$k(x)$ which are trivial on k correspond to the elements of the 'projective line' $k \cup \{\infty\}$ over k.

In the special case $p = x - c \in k[x]$, the integer $v_p(f)$ describes the behaviour near c of the rational function $f \in F = k(x)$ in the following sense: f vanishes at c of order n precisely if $v_p(f) = n > 0$, and f has a pole at c of order $-n$ if, and only if, $v_p(f) = n < 0$.

(d) The completion of $k(x)$ with respect to the topology defined by the valuation v_x as in (c) is the field $k((x))$ of Laurent series over k, and the extended valuation maps $\sum_n a_n x^n$ to $\min\{n \in \mathbb{Z} \mid a_n \neq 0\}$; moreover, the valuation v_∞ leads to the same completion. See 44.11 for these facts.

(e) The non-standard rationals $^*\mathbb{Q}$ and reals $^*\mathbb{R}$ admit valuations that are obtained essentially by factoring out infinitesimals; see Sections 23, 24 (and use 56.5). Compare also 56.16.

Now we describe the valuations on a field F in terms of certain subrings of F.

56.4 Definitions A proper subring R of a field F is called a *valuation ring* of F, if $F = R \cup \{r^{-1} \mid 0 \neq r \in R\}$.

Let v be a valuation on a field F with value group Γ. The ring $R := \{x \in F \mid v(x) \geq 0\}$ meets the requirements of 56.4, because $v : F^\times \to \Gamma$ is not trivial and $v(x) < 0$ implies $v(x^{-1}) = -v(x) > 0$. We call R the *valuation ring of v*.

Then $U := \{x \in F \mid v(x) = 0\}$ is the group of units of this ring R, and the set $M := R \smallsetminus U = \{x \in F \mid v(x) > 0\}$ of non-units is an ideal, the unique maximal ideal of R. Thus the quotient $F_v := R/M$ is a field, the *residue field* of the valuation v. Furthermore $v : F^\times \to \Gamma$ induces a group isomorphism $F^\times/U \to \Gamma$.

The p-adic valuation v_p of the field $\mathbb{Q}$ has the valuation ring $R_p := \{ab^{-1} \mid a \in \mathbb{Z} \wedge b \in \mathbb{Z} \smallsetminus p\mathbb{Z}\}$ with the unique maximal ideal pR_p and the residue field $\mathbb{F}_p$; see 51.7.

56.5 Proposition *Let R be a valuation ring of a field F. Then there exists a valuation v of F such that $R = \{x \in F \mid v(x) \geq 0\}$.*

Proof The set $U = \{r \in R \mid r \neq 0 \wedge r^{-1} \in R\}$ is the group of units of R, and the factor group $\Gamma := F^\times/U$ is not trivial, as $R \neq F$. The definition

$$xU \leq yU \iff x^{-1}y \in R$$

for $x, y \in F^\times$ renders Γ an ordered group; indeed, the relation $\leq$ is a total ordering, because $x^{-1}y \notin R$ implies $y^{-1}x \in R$. A coset $xU \in \Gamma$ is (strictly) positive precisely if $x \in R \smallsetminus U$.

We write Γ additively, and we denote its neutral element U by 0_Γ, to avoid confusion with $0 \in F$. The canonical map $v : F^\times \to \Gamma = F^\times/U$ is a group epimorphism, and $R = \{0\} \cup \{ x \in F^\times \mid v(x) \geq 0_\Gamma \}$. In order to show that v is a valuation, let $x, y \in F^\times$ with $v(x) \leq v(y)$ and $x + y \neq 0$. Then $x^{-1}y \in R$, hence $1 + x^{-1}y \in R$ and $v(1 + x^{-1}y) \geq 0_\Gamma$. We infer that

$$v(x + y) = v(x(1 + x^{-1}y)) = v(x) + v(1 + x^{-1}y) \geq v(x) .$$

This shows that $v(x + y) \geq \min\{v(x), v(y)\}$ for all $x, y \in F^\times$. $\qquad \square$

56.6 Proposition *Two valuations $v : F^\times \to \Gamma$ and $w : F^\times \to \Delta$ of a field F have the same valuation ring if, and only if, there exists an isomorphism $\alpha : \Gamma \to \Delta$ of ordered groups such that $w = \alpha \circ v$.*

In this situation v and w are called *equivalent* (see also 56.11).

Proof If $w = \alpha \circ v$, then v and w have the same valuation ring. Conversely, if v and w have the same valuation ring R, then the group U of units of R is the common kernel of the epimorphisms $v : F^\times \to \Gamma$ and $w : F^\times \to \Delta$. Hence the quotient α of the induced group isomorphisms $\Gamma \leftarrow F^\times/U \to \Delta$ satisfies $w = \alpha \circ v$. Since $v(R \smallsetminus U)$ and $w(R \smallsetminus U)$ are the sets of positive elements of Γ and Δ, respectively, we infer that α is an isomorphism of ordered groups. $\qquad \square$

Extending Theorem 55.4 we now determine all valuations of the field $\mathbb{Q}$ of rational numbers.

56.7 Theorem *Every valuation v of $\mathbb{Q}$ is equivalent to a p-adic valuation $v_p = - \log_p | \ |_p$ for a unique prime number p.*

Proof By 56.1 we have $v(\mathbb{Z}) \geq 0$, and the restriction of v to $\mathbb{Z} \smallsetminus \{0\}$ is not trivial, as v is not trivial on $\mathbb{Q}^\times$. Hence $P := \{ z \in \mathbb{Z} \mid v(z) > 0 \}$ is a non-zero proper ideal of $\mathbb{Z}$, in fact a prime ideal, because $x, y \in \mathbb{Z}$, $v(xy) > 0$ implies $v(x) > 0$ or $v(y) > 0$. Thus $P = p\mathbb{Z}$ for a unique prime number p. In particular, $v(z) = 0$ for all $z \in \mathbb{Z} \smallsetminus p\mathbb{Z}$. By multiplicativity, v and v_p have the same valuation ring $R_p = \{ ab^{-1} \mid a \in \mathbb{Z} \wedge b \in \mathbb{Z} \smallsetminus p\mathbb{Z} \}$, hence they are equivalent by 56.6.

These valuation rings R_p are mutually distinct, because the quotient of R_p modulo its unique maximal ideal pR_p has cardinality p. $\qquad \square$

56.8 Theorem   *Let F be a field and $v : F^\times \to \Gamma$ a valuation with valuation ring R. Then the following conditions are equivalent:*

(i) *R is a maximal subring of F.*

(ii) *R contains only one proper non-zero prime ideal.*

(iii) *The value group Γ is Archimedean.*

Proof (i) implies (ii): Let I be a proper non-zero prime ideal of R. We show that I coincides with the (unique) maximal ideal M of R; clearly $I \subseteq M$. Let $0 \neq a \in M$. Then $a^{-1} \notin R$, hence the ring $R[a^{-1}]$ generated by R and a^{-1} is F, by (i). If $0 \neq b \in I$, then b^{-1} can be written as a finite sum $b^{-1} = \sum_{0 \leq k \leq n} r_k a^{-k}$ with elements $r_k \in R$. We infer that $a^n = bb^{-1}a^n = b \sum_k r_k a^{n-k} \in I$, hence $a \in I$, as I is a prime ideal. This shows that $M \subseteq I$.

(ii) implies (iii): Let $0 < \gamma \in \Gamma$ and $I := \{ a \in F \mid \forall_{n \in \mathbb{N}} v(a) > n\gamma \}$. The defining properties of v imply that I is an ideal of R. In fact, I is a prime ideal of R, because $r_1, r_2 \in R \smallsetminus I$ entails $v(r_i) \leq n_i \gamma$ for $i = 1, 2$ with suitable integers $n_i \in \mathbb{N}$, hence $v(r_1 r_2) \leq (n_1 + n_2)\gamma$ and $r_1 r_2 \notin I$. The preimages of γ under v belong to the maximal ideal of R, but not to I. Hence (ii) implies that $I = \{0\}$, which means that each $a \in F^\times$ satisfies $v(a) \leq n\gamma$ for some $n \in \mathbb{N}$. Thus Γ is Archimedean.

(iii) implies (i): Each $c \in F \smallsetminus R$ satisfies $v(c) < 0$. For each $a \in F$ there exists by (iii) an integer $n \in \mathbb{N}$ such that $nv(c) \leq v(a)$. Hence $v(ac^{-n}) \geq 0$, so $ac^{-n} \in R$. Therefore $a = ac^{-n}c^n$ belongs to the subring $R[c]$ generated by R and c. Thus $R[c] = F$, and (i) holds.   □

56.9 Extension Theorem   *Let v be a valuation of a field F and let E be an extension field of F. Then v has an extension to a valuation of E (with a possibly larger value group).*

Proof Let $R_v = \{ x \in F \mid v(x) \geq 0 \}$ be the valuation ring of v and $M_v = \{ x \in F \mid v(x) > 0 \}$ its maximal ideal. First we construct a valuation ring of E.

We consider all pairs (R, M) consisting of a subring R of E with $R \supseteq R_v$ and a proper ideal M of R with $M \supseteq M_v$, and we order these pairs by double inclusion: $(R, M) \leq (R', M') \Leftrightarrow R \subseteq R' \wedge M \subseteq M'$. Given a chain of such pairs, by taking unions we obtain again a pair of this type (note that always $1 \notin M$). Hence, by Zorn's Lemma, there exists a maximal pair (R, M).

We claim that R is a valuation ring of E. We have $R \neq E$, since $M \supseteq M_v \neq \{0\}$ is a proper ideal of R. Furthermore, we assume that $e \notin R$ and $e^{-1} \notin R$ for some $e \in E$ and aim for a contradiction. Then the ring $R[e]$ generated by R and e is strictly larger than R. If the ideal $M' := \{\sum_i a_i e^i \mid a_i \in M\}$ generated by M in $R[e]$ were proper, then $(R[e], M') > (R, M)$, which contradicts the maximality. Hence $M' = R[e]$, which implies that we have an equation

$$1 = a_0 + a_1 e + \cdots + a_m e^m \qquad\qquad (*)$$

with $a_i \in M$. Similarly, since $R[e^{-1}]$ is strictly larger than R, we have another equation

$$1 = b_0 + b_1 e^{-1} + \cdots + b_n e^{-n}$$

with $b_i \in M$. We may assume that $m \geq n$ by symmetry, and that $m + n$ is chosen to be as small as possible. If we now multiply equation $(*)$ by $1 - b_0$ and substitute for $(1 - b_0)e^m$ from the second equation, we obtain an equation for e of the form $(*)$, but with m replaced by $m - 1$. This is a contradiction, which proves that R is a valuation ring of E.

By construction we have $R_v \subseteq R \cap F$ and $M_v \subseteq M$. Furthermore $F \smallsetminus R_v = (M_v \smallsetminus \{0\})^{-1} \subseteq (M \smallsetminus \{0\})^{-1} = E \smallsetminus R$ is disjoint from R, hence $R_v = R \cap F$. By Proposition 56.5 there exists a valuation w of E with valuation ring R. We have just shown that the valuation ring $R \cap F$ of the restriction $w|_F$ coincides with the valuation ring R_v of v. Using 56.6 we obtain an isomorphism $\alpha : w(F^\times) \to v(F^\times)$ with $v = \alpha \circ w|_F$. By set-theoretic considerations, there exists an ordered abelian group B containing $v(F^\times)$ as an ordered subgroup and an isomorphism $\beta : w(E^\times) \to \mathsf{B}$ extending α. Then $\beta \circ w$ is the desired valuation of E extending v. $\square$

More information on extensions of valuations can be found in the references mentioned at the beginning of this section; see also 56.15 and Exercise 5.

The *topology* τ_v induced on a field F by a valuation $v : F^\times \to \Gamma$ is defined by taking the sets $\{x \in F \mid v(x - a) > \gamma\}$ with $\gamma \in \Gamma$ as a neighbourhood base at $a \in F$. By 13.2(b), τ_v is always a totally disconnected field topology of F.

We say that F is *complete with respect to a valuation* v, if the topological field (F, τ_v) is complete (in the sense of 44.1).

By 56.9 each p-adic valuation of $\mathbb{Q}$ extends to $\mathbb{R}$; this gives (infinitely many) totally disconnected field topologies on $\mathbb{R}$. However, none of the these valuation topologies on $\mathbb{R}$ is locally compact or complete (otherwise $\mathbb{R}$ would contain a copy of the completion $\mathbb{Q}_p$ of $\mathbb{Q}$, which is a contradiction to 54.2).

56.10 Theorem *Let v and v' be valuations of a field F, with respective valuation rings R and R'. Then the following conditions are equivalent.*
  (i) *The valuations v and v' induce the same topology $\tau_v = \tau_{v'}$ on F.*
  (ii) *The subring generated by $R \cup R'$ is a proper subring of F.*

Valuations v, v' which satisfy one (hence both) of the properties in 56.10 are often called *dependent*.

Proof (i) implies (ii): If $\tau_v = \tau_{v'}$, then $R'a \subseteq R$ for some $a \in F^\times$, hence $RR' \subseteq RRa^{-1} \subseteq Ra^{-1}$. The subring S generated by $R \cup R'$ is the additive subgroup generated by RR', and Ra^{-1} is an additive group. We conclude that $S \subseteq Ra^{-1}$. Moreover, $Ra^{-1} = \{\, x \in F \mid v(x) \geq -v(a)\,\}$ is a proper subset of F, hence S is a proper subring.
  (ii) implies (i): The subring S generated by $R \cup R'$ is a valuation ring of F; denote by w the valuation defined by S; see 56.5. It suffices to show that $\tau_v = \tau_w$, because symmetry implies then that also $\tau_{v'} = \tau_w$.
  Since $S \neq F$, we can pick $a \in F \smallsetminus S$. Then $a \notin R$, hence $a^{-1} \in R$. Each non-zero $s \in S$ satisfies $s^{-1}a \notin R$ (otherwise $a = s(s^{-1}a) \in SR = S$, which is a contradiction), hence $a^{-1}s \in R$ and $s = a(a^{-1}s) \in aR$. Thus $R \subseteq S \subseteq aR$ and $bR \subseteq bS \subseteq baR$ for each $b \in F^\times$. This says that τ_v and τ_w have the same neighbourhoods of 0, whence $\tau_v = \tau_w$. $\square$

56.11 Corollary *Let v and v' be valuations of a field F, with respective valuation rings R and R'. Assume that v is a principal valuation, or more generally that the value group of v is Archimedean. Then v and v' induce the same topology $\tau_v = \tau_{v'}$ on F if, and only if, v and v' are equivalent (that is, $R = R'$; see 56.6).*

Proof By 56.8 the valuation ring R is a maximal subring of F, hence the assertion follows from 56.10. $\square$

The following result characterizes the valuation topologies which are locally compact.

56.12 Theorem *Let $v : F^\times \to \Gamma$ be a valuation of a field F. Then the valuation topology τ_v is locally compact if, and only if, $\Gamma \cong \mathbb{Z}$, the residue field F_v is finite, and F is complete with respect to v.*

Proof $R = \{\, a \in F \mid v(a) \geq 0 \,\}$ is the valuation ring of v, and the set $M = \{\, a \in F \mid v(a) > 0 \,\}$ is the maximal ideal of R.
  Assume that τ_v is locally compact. Then (F^+, τ_v) is complete by 43.9. The closed sets aR with $a \in F^\times$ form a neighbourhood base at 0. By local compactness, one of these sets aR is compact, hence $R = a^{-1}aR$

is compact. Since M is an open additive subgroup of R, the quotient $F_v = R/M$ is finite. Each set $I_\alpha := \{\, a \in F \mid v(a) > \alpha \,\}$ with $\alpha > 0$ is an open additive subgroup of R, hence R/I_α is finite. Since $I_\beta/I_\alpha \subseteq R/I_\alpha$ for $0 \leq \beta \leq \alpha$ and since $v : F^\times \to \Gamma$ is surjective, we deduce that each interval $\{\, \beta \in \Gamma \mid 0 \leq \beta \leq \alpha \,\}$ with $\alpha > 0$ is finite. Therefore Γ is Archimedean (as Γ is torsion free) and contains a smallest positive element, which is a generator of Γ (compare the proof of 1.4). This shows that Γ is cyclic.

For the converse implication, we assume that F is complete with respect to a principal valuation $v : F^\times \to \mathbb{Z}$ with finite residue field F_v. It suffices to show that the valuation ring R is compact, because the sets aR with $a \in F^\times$ form a neighbourhood base at 0. The set R is closed in F and therefore complete. Let π be a prime element of v, i.e., $v(F^\times) = \langle v(\pi) \rangle$. Then $M = \pi R$, and multiplication by π induces isomorphisms of $\pi^{n-1} R / \pi^n R$ onto $\pi^n R / \pi^{n+1} R$ for all $n \in \mathbb{N}$. Hence each quotient ring $R/\pi^n R$ with $n \in \mathbb{N}$ is finite. Since $\bigcap_n \pi^n R = \{0\}$ we obtain an injective homomorphism

$$\iota : R \to \textstyle\prod_{n \in \mathbb{N}} R/\pi^n R : r \mapsto (r_n)_n\,, \quad r_n := r + \pi^n R\,,$$

of additive groups (even of rings). Endow each finite group $R/\pi^n R$ with the discrete topology. Then the product group $\prod_n R/\pi^n R$ is compact by Tychonoff's theorem. Moreover, ι is a homeomorphism of R onto its image $\iota(R)$, because the sets $\pi^k R$ form a neighbourhood base of 0 in R and their images $\iota(\pi^k R) = \iota(R) \cap \{\, (x_n)_n \mid x_n = 0 \text{ for } n \geq k \,\}$ form a neighbourhood base of 0 in $\iota(R)$. The image $\iota(R)$ is complete, hence closed in the compact product space. Thus $\iota(R)$ is compact, and R as well. □

The next result says that each field appearing in 56.12 admits only one principal valuation (up to equivalence); more general results can be found in RIBENBOIM 1999 3.W p. 103 and WARNER 1989 32.24.

56.13 Theorem *Let F be a field which is complete with respect to a principal valuation $v : F^\times \to \mathbb{Z}$ with finite residue field. Then v is, up to equivalence, the only principal valuation of F.*

Proof Let R be the valuation ring of v, let M be the maximal ideal of R, let $F_v = R/M$ denote the finite residue field of v, and let p be the characteristic of F_v. Then $F^\times \cong \mathbb{Z} \times R^\times \cong \mathbb{Z} \times F_v^\times \times (1 + M)$; compare the proof of 53.1 and see also Exercise 3. Moreover, M is open and closed in F, hence complete.

Let $m \in M$. The group homomorphism $z \mapsto (1 + m)^z$ of $\mathbb{Z}^+$ into the multiplicative group $1 + M$ is continuous when $\mathbb{Z}$ is provided with the p-adic topology, because $(1 + m)^{p^n} \in 1 + M^{n+1}$ for each $n \in \mathbb{N}$ (this is easily verified by induction on n, using the fact that $p \cdot 1_F \in M$; see also 37.5). As $1 + M$ is complete, the mapping $z \mapsto (1 + m)^z$ extends by 43.23 to a continuous group homomorphism of $\mathbb{Z}_p^+$ into $1 + M$. This implies that the multiplicative group $1 + M$ admits roots of order r for each prime $r \neq p$; see 52.6. (In fact, $1 + M$ becomes a $\mathbb{Z}_p$-module in this fashion; compare WEIL 1967 p. 32.)

Fix any prime number $r > |F_v|$. Then $R^\times = F_v^\times \times (1 + M)$ is the largest r-divisible subgroup of $F^\times \cong \mathbb{Z} \times R^\times$. (Compare the proof of 53.5 for a related argument.)

If R' is the valuation ring of another principal valuation v' of F, then $F^\times \cong \mathbb{Z} \times R'^\times$ (by Exercise 3). Thus the largest r-divisible subgroup $R^\times$ of $F^\times$ is contained in $R'^\times \subseteq R'$. Now R is additively generated by $R^\times = R \smallsetminus M$ (as M is a proper additive subgroup of R), hence $R \subseteq R'$. Since R is a maximal subring of F by 56.8, this implies that $R = R'$, whence v and v' are equivalent (see 56.6). $\qquad\square$

56.14 Corollary *Up to equivalence, the fields $\mathbb{Q}_p$ and $\mathbb{F}_q((x))$ admit only one principal valuation.*

Proof This is a consequence of Theorem 56.13; note that $\mathbb{F}_q((x))$ is complete by 44.11. $\qquad\square$

Let p and q be distinct primes. By 56.9 the q-adic valuation of $\mathbb{Q}$ extends to a valuation w of $\mathbb{Q}_p$. By 56.14, such an extension w cannot be a principal valuation.

56.15 Theorem *Let F be a field which is complete with respect to a principal valuation $v : F^\times \to \mathbb{Z}$ with finite residue field, and let E be a field extension of F of finite degree n. Then v has an extension to a valuation w of E, and the extension w is unique up to equivalence. Moreover, w is again a principal valuation, its residue field is finite, and E is complete with respect to w.*

If $E|F$ is a Galois extension of finite degree n with Galois group G, then $w(a) = v(N(a))/n$ for $a \in E$, where $N : E \to F : a \mapsto \prod_{\gamma \in G} \gamma(a)$ is the norm map of $E|F$.

Proof Extensions of v to E exist by 56.9. For every extension w of v, the valuation topology τ_w renders E a topological field, hence also a topological vector space over the topological field F (with the valuation topology τ_v). Since F is locally compact by 56.12, we infer from 58.6(i) that (E, τ_w) is locally compact. Now 56.12 implies that w is a principal valuation with finite residue field and that E is complete with respect to w. Result 56.13 shows that w is unique up to equivalence.

Thus if $E|F$ is a Galois extension and $\gamma \in G$, then w and $w \circ \gamma$ are equivalent, hence $w = w \circ \gamma$. Therefore $nw(a) = \sum_{\gamma \in G} w(\gamma(a)) = w(\prod_{\gamma \in G} \gamma(a)) = w(N(a))$.

We mention that more direct proofs of more general versions of 56.15 can be found in many books; see COHN 2003a 9.5.3 and 9.2.7, JACOBSON 1989 Section 9.8 Corollary p. 583, WARNER 1989 26.7 p. 260, EBBING-HAUS *et al.* 1991 Chapter 6 Theorem 11, CASSELS 1986 Chapter 7 Theorem 1.1, SERRE 1979 Chapter II §2, RIBENBOIM 1999 5.A p. 127 or the paper LENSTRA–STEVENHAGEN 1989.   □

56.16 Natural ordering valuations Let $(F, <)$ be an ordered field which is not Archimedean; then the prime field $\mathbb{Q}$ is not cofinal in F; compare 11.12. Hence $R = \{a \in F \mid -q \le a \le q$ for some $q \in \mathbb{Q}\}$ is a proper subring of F. In fact, R is a valuation ring of F with the maximal ideal $M = \{a \in F \mid -q \le a \le q$ for each positive $q \in \mathbb{Q}\}$. The corresponding valuation v of F is called the natural valuation of $(F, <)$. The value group $\Gamma = v(F^\times) = F^\times/R^\times$ is the group of Archimedean classes of the additive ordered group F^+, and the residue field $F_v = R/M$ is an Archimedean ordered field. See Section 23 for examples.

An ordered field $(F, <)$ is called exponentially closed, if there exists an isomorphism $e : F^+ \to \{a \in F \mid a > 0\}$ of ordered groups (as in 11.10) with the additional property that $1 + 1/n < e(1) < n$ for some $n \in \mathbb{N}$. ALLING 1962 Theorem 3.1 gives a characterization of all non-Archimedean ordered fields which are exponentially closed, using the natural valuation and Hahn power series (as defined in 64.25).

Exercises

(1) A field F has a valuation (or an absolute value) if, and only if, F is not algebraic over any finite field.

(2) Let k be any field, and let v be a valuation of the simple transcendental extension field $k(x)$. If v is trivial on k, then v is equivalent to one of the valuations v_p or v_∞ defined in 56.3.

(3) Let F be a field with a valuation v and valuation ring R. Show that v is a principal valuation if, and only if, R is a principal ideal domain; in this case $F^\times = \pi^{\mathbb{Z}} \times R^\times$ for each prime element π of v. Assume further that F is complete with respect to v and that the residue field $F_v = R/M$ is finite, where $M = \pi R$ is the maximal ideal of R; then $R^\times \cong F_v^\times \times (1 + M)$.

(4) Let $E|F$ be an algebraic field extension and v a valuation of E. Show that the restriction $v|_F$ is a valuation (i.e. not trivial).

(5) Let F be a field with a valuation $v : F^\times \to \Gamma$, let δ be an arbitrary element of (an ordered group extending) Γ and define w on the polynomial ring $F[x]$ by $w(\sum_j a_j x^j) = \min_j(v(a_j) + j\delta)$, where $a_j \in F$. Show that w extends to a valuation of the field of fractions $F(x)$ of $F[x]$.

(6) A maximal subring R of a field F is a valuation ring of F if, and only if, F is the field of fractions of R, i.e. $F = \{ rs^{-1} \mid r, s \in R, s \neq 0 \}$.

(7) Let k be a field and Γ a non-trivial ordered abelian group. Construct a field F with a valuation $v : F^\times \to \Gamma$ such that $F_v \cong k$.

(8) The field $\mathbb{Q}_p$ is neither a purely transcendental extension nor a finite algebraic extension of any proper subfield.

(9) The skew field $\mathbb{H}$ of Hamilton's quaternions admits no valuation. The p-adic valuation v_p of $\mathbb{Q}$ extends to the rational quaternions only for $p = 2$.

57   Topologies of valuation type

In this section we characterize those ring topologies on a field F which are induced by an absolute value or by a valuation of F; Theorem 57.7 says that these are precisely the topologies of type V (defined in 57.5).

57.1 Definition Let R be a topological ring. A subset $B \subseteq R$ is called *bounded*, if for each neighbourhood U of 0 there exists a neighbourhood V of 0 such that $VB \subseteq U$ and $BV \subseteq U$. This means that B can be made small by multiplication with sufficiently small elements (see also 57.2(iii)). Clearly every subset of a bounded set is bounded.

The topological ring R and also the topology of R is called *locally bounded*, if R contains a non-empty open bounded set; by 57.2(ii) this is equivalent to assuming that some neighbourhood of 0 is bounded.

Each valuation and each absolute value of a field defines a locally bounded field topology; in fact, a set $B \subseteq F$ is bounded with respect to an absolute value $|\ |$ of F precisely if $|B|$ is a bounded subset of $\mathbb{R}$. See Exercise 4 for examples of field topologies which are not locally bounded.

57.2 Lemma *Let R be a topological ring.*

(i) *Each finite and each compact subset of R is bounded.*

(ii) *If $B, B' \subseteq R$ are bounded, then also $B \cup B'$, $B + B'$ and BB' are bounded.*

(iii) *Let R be a field with a non-discrete ring topology. Then $B \subseteq R$ is bounded if, and only if, for each neighbourhood U of 0 there exists an element $a \in R \smallsetminus \{0\}$ with $aB \subseteq U$.*

(iv) *A field R with a ring topology such that R is bounded is either indiscrete or discrete (compare 13.6).*

Proof (i) Let U be a neighbourhood of 0 in R. Since $0 \cdot x = x \cdot 0 = 0$, there exist neighbourhoods V_x and W_x of 0 with $V_x(x + W_x) \subseteq U$ and $(x + W_x)V_x \subseteq U$. Assume that $C \subseteq R$ is compact or finite. Then $C \subseteq \bigcup \{ x + W_x \mid x \in C_0 \}$ for some finite set $C_0 \subseteq C$. Hence $V := \bigcap \{ V_x \mid x \in C_0 \}$ is a neighbourhood of 0 with $VC \subseteq U$ and $CV \subseteq U$.

(ii) is a consequence of the continuity of addition and multiplication.

(iii) If B is bounded, then the condition in (iii) is satisfied, since each neighbourhood of 0 contains a non-zero element a. For the converse, let U be a neighbourhood of 0. By continuity of the multiplication we find a neighbourhood W of 0 with $WW \subseteq U$, and by assumption we find $a \neq 0$ with $aB \subseteq W$. The neighbourhood $V = Wa$ of 0 satisfies $VB = (Va^{-1})(aB) \subseteq WW \subseteq U$.

(iv) If such a field R is not indiscrete, then R contains an open set U with $0 \in U \not\ni 1$; see 13.4. Since $R = aR \not\subseteq U$ for each $a \in R^\times$ we infer from (iii) that R is discrete. $\qquad\square$

57.3 Definition An element a of a topological ring is called *topologically nilpotent*, if the sequence of powers a^n converges to 0.

Non-zero topologically nilpotent elements exist in each field F with a topology defined by an absolute value $|\ | : F \to \mathbb{R}$ or by a valuation $v : F^\times \to \Gamma$ with values in an Archimedean ordered group Γ.

Topologically nilpotent elements play a role in the characterization theorems 57.4 and 57.7.

57.4 Theorem (Shafarevich, Kaplansky) *Let F be a field with a Hausdorff ring topology. Then the following are equivalent.*

(i) *The topology of F is induced by an absolute value $|\ | : F \to \mathbb{R}$.*

(ii) *The set $N := \{ a \in F \mid \lim_{n \to \infty} a^n = 0 \}$ of all topologically nilpotent elements of F is a neighbourhood of 0 with $N \neq \{0\}$, and $(F \smallsetminus N)^{-1}$ is bounded.*

Proof It is easy to see that (i) implies (ii): $N = \{ a \in F \mid |a| < 1 \}$ and $(F \smallsetminus N)^{-1} = \{ a \in F \mid |a| \leq 1 \}$ in the situation of (i), and both sets are bounded neighbourhoods of 0.

Now assume (ii). Then the subset

$$E := \{ a \in F^\times \mid a \notin N \wedge a^{-1} \notin N \} = (F \smallsetminus N) \cap (F \smallsetminus N)^{-1}$$

of $(F \smallsetminus N)^{-1}$ is bounded. We have $N \subseteq \{0\} \cup (F \smallsetminus N)^{-1}$, as $0 \neq a \in N$ and $a^{-1} \in N$ would lead to the contradiction $1 = a^n a^{-n} \to 0 \cdot 0 = 0$ (here we need the Hausdorff property). Hence N is bounded. Moreover,

$$F = N \cup E \cup (N \smallsetminus \{0\})^{-1} \ .$$

The inclusion $NN \subseteq N$ is a consequence of the commutativity of the multiplication. If $a^m \in N$ for some $m \in \mathbb{N}$, then $a \in N$, because $\lim_{n \to \infty} a^{mn} = 0$ entails $\lim_{n \to \infty} a^{mn+j} = 0$ for $0 \leq j < m$, hence $\lim_{n \to \infty} a^n = 0$. This shows that $y^{\mathbb{N}} \subseteq E$ for each $y \in E$; otherwise we find $n \in \mathbb{N}$ with $y^{\pm n} \in N$, hence $y^{\pm 1} \in N$, and we have reached a contradiction.

We claim that $NE \subseteq N$. Since N and E are bounded, NE is bounded by 57.2(ii), hence $aNE \subseteq N$ for some $a \in F^{\times}$ by 57.2(iii). If $x \in N$ and $y \in E$, then $x^n \in aN$ for sufficiently large exponents n, hence $(xy)^n = x^n y^n \in aNy^{\mathbb{N}} \subseteq aNE \subseteq N$ and $xy \in N$.

Now we can show that E is a subgroup of $F^{\times}$: if $x, y \in E$ and $x^{-1}y \notin E$, then $(x^{-1}y)^{\varepsilon} \in N$ for some $\varepsilon \in \{\pm 1\}$, hence $y = x(x^{-1}y) \in xN \subseteq N$ or $x = y(x^{-1}y)^{-1} \in yN \subseteq N$, which is absurd.

The set $R := E \cup N$ is a bounded neighbourhood of 0. From 57.2 we infer that $R + R$ is bounded, and that $(R + R)b \subseteq R$ for some $b \in F^{\times}$.

The factor group $\Gamma := F^{\times}/E$ is not trivial, because $N \neq \{0\}$. The definition

$$xE < yE \Leftrightarrow x^{-1}y \in N$$

for $x, y \in F^{\times}$ renders Γ an ordered group; the relation $<$ is well-defined since $NE = N$, and $<$ is a total ordering, because $x^{-1}y \notin N$ implies $y^{-1}x \in E \cup N$. In fact, $(\Gamma, <)$ is an Archimedean ordered group: if $x, y \in F^{\times}$ and yE is strictly positive, then $y \in N$, hence the elements y^n and $x^{-1}y^n$ converge to 0, which entails $x^{-1}y^n \in N$ and $xE < y^n E$ for all sufficiently large integers n.

According to Theorem 7.8 we can embed the ordered group $(\Gamma, <)$ into the ordered group $(\mathbb{R}^{+}, <)$. Using the automorphisms of $(\mathbb{R}^{+}, <)$ (compare 7.11), we find a monomorphism $\alpha : \Gamma \to \mathbb{R}$ of ordered groups with $\alpha(bE) \leq \log 2$. Now we define the mapping $\varphi : F \to \mathbb{R}$ by

$$\varphi(x) = \exp(-\alpha(xE))$$

for $x \in F^{\times}$ and $\varphi(0) = 0$. Clearly φ is multiplicative, and $\varphi(1) = 1$. We claim that

$$\varphi(x + y) \leq 2 \max\{\varphi(x), \varphi(y)\}$$

for $x, y \in F$. It suffices to consider the case where $\varphi(y) \leq \varphi(x)$. Then $\alpha(yE) \geq \alpha(xE)$ and $x^{-1}y \in N \cup E = R$, hence $(x+y)b = x(1+x^{-1}y)b \in x(R + R)b \subseteq xR$. This implies $\alpha((x + y)E) + \log 2 \geq \alpha((x + y)bE) \geq \alpha(xE)$ and $\varphi(x + y) \leq 2\varphi(x)$. By Artin's trick 55.7, φ is an absolute value of F.

By the assumptions on N, the sets $aN = \{\, x \in F \mid \varphi(x) < \varphi(a) \,\}$ with $0 \neq a \in F$ form a neighbourhood base at 0 for the given topology of F. Therefore φ induces this topology. $\square$

57.5 Definition Let F be a field with a ring topology. One says that F and also its topology are of *type V* (or *locally retrobounded*), if the set $(F \smallsetminus U)^{-1}$ is bounded for each neighbourhood U of 0.

It suffices to check this condition for the elements U of a neighbourhood base at 0 (since subsets of bounded sets are bounded). See also Exercise 5.

Here the symbol V stands for valuation; indeed, the topologies induced by valuations or absolute values are of type V (Exercise 1). Theorem 57.7 essentially says that also the converse statement is true.

57.6 Lemma *Let F be a field. Then each ring topology of type V on F is a locally bounded field topology.*

Proof We may assume that F is a Hausdorff space; otherwise the topology of F is indiscrete by 13.4, hence a locally bounded field topology.

First we show that inversion is continuous. Let U be a neighbourhood of 0. By continuity of addition, there exists a neighbourhood W of 0 with $W + W \subseteq F \smallsetminus \{-1\}$. As $(F \smallsetminus U)^{-1}$ is bounded, there is a neighbourhood $V \subseteq W$ with $V(F \smallsetminus U)^{-1} \subseteq W$ and $(F \smallsetminus U)^{-1}V \subseteq W$. We infer that $-1 \notin V + V(F \smallsetminus U)^{-1}$, hence $1 + v \notin -v(F \smallsetminus U)^{-1}$ and $(1+v)^{-1} = 1 + (1+v)^{-1}(-v) \notin 1 + (F \smallsetminus U)$ for each $v \in V$. This shows that $(1 + V)^{-1} \subseteq 1 + U$. We conclude that inversion is continuous at 1, hence everywhere (see 8.3 or 13.2(a)).

Now we show that F is locally bounded. We can choose a neighbourhood U of 0 with $UU \subseteq F \smallsetminus \{1\}$ by the continuity of multiplication at 0. If $0 \neq x \in U$, then $x^{-1} \notin U$, hence $x \in (F \smallsetminus U)^{-1}$. The set $(F \smallsetminus U)^{-1}$ is bounded, since we have a topology of type V. Therefore, $U \subseteq \{0\} \cup (F \smallsetminus U)^{-1}$ is bounded; see 57.2(i, ii). $\square$

There exist field topologies which are locally bounded, but not of type V; see Exercise 4.

57.7 Theorem (Kowalsky–Dürbaum, Fleischer) *Let F be a field with a Hausdorff ring topology of type V.*

  (i) *If 0 is the only topologically nilpotent element of F and if F is not discrete, then the topology of F is induced by a valuation of F with a value group which is not Archimedean.*

 (ii) *If F contains a non-zero topologically nilpotent element, then the topology of F is induced by an absolute value of F.*

This result says that the non-discrete Hausdorff ring topologies of type V on a field F are precisely the topologies induced by the absolute values of F, together with the valuation topologies of F (valuations with Archimedean value groups define the same topologies as non-Archimedean absolute values; see 56.2).

Proof (i) We shall construct a valuation ring R of F such that R is a bounded neighbourhood of 0. This proves (i), because then the sets aR with $a \in F^\times$ form a neighbourhood base at 0 (see 57.2(iii)), hence the given topology of F coincides with the topology induced by the valuation v associated with R; the value group $v(F^\times)$ cannot be Archimedean (otherwise each element with positive value would be topologically nilpotent).

By 57.6 there exists a bounded neighbourhood V of 0, hence also a neighbourhood W of 0 with $WV \subseteq V$. The set $D := \{\, a \in F \mid aV \subseteq V \,\}$ contains W and hence is a neighbourhood of 0. Moreover, D is bounded, since $D \subseteq Vv^{-1}$ whenever $0 \neq v \in V$, and Vv^{-1} is bounded by 57.2(ii). As F has type V, also $(F \smallsetminus D)^{-1}$ is bounded, hence $U(F \smallsetminus D)^{-1} \subseteq D$ for some neighbourhood U of 0. We pick any non-zero element $c \in U \cap D$, we put $C := \{\, a \in F \mid aD \subseteq c^{-\aleph_0}D \,\}$ and we define R to be the additive subgroup of F generated by C. Note that $c^{-1} \in C$.

Then R is a subring of F, in view of $CC \subseteq C$. Moreover $DD \subseteq D$, hence $D \subseteq C \subseteq R$, whence R is a neighbourhood of 0. By the choice of c we have $(F \smallsetminus R)^{-1} \subseteq (F \smallsetminus D)^{-1} \subseteq c^{-1}D \subseteq c^{-1}C \subseteq CC \subseteq C \subseteq R$. Thus R is a valuation ring of F, provided that $R \neq F$. We complete the proof by showing that R is bounded, which implies $R \neq F$ according to 57.2(iv).

We claim that the set $c^{-\aleph_0}$ is bounded. Since V is bounded, the sets Va with $a \in F^\times$ form a neighbourhood base at 0. If $c^n \in Va$ for some $n \in \mathbb{N}$, then $c^{n+1} \in cVa \subseteq Va$ (as $c \in D$), hence $c^m \in Va$ for all $m \geq n$. By assumption, c is not topologically nilpotent, hence $c^{\mathbb{N}} \cap Va = \emptyset$ for some $a \in F^\times$. Thus $c^{-\mathbb{N}}$ is contained in the set $(F \smallsetminus Va)^{-1}$, which is bounded as F has type V; hence $c^{-\aleph_0}$ is bounded.

By 57.2(ii) also $c^{-\aleph_0}D$ is bounded, hence C is bounded as well in view of $C \subseteq c^{-\aleph_0}Dd^{-1}$ for each non-zero $d \in D$. We infer again from 57.2(ii) that $(C \cup -C) + (C \cup -C)$ is bounded, hence

$$(C \cup -C) + (C \cup -C) \subseteq b^{-1}C$$

for some non-zero element $b \in C$ by 57.2(iii). An easy induction shows that the sum of 2^n sets $C \cup -C$ is contained in $b^{-n}C$, hence $R \subseteq b^{-\aleph_0}C$.

Now $C = C + \{0\} \subseteq b^{-1}C$, hence $bC \subseteq C$, and we infer as in the previous paragraph that $b^{-\aleph_0}$ is bounded (using the bounded set C instead of V). Thus $R \subseteq b^{-\aleph_0}C$ is bounded by 57.2(ii).

(ii) Let $t \in F^\times$ be topologically nilpotent. By 57.6 we find a bounded neighbourhood V of 0. As in the proof of (i) above, we infer that $D := \{ a \in F \mid aV \subseteq V \}$ is a bounded neighbourhood of 0, hence the sets $t^n D$ with $n \in \mathbb{N}$ form a neighbourhood base at 0. Each element of tD is topologically nilpotent, as $DD \subseteq D$. Thus the set N of all topologically nilpotent elements of F is a neighbourhood of 0. Since F has type V, the set $(F \smallsetminus N)^{-1}$ is bounded, and assertion (ii) follows from 57.4.   □

57.8 Corollary   *A non-discrete Hausdorff ring topology τ of a field F is defined by a valuation of F if, and only if, τ has type V and the additive group F^+ contains a bounded open subgroup.*

Proof Each valuation topology of F has type V (by Exercise 1), and $\{ a \in F \mid v(a) > 0 \}$ is a bounded open subgroup of F^+. The converse implication is a consequence of Theorem 57.7, because Archimedean absolute values are excluded if there exists a bounded open subgroup U of F^+: there is an element $u \neq 0$ in U, hence $\mathbb{Z} \cdot 1_F \subseteq Uu^{-1}$ is bounded by 57.2(ii), but $|\mathbb{Z} \cdot 1_F|$ is unbounded for every Archimedean absolute value $|\ |$ of F. The topology of a non-Archimedean absolute value is also induced by a valuation; see 56.2.   □

The field $\mathbb{Q}$ of rational numbers has many field topologies (see Exercise 4 or 13.10), but only few of them are of type V.

57.9 Corollary   *The non-discrete Hausdorff ring topologies of type V on $\mathbb{Q}$ are the usual topology and the p-adic topologies, where p is a prime number.*

Proof This is a consequence of 57.7, 56.7 and 55.4.   □

In fact, all locally bounded field topologies of $\mathbb{Q}$ can be described: each topology of this type is the supremum of finitely many topologies induced by absolute values of $\mathbb{Q}$ (compare Exercise 4). See WARNER 1989 and WIĘSŁAW 1988 for proofs and for more information on locally bounded fields and fields of type V; compare also SHELL 1990 Chapter 4.

According to MUTYLIN 1968 Theorem 5 (see also WIĘSŁAW 1988 9.4), $\mathbb{R}$ and $\mathbb{C}$ are the only locally bounded extension fields of $\mathbb{R}$ that induce on $\mathbb{R}$ the usual topology.

Exercises

(1) Each valuation and each absolute value of a field F induces on F a field topology of type V. The order topology of each ordered field is of type V.

(2) Let F be a field with a non-discrete ring topology, and let U be a bounded neighbourhood of 0. Show that the sets aU with $a \in F^\times$ form a neighbourhood base at 0.

(3) Let τ be a non-discrete Hausdorff ring topology on a field F, let $\infty \notin F$ and let τ_∞ be the topology on $F_\infty := F \cup \{\infty\}$ obtained from τ by adding all sets $F_\infty \setminus B$ where B is closed in F and bounded. Show that τ_∞ is a Hausdorff topology if, and only if, τ is locally bounded, and that τ is of type V if, and only if, the inversion $x \mapsto x^{-1}$, $0 \mapsto \infty$, $\infty \mapsto 0$ is continuous at ∞.

(4) For any non-empty set X of prime numbers, we define a metric d_X on $\mathbb{Q}$ by $d_X(a,b) = \sum_{p \in X} 2^{-p} \cdot |a - b|_p$ (compare 36.6). Show that the topology τ_X defined by d_X is always a field topology of $\mathbb{Q}$, that τ_X is locally bounded precisely if X is finite, and that τ_X has type V precisely if $\operatorname{card} X = 1$.

(5) A topological field F has type V if, and only if, for every neighbourhood W of 0 there exists a neighbourhood U of 0 such that $x, y \in F$ and $xy \in U$ implies that $x \in W$ or $y \in W$.

(6) The completion of a topological field of type V is again a topological field.

(7) Let v be a valuation of a field F, let $a \neq 0$ be a topologically nilpotent element of F, and $R_a := \{ x \in F \mid \forall_{n \in \mathbb{N}} \, nv(x) + v(a) > 0 \}$. Show that R_a is a maximal subring of F, and deduce that the valuation topology of v is also defined by a valuation of F with an Archimedean value group.

58 Local fields and locally compact fields

Here we briefly discuss the concept of local fields, and we explain how one can classify all locally compact skew fields. In Theorem 58.7 we shall see that the local fields together with $\mathbb{R}$ and $\mathbb{C}$ are precisely the locally compact fields which are neither discrete nor indiscrete.

In this section, a locally compact (skew) field is assumed to be endowed with a **non-discrete Hausdorff** topology.

58.1 Local fields Usually a local field is defined as a field F which is complete with respect to a valuation $v : F^\times \to \mathbb{Z}$ with a finite residue field (see Section 56). These fields can be described more explicitly, as follows: either F is isomorphic to the field $\mathbb{F}_q((t))$ of Laurent series over a finite field $\mathbb{F}_q$ (see 64.23 and 44.11), or F is an extension of finite degree of a p-adic field $\mathbb{Q}_p$ (see 56.3(a), 56.15 and 58.2). Indeed, by 56.12 the local fields are precisely the locally compact fields with valuation, and 58.7 gives the assertion (compare also JACOBSON 1989 Theorem 9.16 p. 577 and NEUKIRCH 1992 II.5.2).

A local field F has only one principal valuation $v : F^\times \to \mathbb{Z}$; see 56.13. We call this valuation the *natural valuation* of F. As in 56.1, the prime elements π of F are defined by the condition $v(\pi) = 1$. Each prime element of F generates the unique prime (in fact, maximal) ideal in the natural valuation ring of F.

Many properties of $\mathbb{Q}_p$ can be generalized to arbitrary local fields, as the books by CASSELS 1986 and SERRE 1979 show. However, the multiplicative group $\mathbb{F}_q((t))^\times$ is isomorphic to $\mathbb{Z} \times C_{q-1} \times (\mathbb{Z}_p^+)^{\mathbb{N}}$, where q is a power of the prime p (see WEIL 1967 II.3 Proposition 10 p. 34), in contrast to the analogous result 53.3 on $\mathbb{Q}_p^\times$.

58.2 Field extensions and automorphisms Let F be a local field with its natural valuation v, and let E be a field extension of finite degree n of F. Then E is also a local field and v has a unique extension to E which is the natural valuation w of E; see 56.15.

The index $e = |w(E^\times) : v(F^\times)|$ of the two value groups is called the *ramification index* of $E|F$; this terminology reflects the fact that each prime element of F is a product of e prime elements of E. Furthermore the residue field F_v is canonically embedded into E_w, and the degree $f = [E_w : F_v]$ is called the *residue degree* of $E|F$. These numbers are related via the equation $n = ef$; for proofs see, for example, COHN 2003a 9.5.1, JACOBSON 1989 p. 591, SERRE 1979 p. 29 or BOREVICH–SHAFAREVICH 1966 Chapter 4 §1 Theorem 5 p. 262.

A field extension $E|F$ as considered above is called *unramified*, if $e = 1$; this means that the value groups coincide and that the residue fields have the same degree n as $E|F$. If $f = 1$, then $E|F$ is said to be *totally ramified*; this means that the residue fields coincide and that the value group $v(F^\times)$ has index n in $w(E^\times)$.

Let F be a local field with residue field $\mathbb{F}_q$ (so q is a prime power), and let $n \in \mathbb{N}$. Then F has a unique unramified extension F_n of degree n, namely the splitting field of $x^{q^n} - x$ over F, and $F_n|F$ is a cyclic Galois extension; see REINER 1975 5.10 and 5.11, SERRE 1979 III §5 Theorem 2 p. 54 or WEIL 1967 I §4 Corollaries 2 and 3 p. 18/19. If $F = \mathbb{F}_q((t))$, then $F_n = \mathbb{F}_{q^n}((t))$.

Let F be a local field, and let π be a prime element of F. Adjoining to F a root of the polynomial $x^n - \pi$ gives a totally ramified field extension of degree n over F, for any $n \in \mathbb{N}$. In fact, the totally ramified extensions of F are precisely the extensions generated by roots of Eisenstein polynomials over F, i.e., polynomials $x^n + \sum_{i=0}^{n-1} a_i x^i$ with $v(a_0) = 1$ and $v(a_i) \geq 1$ for all i; see SERRE 1979 I §6(ii) p. 19, CASSELS 1986 Theorem 7.1 p. 133, RIBENBOIM 1999 4.H p. 116, LANG 1970 Chapter II Proposition 11 or ROBERT 2000 2.4.2.

Each local field F of characteristic $p \neq 0$ is isomorphic to $\mathbb{F}_q((t))$ for some power q of p; indeed, each prime element π of F is transcendental over the prime field $\mathbb{F}_p$ (otherwise π would be a root of unity), hence $\mathbb{F}_p(\pi) \cong \mathbb{F}_p(x)$ with the valuation v_x as in 56.3(c). Therefore F contains the completion $\mathbb{F}_p((\pi)) \cong \mathbb{F}_p((x))$ (compare 43.9 and 44.11(3)), and the extension $F|\mathbb{F}_p((\pi))$ is finite-dimensional (of degree at most $[F_v : \mathbb{F}_p]$) and unramified, hence $F \cong \mathbb{F}_p((x))_n = \mathbb{F}_{p^n}((t))$; see also RIBENBOIM 1999 Chapter 7 Theorem 1 p. 196.

One can show that $\mathbb{Q}_p$ has only finitely many extensions of given degree (see 54.3 for the case of quadratic extensions); indeed, the Eisenstein polynomials of fixed degree form a compact topological space $X \subset \mathbb{Q}_p[x]$, and by Krasner's Lemma the isomorphism type of the field $\mathbb{Q}_p[x]/(f)$ is a locally constant function of $f \in X$; see ROBERT 2000 3.1.6, WEIL 1967 p. 207f, REINER 1975 33.8 or LANG 1970 Chapter II Proposition 14. For local fields of positive characteristic, the analogous statement is not true; see Exercise 3. Enumerations of extensions of $\mathbb{Q}_p$ of low degree can be found in PAULI–ROBLOT 2001 and KLAAS *et al.* 1997 IX; see also the online database under http://math.asu.edu/~jj/localfields/.

Theorem 56.13 implies that each field automorphism of a local field is an isometry with respect to the natural valuation, hence continuous. The automorphism group G of a finite field extension of $\mathbb{Q}_p$ fixes each element of $\mathbb{Q}$ and of $\mathbb{Q}_p = \overline{\mathbb{Q}}$, hence G is finite. The Galois groups of Galois extensions of local fields are always soluble; see SERRE 1979 IV §2 Corollary 5 p. 68 or RIBENBOIM 1999 9.Q p. 254. Thus each polynomial equation with coefficients in $\mathbb{Q}_p$ is solvable by radicals over $\mathbb{Q}_p$.

Let q be a power of the prime p. The automorphism group G of $\mathbb{F}_q((t))$, which coincides with the automorphism group of the power series ring $\mathbb{F}_q[[t]]$ by 56.14, is notoriously complicated. The kernel N of the action of G on $\mathbb{F}_q[[t]]/(t^2)$ is called the Nottingham group. The quotient G/N is finite, and N may be described as the set of all power series $t + \sum_{i>1} a_i t^i$ where $a_i \in \mathbb{F}_q$, with substitution as the group operation. This group N is a finitely generated pro-p-group which contains every finite p-group and every finitely generated pro-p-group as a closed subgroup; see LEEDHAM–GREEN–McKAY 2002 12.4.11, 12.4.16.

Note that $\mathbb{F}_q((t))$ admits many proper field endomorphisms; for example, one can map t to t^k for each $k \geq 2$.

58.3 Local–global principles The fields $\mathbb{R}$ and $\mathbb{Q}_p$, p a prime, are the completions of $\mathbb{Q}$ with respect to the absolute values of $\mathbb{Q}$; see 55.4, 44.11, 51.4. Vaguely speaking, a local–global principle relates the behaviour of mathematical objects over $\mathbb{Q}$ to their behaviour over $\mathbb{R}$ and over the local fields $\mathbb{Q}_p$ for all primes p. (The finite extensions of $\mathbb{Q}$ and of $\mathbb{F}_p(x)$ are often called global fields.) For example, a rational number a is a square in $\mathbb{Q}$ if, and only if, a is a square in $\mathbb{R}$ (i.e. $a \geq 0$) and in $\mathbb{Q}_p$ for each prime p (just observe that the p-adic value $v_p(a)$, as defined in 56.3(a), is even if a is a square in $\mathbb{Q}_p$).

By considering the quadratic form $ax^2 - y^2$, we see that this example is a very special case of the following famous theorem of Hasse–Minkowski: a quadratic form $f = \sum_i a_i x_i^2$ with rational coefficients a_i is isotropic (as defined in Section 54) over $\mathbb{Q}$ if, and only if, f is isotropic over $\mathbb{R}$ and over $\mathbb{Q}_p$ for each prime p (in fact, it suffices to consider only finitely many primes p). See SERRE 1973, BOREVICH–SHAFAREVICH 1966, SCHARLAU 1985, CASSELS 1978 or LAM 2005 for proofs and extensions. Furthermore, Hensel's Lemma may be regarded as a local–global principle (see the introduction to Chapter 5).

For Diophantine equations of higher degree (and for systems of more than two quadratic forms), such a general local–global principle does not hold. For example, the cubic form $3x^3 + 4y^3 + 5z^3$ is isotropic over $\mathbb{R}$ and over $\mathbb{Q}_p$ for each prime p, but not over $\mathbb{Q}$ (Selmer 1951; see CASSELS 1986 Chapter 10 Lemma 9.1 or BOREVICH–SHAFAREVICH 1966 Chapter 1 7.6). For more examples see Exercise 4 and CASSELS 1986 Chapter 4 3bis. Still, the failure of a local–global principle can sometimes be measured (via a Galois cohomological reformulation), thus salvaging some connection between local and global behaviour; see the survey by MAZUR 1993.

58.4 Locally compact skew fields In the rest of this section we consider the classification of all locally compact skew fields. The connected skew fields of this type are $\mathbb{R}$, $\mathbb{C}$ and $\mathbb{H}$; see 13.8 or 58.11. The fields $\mathbb{F}_q((x))$ and $\mathbb{Q}_p$, endowed with their natural valuation topologies, are examples of locally compact disconnected fields (by 44.11 and 51.10); more generally, each local field is also an example (by 13.2(c) or 56.12). In fact, 13.2(c) shows that each skew field with finite (right or left) dimension over a local field F is a locally compact skew field with respect to the product topology obtained from F.

The classification results 58.7 and 58.9 say that these remarks describe all locally compact skew fields.

There are two approaches to achieve this classification. The first approach applies to locally compact skew fields F which are totally disconnected: the study of the compact (open) subrings of such a skew field F shows that the (unique) maximal compact subring R of F is a valuation ring (in a non-commutative sense); see JACOBSON 1989 Section 9.13 for a good exposition. In fact, the maximal ideal of R is the set $M = \{ x \in F \mid x^n \to 0 \}$ of all topologically nilpotent elements of F, and $R = F \smallsetminus (M \smallsetminus \{0\})^{-1}$.

Another approach, which can be found in WEIL 1967 Chapter I (and in BOURBAKI 1972 VI.9, ROBERT 2000 Appendix to Chapter 2), treats the connected and the totally disconnected case simultaneously; here one uses a Haar measure of the additive group F^+ for a direct construction of an absolute value on F, as follows.

58.5 Lemma *Let F be a locally compact skew field. Then there exists a unique function $| \ | : F \to \mathbb{R}$ such that*

$$\mu(aX) = |a|\mu(X)$$

for each $a \in F$, each compact subset $X \subseteq F$ and each Haar measure μ of F^+.

(i) *This function $| \ | : F \to \mathbb{R}$ is multiplicative and continuous, and $|1| = 1$.*

(ii) *The balls $B_r := \{ x \in F \mid |x| \leq r \}$ with $0 < r \in \mathbb{R}$ are compact and form a neighbourhood base at 0 for F.*

(iii) *Some power $| \ |^s$ of $| \ |$ is an absolute value of F inducing the given locally compact topology of F.*

(iv) *Each discrete sub-skew-field F_0 of F is finite.*

Proof Recall that F carries a non-discrete Hausdorff topology. By definition, a Haar measure of F^+ is a translation invariant measure μ defined on all Borel subsets of F such that $\mu(X) > 0$ if $X \neq \emptyset$ is open in F. Each Haar measure is regular in the sense of 10.2 on (the σ-algebra generated by the) compact sets; see HALMOS 1950 §64 Theorem I p. 288 and p. 230. Like each locally compact group, F^+ admits a Haar measure μ, and μ is determined uniquely up to a positive factor; see HALMOS 1950 §58 Theorem B p. 254 and §60 Theorem C p. 263. (For constructions of the Haar measure without using the axiom of choice see HEWITT–ROSS 1963 §15, LOOMIS 1945 and BREDON 1963.) Moreover, $\mu(X) < \infty$ for each compact subset X of F; see HALMOS 1950 p. 255 or HEWITT–ROSS 1963 15.8 and 11.24.

Since $x \mapsto ax$ is an automorphism of the topological group F^+ for $0 \neq a \in F$, the assignment $X \mapsto \mu(aX)$ is also a Haar measure of F^+, hence $\mu(aX) = r\mu(X)$ with a positive real factor $r = |a|$, which is independent of X and μ.

(i) The function $|\ |$ is multiplicative, as an immediate consequence of its definition, and $|1| = 1$. Now we show that $|\ |$ is upper semicontinuous: let C be a compact neighbourhood of 0 in F, let $a \in F$ and $\varepsilon > 0$. Since μ is regular (10.2), there exists an open set U in F with $aC \subseteq U$ and $\mu(U) \le \mu(aC) + \varepsilon$. Furthermore we find a neighbourhood V of a with $VC \subseteq U$. Each $x \in V$ satisfies $xC \subseteq U$, hence

$$|x| = \mu(xC)/\mu(C) \le \mu(U)/\mu(C) \le |a| + \varepsilon/\mu(C)\ .$$

Therefore $|\ |$ is upper semicontinuous.

In particular, $|\ |$ is continuous at 0. Furthermore $|x| = |x^{-1}|^{-1}$ for $x \ne 0$, hence $|\ |$ is also lower semicontinuous everywhere on $F^\times$, and therefore continuous on $F^\times$.

(ii) Let V be a compact neighbourhood of 0 in F. There exists a compact neighbourhood W of 0 such that $VW \subseteq V$, as V is bounded by 57.2(i). Since 0 is not isolated in $V \cap W$ and $|\ |$ is continuous, we can choose $a \in V \cap W$ such that $0 < |a| < 1$. Then $a^n \in V$ for all $n \ge 1$, by induction on n. From (i) we infer that 0 is the only accumulation point of the sequence of powers a^n. This sequence is contained in the compact set V, hence it has the limit 0.

Now we show that each ball B_r is compact. Let $x \in B_r \smallsetminus V$. As $a^n x$ converges to 0, there exists a smallest integer $n \in \mathbb{N}$ with $a^n x \in V$. Thus $a^n x \in V \smallsetminus aV$, which implies $|a|^n r \ge |a^n x| \ge \inf |V \smallsetminus aV| > 0$. This gives an upper bound N for n which is independent of x. We conclude that B_r is contained in the compact set $V \cup \bigcup_{n \le N} a^{-n} V$. Furthermore B_r is closed by (i), hence compact.

In order to show that these balls form a neighbourhood base at 0, we proceed indirectly and consider a neighbourhood U of 0 such that $B_{1/n} \not\subseteq U$ for each $n \in \mathbb{N}$. We obtain a sequence $b_n \in B_{1/n} \smallsetminus U$, which has an accumulation point b in the compact set B_1. Each ball $B_{1/n}$ contains $b_n, b_{n+1}, b_{n+2}, \ldots$ (since these balls form a chain), hence $b \in B_{1/n}$. Thus $b \in \bigcap_{n \in \mathbb{N}} B_{1/n} = \{0\}$ and $b = 0$, which is a contradiction, since $b_n \notin U$ for all $n \in \mathbb{N}$.

(iii) Let $M := \max |1 + B_1|$. The multiplicativity of $|\ |$ implies that $|x + y| \le M \max\{|x|, |y|\}$ for all $x, y \in F$, since $|x + y| = |1 + yx^{-1}||x|$ and $yx^{-1} \in B_1$ for $0 \ne |x| \ge |y|$. (If $M = 1$, then $|\ |$ is an ultrametric absolute value; compare 55.3.)

Now choose $s \in \mathbb{R}$ such that $M^s \le 2$. Then the multiplicative function $\varphi = |\ |^s$ satisfies $\varphi(x + y) \le 2 \max\{\varphi(x), \varphi(y)\}$. By Artin's trick 55.7, this function φ is an absolute value of F, and by (ii) it induces the given topology on F.

(iv) Since F_0 is discrete, we infer $|F_0| = \{0, 1\}$ from (iii). Moreover, the additive group F_0 is closed in F: if a sequence of elements $a_n \in F_0$ converges to $a \in F$, then $a_{n+1} - a_n \in F_0$ converges to 0 and is finally constant, hence $a \in F_0$. Thus F_0 is a closed, discrete subset of the compact ball B_1, hence finite. □

The following proposition is a basic principle of functional analysis.

58.6 Proposition *Let V be a Hausdorff topological vector space over a locally compact skew field F. Then the following hold.*

  (i) *Each finite-dimensional subspace S of V is closed in V and carries the product topology obtained from any isomorphism with F^d, where $d = \dim S$.*

 (ii) *If V is locally compact, then V has finite dimension over F.*

Proof To fix the notation, we consider V as a left vector space over F (with scalars on the left). For vector spaces V with countable neighbourhood bases, the nets appearing in the proof for (i) can be replaced by sequences.

(i) We use the following property of F, which is a consequence of 58.5(i,ii): if $(a_n)_n$ is a net in F such that the net $(|a_n|)_n$ converges to ∞ in the one-point compactification of $\mathbb{R}$, then the net $\left(a_n^{-1}\right)_n$ converges to $0 \in F$; this means that the inversion $a \mapsto a^{-1}$ on $F^\times$ extends to a continuous involution on the one-point compactification $F \cup \{\infty\}$. (Thus 58.6 holds for complete fields F of type V, by Exercise 3 of Section 57.)

First we show by induction on d that each d-dimensional subspace of V is closed in V; this is obvious if $d = 0$. Let $d > 0$, let H be a subspace of dimension $d - 1$ and $v \in V \smallsetminus H$. By induction, H is closed in V. In order to show that $\langle H, v \rangle = H \oplus Fv$ is closed in V, we consider a convergent net $h_n + a_n v \to w \in V$ with $h_n \in H$ and $a_n \in F$; we have to show that $w \in \langle H, v \rangle$. If the real numbers $|a_n|$ are unbounded, then the scalars a_n^{-1} accumulate at 0, hence the vectors $-a_n^{-1}(h_n + a_n v) + v = -a_n^{-1} h_n \in H$ accumulate at $0 \cdot w + v = v \notin H$, a contradiction. This contradiction shows that the real numbers $|a_n|$ are bounded, hence the scalars a_n accumulate at some $a \in F$ (by 58.5(ii)), and the vectors $(h_n + a_n v) - a_n v = h_n \in H$ accumulate at $w - av$. We obtain $w - av \in H$ and $w \in \langle H, v \rangle$.

Let H be a closed subspace of V and $v \in V \smallsetminus H$. We claim that the linear bijection

$$\lambda : H \times F \to \langle H, v \rangle : (h, a) \mapsto h + av$$

is a homeomorphism (then an easy induction completes the proof of (i)).

Clearly λ is continuous; it suffices to verify the continuity of λ^{-1} at 0. For this we consider a net $h_n + a_n v$ converging to 0 with $h_n \in H$ and $a_n \in F$. We have to show that $a_n \to 0$ (which entails $h_n \to 0$). Proceeding indirectly, we assume that the scalars a_n do not converge to 0. Then the real numbers $|a_n^{-1}|$ do not converge to ∞, hence they have a bounded subnet. Thus the scalars a_n^{-1} have an accumulation point b in F (by 58.5(ii)). We infer that the vectors $-a_n^{-1}(h_n + a_n v) + v = -a_n^{-1}h_n \in H$ accumulate at $b \cdot 0 + v = v \notin H$, a contradiction, as H is closed in the Hausdorff space V.

(ii) Let C be a compact neighbourhood of 0 in V, and pick $a \in F$ with $0 < |a| < 1$. Then the powers a^n converge to 0, and the sets $a^n C$ form a countable neighbourhood base of 0 in V (by 58.5(i, ii)). The compact set C is covered by finitely many sets $v_i + aC$, $1 \le i \le d$, where $v_i \in V$. Let $S = \langle v_1, \ldots, v_d \rangle$ be the (finite-dimensional) subspace generated by $v_1, \ldots, v_d$. Then $C \subseteq S + a^k C$ for each $k \in \mathbb{N}$, as an easy induction on k shows. Let $v \in V$. The vectors $a^n v$ converge to 0, hence for all sufficiently large n we have $a^n v \in C \subseteq S + a^{2n}C$ and $v \in S + a^n C$. This shows that S is dense in V, and (i) implies $S = V$. $\square$

If F is commutative, then the crucial fact 58.5(iii) that the topology of F is induced by an absolute value may also be inferred from Theorem 57.4; see WIĘSŁAW 1988 6.2 Lemma 3 p. 153. This suffices for the following two results.

58.7 Theorem   *Each locally compact field F is a field extension of finite degree of $\mathbb{R}$, $\mathbb{Q}_p$ or $\mathbb{F}_p((x))$, where p is a prime. Thus the locally compact fields are precisely the fields $\mathbb{R}$, $\mathbb{C}$ and the local fields.*

Proof By 58.5(iii), the topology of F is described by an absolute value φ of F.

First we consider the case where F has characteristic zero. Then $\mathbb{Q} \subseteq F$, and the restriction $\varphi|_\mathbb{Q}$ is not trivial by 58.5(iv), hence an absolute value of $\mathbb{Q}$. Using Ostrowski's enumeration (55.4) of all absolute values of $\mathbb{Q}$, we conclude that $\varphi|_\mathbb{Q}$ is a power of the usual absolute value or of a p-adic absolute value. The topological closure $\overline{\mathbb{Q}}$ of $\mathbb{Q}$ in F is complete, hence isomorphic to $\mathbb{R}$ or $\mathbb{Q}_p$ as a topological field; see 44.11. Lemma 58.6 shows that F has finite dimension over $\overline{\mathbb{Q}}$.

Now we consider the case where F has characteristic $p > 0$. Then $\mathbb{F}_p \subseteq F$. We can choose $x \in F$ with $0 < \varphi(x) < 1$. Then x is not a root of unity, hence x is transcendental over $\mathbb{F}_p$, and the restriction of φ to $\mathbb{F}_p(x)$ is obtained from the degree valuation of the field $\mathbb{F}_p(x)$

of rational functions; see 56.2 and 56.3(c). The topological closure of $\mathbb{F}_p(x)$ is complete (43.9), hence isomorphic to the Laurent series field $\mathbb{F}_p((x))$; see 44.11. Lemma 58.6 shows that F has finite dimension over the closure of $\mathbb{F}_p(x)$. □

58.8 Corollary *The real field* $\mathbb{R}$ *is the only locally compact field which is formally real. The complex field* $\mathbb{C}$ *is the only locally compact field which is algebraically closed.*

Proof Both statements follow from 58.7, because $\mathbb{Q}_p$ is not formally real by 54.2, and a local field is never algebraically closed; see 58.2. □

A locally compact ring topology on a (skew) field is in fact a (skew) field topology; see 62.4 (or 13.4) and 62.2.

Thus Corollary 58.8 together with 6.4 shows that the abstract field $\mathbb{R}$ has only one locally compact ring topology (apart from the discrete and the indiscrete topology), namely the usual one, and that the (non-trivial) locally compact ring topologies of $\mathbb{C}$ are the images of the usual topology under field automorphisms. This means that the connectedness assumptions in 13.9 can be replaced by non-discreteness. Similarly, each local field has only one non-trivial locally compact ring topology (by 58.5, 56.12 and 56.13).

58.9 Theorem *Let* F *be a locally compact skew field. Then the centre* Z *of* F *is not discrete, hence* Z *is isomorphic to* $\mathbb{R}$, $\mathbb{C}$ *or to a local field, and* F *has finite dimension over* Z.

Proof Let P be the prime field of F. There exists an element $a \in F$ with $|a| \neq 0, 1$; see 58.5. The field $P(a)$ is not discrete, hence F has finite (right or left) dimension n over the topological closure $\overline{P(a)}$ by 58.6. Clearly $\overline{P(a)}$ is a field (not necessarily contained in Z). Now a general algebraic result on skew fields implies that F has dimension at most n^2 over its centre Z; see COHN 1995 3.1.4 p. 95. As F is infinite, we infer that Z is infinite, hence not discrete by 58.5(iv). This shows that Z is one of the fields appearing in 58.7.

As an alternative, one can prove directly that the centre Z contains an element a with $|a| \neq 0, 1$; see WEIL 1967 I §4 Proposition 5 p. 20, BOURBAKI 1972 VI.9.3 or WARNER 1989 27.3 and 27.5. □

There remains the algebraic problem to classify all skew fields F of finite dimension over a centre Z as in 58.9. If $Z \in \{\mathbb{R}, \mathbb{C}\}$, then $F = \mathbb{H}$ is the only proper skew field arising, by a classical result of Frobenius (see, for example, JACOBSON 1985 7.7, EBBINGHAUS *et al.* 1991 Chapter 8

§2, PALAIS 1968 or 58.11 below). For local fields Z, this problem was solved by Hasse in 1931; for each local field Z, there are infinitely many possibilities for F, to be described now.

58.10 Cyclic algebras Let Z be any field, and let $E|Z$ be a cyclic Galois extension of degree $n \in \mathbb{N}$. Choose a generator γ of the Galois group $\mathrm{Gal}_Z E$ and $c \in Z^\times$. The *cyclic algebra* $[E|Z; \gamma, c]$ determined by these data is the associative Z-algebra defined by the following properties: it is an n-dimensional (left) vector space $\bigoplus_{i=0}^{n-1} Eb^i$ over E with a basis $1, b, \ldots, b^{n-1}$ consisting of powers of some element b, and the multiplication satisfies the rules $b^n = c$ and $bx = x^\gamma b$ for all $x \in E$.

Denote by $N_{E|Z} \leq Z^\times$ the group of all non-zero norms of $E|Z$. One can show that $[E|Z; \gamma, c]$ is always a simple algebra of dimension n^2 over its centre Z, that $[E|Z; \gamma, c]$ is Z-isomorphic to $[E|Z; \gamma, c']$ precisely if $cN_{E|Z} = c'N_{E|Z}$, and that $[E|Z; \gamma, c]$ is a skew field if n is the order of $cN_{E|Z}$ in the norm factor group $Z^\times/N_{E|Z}$; see REINER 1975 30.4 and 30.7, JACOBSON 1989 8.5 or SCHARLAU 1985 Chapter 8 §12.

For example, the cyclic algebra $[\mathbb{C}|\mathbb{R}; \gamma, c]$ with $c \in \mathbb{R}^\times$ is isomorphic to the skew field $\mathbb{H}$ of Hamilton's quaternions if $c < 0$, and isomorphic to the ring $\mathrm{End}\,\mathbb{R}^2 = \mathbb{R}^{2\times2}$ of all real 2×2 matrices if $c > 0$.

For local fields Z there are many possibilities. The unique unramified extension Z_n of degree n over Z is a cyclic Galois extension of Z (see 58.2), the norm factor group $Z^\times/N_{Z_n|Z}$ is cyclic of order n, and each prime element π of Z represents a generator of $Z^\times/N_{Z_n|Z}$ (see JACOBSON 1989 p. 610 or REINER 1975 14.1). Let q be the order of the finite residue field of Z, and let $\gamma \in \mathrm{Gal}_Z Z_n$ be the distinguished generator which induces the Frobenius automorphism $x \mapsto x^q$ on the finite residue field of Z_n (equivalently, $\zeta^\gamma = \zeta^q$ for some root of unity $\zeta \in Z_n$ of order $q^n - 1$). Then the cyclic algebra

$$Z_{n,r} := [Z_n|Z; \gamma, \pi^r]$$

is a skew field (of dimension n^2 over its centre Z) if the integer r is prime to n. The $\varphi(n)$ integers $r \in \{1, \ldots, n\}$ prime to n give $\varphi(n)$ skew fields $Z_{n,r}$ which are mutually not isomorphic as Z-algebras. We remark that the skew field $Z_{n,r}$ contains *each* field extension of Z of degree n as a maximal subfield; see SERRE 1979 XIII.3 Corollary 3 p. 194 or REINER 1975 31.11.

Choosing for γ another generator leads to the same skew fields $Z_{n,r}$, since $[E|Z; \gamma^r, c^r]$ is Z-isomorphic to $[E|Z; \gamma, c]$ for each integer r prime to n; see REINER 1975 30.4(i) or SCHARLAU 1985 Chapter 8 12.4.

58.11 Theorem (Pontryagin, Jacobson, Hasse) *Let F be a locally compact skew field. Then F is isomorphic to a cyclic algebra over a locally compact field. More precisely, F is isomorphic to $\mathbb{R}$, $\mathbb{C}$, $\mathbb{H}$ or to a cyclic algebra $Z_{n,r}$, where Z is a local field, $n \in \mathbb{N}$ and $r \in \{1, \ldots, n\}$ is prime to n.*

Proof By Theorem 58.9, such a skew field F has finite dimension over its centre Z. General results on skew fields say that this dimension is a square n^2, and that the maximal subfields of F are precisely the fields E with $Z \leq E \leq F$ and $[E : Z] = n$; see, for example, COHN 2003b 5.1.12 or REINER 1975 Theorem 7.15.

If $Z = \mathbb{C}$, then $F = Z$, as $\mathbb{C}$ is algebraically closed. If $Z = \mathbb{R} \neq F$, then $E \cong \mathbb{C}$ and $n = 2$. The complex conjugation $\gamma \in \operatorname{Aut} E$ is induced by an inner automorphism of F, by the Skolem–Noether Theorem (see JACOBSON 1989 Theorem 4.9 p. 222). Thus we find $b \in F^\times$ with $bx = x^\gamma b$ for each $x \in E$. We have $F = E \oplus Eb$, as $b \notin E$. Furthermore b^2 commutes with E and b, so $b^2 = c \in Z^\times = \mathbb{R}^\times$. If $c > 0$, then $c = r^2$ with $r \in \mathbb{R}$, hence $(b + r)(b - r) = b^2 - c = 0$ and $b = \pm r \in Z$, a contradiction. This shows that $c < 0$, thus the cyclic algebra $F = [E|Z; \gamma, c]$ is isomorphic to $\mathbb{H}$.

Now let Z be a local field. In this case, the crucial step is to prove the existence of a maximal subfield E of F such that the extension $E|Z$ is unramified; for this we refer to the two proofs given in SERRE 1979 XII §1 and §2. Now $E|Z$ is a cyclic Galois extension (compare 58.2), and the Skolem–Noether Theorem yields an element $b \in F^\times$ such that conjugation by b induces on E a generator γ of the Galois group $\operatorname{Gal}_Z E$. This implies that the elements $1, b, \ldots, b^{n-1}$ are linearly independent over E (compare JACOBSON 1989 Theorem 8.8 p. 478, SCHARLAU 1985 Chapter 8 Theorem 12.2 p. 317, or COHN 1995 Theorem 3.5.5 p. 124). For dimension reasons we obtain $F = \bigoplus_{i=0}^{n-1} Eb^i$. Furthermore, the element b^n commutes with E and b, hence $b^n \in Z$. This shows that F is a cyclic algebra $[E|Z; \gamma, b^n]$ which is isomorphic to one of the skew fields $Z_{n,r}$ described in 58.10.

Other proofs of Theorem 58.11 can be found in WEIL 1967 I §4 Proposition 5 p. 20 and JACOBSON 1989 Theorem 9.21 p. 607.   □

We remark that the locally compact (skew) field topology of a skew field F as in Theorem 58.11 is uniquely determined (as a consequence of the results 58.9, 13.9 and 58.6) except if $F = \mathbb{C}$ (compare 14.11 and 14.12).

58.12 Brauer groups   For an arbitrary field Z, we denote by $B(Z)$ the set of all Z-isomorphism types of skew fields D of finite dimension over their centre Z. One can make $B(Z)$ into an abelian group, using the tensor product:

$$D_1 + D_2 = D_3 \iff D_1 \otimes_Z D_2 \cong D_3^{k \times k} \text{ for some } k \in \mathbb{N}.$$

Then $Z \in B(Z)$ is the neutral element, and the additive inverse of D is $-D = D^{\mathrm{op}}$, the skew field with the opposite multiplication, as $D \otimes_Z D^{\mathrm{op}} \cong Z^{k \times k}$ with $k = \dim_Z D$. The Brauer group $B(Z)$ is always a torsion group. See COHN 2003a 5.4, COHN 2003b 5.2 and 5.5.4, JACOBSON 1989 4.7 and Theorem 8.12, or REINER 1975 Section 28 and Theorem 29.22 for details and for more elegant descriptions.

If Z is algebraically closed, then $B(Z) = \{0\}$, and the same holds for each finite field Z by Wedderburn's theorem. The Brauer group $B(\mathbb{R}) = \{\mathbb{R}, \mathbb{H}\}$ of the real numbers is a cyclic group of order 2, by a result of Frobenius (compare 58.11), and the same is true for every real closed field (see 12.10).

Now let Z be a local field. Then

$$B(Z) = \{ Z_{n,r} \mid n \in \mathbb{N}, r \in \{1, \dots, n\} \text{ is prime to } n \}$$

by 58.11. One can show that $Z_{n,r} + Z_{n,s} = Z_{n,r+s}$ in $B(Z)$; see REINER 1975 Theorem 30.4 or SCHARLAU 1985 Chapter 8 Theorem 12.7. This implies that the so-called Hasse invariant, i.e., the mapping

$$B(Z) \to \mathbb{Q}/\mathbb{Z} : Z_{n,r} \mapsto \tfrac{r}{n} + \mathbb{Z} \in \mathbb{Q}/\mathbb{Z},$$

is an isomorphism $B(Z) \cong \mathbb{Q}/\mathbb{Z}$, for each local field Z; see REINER 1975 Theorem 30.4, JACOBSON 1989 Theorem 9.22 or SERRE 1979 XIII §3.

For global fields like $\mathbb{Q}$, the situation is slightly more complicated, there is an exact sequence

$$0 \to B(\mathbb{Q}) \to B(\mathbb{R}) \oplus \bigoplus_p B(\mathbb{Q}_p) \to \mathbb{Q}/\mathbb{Z} \to 0,$$

which originates from a local–global principle for the splitting of skew fields and from a non-trivial relation between the Hasse invariants of the completions $D \otimes \mathbb{Q}_p$ of a skew field $D \in B(\mathbb{Q})$. One can show that each skew field of finite dimension over $\mathbb{Q}$ is a cyclic algebra (over a finite extension Z of $\mathbb{Q}$; compare 58.10); see REINER 1975 Section 32 or WEIL 1967 XIII §3 and §6.

Exercises

(1) Show that the polynomial $f = (x^p - 1)/(x - 1)$ is irreducible over $\mathbb{Q}_p$ and defines a totally ramified extension of $\mathbb{Q}_p$ of degree $p - 1$.

(2) Let $c \in \mathbb{Q}_p$ be a non-square and $F = \mathbb{Q}_p(\sqrt{c})$. Show that $F|\mathbb{Q}_p$ is unramified precisely if $v_p(a^2 - c)$ is even for each $a \in \mathbb{Q}_p$. Deduce that $\mathbb{Q}_2(\sqrt{3})|\mathbb{Q}_2$ is ramified, and that $\mathbb{Q}_2(\sqrt{5})|\mathbb{Q}_2$ is unramified.

(3) Show that $\mathbb{F}_2((t))$ has infinitely many separable quadratic extensions (consider the splitting fields of polynomials $x^2 + t^n x + t$ with $n \in \mathbb{N}$).

(4) Show that the polynomial $(x^2 - 2)(x^2 - 17)(x^2 - 34)$ has a root in $\mathbb{Q}_p$ and in $\mathbb{F}_p$ for each prime p (but not in $\mathbb{Q}$).

(5) Let $E|Z$ be a Galois extension of degree 2, let γ be the generator of $\mathrm{Gal}_Z E$ and $c \in Z^\times$. Show that the cyclic algebra $[E|Z; \gamma, c]$ is isomorphic to the Z-algebra which consists of all matrices

$$\begin{pmatrix} x & cy^\gamma \\ y & x^\gamma \end{pmatrix}$$

with $x, y \in E$. Show directly that these matrices form a skew field precisely if $c \neq xx^\gamma$ for all $x \in E$.

(6) Let $Z = \mathbb{F}_q((x))$ for some prime power q, and let $r \in \{1, \dots, n\}$ be prime to n. Show that the cyclic algebra $[Z_n|Z; \gamma^r, x]$ as in 58.10 is isomorphic to the skew Laurent series field $\mathbb{F}_{q^n}((t))$ with the multiplication rules $t^n = x$ and $ta = a^{q^r} t$ for $a \in \mathbb{F}_{q^n}$.

6

Appendix

The first section of the appendix collects a few facts on ordinal and cardinal numbers. Then we deal with topological groups, and we summarize the duality theory of locally compact abelian groups. Finally we present basic facts and constructions of field theory.

61 Ordinals and cardinals

This section is based on the system NBG (von Neumann–Bernays–Gödel) of set theory, having as its primitive notions *classes* as objects, a predicate *set* for certain classes, and the *element* relation $\in$; see Dugundji 1966 Chapter I §8, Rubin 1967, or Smullyan–Fitting 1996. Each element of a class is a set, and so is each subclass of a set. The class of all sets is partially ordered by the relation $\in$. The *axiom of foundation* stipulates that each $\in$-descending *sequence* of sets is finite.

In this system, the class $\mathcal{O}$ of all *ordinals* or *ordinal numbers* consists of all sets which are linearly ordered by the relation $\in$, formally

$$\nu \in \mathcal{O} \rightleftharpoons \underset{\sigma \in \nu}{\forall} \underset{\xi}{\forall} [(\xi \in \sigma \Rightarrow \xi \in \nu) \wedge (\xi \in \nu \Rightarrow \xi \in \sigma \vee \xi = \sigma \vee \sigma \in \xi)] \, .$$

By the axiom of foundation, each ordinal number is even well-ordered (which means that every non-empty subset has a smallest element). Any well-ordered set is order isomorphic to a unique ordinal. The *well-ordering principle* says that every set can be well-ordered, it implies that there are ordinal numbers of arbitrary cardinality.

Note that the well-ordering principle is equivalent to the *axiom of choice* (any Cartesian product of non-empty sets is non-empty) and to *Zorn's lemma* (if each chain in a partially ordered set S has an upper bound in S, then there exists at least one maximal element in S). This is an immediate consequence of *Hausdorff's maximal chain principle* (each chain in S is contained in a maximal chain). Conversely, the maximal chain principle can be obtained by applying Zorn's lemma to the set of all subchains of S, ordered by inclusion. The well-ordering principle also follows easily: Consider the set $\mathcal{S}$ of all injective maps $\varphi : \xi \to M$ where ξ is an ordinal number. Write $\varphi \leq \psi$ if ψ is an extension of φ. Any chain in $\mathcal{S}_<$ has a common extension, i.e., an upper bound. By Zorn's lemma there is a maximal element $\sigma : \mu \to M$ in $\mathcal{S}$. The map σ is necessarily surjective and σ carries the well-ordering from μ to M. A proof of the well-ordering principle from the axiom of choice is given in DUGUNDJI 1966 Chapter II §2. For a full account of the axiom of choice and related topics see RUBIN–RUBIN 1985.

61.1 *If $\sigma \in \nu \in \mathcal{O}$, then $\sigma \in \mathcal{O}$.*

Proof If $\xi \in \eta \in \sigma$, then successively $\eta \in \nu$ and $\xi \in \nu$. The possibilities $\xi \in \eta \in \sigma \in \xi$ or $\xi \in \eta \in \sigma = \xi$ contradict the axiom of foundation. Hence $\xi \in \sigma$.

If $\xi, \eta \in \sigma$, then $\xi, \eta \in \nu$ and therefore $\xi \in \eta$ or $\xi = \eta$ or $\eta \in \xi$. $\quad\square$

61.2 *If $\mu, \nu \in \mathcal{O}$ and $\mu \subset \nu$ (proper inclusion), then $\mu \in \nu$.*

Proof Let α be the $\in$-minimal element of the non-empty set $\nu \smallsetminus \mu$. We show that $\alpha = \mu$: in fact, $\xi \in \alpha \Rightarrow \xi \in \mu$ by minimality of α. Conversely, $\xi \in \mu \Rightarrow \xi \in \alpha$, because $\alpha \notin \mu$ and hence $\xi \neq \alpha$ and $\alpha \notin \xi$. $\quad\square$

61.3 *Any non-empty class $\mathcal{M} \subseteq \mathcal{O}$ has a minimum $\bigcap \mathcal{M} \in \mathcal{O}$.*

Proof $\bigcap \mathcal{M} = \delta$ is a set. If $\xi \in \sigma \in \delta$, then $\sigma \in \mu$ and hence $\xi \in \mu$ for each $\mu \in \mathcal{M}$. This means that $\xi \in \delta$. Similarly, it follows that δ is linearly ordered by $\in$. Thus, $\delta \in \mathcal{O}$. If $\delta \subset \mu$ for each $\mu \in \mathcal{M}$, then $\delta \in \delta$ by 61.2, but this contradicts the axiom of foundation. $\quad\square$

61.4 Corollary *The class $\mathcal{O}$ itself is well-ordered by $\in$.*

Proof If $\mu, \nu \in \mathcal{O}$, then $\mu \cap \nu = \min \{\mu, \nu\}$, thus $\mathcal{O}$ is linearly ordered. $\quad\square$

61.5 *If $\mathcal{M}$ is a subset of $\mathcal{O}$, then $\bigcup \mathcal{M} = \sup \mathcal{M} \in \mathcal{O}$.*

Proof (a) Because $\mathcal{M}$ is a set, so is $\sigma = \bigcup \mathcal{M}$ (by an axiom of NBG). If $\xi \in \eta \in \sigma$, then $\eta \in \mu$ for some $\mu \in \mathcal{M}$. Consequently, $\xi \in \mu$ and then $\xi \in \sigma$. If $\xi, \eta \in \sigma$, then there are $\mu, \nu \in \mathcal{M}$ such that $\xi \in \mu$ and $\eta \in \nu$. Since $\mathcal{M}$ is linearly ordered, we may assume that $\mu \subseteq \nu$. By the definition of $\mathcal{O}$ it follows that ξ and η are comparable, and $\sigma \in \mathcal{O}$.

(b) If $\mu \in \mathcal{M}$, then $\mu \subseteq \sigma$ and hence $\mu = \sigma$ or $\mu \in \sigma$. Therefore, σ is an upper bound, even a least upper bound, since $\xi \in \sigma$ implies $\xi \in \mu$ for some $\mu \in \mathcal{M}$. $\qquad\square$

61.6 Explicit description   *Each ordinal number is the set of all its predecessors. The immediate successor of ν is $\nu + 1 := \nu \cup \{\nu\}$.* $\qquad\square$

61.7 Small ordinals   The first ordinal numbers are $0 = \emptyset$, $1 = \{0\}$, $2 = \{0, 1\}$, and the subsequent *finite* numbers. The smallest infinite ordinal number $\omega = \omega_0$ is the set of all finite ordinals, it is followed by $\omega + 1$, $\omega + 2$, ..., $\omega + \omega = \omega \cdot 2$, ..., $\omega \cdot 3$, ..., $\omega \cdot \omega = \omega^2$, ..., ω^3, ..., $\omega^\omega = {}^2\omega$, ..., $\omega^{\omega^\omega} = {}^3\omega$, ..., ${}^\omega\omega$, ... All these ordinal numbers are countable, because a (countable) sequence of countable ordinals has a countable upper bound. The first uncountable ordinal number $\Omega = \omega_1$ is the set of all countable ordinals. $\qquad\square$

61.8 Cardinal numbers   A class of bijectively equivalent sets may be considered as a cardinal number. It is more convenient, however, to select a canonical element of the class to denote the cardinal number: in each class of bijectively equivalent ordinals there is a smallest one, the so-called initial ordinal, or *cardinal*, of this class. The infinite cardinals form a well-ordered subclass $\mathcal{C}$ of $\mathcal{O}$, and there is a (unique) order isomorphism $(\alpha \mapsto \omega_\alpha) : \mathcal{O} \to \mathcal{C}$. In contexts where the relevant property is *size* rather than *ordering*, the ordinal ω_α is usually designated by $\aleph_\alpha$.

The *cardinality* (or cardinal number) card M of a set M is defined to be the smallest element $\alpha \in \mathcal{O}$ such that there exists an injective map $\mu : M \to \alpha$. Thus $\aleph_0 = \text{card}\,\mathbb{N}$. Moreover $M \subseteq N$ implies card $M \leq$ card N.

For more details see DUGUNDJI 1966 Chapter II §7, RUBIN 1967, BACHMANN 1967, LÉVY 1979, or CIESIELSKI 1997. Ordinals and cardinals in the system ZF of Zermelo and Fraenkel are presented in Chapter 1 of HOLZ *et al.* 1999. We mention the Theorem of Cantor–Bernstein (or Schröder–Bernstein): *If there are injections $\varphi : X \to Y$ and $\psi : Y \to X$, then there exists also a bijection $\sigma : X \to Y$* (for proofs see RAUTEN-BERG 1987).

61.9 Addition and multiplication   We define card M + card N = card $M \cup N$ if $M \cap N = \emptyset$, and card $M \cdot$ card N = card $M \times N$.

61.10 Powers   We denote by M^X the set of all maps $\varphi : X \to M$, as usual. Exponentiation of cardinal numbers is defined by card $M^{\text{card}\,X}$ = card M^X. Note that card $2^X = 2^{\text{card}\,X}$.

Attention: the least upper bound ω^ω of all ordinal numbers ω^n with finite n is to be distinguished from the set $\aleph_0^{\aleph_0}$; the first one is countable, the other is not.

61.11 Theorem *If* $\operatorname{card} M > 1$ *and* $X \neq \emptyset$, *then* $\operatorname{card} M^X > \operatorname{card} X$.

Proof Obviously, $\operatorname{card} X \leq \operatorname{card} M^X$. Assume that there is a bijective map $(x \mapsto \varphi_x) : X \to M^X$. For each $x \in X$ choose an element $\psi(x) \in M$ with $\psi(x) \neq \varphi_x(x)$; this requires the axiom of choice. Then $\psi \neq \varphi_x$ for any x, a contradiction. □

The following theorem plays an important role in determining the cardinality of several sets constructed in this book.

61.12 Squares *For* $\alpha \in \mathcal{O}$ *we have* $\aleph_\alpha^2 = \aleph_\alpha$.

Proof (a) One has $\aleph_0^2 = (\operatorname{card} \omega)^2 = \operatorname{card}(\omega \cdot \omega) = \aleph_0$ by 61.7.
(b) It can easily be verified (compare SMULLYAN–FITTING 1996 Chapter 9 §8) that the definition $(\alpha, \beta) \prec (\gamma, \delta) \leftrightharpoons$

$$(\alpha \cup \beta < \gamma \cup \delta) \vee (\alpha \cup \beta = \gamma \cup \delta \wedge \alpha < \gamma) \vee (\alpha \cup \beta = \gamma \cup \delta \wedge \alpha = \gamma \wedge \beta < \delta)$$

yields a well-ordering $\prec$ of $\mathcal{O} \times \mathcal{O}$. Since each well-ordered set is order isomorphic to a unique ordinal, the relation $\prec$ determines an injective map $\kappa : \mathcal{O} \times \mathcal{O} \to \mathcal{O}$.
(c) Assume that there is a least α such that $\aleph_\alpha^2 > \aleph_\alpha$. Then $\alpha > 0$ by step (a). We have $\omega_\alpha \times \omega_\alpha = \bigcup_{\mu \in \omega_\alpha} \mu \times \mu$. From the minimality of α we infer that $\operatorname{card} \mu \times \mu = \operatorname{card} \mu$ and hence $\kappa(\mu \times \mu) \in \omega_\alpha$. Therefore, $\kappa(\omega_\alpha \times \omega_\alpha) = \bigcup_{\mu \in \omega_\alpha} \kappa(\mu \times \mu) \subseteq \omega_\alpha$ and $\aleph_\alpha^2 = \aleph_\alpha$ contrary to the assumption.

For a different proof see DUGUNDJI 1966 Chapter II Theorem 8.5; compare also BACHMANN 1967 §28 and DEISER 2005. □

61.13 Corollary (a) *If* $S = \bigcup_{\iota \in K} M_\iota$, *where* $\operatorname{card} K \leq \aleph_\alpha$ *and* $\operatorname{card} M_\iota \leq \aleph_\beta$ *for each* $\iota \in K$, *then* $\operatorname{card} S \leq \aleph_\alpha \cdot \aleph_\beta = \aleph_{\max\{\alpha,\beta\}}$.
(b) *For each finite* $n > 0$ *we have* $\aleph_\alpha^n = \aleph_\alpha$, *and* $\aleph^{\aleph_\alpha} = (2^{\aleph_0})^{\aleph_\alpha} = 2^{\aleph_0 \aleph_\alpha} = 2^{\aleph_\alpha}$. □

61.14 Corollary *If* M *is an infinite set, and if* $\mathfrak{F}$ *consists of finite sequences in* M (*or of finite subsets of* M), *then* $\operatorname{card} \mathfrak{F} \leq \operatorname{card} M$.

Proof If $\operatorname{card} M = \aleph_\alpha$, then there are $\aleph_\alpha^n = \aleph_\alpha$ sequences of length n in M, and at most $\aleph_\alpha$ subsets of size n. Hence $\operatorname{card} \mathfrak{F} \leq \aleph_0 \cdot \aleph_\alpha = \aleph_\alpha$. □

61.15 Theorem *If $\beta \leq \alpha + 1$ then $\aleph_\beta^{\aleph_\alpha} = 2^{\aleph_\alpha}$.*

Proof We have $\aleph_\beta \leq \aleph_{\alpha+1} \leq 2^{\aleph_\alpha}$ by 61.11. Hence $2^{\aleph_\alpha} \leq \aleph_\beta^{\aleph_\alpha} \leq (2^{\aleph_\alpha})^{\aleph_\alpha} = 2^{\aleph_\alpha}$. □

61.16 Corollary *If $\mathrm{Sym}\,M$ denotes the group of all permutations of M (i.e. bijections $M \to M$), then $\mathrm{card}\,\mathrm{Sym}\,M = 2^{\mathrm{card}\,M}$. Similarly, $\mathrm{card}\{\, S \mid S \subseteq M \wedge \mathrm{card}\,S = \mathrm{card}\,M \,\} = 2^{\mathrm{card}\,M}$.*

Proof Let $c := \mathrm{card}\,M$ and partition M into c pairs. There are 2^c permutations of order 2 mapping each pair to itself. Therefore, $2^c \leq \mathrm{card}\,\mathrm{Sym}\,M \leq c^c = 2^c$.

Partition M into two subsets A and $M \smallsetminus A$ of cardinality c. Then $2^c = \mathrm{card}\{\, S \mid A \subseteq S \subseteq M \,\} \leq \mathrm{card}\{\, S \mid S \subseteq M \wedge \mathrm{card}\,S = c \,\} \leq 2^c$. □

61.17 The continuum problem By the well-ordering principle, we have $\mathrm{card}\,\mathbb{R} = 2^{\aleph_0} = \aleph_\alpha$ for some ordinal number α. Cantor's theorem 61.11 shows that $\alpha > 0$, but otherwise does not give any clue how the value of α might be determined. See BAUMGARTNER–PRIKRY 1977 for a survey of the history of this so-called *continuum problem*. Cantor and Hilbert conjectured that $2^{\aleph_0} = \aleph_1$ (continuum hypothesis, CH). This is the first case of the *generalized continuum hypothesis* (GCH) which asserts that $2^{\aleph_\alpha} = \aleph_{\alpha+1}$ for each $\alpha \in \mathcal{O}$. See also STILLWELL 2002.

Without using ordinals, GCH may be expressed as follows: if $\mathfrak{m}$, $\mathfrak{n}$ denote classes of infinite bijectively equivalent sets, then $\mathfrak{m} \leq \mathfrak{n} < 2^{\mathfrak{m}}$ entails that $\mathfrak{m} = \mathfrak{n}$. This form of GCH implies the axiom of choice; compare SPECKER 1954, GILLMAN 2002 or COHEN 1966 Chapter 4 §12 p. 148ff.

By the work of GÖDEL 1940 and COHEN 1966 it has become clear that the continuum hypothesis can neither be proved nor disproved on the basis of the usual axioms of set theory. More precisely, if the system NBG including the axiom of choice is consistent, then it is not possible to derive a contradiction by assuming in addition the continuum hypothesis or its negation. Specifically, each assertion $2^{\aleph_0} = \aleph_n$ with $0 < n \in \omega$ is consistent with the usual set theory, but not $2^{\aleph_0} = \aleph_\omega$. Analogously, the continuum hypothesis is independent of the axiom system ZF of Zermelo and Fraenkel; compare WOODIN 2001. Moreover, GCH is consistent with ZF as well as with NBG; see the survey by MARTIN 1976 or the more recent treatment of the continuum problem by SMULLYAN–FITTING 1996, in particular p. 185 and Chapter 19 §6.

Exercises

(1) Show that Ω^2 is well-ordered by the lexicographic ordering.

(2) Show that $\mathbb{R}$ has only $\aleph = \operatorname{card} \mathbb{R}$ countable subsets.

(3) Prove the following theorem of König: if $\operatorname{card} A_\iota < \operatorname{card} B_\iota$ for $\iota \in K$, then $\sum_{\iota \in K} \operatorname{card} A_\iota < \prod_{\iota \in K} \operatorname{card} B_\iota$.

(4) Use König's theorem to prove that $2^{\aleph_0} \neq \aleph_\omega$.

62 Topological groups

The additive and multiplicative groups of the fields considered in this book are topological groups of a very special nature. For better understanding and in order to avoid repetition of arguments, these groups ought to be looked at in the context of topological groups in general. We introduce a few basic notions of the theory and collect some simple facts. For more details see PONTRYAGIN 1986, BOURBAKI 1966 Chapter III, HEWITT–ROSS 1963 Chapter II or STROPPEL 2006.

62.1 Definitions Assume that $(G, \cdot)$ is a group and that (G, τ) is a topological space. Then $G = (G, \cdot, \tau)$ is called a *semi-topological group* if multiplication is continuous in each variable separately (i.e., if left and right multiplication with a fixed element are homeomorphisms).

G is said to be a *para-topological group* if multiplication is continuous in both variables simultaneously. If, moreover, inversion is also continuous, then G is a *topological group*.

Remarks $(\mathbb{R}, +)$ with Sorgenfrey's topology (5.73 or 24.17) is a para-topological, but not a topological group. Any infinite group with the *cofinite* topology (each neighbourhood has a finite complement) is semi-topological, but not para-topological.

62.2 Proposition *A regular, locally compact semi-topological group is a topological group.*

This is Theorem 2 in ELLIS 1957. The conclusion is true, in fact, under rather weak hypotheses; see WU 1962 and SOLECKI–SRIVASTAVA 1997. □

The set $V_\varepsilon = \{ (x, y) \mid x, y \in \mathbb{R} \wedge |y - x| < \varepsilon \}$ is a surrounding of the diagonal in $\mathbb{R} \times \mathbb{R}$. The filter (see 21.1) generated by all these surroundings (with $\varepsilon > 0$) is a uniformity $\mathfrak{U}$ on $\mathbb{R}$. A function $f : \mathbb{R} \to \mathbb{R}$ is uniformly continuous if $f^{-1}(\mathfrak{U}) \subseteq \mathfrak{U}$. These notions can be generalized to arbitrary topological groups as follows.

62.3 If U denotes a neighbourhood of 1 in the topological group G, then the *surroundings* $V_U = \{\,(x,y) \in G \times G \mid yx^{-1} \in U\,\}$ of the diagonal in $G \times G$ generate a uniformity on G, the *right uniformity*; it makes right and left multiplications of G uniformly continuous (see BOURBAKI 1966 Chapter III §3 no.1, or HEWITT–ROSS 1963 Chapter II §4). Inversion interchanges the right and the left uniformity. For a fuller treatment of uniform structures on groups see ROELCKE–DIEROLF 1981.

62.4 Regularity *If G is a topological group and if $\mathfrak{V}$ denotes the filter of all neighbourhoods of 1, then $\bigcap \mathfrak{V} = \{1\}$ implies that G is a T_1-space (i.e., points are closed in G), and then G is even completely regular (but not necessarily normal).*

Proof By homogeneity, each point has a neighbourhood that does not contain the element 1. Hence $G \smallsetminus \{1\}$ is open. Because the group operations are continuous, there are arbitrarily small neighbourhoods of the form UU^{-1} in $\mathfrak{V}$, and $\overline{U} \subseteq UU^{-1}$ shows that G is regular; see also BOURBAKI 1966 Chapter II §1 Proposition 3; according to BOURBAKI 1966 Chapter IX §1 no.5, the space G is even completely regular. The existence of non-normal regular topological groups has been shown by MARKOV 1945; compare HEWITT–ROSS 1963 8.10–12.　　□

62.5 Subgroups *Let G be a topological group. If H is a subgroup of the group G, and if $\overline{H}$ is the topological closure of H in G, then $\overline{H}$ is also a subgroup of G.*

Proof The map $\gamma = ((x,y) \mapsto xy^{-1})$ is continuous on $G \times G$. Since $\overline{H} \times \overline{H} = \overline{H \times H}$ and H is a group, $\gamma(\overline{H \times H}) \subseteq \overline{\gamma(H \times H)} \subseteq \overline{H}$.　　□

62.6 Quotients Let H be a subgroup of a topological group G. The quotient topology on the coset space $G/H = \{\,xH \mid x \in G\,\}$ is characterized by the property that the canonical projection $\eta : G \to G/H :$ $x \mapsto xH$ is continuous and open. In fact, if U is open in G, then so is the preimage $UH = \bigcup_{h \in H} Uh$ of $\eta(U)$, and this means that $\eta(U)$ is open in G/H; see also HEWITT–ROSS 1963 Chapter II, 5.15–17.

The adequate notion of a substructure of a topological group is that of a closed subgroup rather than just a subgroup.

62.7 *Any open subgroup H of a topological group is also closed.*

Proof The complement of H is a union of cosets of H, hence open.　　□

From now on, we assume that all **topological groups** are T_1-spaces.

62.8 Proposition *If H is a closed subgroup of a topological group G, then G/H is a Hausdorff space. In fact, G/H is even regular.*

Proof If x is in the open complement of H, then, by continuity of the group operations, there are neighbourhoods U of 1 and V of x such that $U^{-1}V \cap H = \emptyset = VH \cap UH$, and 62.6 shows that $\eta(V)$ and $\eta(U)$ are disjoint neighbourhoods of x and 1, respectively. For the last assertion, see HEWITT–ROSS 1963 Chapter II 5.21. □

62.9 Metric *If G is a topological group with a countable neighbourhood base at 1, then G is metrizable, and there exists a metric which is invariant under right multiplications.*

For *proofs* see BOURBAKI 1966 Chapter IX §3 or HEWITT–ROSS 1963 Chapter II §8.

62.10 Normality *A locally compact group is paracompact and hence is a normal space.*

Proof A topological space is called paracompact if each open cover has a locally finite refinement. If V is a compact symmetric ($V = V^{-1}$) neighbourhood of 1 in G, then $H = \bigcup_{n \in \mathbb{N}} V^n$ is an open subgroup of G. Being a countable union of the compact sets V^n, the group H is a Lindelöf space and hence is paracompact; see DUGUNDJI 1966 Chapter VIII Theorem 6.5. Obviously, G, as a union of disjoint cosets of H, is also paracompact. By a standard result, paracompact spaces are normal; compare DUGUNDJI 1966 Chapter VIII Theorem 2.2. □

62.11 Factor groups *If N is a closed normal subgroup of the topological group G, then G/N is a topological group.*

Proof Recall from 62.6 that the canonical epimorphism $\eta : G \to G/N$ is continuous and open. If $UV^{-1} \subseteq W$ for open subsets $U, V, W \subseteq G$, then $\eta(U)\eta(V)^{-1} \subseteq \eta(W)$. This shows the continuity of the group operations in G/N; see also HUSAIN 1966 §24. □

62.12 Proposition *If K is a connected closed subgroup of the topological group G, then G is connected if, and only if, G/K is connected.*

Proof If G is connected, so is its continuous image G/K. If G is not connected, then there exists a continuous map φ of G onto the discrete space $\{0,1\}$, and φ is constant on each (connected) coset xK. Hence

φ induces a continuous surjection $\psi : G/K \to \{0,1\}$, and G/K is not connected. $\qquad\qquad\qquad\qquad\qquad\qquad\qquad\qquad\qquad\qquad\qquad\qquad\square$

62.13 Connected component   If G is a topological group and K is the largest connected subset of G containing the element 1, then K is a closed normal subgroup of G, the *connected component*, and G/K is totally disconnected. If G is locally compact, then $\dim G/K = 0$ (compare Exercise 3).

In fact, $K \times K$ is connected, hence its continuous image KK^{-1} is contained in K, and K is invariant under inner automorphisms. Moreover, K is closed in G; see DUGUNDJI 1966 Chapter V Theorem 1.6. By 62.12, the connected component of G/K consists of one element, and this implies that G/K is totally disconnected. For the last claim see HOFMANN–MORRIS 1998 E8.6. $\qquad\qquad\qquad\qquad\qquad\qquad\qquad\qquad\square$

Finally, we mention a fundamental structure theorem.

62.14 Theorem (Mal'cev–Iwasawa)   *Let G be a locally compact, connected topological group.*
(1) *Then there exists a maximal compact subgroup C, and each maximal compact subgroup of G is connected and conjugate to C. Moreover, each compact subgroup of G is contained in a maximal one.*
(2) *There are closed subgroups $S_\mu \cong \mathbb{R}$ such that the multiplication map*

$$(k, s_1, \ldots, s_m) \mapsto ks_1 \ldots s_m : C \times S_1 \times \ldots \times S_m \to G$$

is a homeomorphism. In particular, $G \approx C \times \mathbb{R}^m$.

For a *proof* see IWASAWA 1949 Theorem 13, compare also HOFMANN–TERP 1994. The result has first been obtained for Lie groups. It can then be extended to general locally compact, connected groups, because each of these has arbitrarily small compact normal subgroups with Lie factor groups.

Exercises

(1) If inversion in a topological group is uniformly continuous with respect to the right uniformity, then multiplication is uniformly continuous in two variables.

(2) If H is a locally compact subgroup of the regular topological group G, then H is closed in G.

(3) A topological space is called 0-dimensional, if each point has arbitrarily small open and closed neighbourhoods. Show that every locally compact, 0-dimensional group has arbitrarily small compact open subgroups.

63   Locally compact abelian groups and Pontryagin duality

Each locally compact abelian (or LCA) group A has a dual A^*, which is again an LCA group. The Pontryagin–van Kampen Theorem asserts that A^{**} is canonically isomorphic to A. Therefore, each question concerning A is reflected in a question on A^*. Usually, one of the two problems is easier to deal with than the other. This makes Pontryagin duality the most important tool in the structure theory of LCA groups. For an introduction to duality theory see PONTRYAGIN 1986 or MORRIS 1977. A full account is given in HEWITT–ROSS 1963 Chapter VI; see also the elegant exposition in HOFMANN–MORRIS 1998 Chapter 7, or STROPPEL 2006.

In this section, **all groups are commutative.**

63.1 Characters Let A be an abelian topological group, and denote by $\mathbb{T} = \mathbb{R}/\mathbb{Z}$ the torus group; compare 8.7. A *character* of A is a continuous homomorphism $\chi : A \to \mathbb{T}$. It is convenient to denote the values of χ in $\mathbb{T}$ by $\langle x, \chi \rangle$ rather than by $\chi(x)$. Under pointwise addition, the characters of A form a group A^*.

63.2 The *compact-open topology* on A^* is generated by all sets

$$W(C, U) = \{ \chi \in A^* \mid \langle C, \chi \rangle \subseteq U \} \,,$$

where C is compact and U is open in $\mathbb{T}$; the sets $W(C, U)$ with $0 \in U$ form a neighbourhood base at 0. The group A^* will always be equipped with the compact-open topology. With this topology, A^* is a topological group.

63.3 Adjoint *Any continuous homomorphism* $\gamma : A \to B$ *induces a continuous 'adjoint' homomorphism* $\gamma^* : B^* \to A^*$ *defined by the condition* $\langle a, \gamma^* \beta \rangle = \langle a\gamma, \beta \rangle$. *Obviously,* γ^* *is injective, if* γ *is surjective.*

Continuity of γ^* follows from $\langle C, \gamma^* \beta \rangle = \langle C\gamma, \beta \rangle$ and the fact that $C\gamma$ is compact if C is.

63.4 Natural homomorphism For each $a \in A$, the evaluation map $\delta_a = (\chi \mapsto \langle a, \chi \rangle) : A^* \to \mathbb{T}$ is a character of A^*, and $\delta_A = (a \mapsto \delta_a) : A \to A^{**}$ is a homomorphism. If δ_A is an isomorphism of topological groups, then A is called reflexive.

Note that the map δ_a is continuous: if $\Omega = W(a, U)$, then $\langle a, \Omega \rangle \subseteq U$ by definition.

63.5 *If A is compact, then A^* is discrete, and if A is discrete, then A^* is compact.*

Proof If A is compact, then $W(A, U)$ is open, and if U is small, then $W(A, U) = \{0\}$ because $\mathbb{T}$ has no small subgroups. If A is discrete, then the topology on A^* is the topology of pointwise convergence, and A^* is closed in the compact space $\mathbb{T}^A$.   □

63.6 Theorem *Let A be an LCA group.*
 (a) *Let C be a compact neighbourhood of 0 in A. If U is a sufficiently small neighbourhood of 0 in $\mathbb{T}$, then the closure of $W(C, U)$ is compact.*
 (b) *The character group A^* is locally compact.*

The *proof* uses Ascoli's theorem; for details see MORRIS 1977 Theorem 10, HOFMANN–MORRIS 1998 Theorem 7.7(ii) or STROPPEL 2006 Theorem 20.5.

The duality theorem 63.27 asserts that A is reflexive, i.e., that δ_A is injective and surjective, continuous and open, if A is locally compact. These four properties of δ_A will be discussed separately.

A topological space A is said to be a *k-space* if it has the following property: $O \subseteq A$ is open in A if, and only if, $O \cap C$ is open in C for each compact subspace C of A.

In particular, A is a k-space, if each point of A has a countable neighbourhood base. (Use the fact that a convergent sequence together with its limit point is compact, or see DUGUNDJI 1966 XI.9.3.)

63.7 Continuity *If the underlying space of the topological group A is a k-space, in particular, if A is locally compact, then $\delta_A : A \to A^{**}$ is continuous.*

A *proof* is given in HOFMANN–MORRIS 1998 Theorem 7.7(iii). For LCA groups, the proof is easier; see HEWITT–ROSS 1963 24.2, PONTRYAGIN 1986 Definition 37 or STROPPEL 2006 Lemma 20.10.

63.8 *As topological groups, $(A \oplus B)^*$ and $A^* \oplus B^*$ are isomorphic.*

63.9 Corollary *If A and B are reflexive, so is $A \oplus B$.*

63.10 Vector groups The characters of $\mathbb{R}$ are given by $x \mapsto ax + \mathbb{Z}$ with $a \in \mathbb{R}$. Hence $\mathbb{R}$ is self-dual, moreover, $\mathbb{R}^n$ is self-dual for any $n \in \mathbb{N}$; see 8.31a (or Exercise 5 of Section 52).

63.11 Definition We say that A has *enough characters*, if for each $a \in A$, $a \neq 0$, there is a character $\chi \in A^*$ such that $\langle a, \chi \rangle \neq 0$, i.e. if the characters separate the elements of A. The following fact is obvious.

63.12 Injectivity   *The natural map* $\delta_A : A \to A^{**}$ *is injective if, and only if, A has enough characters.*

63.13 Theorem   *Each compact group has enough characters.*

For a *proof* see PONTRYAGIN 1986 Theorems 31 and 33 or HOFMANN–MORRIS 1998 Corollary 2.31. The case of general LCA groups will be reduced to the compact case; see 63.18.

63.14 Splitting Theorem   *Any LCA group A is isomorphic as a topological group to $\mathbb{R}^n \oplus H$, where H has a compact open subgroup.*

Proof HOFMANN–MORRIS 1998 Theorem 7.57; see also PONTRYAGIN 1986 §39.   □

63.15 Lemma   *If B is an open subgroup of A, then any character $\beta \in B^*$ is a restriction $\alpha|_B$ of some character $\alpha \in A^*$. In other words, if $\gamma : B \to A$ is injective and open, then the adjoint $\gamma^* : A^* \to B^*$ is surjective.*

Proof The divisible group $\mathbb{T}$ is injective; see 1.23. This means that β can be extended to a homomorphism $\alpha : A \to \mathbb{T}$, and α is continuous, because B is open in A.   □

63.16 Corollary   *Each discrete group D has enough characters.*

Proof Obviously, each cyclic subgroup of D has enough characters and is open.   □

63.17 Corollary   *If H has a compact open subgroup K, then H has enough characters.*

Proof There is a canonical projection $\kappa : H \to D$ onto the discrete group $D = H/K$. If $c\kappa \neq 0$, then $\langle c\kappa, \alpha \rangle = \langle c, \kappa^*\alpha \rangle \neq 0$ for some $\alpha \in D^*$ and $\kappa^*\alpha \in H^*$. If $0 \neq c \in K$, then $\langle c, \gamma \rangle \neq 0$ for some γ in the restriction $H^*|_K$.   □

63.18   *If A is locally compact, then the natural map $\delta_A : A \to A^{**}$ is injective.*

Proof This is an immediate consequence of 63.9–17. See also STROPPEL 2006 Proposition 20.12.   □

63.19 Lemma   *Every finitely generated discrete group A is reflexive.*

Proof By the fundamental theorem on finitely generated abelian groups (HOFMANN–MORRIS 1998 Theorem A1.11 or most books on abelian groups), A is a direct sum of cyclic groups. Each finite cyclic group is self-dual, and $\mathbb{Z}$ and $\mathbb{T}$ are character groups of each other. Thus 63.9 implies the assertion. $\qquad\square$

63.20 Theorem *If A is compact, then δ_A is surjective, and hence A is reflexive.*

Proof If A is compact, then $D = A^*$ is discrete, and $C = D^*$ is compact again. Because of 63.13, the group A may be considered as a subgroup of C. By the very definition, A separates the points of D: for each $x \in D$, $x \neq 0$, there is some $a \in A$ with $\langle a, x \rangle \neq 0$. If suffices to show that A is dense in C, and this means that each non-empty open set O in C of the form $O = W(F, U)$ contains an element of A. Here, F is compact in D, hence F is finite. According to the previous lemma, F generates a reflexive subgroup G of D. The adjoint $\kappa : D^* \to G^*$ of the inclusion map $G \hookrightarrow D$ is surjective by 63.15. The restriction $A|_G = A\kappa$ separates the elements of G. Assume that $A\kappa \neq G^*$. Since $G^*/A\kappa$ is compact, 63.13 implies that there is some $x \in G^{**} = G$ with $x \neq 0$ and $\langle A\kappa, x \rangle = \langle A, x \rangle = 0$, which is a contradiction. Therefore, $A\kappa = G^*$. If $w \in O$, then $w|_G \in O\kappa$, and there is some $a \in A$ such that $a\kappa = w|_G$. Then $\langle a, F \rangle = \langle w, F \rangle \subseteq U$ and $a \in O$. As A is compact and δ_A is a continuous bijection, δ_A is a homeomorphism. $\qquad\square$

63.21 Corollary *If D is discrete, then δ_D is surjective, hence D is reflexive.*

Proof The group $C = D^*$ is compact by 63.5. If the discrete group $S = D^{**}/D$ is not trivial, then some element $c \in C^{**} = C$ induces a non-trivial character on S. This means that $\langle c, D^{**} \rangle \neq 0$ and $\langle c, D \rangle = 0$, but the last condition implies $c = 0$, which is a contradiction. $\qquad\square$

63.22 Annihilators If L is a subgroup of the topological abelian group A and if Λ is a subgroup of A^*, then

$$L^{\perp} = \{ \chi \in A^* \mid \langle L, \chi \rangle = 0 \} \quad \text{and} \quad \Lambda^{\perp} = \{ a \in A \mid \langle a, \Lambda \rangle = 0 \}$$

are called the *annihilators* of L and of Λ, respectively.

63.23 Lemma *Assume that K is a compact open subgroup of H, and let κ denote the canonical projection of H onto the discrete group*

$D = H/K$. Then $(\varphi \mapsto \kappa\varphi) : D^* \to K^\perp$ is an isomorphism of topological groups, and $K^\perp$ is a compact open subgroup of H^*.

This can be *proved* by direct verification; see HOFMANN–MORRIS 1998 7.13.

63.24 Lemma *Assume again that K is a compact open subgroup of H. By 63.15, the inclusion $\iota : K \to H$ has a surjective adjoint $\iota^* : H^* \to K^*$ with kernel $K^\perp$. Hence each $x \in K^{\perp\perp}$ induces a character on $H^*/K^\perp \cong K^*$. Because of 63.20, there is an element $z_x \in K$ such that $\langle x, \xi \rangle = \langle z_x, \iota^*\xi \rangle = \langle z_x \delta_H, \xi \rangle$ for each $\xi \in H^*$. Consequently, $K\delta_H = K^{\perp\perp}$.* □

63.25 Surjectivity *If A is locally compact, then the natural map δ_A is surjective.*

Proof By the Splitting Theorem, it suffices to prove surjectivity for a group H having a compact open subgroup K. Let $\delta = \delta_H$ denote the natural map $H \to H^{**}$. The last two lemmas imply that $K^\perp$ is compact and open in H^* and that $K^{\perp\perp} = K\delta \le H\delta = G \le H^{**}$, moreover, $K^{\perp\perp}$ is compact and open in H^{**}. Thus G is an open subgroup of H^{**}.

If $G < H^{**}$, then there is a non-trivial character $\varphi : H^{**} \to \mathbb{T}$ with $\langle G, \varphi \rangle = 0$. In particular, $\langle K^{\perp\perp}, \varphi \rangle = 0$. Lemma 63.24, applied to $K^\perp \hookrightarrow H^*$ instead of $K \hookrightarrow H$, gives an element $\zeta \in K^\perp$ such that $\langle x, \varphi \rangle = \langle x, \zeta \rangle$ for all $x \in H^{**}$. It follows that $\langle H\delta, \zeta \rangle = \langle H, \zeta \rangle = 0$, hence $\zeta = 0$ and $\varphi = 0$. This contradiction shows that $G = H^{**}$. □

63.26 *If A is locally compact, then the natural map $\delta_A : A \to A^{**}$ is open.*

Proof Again, it suffices to consider a group H with a compact open subgroup K. The map $\delta = \delta_H$ is bijective and continuous, hence the restriction $\delta|_K$ is a homeomorphism onto the image $K\delta = K^{\perp\perp}$, and this image is an open subgroup of H^{**}, as we have seen in the last proof. Since δ is a group isomorphism, it follows that δ is open in a neighbourhood of each point. □

Combined, Theorems 63.7, 63.18, 63.25, and 63.26 can be stated as

63.27 Pontryagin duality *If A is a locally compact abelian group, then A is reflexive, i.e., the natural homomorphism $\delta_A : A \to A^{**}$ is a homeomorphism and hence also a group isomorphism. Therefore, A^{**} and A can be identified via δ_A; the groups A and A^* are then duals of each other.*

Note   One should keep in mind that for the crucial steps 63.6, 63.7, 63.13, and 63.14 no proofs have been given, only convenient references. See also STROPPEL 2006 Theorem 22.6.

63.28 Evaluation   *If A is locally compact, then $(x, \xi) \mapsto \langle x, \xi \rangle$ is a continuous map $\omega : A \times A^* \to \mathbb{T}$.*

Proof Only continuity of ω at $(0, 0)$ has to be shown. Let V be a compact neighbourhood of 0 in A, and put $\Omega = W(V, U)$. Then $\langle V, \Omega \rangle \subseteq U$ by the very definition of Ω. $\square$

Applications

63.29 Proposition   *If C is a connected subgroup of A and Γ is a compact subgroup of A^*, then $\langle C, \Gamma \rangle = 0$.*

Proof Let $\iota : \Gamma \hookrightarrow A^*$ denote the inclusion. Its adjoint $\iota^* : A \to \Gamma^*$ is a continuous homomorphism into a discrete group, and $C\iota^*$ is a connected subgroup of Γ^*. Hence $C\iota^* = 0$ and $\langle C, \iota\Gamma \rangle = \langle C\iota^*, \Gamma \rangle = 0$. $\square$

63.30 Theorem   *Let A be an LCA group. Then A is connected if, and only if, A^* has no compact subgroups other than $\{0\}$.*

Proof If A is connected and Γ is a compact subgroup of A^*, then $\langle A, \Gamma \rangle = 0$ by the last proposition, and this means that $\Gamma = 0$.

For the converse, let C denote the connected component of A. By 62.13 and Exercise 3 of Section 62, the factor group A/C is 0-dimensional and has arbitrarily small compact open subgroups S/C, where $C \leq S$ and S is open in A. If A is not connected, then $A/C \neq 0$ and $A/S \neq 0$ for a suitable choice of S. Consider the canonical projection κ of A onto the discrete group $A/S = D$. Its adjoint $\kappa^* : D^* \to A^*$ is injective (63.3) and maps the compact group D^* homeomorphically onto a non-trivial subgroup of A^*. This proves the converse. $\square$

An abelian group A is divisible (resp. torsion free) if each homomorphism $A \to A : x \mapsto n \cdot x$ with $n \in \mathbb{N}$ is surjective (resp. injective).

63.31 Proposition   *If A is divisible, then A^* is torsion free.*

This is an immediate consequence of 63.3 and the identity $\langle n \cdot a, \chi \rangle = \langle a, n \cdot \chi \rangle$. The converse does not hold in general; see HEWITT–ROSS 1963 24.44. The following is true, however.

63.32 Theorem *If A is a torsion free abelian group, and if A is compact or discrete, then A^* is divisible.*

Proof Put $\Lambda = n \cdot A^*$. By 63.5, the group A^* is discrete or compact. Hence Λ is closed in A^*. We have $\Lambda^\perp = \{\, a \in A \mid \langle n \cdot a, A^* \rangle = 0 \,\} = 0$, since A is torsion free. The adjoint of the quotient map $A^* \to A^*/\Lambda$ yields an injection $(A^*/\Lambda)^* \to \Lambda^\perp = 0$, and $n \cdot A^* = A^*$.

If A is discrete, the proof follows also immediately from 63.15. $\qquad \square$

Exercises

(1) Let $0 \to A \to B \to C \to 0$ be a *proper* exact sequence of LCA groups (i.e., $A \to B$ is an embedding and $B \to C$ is a quotient morphism with kernel A). Then $C^* \cong A^\perp \le B^*$ and the adjoint sequence $0 \to C^* \to B^* \to A^* \to 0$ is also exact and proper.

(2) Let D denote the group $(\mathbb{R}, +)$, taken with the discrete topology. Show that the adjoint ε^* of the identity mapping $\varepsilon : D \to \mathbb{R}$ is not surjective. In fact, $\operatorname{card} \mathbb{R}^* < \operatorname{card} D^*$.

(3) Show that a compact abelian group A is connected if, and only if, it is divisible.

64   Fields

This section deals with fields and field extensions. Most proofs are easily accessible in the literature, there are many algebra books with chapters on field theory. We mention COHN 2003a and JACOBSON 1985, 1989; see also MORANDI 1996, LANG 1993, BOURBAKI 1990.

64.1 Definition A *field* is a set F with two binary operations, addition $+$ and multiplication, such that the following axioms hold.
 (a) $(F, +)$ is a commutative group with neutral element 0.
 (b) $F^\times = F \smallsetminus \{0\}$ is a commutative group with respect to multiplication (with neutral element 1).
 (c) Addition and multiplication are related by the distributive law
    $(a + b)c = ac + bc$.

Let F be a subfield of a field E; then we speak of the *field extension* $E|F$. Clearly E is a vector space over F; we call the dimension $\dim_F E$ of this vector space the *degree* of $E|F$ and denote it by $[E : F]$. An extension $E|F$ is called *finite*, if its degree $[E : F]$ is finite. The following simple observation is very useful.

64.2 Degree formula *Let $D|E$ and $E|F$ be field extensions. Then $[D : F] = [D : E][E : F]$.*

64.3 Quotients of rings *Let M be an ideal of a commutative ring R. Then the quotient ring R/M is a field if, and only if, M is a maximal ideal in R.*

Proof If R/M is a field, then the trivial ideal is maximal in R/M, hence M is maximal in R. Conversely, let M be maximal. If $a \in R \smallsetminus M$, then the ideal $aR + M$ coincides with R. Thus there exists an element $b \in R$ with $1 \in ab + M$, and this means that $b + M$ is the inverse of $a + M$ in R/M.   □

64.4 Prime fields Each field F has a smallest subfield P, which is the intersection of all subfields of F. One calls P the *prime field* of F. If $P \cong \mathbb{Q}$, then F is said to have *characteristic* zero, and one writes $\operatorname{char} F = 0$. Otherwise $P \cong \mathbb{F}_p = \mathbb{Z}/p\mathbb{Z}$ for some prime number p, and then F has the characteristic $\operatorname{char} F = p$.

64.5 Definitions Let $E|F$ be a field extension. An element $a \in E$ is called *algebraic* over F, if $f(a) = 0$ for some polynomial $f \neq 0$ with coefficients in F. By $F[x]$ (or $F[t]$) we denote the polynomial ring over F (compare 64.22), and we call a polynomial *monic* if its leading coefficient is 1. The monic polynomial $f \in F[x]$ of minimal degree with $f(a) = 0$ is uniquely determined; it is called the *minimal polynomial* of a over F. Such a minimal polynomial is irreducible in $F[x]$.

The extension $E|F$ is called *algebraic* (and the field E is called algebraic over F), if each element of E is algebraic over F.

Each algebraic extension $E|F$ satisfies $\operatorname{card} E \leq \max\{\aleph_0, \operatorname{card} F\}$; this is a consequence of 61.14: we have $\operatorname{card} F[x] = \max\{\aleph_0, \operatorname{card} F\}$, and each polynomial has only finitely many roots.

For $a \in E$ we denote by $F[a]$ the subring generated by $F \cup \{a\}$, and $F(a)$ denotes the subfield generated by $F \cup \{a\}$ (which is the field of fractions of $F[a]$). For $A \subseteq E$, the ring $F[A]$ and the field $F(A)$ are defined analogously.

64.6 Algebraic elements *Let $E|F$ be a field extension. An element $a \in E$ is algebraic over F if, and only if, $F[a]$ has finite dimension n as a vector space over F.*

In this case, $F[a] = F(a)$ is a field, the minimal polynomial $g \in F[x]$ of a has degree n, the evaluation map $F[x] \to F(a) : f \mapsto f(a)$ is a surjective ring homomorphism, and its kernel is the ideal $(g) := gF[x]$.

Proof $F[a]$ is the vector space over F generated by $1, a, a^2, \dots$. This vector space has finite dimension if, and only if, the elements $1, a, \dots, a^k$ are linearly dependent for some $k \in \mathbb{N}$, and this means that a is algebraic over F.

If this holds, then g is the monic polynomial of minimal degree in the ideal $I = \{ f \in F[x] \mid f(a) = 0 \}$. Since $F[x]$ is a principal ideal domain, we infer that $I = (g)$. The vector space $F[x]/I$ has the basis $\{ x^i + I \mid 0 \le i < \deg g \}$. Now $F[x]/I \cong F[a]$, hence $\deg g = n$.

Since g is irreducible, $I = (g)$ is a maximal ideal in the principal ideal domain $F[x]$. By 64.3, the quotient $F[x]/I \cong F[a]$ is a field, hence $F[a] = F(a)$. □

64.7 Transitivity *If $D|E$ and $E|F$ are algebraic field extensions, then $D|F$ is algebraic.*

Proof Let $a \in D$. Then $a^n + \sum_{i=0}^{n-1} c_i a^i = 0$ for suitable elements $c_i \in E$ and $n \in \mathbb{N}$. By 64.2, the field $C = F(c_0, c_1, \dots, c_{n-1})$ is a finite extension of F, hence $[F(a) : F] \le [C(a) : F] = [C(a) : C][C : F] \le n \cdot [C : F]$ is finite. □

64.8 Splitting fields Let $f \in F[x]$ be a non-zero polynomial. Choose an irreducible factor g of f. By 64.3, $E = F[x]/gF[x]$ is a field, and we consider F as a subfield of E by identifying $c \in F$ with $c + gF[x] \in E$. Then f has a root in E, namely $x + gF[x]$. (For example, $f = x^2 + 1$ has the root $x + f\mathbb{R}[x]$ in $\mathbb{C} = \mathbb{R}[x]/f\mathbb{R}[x]$.)

Repeated application of this construction gives a finite extension $E|F$ such that f is a product of linear factors in $E[x]$. If E is minimal with this property, i.e., if E is generated by F together with all roots of f, then E is called a *splitting field* of f over F. The splitting field of f over F is uniquely determined up to isomorphism; see COHN 2003a 7.2.3, JACOBSON 1985 Section 4.3, LANG 1993 V.3.1. More generally, there exists a splitting field for any set of non-zero polynomials over F.

Let q be a power of a prime p. A finite field F with card $F = q$ is a splitting field of the polynomial $x^q - x$ over $\mathbb{F}_p = \mathbb{Z}/p\mathbb{Z}$ (in fact, F consists just of all roots of $x^q - x$). Thus a finite field is determined up to isomorphism by its cardinality.

64.9 Definition An algebraic field extension $E|F$ is called *normal*, if each irreducible polynomial $f \in F[x]$ that has some root in E splits in $E[x]$ into a product of linear factors.

64.10 Theorem *An algebraic extension $E|F$ is normal precisely if E is a splitting field of some set of non-zero polynomials over F.*

Indeed, if $E|F$ is normal, then E is a splitting field of the set of minimal polynomials of elements in E. For the converse see COHN 2003a 7.2.4, JACOBSON 1989 Theorem 8.17 or LANG 1993 V.3.3.

64.11 Definition A polynomial is said to be *separable* if all its roots (in some splitting field) are simple. Let $E|F$ be a field extension. An element $a \in E$ is called *separable* over F, if a is algebraic over F and the minimal polynomial of a over F is separable. The extension $E|F$ is called *separable* if every element of E is separable over F.

Each multiple root of a polynomial $f \in F[x]$ is also a root of the derivative f'. If f is irreducible and char $F = 0$, then $f' \neq 0$ cannot divide f, hence f' and f have no common factor. This shows that each algebraic extension of fields with characteristic 0 is separable.

64.12 Primitive elements *Let $E|F$ be a finite separable extension. Then $E|F$ is simple, i.e., $E = F(a)$ for a suitable 'primitive' element $a \in E$.*

This is proved in COHN 2003a 7.9.2, JACOBSON 1985 Section 4.14 and LANG 1993 V.4.6.

64.13 Definition A field F is called *algebraically closed*, if F has no proper algebraic extension; this means that each irreducible polynomial in $F[x]$ has degree 1, or equivalently, that each non-zero polynomial in $F[x]$ has a root in F. An extension field $F^\natural$ of a field F is called an *algebraic closure* of F, if $F^\natural|F$ is algebraic and $F^\natural$ is algebraically closed.

64.14 Theorem *Every field F has an algebraic closure $F^\natural$.*

For the proof, one uses Zorn's Lemma (or some equivalent principle) to construct a maximal algebraic extension of F; such an extension is algebraically closed by 64.7. See JACOBSON 1989 Section 8.1 or MORANDI 1996 3.14 for the details, and COHN 2003a 7.3.4, 11.8.3 for other construction methods. These references show moreover that the algebraic closure $F^\natural$ of a field F is uniquely determined up to isomorphism. This fact is also a consequence of the following result.

64.15 Theorem *Let $F^\natural$ be an algebraic closure of the field F, let $\alpha : F \to F$ be a field automorphism of F, and let $E|F$ be an algebraic extension. Then α has an extension to a monomorphism $E \to F^\natural$ of fields. If E is algebraically closed, then any such extension is an isomorphism of E onto $F^\natural$.*

For a proof see LANG 1993 V.2.8, BOURBAKI 1990 V §4 or MORANDI 1996 3.20, 3.22. Using (an equivalent of) Zorn's Lemma in the proofs of 64.14 and 64.15 is unavoidable, because some models of ZF set theory contain fields F without containing an algebraic closure of F; see LÄUCHLI 1962.

64.16 Cardinality of the algebraic closure *We have* card $F^\natural = \max\{\aleph_0, \operatorname{card} F\}$.

For the proof see the inequality in 64.5 and note that a finite field E is not algebraically closed (as $1 + \prod_{a \in E}(x - a)$ has no root in E).

64.17 Definitions Let $E|F$ be a field extension and let $\Gamma \le \operatorname{Aut} E$ be a group of automorphisms of E. Then

$$\operatorname{Aut}_F E := \{\, \alpha \in \operatorname{Aut} E \mid \alpha(a) = a \text{ for all } a \in F \,\}$$

is a subgroup of $\operatorname{Aut} E$, and $\operatorname{Fix}_E \Gamma := \{\, a \in E \mid \alpha(a) = a \text{ for all } \alpha \in \Gamma \,\}$ is a subfield of E.

An algebraic extension $E|F$ is said to be a *Galois extension* if $F = \operatorname{Fix}_E \operatorname{Aut}_F E$, and then $\operatorname{Gal}_F E := \operatorname{Aut}_F E$ is called the *Galois group* of $E|F$. An algebraic extension $E|F$ is a Galois extension if, and only if, $E|F$ is normal and separable; see COHN 2003a 7.6.1, 11.8.4.

64.18 Main Theorem of Galois Theory *Let $E|F$ be a finite Galois extension and $\Gamma = \operatorname{Gal}_F E$. Then $|\Gamma| = [E : F]$, and the following holds.*
  (i) *The map $\Delta \mapsto \operatorname{Fix}_E \Delta$ is a lattice anti-isomorphism of the subgroup lattice of Γ onto the lattice $\{\, D \mid F \le D \le E \,\}$ of all fields between F and E.*
  (ii) *Let $\Delta \le \Gamma$ and $D = \operatorname{Fix}_E \Delta$. Then $E|D$ is a Galois extension with Galois group Δ, and the degree $[D : F]$ is the index of Δ in Γ. The extension $D|F$ is a Galois extension precisely if Δ is a normal subgroup of Γ, and then $\operatorname{Gal}_F D \cong \Gamma/\Delta$.*

64.19 Transcendental elements Let $E|F$ be a field extension. An element $t \in E$ is said to be *transcendental* over F, if t is not algebraic over F. In this case, $F[t]$ is isomorphic to the polynomial ring $F[x]$ (via evaluation as in 64.6), and $F(t)$ is isomorphic to the field $F(x)$ of rational functions (which is the field of fractions of $F[x]$).

Such a field $F(t)$ has many automorphisms: each of the substitutions $t \mapsto t^{-1}$ and $t \mapsto at + b$ with $a \in F^\times$, $b \in F$ yields a unique automorphism of $F(t)$ fixing all elements of F. Hence $\operatorname{Aut}_F F(t)$ contains the group $\operatorname{PGL}_2 F$ (which is generated by these substitutions). In fact,

$\mathrm{Aut}_F F(t) = \mathrm{PGL}_2 F$ by a result related to Lüroth's Theorem; see COHN 2003a 11.3.3 or JACOBSON 1989 Section 8.14.

We claim that $[F(t) : F] = \max\{\aleph_0, \mathrm{card}\, F\} = \mathrm{card}\, F(t)$ if t is transcendental over F. Indeed, the elements $(t-a)^{-1}$ with $a \in F$ are linearly independent over F by Exercise 5 below; the other necessary estimates are obvious.

64.20 Algebraic independence and transcendency bases Let $E|F$ be a field extension. A subset $T \subseteq E$ is called *algebraically independent* over F, if each $t \in T$ is transcendental over $F(T \smallsetminus \{t\})$. Then the field $F(T)$ consists of all fractions of polynomials in finitely many variables taken from T; such a field $F(T)$ is called a *purely transcendental extension* of F.

If $T \neq \emptyset$ is algebraically independent over F, then

$$[F(T) : F] = \max\{\aleph_0, \mathrm{card}\, F, \mathrm{card}\, T\} = \mathrm{card}\, F(T) \ .$$

Indeed, if T is finite, then the results 64.2, 64.19 and 64.13b imply that $[F(T) : F] = \max\{\aleph_0, \mathrm{card}\, F\}^{\mathrm{card}\, T} = \max\{\aleph_0, \mathrm{card}\, F\}$; now the first equation for $[F(T) : F]$ follows from this in view of 61.13a, 61.14 and $F(T) = \bigcup\{ F(X) \mid X \subseteq T, X \text{ is finite}\}$. Again by 61.13b and 61.14 we have $\mathrm{card}\, F(T) = \max\{\mathrm{card}\, F, [F(T) : F]\} = [F(T) : T]$.

By Zorn's Lemma, there exists a maximal set $T \subseteq E$ which is algebraically independent over F, a so-called *transcendency basis* of $E|F$; then E is algebraic over $F(T)$. By a result due to STEINITZ 1910, any two transcendency bases of E over F have the same cardinality; see COHN 2003a 11.2.1, JACOBSON 1989 Section 8.12, MORANDI 1996 19.15 or BOURBAKI 1990 V §14 Theorem 3. This cardinality is called the *transcendency degree* $\mathrm{trdeg}(E|F)$ of $E|F$.

For every infinite field F, we have $\mathrm{card}\, F = \max\{\aleph_0, \mathrm{trdeg}(F|P)\}$, where P is the prime field of F.

64.21 Theorem (Steinitz) *Two algebraically closed fields F, F' with prime fields P, P' are isomorphic if, and only if, $\mathrm{char}\, F = \mathrm{char}\, F'$ and $\mathrm{trdeg}(F|P) = \mathrm{trdeg}(F'|P')$.*

Such a field F is an algebraic closure of $P(T)$ for each transcendency basis T of F over P; hence 64.21 follows from the fact that $\mathrm{trdeg}(F|P)$ is well-defined (64.20). Note that $\mathrm{trdeg}(F|P) = [F : P] = \mathrm{card}\, F$ if F is uncountable.

Each permutation of T induces an automorphism of $P(T)$, which extends to F by 64.15. This shows that algebraically closed fields have many automorphisms (see 52.10 for $\mathbb{Q}^\natural$ and $\mathbb{F}_p^\natural$). See also Exercise 4 and Theorem 14.11.

Now we define some rings and fields which consist of various types of formal power series.

64.22 Power series and polynomials Let F be a field. We endow the set $F^{\mathbb{N}_0}$ of all maps (sequences) $f : \mathbb{N}_0 \to F$ with pointwise addition and with the convolution product fg defined by

$$(fg)(n) = \sum_{i=0}^n f(i)g(n-i) \ .$$

Then $F^{\mathbb{N}_0}$ is a commutative ring (and an F-algebra) with the unit element δ_0 defined by $\delta_0(0) = 1$ and $\delta_0(n) = 0$ for $n \neq 0$.

We claim that each element $f \in F^{\mathbb{N}_0}$ with $f(0) \neq 0$ is invertible in this ring. Indeed, the map g defined recursively by $g(0) = f(0)^{-1}$ and $g(n) = -f(0)^{-1} \sum_{i=1}^n f(i)g(n-i)$ for $n > 0$ satisfies the equation $fg = \delta_0$.

Now we switch to the usual notation. For $n \in \mathbb{N}_0$ let $\delta_n \in F^{\mathbb{N}_0}$ be defined by $\delta_n(n) = 1$ and $\delta_n(m) = 0$ for $n \neq m$. The special element $t := \delta_1$ is called an indeterminate, it satisfies $t^n = \delta_n$ for $n \in \mathbb{N}_0$. The F-subalgebra of $F^{\mathbb{N}_0}$ generated by t is the *polynomial ring* $F[t] = \{\sum_{i=0}^n a_i t^i \mid n \in \mathbb{N}_0, a_i \in F\}$; it consists of those elements $f \in F^{\mathbb{N}_0}$ which have finite support $\operatorname{supp}(f) := \{n \in \mathbb{N}_0 \mid f(n) \neq 0\}$, and $\deg F := \max \operatorname{supp}(f)$ is the *degree* of a polynomial $f \neq 0$.

Often an arbitrary element $f \in F^{\mathbb{N}_0}$ is written formally as the power series $f = \sum_{n \geq 0} f(n)t^n$, and the ring $F^{\mathbb{N}_0}$ is written as the (formal) *power series ring*

$$F[[t]] = \left\{ \textstyle\sum_{n \geq 0} a_n t^n \mid a_n \in F \right\} \ .$$

The group of units of $F[[t]]$ is the direct product of $F^\times$ and of the multiplicative group $1 + tF[[t]]$.

The notation of these ring elements as series has a topological meaning: in the product topology of $F[[t]] = F^{\mathbb{N}_0}$ obtained from the discrete topology on F, the series f is the limit of the polynomials $\sum_{n=0}^N f(n)t^n$ with $N \in \mathbb{N}$. Moreover, $F[[t]]$ is the completion of the polynomial ring $F[t]$ with respect to this topology; see 44.11. $\square$

64.23 Laurent series Let F be a field. We denote by

$$L = \left\{ f \in F^{\mathbb{Z}} \mid \exists_{k \in \mathbb{Z}} \, \forall_{n < k} \, f(n) = 0 \right\}$$

the set of all mappings $f : \mathbb{Z} \to F$ such that the support $\operatorname{supp}(f) =$

$\{n \in \mathbb{Z} \mid f(n) \neq 0\}$ is bounded below. For $0 \neq f \in L$ we define $v(f) = \min \mathrm{supp}(f) \in \mathbb{Z}$, and $v(0) := \infty$. This set L is a commutative ring with pointwise addition and with the convolution product fg defined by the finite sums

$$(fg)(n) = \sum \{f(i)g(j) \mid i+j = n, \, i \geq v(f), \, j \geq v(g)\} \, .$$

We claim that L is in fact a field. The mapping $v : L \to \mathbb{Z} \cup \{\infty\}$ satisfies $v(fg) = v(f) + v(g)$ and $v(f+g) \geq \min\{v(f), v(g)\}$ for $f, g \in L$ (compare 56.1). Therefore $L_0 := \{f \in L \mid v(f) \geq 0\}$ is a subring of L, and we identify L_0 with the power series ring $F[[t]]$ from 64.22 via the isomorphism given by the restriction $f \mapsto f|_{\mathbb{N}_0}$. In particular, we consider $t = \delta_1$ as an element of L with $v(t) = 1$. Each non-zero element $f \in L$ can be written as $f = t^n g$, where $n = v(f)$ and $g \in L$ with $v(g) = 0$. The element $\delta_{-1} \in L$ (defined by $\delta_{-1}(-1) = 1$ and $\delta_{-1}(z) = 0$ for $z \neq -1$) is the inverse of $t = \delta_1$, and g is invertible in $L_0 = F[[t]]$ by 64.22. Hence f is invertible in L, and L is a field. In fact, we have shown that L is the field of fractions of the power series ring $L_0 = F[[t]]$. We remark that v is a valuation of L with valuation ring L_0 and residue field F; see Section 56.

Now we switch to the usual notation. We have $L = \bigcup_{n \geq 0} t^{-n} L_0$, hence we write each $f \in L$ formally as the Laurent series $f = \sum_{n \geq v(f)} f(n) t^n$ (compare 64.22), and we write the field L as the (formal) *Laurent series field*

$$F((t)) := L = \left\{ \sum_{n \geq z} a_n t^n \mid z \in \mathbb{Z}, a_n \in F \right\} \, .$$

In this notation, the multiplicative group of $F((t))$ has the direct decomposition $F((t))^\times = F^\times \times t^{\mathbb{Z}} \times (1 + tF[[t]])$.

Again the series notation has a topological meaning: with respect to the valuation topology induced by v (see 13.2b), f is the limit of the Laurent polynomials $\sum_{n=v(f)}^{N} f(n) t^n$ with $v(f) \leq N \in \mathbb{Z}$; note that F is discrete in this topology. Moreover, $F((t))$ is the completion with respect to this valuation topology of the field $F(t)$ of rational functions; see 44.11. A Laurent series f belongs to $F(t)$ precisely if the coefficients $f(n)$ finally satisfy a linear recurrence relation; see RIBENBOIM 1999 3.1.N p. 92.

64.24 Puiseux series Let F be a field. We define the set $P \subseteq F^{\mathbb{Q}}$ by

$$P = \bigcup_{n \in \mathbb{N}} P_n, \quad \text{where } P_n = \left\{ f \in F^{\mathbb{Q}} \mid \exists_{k \in \mathbb{N}} \, \mathrm{supp}(f) \subseteq n^{-1}(\mathbb{N} - k) \right\} \, .$$

Thus P consists of all maps $f : \mathbb{Q} \to F$ with supports which are contained in a cyclic subgroup $n^{-1}\mathbb{Z}$ of $(\mathbb{Q}, +)$ and bounded below. We endow P with pointwise addition and with the convolution product fg, where $(fg)(r)$ with $r \in \mathbb{Q}$ is defined as in 64.23, except that now the summation extends over the finitely many pairs $(i, j) \in \operatorname{supp}(f) \times \operatorname{supp}(g)$ with $i + j = r$.

The bijection $P_n \to F((t)) : f \mapsto \hat{f}$ defined by $\hat{f}(z) = f(z/n)$ for $z \in \mathbb{Z}$ is a ring isomorphism, hence P_n is a field by 64.23. Therefore also $P = \bigcup_n P_{n!}$ is a field. We remark that the map $v : P \to \mathbb{Q} \cup \{\infty\}$ defined as in 64.23 is a valuation of P with value group $\mathbb{Q}$ and residue field F.

Similarly as in 64.22, we define $\delta_q \in P$ for $q \in \mathbb{Q}$ by $\delta_q(q) = 1$ and $\delta_q(x) = 0$ for $q \neq x \in \mathbb{Q}$; then $\delta_q \delta_r = \delta_{q+r}$ for $q, r \in \mathbb{Q}$. We write $t := \delta_1$ and $t^q := \delta_q$; then $(t^{1/n})^n = t$ for $n \in \mathbb{N}$. The isomorphism $P_n \to F((t))$ described above maps $t^{1/n} \in P_n$ to $t \in F((t))$; we identify via this isomorphism and write $F((t^{1/n}))$ for P_n, and $F[[t^{1/n}]]$ for the valuation ring $\{ f \in P_n \mid v(f) \geq 0 \}$ of P_n. Moreover we write

$$F((t^{1/\infty})) := P = \bigcup_{n \in \mathbb{N}} F((t^{1/n})) .$$

By 64.23 each element $f \in P_n = F((t^{1/n}))$ can be written as $f = \sum_{i \geq N} a_i t^{i/n}$ with $N \in \mathbb{Z}$ and $a_i \in F$ (these infinite sums converge in the valuation topology defined by v). Such a series is called a *Puiseux series*, and $F((t^{1/\infty}))$ is called the *Puiseux series field* over F.

The multiplicative group of $F((t^{1/\infty}))$ has the direct decomposition $F((t^{1/\infty}))^\times = F^\times \times t^{\mathbb{Q}} \times \{1 + f \mid v(f) > 0\}$, and $\{1 + f \mid v(f) > 0\} = \bigcup_{n > 0}(1 + t^{1/n} F[[t^{1/n}]])$.

Let F be an algebraically closed of characteristic 0. Then the Newton–Puiseux theorem says that $F((t^{1/\infty}))$ is an algebraic closure of the Laurent series field $F((t))$; see, for example, SERRE 1979 IV Proposition 8 p. 68, RIBENBOIM 1999 7.1.A p. 186 or VÖLKLEIN 1996 Theorem 2.4.

This implies that $\mathbb{C}((t^{1/\infty}))$ is isomorphic to $\mathbb{C}$; see 64.21 and observe that $\operatorname{trdeg}(\mathbb{C}((t^{1/\infty}))|\mathbb{Q}) \leq \operatorname{card}\mathbb{C}((t^{1/\infty})) \leq \operatorname{card}\mathbb{C}^{\mathbb{Q}} = \operatorname{card}\mathbb{C} = \operatorname{trdeg}(\mathbb{C}|\mathbb{Q})$; compare also 64.25 and 14.9.

64.25 Hahn power series Let F be a field and let Γ be an ordered abelian group. On the set

$$F((\Gamma)) := \{ f \in F^\Gamma \mid \operatorname{supp}(f) \text{ is a well-ordered subset of } \Gamma \}$$

one can define an addition and a multiplication as in 64.24, and this turns $F((\Gamma))$ into field, the field of *Hahn power series* over Γ. The elements of $F((\Gamma))$ with finite support form a subring, the group ring $F[\Gamma]$, and the elements of $F((\Gamma))$ with countable support form the subfield

$$F((\Gamma))_1 := \{\, f \in F((\Gamma)) \mid \operatorname{card} \operatorname{supp}(f) \leq \aleph_0 \,\};$$

see DALES–WOODIN 1996 2.7 and 2.15, PRIESS-CRAMPE 1983 II §5, SHELL 1990 Appendix B.7, NEUMANN 1949a or ENGLER–PRESTEL 2005 Exercise 3.5.6 for details, and RIBENBOIM 1992 for generalizations.

As in 64.24, one has the valuation $v : F((\Gamma)) \to \Gamma \cup \{\infty\}$ defined by $v(f) = \min \operatorname{supp}(f)$ for $f \neq 0$ and $v(0) = \infty$, with value group Γ and residue field F.

Such a field $F((\Gamma))$ or $F((\Gamma))_1$ is algebraically closed if, and only if, F is algebraically closed and Γ is divisible; see DALES–WOODIN 1996 2.15, RIBENBOIM 1992 5.2 or PRIESS-CRAMPE 1983 p. 52 Satz 6. This implies that $\mathbb{C}((\mathbb{Q})) \cong \mathbb{C} \cong \mathbb{C}((\mathbb{Q}))_1$; see 64.21.

Using 12.10 we infer that the fields $\mathbb{R}((\Gamma))$ and $\mathbb{R}((\Gamma))_1$ are real closed for each divisible ordered abelian group Γ. The unique ordering of these fields is described by $f > 0 \Leftrightarrow f(v(f)) > 0$.

Exercises

(1) Show that each finite subgroup G of the multiplicative group of a field is cyclic. In particular, the multiplicative group of every finite field is cyclic.

(2) Let $E|F$ be a Galois extension with Galois group Γ. Then the minimal polynomial of $a \in E$ over F is the product $\prod_{b \in B}(x - b)$, where $B = \{\, \gamma(a) \mid \gamma \in \Gamma \,\}$ is the orbit of a under Γ.

(3) Let E be a splitting field of the polynomial $f = x^5 - 16x + 2$ over $\mathbb{Q}$. Show that $\operatorname{Gal}_\mathbb{Q} E$ induces the full symmetric group on the five roots of f.

(4) Every algebraically closed field F has precisely $2^{\operatorname{card} F}$ field automorphisms.

(5) If t is transcendental over the field F, then the rational functions $(t - a)^{-1}$ with $a \in F$ are linearly independent over F.

(6) Let $q = p^n$ be a power of the prime p. Show that the automorphism group of the finite field $\mathbb{F}_q$ of cardinality q is the cyclic group of order n which is generated by the Frobenius automorphism $a \mapsto a^p$.

Hints and solutions

1 The additive group of real numbers

(1) Every element $x \in B'$ is a linear combination of some finite set $B_x \subseteq B$. Note that $B = \bigcup_{x \in B'} B_x$ and apply 61.13a.

(2) If $a \in A$ has order $p^k r$, where r is not divisible by the prime p, then $up^k + vr = 1$ for suitable integers u, v (see 1.5), and $a = vr \cdot a + up^k \cdot a$. The first summand has order p^k, the order of the other summand is prime to p. The assertion follows by induction.

(3) We have $\mathbb{R}^+ \times \mathbb{Z}^n < \mathbb{R}^+ \times \mathbb{Q}^n \cong \mathbb{R}^+$ by 1.17. The group $\mathbb{R}^+$ is divisible, but $\mathbb{Z}^n$ and hence $\mathbb{R}^+ \times \mathbb{Z}^n$ are not.

(4) If B is a basis of $\mathbb{R}$ over $\mathbb{Q}$, then each map $\lambda : B \to \mathbb{Q}$ defines a linear form. The kernel H of λ is a hyperplane, and λ is determined by H up to a scalar in $\mathbb{Q}^\times$. Hence $\aleph_0 \cdot \operatorname{card} \mathcal{H} = \operatorname{card} \mathbb{Q}^B = \aleph_0^{\aleph} = 2^{\aleph}$ by 1.12 and 61.15.

(5) $\frac{1}{2}(\mathcal{C} + \mathcal{C}) = \{ \sum_{\nu=1}^\infty c_\nu 3^{-\nu} \mid c_\nu \in \{0, 1, 2\} \} = [0, 1]$ generates $\mathbb{R}^+$ as a group.

(6) See 35.10.

(7) If M is a maximal subgroup of F^+, then the quotient F/M is an abelian group without non-trivial proper subgroups, hence cyclic of prime order p. The divisibility of F^+ implies that $F = pF \subseteq M$, a contradiction.

(8) Each such subgroup of $\mathbb{R}^+$ is isomorphic to $\mathbb{Z}^2$. Any automorphism of odd order of $\mathbb{Z}^2$ is given by an integer matrix with determinant 1, since $\operatorname{Aut} \mathbb{Z}^2 = \operatorname{GL}_2\mathbb{Z}$. If $X \in \operatorname{SL}_2\mathbb{R}$ has trace t and finite order $n > 1$, then $X^2 = tX - E$ and $X^3 = (t^2 - 1)X - tE$. Each complex eigenvalue λ of X satisfies $|\lambda| = 1$. This implies $|t| \leq 2$. In the case $t = 2$, induction shows $X^n = nX - (n-1)E$, and $X^n = E$ implies $X = E$. If $t = -2$, then $-X = E$ by the last argument. $|t| = 1$ yields $X^3 = \pm E$, and $t = 0$ gives $X^2 = -E$. In particular, $n \neq 5$.

Alternative solution: Every automorphism of $\mathbb{Z}^2$ is described by an integer 2×2 matrix M. If M has order 5, then 1 is not an eigenvalue of M, hence M is annihilated by the cyclotomic polynomial $\Phi_5 = (x^5 - 1)/(x - 1)$. Now Φ_5 is irreducible over $\mathbb{Q}$ (apply Eisenstein's criterion, which is due to SCHÖNEMANN 1846, to $\Phi_5(x + 1)$), hence Φ_5 divides the characteristic polynomial of M, which has degree 2, a contradiction.

(9) By 1.17, $\mathbb{R}/\mathbb{Q} \cong \mathbb{R}$.

2 The multiplication of real numbers, with a digression on fields

(1) As $\mathbb{Q}(T) = \bigcup \{ \mathbb{Q}(S) \mid S \subseteq T \wedge \operatorname{card} S < \aleph_0 \}$, we may assume that T is finite. Since $\mathbb{Q}(S \cup \{t\}) = \mathbb{Q}(S)(t)$, the claim can then be proved by induction. If F is any ordered field, then there is a unique ordering for the polynomial ring $F[t]$ such that $F < t$. This ordering yields the desired ordering of the field of fractions $F(t)$.

(2) Every involution of the rational vector space $\mathbb{R}$ yields such an extension. An isomorphism between two extensions maps the copies of $\mathbb{R}$ onto each other, since $\mathbb{R}^+$ has no subgroup of index 2. Involutions defining isomorphic extensions have eigenspaces of the same dimension; hence we obtain infinitely many extensions. If an extension of $\mathbb{R}^+$ by C_2 is commutative, then it is isomorphic to $\mathbb{R} \oplus C_2$ by 1.23 and 1.22.

3 The real numbers as an ordered set

(1) The chain $\check{\mathbb{R}}$ is order complete, but neither strongly dense in itself nor strongly separable.

(2) If $\varphi : \mathbb{R}^2_{\text{lex}} \to \mathbb{R}_<$ is injective and order-preserving, then each set $\{t\} \times \mathbb{R}$ is mapped onto a connected set, hence onto an interval, but any family of pairwise disjoint open intervals in $\mathbb{R}$ is at most countable (Souslin's condition).

4 Continued fractions

(1) Consider the continued fraction $[1; 1, 1, 1, \dots]$ for the golden ratio $\gamma = (\sqrt{5} + 1)/2$. The approximating fractions have the form $f_{\nu+1}/f_\nu$ with $f_{\nu+1} = f_\nu + f_{\nu-1}$; the $f_\nu = 1, 1, 2, 3, 5, 8, \dots$ are the Fibonacci numbers. The quotients $f_{\nu+1}/f_\nu$ converge to $\gamma > 8/5$, and $(8/5)^2 > 5/2$. Therefore, $f_6/f_5 = 13/8 > 8/5$, $f_{4+\nu} \geq 5 \cdot (8/5)^\nu$ and $f_{2+\nu} f_{3+\nu} \geq 6 \cdot (5/2)^\nu$. By induction, $q_\nu = c_\nu q_{\nu-1} + q_{\nu-2} \geq f_\nu$, and the claim follows.

(2) Use $p_\nu = c_\nu p_{\nu-1} + p_{\nu-2}$ and $q_\nu = c_\nu q_{\nu-1} + q_{\nu-2}$.

(3) Let ν be even. Then $p_\nu/q_\nu < \zeta < p_{\nu+1}/q_{\nu+1}$ and $p_{\nu+1}/q_{\nu+1} - p_\nu/q_\nu = (q_\nu q_{\nu+1})^{-1} < \frac{1}{2}(q_\nu^{-2} + q_{\nu+1}^{-2})$, since the geometric mean is smaller than the arithmetic mean.

5 The real numbers as a topological space

(1) A is open (closed) in X and in Y if there are open (closed) subsets U, V in Z such that $A = U \cap X = V \cap Y$. Put $U \cap V = W$. It follows that $A = W \cap X = W \cap Y = W \cap (X \cup Y) = W \cap Z = W$ is open (closed) in Z.

(2) The set $\dot{\mathbb{Q}}$ is dense in $\check{\mathbb{R}}$, and $\check{\mathbb{R}}$ is separable. Each interval of the form $]\grave{a}, \acute{c}[= [\acute{a}, \grave{c}]$ is open and closed. Hence $\check{\mathbb{R}}$ is not connected and does not have a countable basis; in particular, $\check{\mathbb{R}}$ is not metrizable. Being orderable, the space $\check{\mathbb{R}}$ is normal, and order-completeness implies that $\check{\mathbb{R}}$ is locally compact.

(3) (a) Obviously $X = X_{\text{iii}}$ is a Hausdorff space and $\mathbb{Q}$ is dense in X. If X is a disjoint union of two proper open subsets, then there is also a partition of X into open subsets A and B such that A has a least upper bound s. As in the case of the ordinary reals, $s \notin A$ and each interval $]u, s]$ contains elements from $A \cap \mathbb{Q}$. This contradiction shows that X and the homeomorphic spaces $] \, , x[$ and $]x, \, [$ are connected, but X is not locally connected because each open interval in $\mathbb{Q}$ is open in X.

(b) The remaining open sets form a topology on $X = X_{ii}$ and $\mathbb{Q}$ is dense in X. By definition, the complement of $\{0\}$ is not open. Suppose that X is partitioned into two open subsets A and B such that $1 \in B$. Since $[1, \ [$ is connected, one has $] \ , -1] \subseteq A$ and $[1, \ [\subseteq B$, but then $0 \in A \cap B$. Therefore, X is connected. Similarly, X is locally connected and each point separates X into two connected subsets.

(c) A non-empty intersection of two basic open sets is another one. Hence $X = X_{iv}$ is a topological (Hausdorff) space. The countable set $\mathbb{N} \times ([0, 1[\cap \mathbb{Q})$ is dense in X and X is separable. Being a union of the connected space $\mathbb{N}$ and connected 'bristles' each of which meets $\mathbb{N}$ the space X is connected. Since $\mathbb{N}$ is locally connected, X has a basis of connected open sets. By 5.22, each point $(n, 0)$ has a connected complement in $\mathbb{N} \times \{0\}$. Therefore, condition (iv') is violated.

(4) Any interval with rational endpoints is open and closed in $\mathfrak{J}$. Hence any permutation of a finite number of pairwise disjoint intervals with rational endpoints is induced by a homeomorphism of $\mathfrak{J}$, and n distinct points can be mapped by a homeomorphism to n other points whenever the two sets are disjoint.

(5) $T_{1000\ldots}$ and $T_{0111\ldots}$ define the same vertex of T. A similar remark applies whenever two binary expansions represent the same number. The diameter of $T_{c|\nu}$ converges to 0 as ν increases. Hence φ is well-defined and continuous. Obviously, φ is surjective. This curve φ is named after PÓLYA 1913. According to LAX 1973, the map φ is nowhere differentiable; for a triangle with sides of length 9, 40, 41 however, the analogous construction yields a map ψ such that $\psi'(t) = 0$ almost everywhere; see also PRACHAR–SAGAN 1996. More on space-filling curves can be found in SAGAN 1994.

(6) A typical example is $S = \{ x \in \mathbb{R} \mid x = 0 \vee |x| > 1 \}$.

(7) By 4.11, the chain $\mathfrak{J}$ of irrational numbers admits a complete metric.

(8) Assume that $\varphi : \mathbb{L} \to \mathbb{L}$ is continuous and increasing. Put $c_0 = 0$ and $c_{\kappa+1} = \varphi(c_\kappa)$. The set $\{ c_\kappa \mid \kappa < \omega \}$ is bounded. Hence the c_κ converge to some $c \in \mathbb{L}$ and $\varphi(c) = c$.

(9) Each finitely generated Boolean algebra is finite and isomorphic to the power set 2^N of the finite set N of atoms (compare COHN 2003a 3.4.5 or KOPPELBERG 1989 Corollary 2.8 p. 30).

List the elements of the given algebras $A = \{0, 1, a_1, a_2, \ldots\}$ and $B = \{0, 1, b_1, b_2, \ldots\}$. Then there is an isomorphism φ of the algebra A_n generated by $a_1, a_2, \ldots, a_n$ into B. Choose the first element b in B which is not in $\varphi(A_n)$. Since the algebras are supposed to have no atoms, each interval of A_n contains further elements of A. Therefore there is a first element $a \in A$ such φ has an extension $\varphi' : \langle A_n, a \rangle \to B$ with $\varphi'(a) = b$. Now interchange the roles of A and B and proceed in the same way, so that both sequences will be exhausted.

Compare also HODGES 1993 p. 100.

(10) If $q \in \mathbb{Q}^2$, then each straight line through q with rational slope $r \neq 0$ is contained in S. If $u, v \in \mathfrak{J}$, then there are increasing sequences of rational numbers a_ν converging to u and b_ν converging to v with decreasing slopes $(b_{\nu+1} - b_\nu)/(a_{\nu+1} - a_\nu)$. Hence there is an increasing continuous map $\varphi : [0, 1] \to S$ (which is piecewise linear on each interval $[0, \ell]$ with $\ell < 1$) such that $\varphi(0) = (a_1, b_1)$ and $\varphi(1) = (u, v)$.

6 The real numbers as a field

(1) Well-order a maximal set T of algebraically independent elements of $\mathbb{R}$. Note that any extension of a countable field by a countable set is itself countable by 61.14. Therefore, T is uncountable. Choose t_ν as the first element in T such that $t_\nu \notin \mathbb{Q}(\{\, t_\kappa \mid \kappa < \nu \,\}) = F_\nu$. Then the F_ν form a chain of subfields of $\mathbb{R}$ of length at least $\Omega = \omega_1$ (compare 61.7).

(2) Clearly $\mathbb{Z}[[x]]^\flat$ is a subring of $\mathbb{Z}[[x]]$. If $f = \sum_n f_n x^n \in \mathbb{Z}[[x]]^\flat$ and $|f_n| \leq Cn^k$ for all $n \in \mathbb{N}$, then the series for $f(1/2)$ converges absolutely in $\mathbb{R}$. Thus the evaluation $f \mapsto f(1/2)$ is a surjective ring homomorphism $\mathbb{Z}[[x]]^\flat \to \mathbb{R}$. The ideal generated by $1 - {}^\cdot 2x$ is contained in the kernel of this evaluation. Let $f(1/2) = 0$ and define $g \in \mathbb{Z}[[x]]$ by $g = f \cdot (1 - 2x)^{-1} = f \cdot \sum_n 2^n x^n$; we show now that the coefficients g_n of g are bounded by some polynomial in n.

We have $g_n = \sum_{j=0}^n f_j 2^{n-j} = 2^n\bigl(f(1/2) - \sum_{j>n} f_j 2^{-j}\bigr) = -2^n \sum_{j>n} f_j 2^{-j}$, thus $|g_n| \leq C\sum_{j=n+1}^{2n} j^k 2^{n-j} + C\sum_{j>2n} j^k 2^{n-j} \leq C(2n)^k + C\sum_{j>2n} j^k 2^{n-j}$. Hence it suffices to show that $\sum_{j>2n} j^k 2^{n-j} = \sum_{j\geq 1}(2n+j)^k 2^{-n-j}$ is bounded above by a constant which depends only on k and not on n.

The function $h(x) := (2x+j)^k 2^{-x}$ with fixed values $j, k \in \mathbb{N}$ is decreasing for $x \geq x_0 := k/(\ln 2) - j/2$ and increasing for $1 \leq x \leq x_0$, hence we have $(2n+j)^k 2^{-n} \leq h(x_0) = (2k/(\ln 2))^k 2^{-x_0} \leq (4k)^k 2^{j/2}$ for $j, k, n \in \mathbb{N}$, in view of $\ln 2 > 1/2$. This implies $\sum_{j\geq 1}(2n+j)^k 2^{-n-j} \leq (4k)^k \sum_{j\geq 1} 2^{-j/2} = (4k)^k/(\sqrt{2}-1)$, which does not depend on n.

7 The real numbers as an ordered group

(1) Monotonicity being obvious, $\mathbb{R}^2_{\mathrm{lex}}$ is an ordered group. It is not Archimedean, since $n \cdot (0,1) = (0,n) < (1,0)$.

(2) Assume that $\varphi : A \to \mathrm{Aut}_< B : a \mapsto \varphi_a$ is a homomorphism into the group of order-preserving automorphisms of B. Then the set $A \times B$ with the lexicographic ordering and the operation $(a,b) \oplus (c,d) = \bigl(a+c, b+\varphi_a(d)\bigr)$ is an ordered group, because $d < d'$ implies $\varphi_a(d) < \varphi_a(d')$. It is not Archimedean for the same reason as in Exercise 1.

(3) The additive group $\mathbb{Z}\log 2 + \mathbb{Z}\log 3$ is dense in $\mathbb{R}$, since $\log 3/\log 2 \notin \mathbb{Q}$; see 2.2, 1.6 and 35.10.

(4) By 7.12, the automorphisms α in question are the maps $\alpha = \varphi_r$ where r is a positive real number r such that $rA = A$. In particular, $\alpha(1) = r = m + nc$ and $\alpha(c) = rc = mc + nc^2 = p + qc$ with $m, n, p, q \in \mathbb{Z}$. If c is not quadratic over $\mathbb{Q}$, then $n = 0$ and $\alpha = \varphi_r = \varphi_m$, hence $m = 1$ and $\alpha = \mathrm{id}$.

Let c be quadratic over $\mathbb{Q}$. Then φ_r is an endomorphism of the ordered group A precisely if $0 < r \in A$ and $rc \in A$, i.e., for $0 < r \in A \cap Ac^{-1}$. We have $A \cap Ac^{-1} = (\mathbb{Z} \oplus \mathbb{Z}c) \cap (\mathbb{Z} \oplus \mathbb{Z}c^{-1}) = \mathbb{Z} \oplus (\mathbb{Z}c \cap Ac^{-1}) = \mathbb{Z} \oplus \mathbb{Z}kc$ where $k \in \mathbb{N}$ is the smallest integer such that $kc^2 \in A$, hence $A \cap Ac^{-1} = \mathbb{Z}[kc]$ is the ring generated by the real algebraic integer kc. By Dirichlet's unit theorem, the group of units of $\mathbb{Z}[kc]$ is the direct product of $\{\pm 1\}$ with an infinite cyclic group (which is generated by the fundamental unit $u > 1$ obtained from solving some Pell equation); see BOREVICH–SHAFAREVICH 1966 Chapter 2, 4.3 Theorem 5 (for example, $\mathbb{Z}[\sqrt{5}]$ has the fundamental unit $u = 2 + \sqrt{5}$). Thus the positive units of $\mathbb{Z}[kc]$, hence also the automorphisms of the ordered group A, form an infinite cyclic group.

8   The real numbers as a topological group

(1) Cantor's middle third set $\mathcal{C}$ is a meagre null set, and so is $S = \mathbb{Z} + \mathcal{C}$. From $\mathcal{C} + \mathcal{C} = [0, 2]$ (see Section 1, Exercise 5) it follows that $S + \mathcal{C} = \mathbb{R}$.

(2) This follows directly from the definitions.

(3) (a) Any proper subgroup of $\mathbb{R}$ is totally disconnected, because each connected subset of $\mathbb{R}$ is an interval and generates all of $\mathbb{R}$. A proper vector subgroup A is dense in $\mathbb{R}$; hence a compact neighbourhood contains an interval and generates $\mathbb{R}$.

(b) By the definition of the Tychonoff topology (or the universal property of topological products) the direct product of any family of topological groups is itself a topological group.

(4) By 8.26, the endomorphism ring $\mathsf{P} = \mathrm{End}_c\,\mathbb{R}$ consists of all maps $\varphi_r = (x \mapsto rx)$ with $r \in \mathbb{R}$. Clearly $\varphi : \mathbb{R} \to \mathsf{P} : r \mapsto \varphi_r$ is an isomorphism of the additive groups, and $\varphi_{rs} = \varphi_r \circ \varphi_s$, hence φ is also multiplicative. A typical neighbourhood (C, U) in P is $\{\rho \in \mathsf{P} \mid \rho(C) \subseteq U\}$, where C is compact and U is open in $\mathbb{R}$. Continuity of φ^{-1} follows from the fact that φ^{-1} maps $(1, U)$ to U. If $rC \subseteq U$, then $VC \subseteq U$ for some neighbourhood V of r, and φ is also continuous.

9   Multiplication and topology of the real numbers

(1) Each automorphism α of $\mathbb{R}^\times$ maps the connected component $\mathbb{R}_{\mathrm{pos}}$ of $\mathbb{R}^\times$ onto itself. By 8.24, α induces on $\mathbb{R}_{\mathrm{pos}}$ a map $x \mapsto x^a$ with $a \neq 0$. Moreover, $(-1)^\alpha = -1$ since -1 is the only involution in $\mathbb{R}^\times$, and $(-x)^\alpha = -x^\alpha$.

(2) Obviously, each motion in $\mathbb{M}$ preserves the distance d. Conversely, if a map φ of $\mathbb{R}$ satisfies $|\varphi(x) - \varphi(y)| = |x - y|$ for all $x, y \in \mathbb{R}$, then $\varphi_a = x \mapsto \varphi(x) - a$ has the same property. Putting $a = \varphi(0)$, we get $|\varphi_a(x)| = |x| = \pm x$. By continuity, the sign is constant and $\varphi \in \mathbb{M}$.

(3) The orientation preserving elements in the groups in question form a (normal) subgroup A of index 2. Each element in the complement of A has order 2 and generates a factor C_2 of a semidirect decomposition.

10   The real numbers as a measure space

(1) Closed sets, and points in particular, are contained in $\mathfrak{O}_\delta$. Each countable set is in $\mathfrak{O}_{\delta\sigma} = \mathfrak{O}_1$ but not in $\mathfrak{O}$ (because intervals are uncountable).

(2) Let $\mathbb{Q} = \{r_\mu \mid \mu \in \mathbb{N}\}$ and choose $O_{\nu|\kappa}$ as a $2^{-\kappa}$-neighbourhood of r_{ν_1}.

(3) If $k \in \mathbb{N}$ and $k > 1$, then $k \cdot S \subset S$ and $k \cdot \overline{\lambda}(S) = \overline{\lambda}(k \cdot S) \leq \overline{\lambda}(S)$. Hence $\overline{\lambda}(S)$ is either 0 or ∞. A subspace of finite codimension has only countably many cosets in $\mathbb{R}$. The claim follows as in 10.9 and 10.10.

(4) If $\mathfrak{T} \subset \mathfrak{M}$ and $\operatorname{card}\mathfrak{T} \leq \aleph$, then an argument analogous to 10.15 shows that $\mathfrak{S} \cup \mathfrak{T}$ generates a σ-field of cardinality at most $\aleph$.

(5) Let H denote a hyperplane in the rational vector space $\mathbb{R}$. Put $C = \mathcal{C}_2 \cap H$ and $B = ([0, 1] \smallsetminus \mathcal{C}_2) \cap H$. Then $\overline{\lambda}(B) = 1/2$ and $\overline{\lambda}(B \cup C) = 1$ by the remark following 10.11. Subadditivity of $\overline{\lambda}$ gives $\overline{\lambda}(C) \geq 1/2$, but $\underline{\lambda}(C) = 0$ by 10.8.

(6) The first claim follows from 10.15. Consider now a non-measurable set $C \subset \mathcal{C}_2$ as in Exercise 5 and a homeomorphism φ of $\mathbb{R}$ which maps $\mathcal{C}_2$ onto

Cantor's middle third set $\mathcal{C}$. Then $\varphi(\mathcal{C})$ is a null set and therefore measurable, but its preimage $\mathcal{C}$ is not measurable.

(7) For a given sequence $\nu \in \mathbb{N}^{\mathbb{N}}$, choose $\mu \in \overline{A} = A$ such that $|\nu_1 - \mu_1| \leq |\nu_1 - \xi_1|$ for all $\xi = (\xi_\iota)_\iota \in A$ and, inductively, $|\nu_\kappa - \mu_\kappa| \leq |\nu_\kappa - \xi_\kappa|$ for all $(\mu_1, \ldots, \mu_{\kappa-1}, \xi_\kappa, \xi_{\kappa+1} \ldots) \in A$. Whenever there are two possibilities, select the smaller one as μ_κ. Put $\rho(\nu) = \mu$. Then $\rho|_A = \mathrm{id}_A$, and ρ is continuous because μ_κ depends only on $(\nu_0, \ldots, \nu_\kappa)$.

11 The real numbers as an ordered field

(1) Let $F_0 = \mathbb{Q}$ and define inductively $F_{2\nu+1} = F_{2\nu}(\exp F_{2\nu})$ and $F_{2\nu+2} = F_{2\nu+1}(\log(F_{2\nu+1} \cap \mathbb{R}_{\mathrm{pos}}))$. Then each F_ν is countable, and $F = \bigcup_{\nu \in \mathbb{N}} F_\nu$ has the required property.

(2) If $a > 0$, then $F < at + b$, but if $c \neq 0$ or $a < 0$, then $F < (at + b)/(ct + d)$ does not hold.

(3) By 11.14, the field $\mathbb{R}$ with the given structure embeds as a subfield into $\mathbb{R}$ with the usual structure. The subfield is order complete because its ordering is isomorphic to the usual one, hence it contains the closure of $\mathbb{Q}$.

(4) If F^+ is Archimedean, then F embeds into $\mathbb{R}$ by 11.14, hence the multiplicative group of positive elements is Archimedean. (Alternatively, $a > 0$ implies that $(1 + a)^n > n \cdot a$ is not bounded above.) Conversely, if $\mathbb{N} \leq k$ in F, then $2^n \leq k$ and the multiplication is not Archimedean.

12 Formally real and real closed fields

(1) For $f, g \in F[t] \smallsetminus \{0\}$, we call the rational function f/g positive if the leading coefficient of the product fg is positive with respect to $<$. This defines an ordering on $F(t)$; on $\mathbb{R}(t)$ this is the ordering $P_{\infty,+}$ introduced in Exercise 3 below. For the uniqueness we observe that the condition $F < t$ determines the extension uniquely on the polynomial ring $F[t]$, and then also on the field $F(t)$ of fractions of $F[t]$.

(2) Let $<$ be the unique ordering of $\mathbb{R}$. The field $\mathbb{Q}(\sqrt{2})$ admits the field automorphism α defined by $\alpha(a + b\sqrt{2}) = a - b\sqrt{2}$ for $a, b \in \mathbb{Q}$. Hence $\mathbb{Q}(\sqrt{2})$ has the two domains of positivity $P_1 = \{a + b\sqrt{2} \mid a, b \in \mathbb{Q}, a + b\sqrt{2} > 0\}$ and $P_2 = \alpha(P_1) = \{a + b\sqrt{2} \mid a, b \in \mathbb{Q}, a - b\sqrt{2} > 0\}$.

Now let P be any domain of positivity of $\mathbb{Q}(\sqrt{2})$. By applying α if necessary, we may assume that $\sqrt{2} \in P$; then we have to show that $P = P_1$. Since P and P_1 are subgroups of index 2 in $\mathbb{Q}(\sqrt{2})^\times$, it suffices to prove the inclusion $P \subseteq P_1$. We know that $P \cap \mathbb{Q}$ is the set of positive rational numbers by 11.7.

Proceeding indirectly, we assume that there are rational numbers a, b with $a + b\sqrt{2} \in P$ and $a + b\sqrt{2} < 0$. If $b < 0$, then $-b\sqrt{2} \in P$, hence $a \in P, a > 0$ and $a - b\sqrt{2} > 0$. Thus the rational number $a^2 - 2b^2 = (a + b\sqrt{2})(a - b\sqrt{2}) \in P$ is negative, a contradiction. The case $a < 0 < b$ is treated analogously.

(3) $P_{\infty,+}$ is the domain of positivity of the ordering described in Exercise 1. Applying the field automorphisms of $\mathbb{R}(t)$ given by $t \mapsto -t$ or $t \mapsto \pm t^{-1} + r$ (compare 64.19), we infer that also $P_{\infty,-}$ and $P_{r,\pm}$ are domains of positivity.

Let P be any domain of positivity of $\mathbb{R}(t)$. The corresponding ordering $<$ induces on $\mathbb{R}$ the usual ordering (11.7). We can achieve that $\mathbb{R} < t$ by applying the field automorphisms just mentioned. Now Exercise 1 yields $P = P_{\infty,+}$.

(4) The field $\mathbb{C}((t^{1/\infty}))$ is algebraically closed by the Newton–Puiseux theorem; see 64.24. Because $\mathbb{C}((t^{1/\infty})) = \mathbb{R}((t^{1/\infty}))(\sqrt{-1})$, the field $\mathbb{R}((t^{1/\infty}))$ is real closed by 12.10. In the unique ordering $<$ of $\mathbb{R}((t^{1/\infty}))$, we have $\mathbb{N} < t^{-1}$, since $t = (t^{1/2})^2$ and $1 - nt = \left(\sum_{k \geq 0} \binom{1/2}{k} (-n)^k t^k \right)^2$ are squares in $\mathbb{R}((t^{1/\infty}))$ for each $n \in \mathbb{N}$.

(5) We show that each element $s \in P$ is totally real: There exists a sequence of quadratic field extensions $F_{\nu+1} = F_\nu(\sqrt{1 + c_\nu^2})$ such that $F_0 = \mathbb{Q}$ and $s = a + b\sqrt{1 + c^2}$ with $a, b, c \in F_n$. By induction we may assume that a, b, c are totally real. If $F_n(s)$ is contained in the Galois extension H of $\mathbb{Q}$, and if $\Gamma = \operatorname{Aut} H$, then $\{\, s^\gamma \mid \gamma \in \Gamma \,\}$ is the set of all roots of the minimal polynomial of s over $\mathbb{Q}$ (see Section 64, Exercise 2). For every $\gamma \in \Gamma$ the elements a^γ, b^γ und c^γ are real. Hence $(1 + c^2)^\gamma = 1 + (c^\gamma)^2 > 0$ and s^γ is real.

For the converse, let a be totally real. Then the minimal polynomial of a^2 has only positive roots. Hence a^2 is a sum of squares; see JACOBSON 1989 Theorem 11.7.

(6) Let M be a maximal subfield of $\mathbb{R}$. Then $\mathbb{R} = M(a)$ for any $a \in \mathbb{R} \smallsetminus M$, and a is algebraic over M, otherwise $M < M(a^2) < \mathbb{R}$. Thus $[\mathbb{R} : M]$ and $[\mathbb{C} : M] = 2[\mathbb{R} : M]$ are finite, and 12.15 implies that $2 = [\mathbb{C} : M] = 2[\mathbb{R} : M]$, hence $\mathbb{R} = M$.

(7) Proceed as in Section 6, Exercise 1, and note that the algebraic closure of any countable field is countable by 64.16.

(8) Let $p = \operatorname{char} F$. First we show that F is perfect, that is, either $p = 0$ or $F = F^p$ for $p > 0$. Indeed, if $p > 0$ and $a \in F \smallsetminus F^p$, then each polynomial $f = x^{p^k} - a$ with $k \in \mathbb{N}$ is irreducible over F: if r is a root of f in some splitting field, then $f = (x - r)^{p^k}$; hence any non-trivial factor of f in $F[x]$ is of the form $c(x - r)^d$ with $c \in F^\times$ and $0 < d < p^k$. Then F contains r^d and $r^{p^k} = a$, hence by Bezout's theorem 1.5 also r^g with $g := \gcd(d, p^k)$. From $g < p^k$ and $(r^g)^{p^k/g} = a$ we infer that $a \in F^p$, a contradiction.

As F is perfect, every finite extension of F is separable (see the proof of 12.15), hence simple (64.12). Thus the degrees of the finite extensions of F are bounded, and there is a finite extension E of F of maximal degree. By maximality, E is algebraically closed. Now apply 12.15.

(9) If a field F is Euclidean, then F is Pythagorean with a unique ordering; see 12.8. Conversely, assume that F is Pythagorean with a unique ordering. By 12.5, the unique domain of positivity P of F consists of all non-zero sums of squares of F, and P is the set of all non-zero squares of F since F is Pythagorean. Hence F is Euclidean.

13 The real numbers as a topological field

(1) The inclusions $a + n\mathbb{Z} - (b + n\mathbb{Z}) \subseteq a - b + n\mathbb{Z}$ for $a, b \in \mathbb{Q}$, $n \in \mathbb{N}$ show that the non-zero ideals of $\mathbb{Z}$ yield a topology τ on $\mathbb{Q}$ such that $\mathbb{Q}^+$ is a topological group. For $p, s \in \mathbb{Z}$ and $q, r, n \in \mathbb{N}$ we have $(pr^{-1} + nrs\mathbb{Z})(qs^{-1} + nrs\mathbb{Z}) \subseteq pq(rs)^{-1} + n\mathbb{Z}$, hence τ is a ring topology of $\mathbb{Q}$. However, τ is not a field topology, because the set $(1 + 2\mathbb{Z})^{-1}$ does not contain any coset $1 + n\mathbb{Z}$ with $n \in \mathbb{N}$.

Let τ be any ring topology of $\mathbb{Q}$ which is not a field topology, and let B be any Hamel B basis of $\mathbb{R}$; compare 1.13. Endow $\mathbb{Q}^B$ with the product topology obtained from τ, and consider $\mathbb{R}$ as a subspace of $\mathbb{Q}^B$. Then $\mathbb{R}$ is a topological ring (since the multiplication is bilinear), but not a field (because $\mathbb{Q}$ retains its topology τ).

(2) By 8.15 and 13.6 the topological additive group is isomorphic to $\mathbb{R}$, and then 13.7 gives the assertion.

(3) Since $\mathbb{Q}$ is dense in $\mathbb{R}$, a continuous isomorphism $\mathbb{Q}(s) \to \mathbb{Q}(t)$ is the identity, whence $\mathbb{Q}(s) = \mathbb{Q}(t)$. This equation means that $s = (at + b)/(ct + d)$ for some invertible matrix $\left(\begin{smallmatrix} a & b \\ c & d \end{smallmatrix}\right)$; see COHN 2003a 11.3.3 or JACOBSON 1989 Section 8.14.

(4) If T is a transcendency basis, so is $\{t - r_t \mid t \in T\}$ for any choice of numbers $r_t \in \mathbb{Q}$.

(5) Use $x^2 = 1 + 2/((x-1)^{-1} - (x+1)^{-1})$ and $4xy = (x+y)^2 - (x-y)^2$.

(6) Let $c = \begin{pmatrix} a & -\overline{b} \\ b & \overline{a} \end{pmatrix} \in \mathbb{H}$. Then $c \neq 0$ implies $\det c = a\overline{a} + b\overline{b} > 0$, and c is invertible. Hence $\mathbb{H}$ is a division algebra. Put $i = \left(\begin{smallmatrix} & 1 \\ -1 & \end{smallmatrix}\right)$ and $j = \left(\begin{smallmatrix} & -1 \\ 1 & \end{smallmatrix}\right)$. If $ci = ic$, then $b = 0$; if $cj = jc$, then $\overline{a} = a$ and $\overline{b} = b$. Therefore, the centralizer of i and j consists of the (real) diagonal matrices in $\mathbb{H}$ and this is also the centre of $\mathbb{H}$. Moreover, $\mathbb{H} \cong \mathbb{R}^4$ and $\det c = |a|^2 + |b|^2$ is the square of the Euclidean norm; see also 34.17.

(7) By induction, $c^{c_n} < 2$. For $x < 2$ we have $x < c^x$ because $x^{-1} \log x$ is strictly increasing on the open interval $(1, e)$. Hence the sequence $(c_n)_n$ converges to a number $s \leq 2$, and $s = c^s = 2$.

(8) The contribution of finitely many elements a_n to Ca is negligible. If $|a_n - s| < \varepsilon$ for all n, then $|c_n - s| < \varepsilon$ for all n. If $a_n = (-1)^n \cdot (2n + 1)$, then Ca does not converge, but CCa does.

(9) If $\prod_n (1 - a_n) > 0$, then $1 + \sum_n a_n < \prod_n (1 + a_n) < \prod_n (1 - a_n)^{-1} < \infty$. Conversely, assume that $\sum_n a_n < \infty$. Discarding finitely many elements of the sequence, we may assume that $\sum_n a_n < 1/2$. Then $\prod_n (1 - a_n)^{-1} < \prod_n (1 + 2a_n) < \prod_n \exp(2a_n) = \exp(2\sum_n a_n) < e$.

(10) If the field F is arcwise connected, then there is also an arc C from 0 to 1 (by homogeneity). Let W be an arbitrary neighbourhood of 0. For each $c \in C$ there are neighbourhoods U_c of c and V_c of 0 such that $U_c V_c \subseteq W$ since multiplication is continuous. The compact arc C is covered by finitely many of the sets U_c. The intersection V of the (finitely many) corresponding sets V_c is a neighbourhood of 0 with $V \subseteq CV \subseteq W$, and CV is arcwise connected because Cv is an arc from 0 to v.

(11) Let $T \subseteq \mathbb{R}$ be a transcendency basis over $\mathbb{Q}$; then $\operatorname{card} T = \operatorname{card} \mathbb{R}$ (see 64.20, 64.5). As mentioned in 13.10, there exists an arcwise connected Hausdorff field topology on $\mathbb{Q}(T)$ such that the subfield $\mathbb{Q}$ is discrete. Let B be a basis of the vector space $\mathbb{R}$ over $\mathbb{Q}(T)$, and consider $\mathbb{R}$ as a subspace of the product space $\mathbb{Q}(T)^B$. This gives a Hausdorff ring topology on $\mathbb{R}$ (the multiplication is bilinear), which is arcwise connected (as T is homeomorphic to $\mathbb{R}$) and distinct from the usual topology of $\mathbb{R}$ (since $\mathbb{Q}$ is discrete).

Applying the result of Gelbaum, Kalisch and Olmsted mentioned in 13.10 to the strange ring topology of $\mathbb{R}$ described in the previous paragraph, we obtain a field topology of $\mathbb{R}$ which is again arcwise connected. This field topology is distinct from the usual topology of $\mathbb{R}$, because T is connected in the new topology, but totally disconnected in the usually topology (since every interval contains infinitely many rational numbers).

14 The complex numbers

(1) A discrete subgroup $D < \mathbb{C}^+$ contains an element a of minimal absolute value $|a| > 0$. Let $b \in D \smallsetminus \mathbb{Z}a$ such that the triangle with vertices $0, a, b$ has minimal positive area; then $D = \mathbb{Z}a + \mathbb{Z}b$. Compare also 8.6.

(2) If C is a subgroup of $\mathbb{C}^\times$ and $c \in C \smallsetminus \mathbb{S}_1$, we may assume that $|c| > 1$. Then the powers c^n converge to ∞ and C is not compact.

(3) The mapping $z \mapsto (z/|z|, \log|z|) : \mathbb{C}^\times \to \mathbb{S}_1 \times \mathbb{R}$ is an isomorphism of topological groups. The compact image $\varphi(\mathbb{S}_1)$ is trivial by 8.6. The map $t \mapsto \varphi(e^t)$ is a continuous homomorphism of $\mathbb{R}^+$ into $\mathbb{C}^+$, hence $\mathbb{R}$-linear, which gives the assertion.

(4) Each G-conjugacy class in N is connected and discrete, hence a singleton.

(5) Otherwise $\mathbb{R}$ or $\mathbb{C}$ would be of the form $F(t)$ where t is transcendental over the subfield F. Such a field $F(t)$ has always automorphisms of order 3, for example the F-linear automorphism determined by $t \mapsto (1 - t)^{-1}$; see 64.19. This contradicts 6.4 and 14.13(i).

(6) If a is contained in some maximal subfield M, then M is real closed by Exercise 5 and 12.15, hence $\mathbb{Q}(a)$ is formally real. Conversely, assume that $\mathbb{Q}(a)$ is formally real. Let T be a transcendency basis (64.20) of $\mathbb{C}$ over $\mathbb{Q}(a)$. Then $\mathbb{Q}(a, T)$ is formally real; see Exercise 1 of Section 12. Proposition 12.16 yields a real closed field M with $\mathbb{Q}(a, T) \subseteq M \subset \mathbb{C}$. From 12.10 we infer that $\mathbb{C} = M(\sqrt{-1})$, hence M is a maximal subfield of $\mathbb{C}$.

(7) The set $\Delta = \{ (a + b, a + \zeta b, a + \zeta^2 b) \mid a \in \mathbb{C},\ b \in \mathbb{C}^\times \}$ is the orbit of the equilateral triangle $(1, \zeta, \zeta^2)$ under the group $\mathrm{Aff}\,\mathbb{C}$ of similarity transformations; see 14.17. Thus Δ is the set of all equilateral triangles.

By 14.11 and 14.12 there exist many discontinuous field automorphisms α of $\mathbb{C}$. The set $\{\zeta, \zeta^2\}$ of all roots of unity of order 3 is invariant under α, hence Δ is invariant under α. Now $\mathbb{R}$ is not invariant under α (otherwise α would be $\mathbb{R}$-linear by 6.4, hence continuous), but $\mathbb{R}$ can be defined in terms of $0, 1$ and the α-invariant set $\{i, -i\}$ using any one of the mentioned geometric notions; for example, $\mathbb{R} = \{ x \in \mathbb{C} \mid 0, 1, x \text{ are collinear} \} = \{ x \in \mathbb{C} \mid (x, i) \text{ is congruent to } (x, -i) \}$.

(8) The extension $E_1 E_2 | F$ is finite by 64.2. It is a Galois extension, since E_i is the splitting field of some set X_i of separable polynomials over F (compare 64.17), and $E_1 E_2$ is the splitting field of $X_1 \cup X_2$. The Galois group of $E_1 E_2 | F$ is solvable, because it embeds into the direct product $\mathrm{Gal}_F E_1 \times \mathrm{Gal}_F E_2$, via restrictions to E_i.

(9) Every element g of a free group can be written uniquely in the form $g = x_1^{e_1} \cdots x_n^{e_n}$ where $x_1, \ldots, x_n$ are free generators with $x_i \neq x_{i+1}$ for $1 \leq i < n$ and $0 \neq e_i \in \mathbb{Z}$. If $g^2 = 1$, then $g = g^{-1} = x_n^{-e_n} \cdots x_1^{-e_1}$, hence $x_i = x_{n+1-i}$ and $e_i = -e_{n+1-i}$ by uniqueness. If $n = 2k - 1$ is odd, then $e_k = -e_k$, a contradiction to $e_k \neq 0$. If $n = 2k > 0$, then $x_k = x_{k+1}$, again a contradiction. Hence $n = 0$ and $g = 1$.

Similar arguments with normal forms show that free groups are torsion free; see COHN 2003b 3.4.1, JACOBSON 1989 2.13 p. 89 or LANG 1993 I.12.6 p. 74.

21 Ultraproducts

(1) (a) We have char $E = p$; (b) every element of E is a sum of two squares; (c) if p is odd, then -1 is a square in E if, and only if, $p \equiv 1 \bmod 4$ or $2\mathbb{N} \in \Psi$; (d) card $E = \aleph$; (e) $\mathbb{F}_{p^\nu}^\natural = \mathbb{F}_p^\natural$ and $E^\natural < (\mathbb{F}_p^\natural)^\Psi$ for analogous reasons as in 21.8(f).

(2) If $M \in \Psi$ and $s_\nu < t_\nu$ for $\nu \in M$, then also $h(s_\nu) < h(t_\nu)$ for $\nu \in M$. Hence h^Ψ is strictly increasing. Because h is surjective, so is h^Ψ.

(3) The image is $[-1, 1]^\Psi = \{ x \in \mathbb{R}^\Psi \mid -1 \le x \le 1 \}$.

22 Non-standard rationals

(1) If $\overline{x} = (x_\nu)_\nu/\Psi \in \mathbb{Q}^\natural$, then there exists a polynomial f with rational coefficients such that $f(x_\nu) = 0$ for almost all ν. This equation has only finitely many solutions, and 21.2(c) shows that $x_\nu = c \in \mathbb{Q}^\natural$ for almost all ν. If $\overline{x} \in \mathbb{Q}^\Psi$, then $x_\nu \in \mathbb{Q}$ for all ν.

(2) If $2\mathbb{N} \in \Psi$, then $s_\nu{}^2 < 2$ for almost all $\nu \in \mathbb{N}$. If $2\mathbb{N} \notin \Psi$, then the set of all odd numbers belongs to Ψ and $s_\nu{}^2 > 2$ for almost all ν.

(3) By definition, $\pi(p_\nu) = \nu$. The prime number theorem implies $\nu \cdot \log p_\nu < 2p_\nu$ for almost all $\nu \in \mathbb{N}$ and hence $k \cdot \nu < p_\nu$ for each $k \in \mathbb{N}$ and allmost all ν. On the other hand, $2\pi(n) > n/\log n$ and $2k \cdot \log n < \sqrt{n}$, hence $\pi(n)^2 > k^2 \cdot n$ for each $k \in \mathbb{N}$ and almost all n.

23 A construction of the real numbers

(1) If $a, b \in R$ and $0 < a \ll b < m \in \mathbb{N}$, then $a \in N$.

(2) Being a Euclidean field is a first-order property. Hence $^*E$ is a Euclidean field. Obviously, $^*\mathbb{Q} < {}^*E$, and $K \le {}^*E$. Each element of K is algebraic over the field $^*\mathbb{Q}$. Put $c_0 = 2$ and $c_{\nu+1} = \sqrt{c_\nu}$. Then $(c_\nu)_\nu$ represents an element $\overline{c} \in {}^*E$, and $\overline{c}$ is not of finite degree over $^*\mathbb{Q}$. Therefore, $K < {}^*E$.

(3) Similarly, $^*P$ is a Pythagorean field. If $\overline{x} \in {}^*P$, then $\overline{x}^4 \ne 2$ since $\sqrt[4]{2} \notin P$; see Exercise 5 of Section 12. Consequently, $^*P < {}^*E$.

24 Non-standard reals

(1) For $x = (x_\nu)_\nu \in \mathbb{R}^\mathbb{N}$, put $Cx = \{\nu \mid x_\nu = 0\}$. Then $x \in M \Leftrightarrow Cx \in \Psi$, and $A \in \Psi$ if, and only if, $Cx = A$ for some $x \in M$.

(2) Surjectivity is obvious from the definition $^*\exp \overline{x} = \overline{(\exp x_\nu)_\nu}$. This shows also that the kernel of $^*\exp$ is $^*\mathbb{Z}2\pi i$.

(3) A positive element b can be represented by a sequence $(b_\nu)_\nu$, where $b_\nu > 0$ for all $\nu \in \mathbb{N}$. Then $b^a = (e^{a_\nu \log b_\nu})_\nu/\Psi$.

(4) The conjugation ι of $^*\mathbb{C}$ fixes exactly the elements of $^*\mathbb{R}$. By 24.6 there is an isomorphism $\alpha : {}^*\mathbb{C} \cong \mathbb{C}$, and $F = {}^*\mathbb{R}^\alpha$ is the fixed field of the involution $\iota^\alpha = \alpha^{-1}\iota\alpha$.

(5) The map $x \mapsto 1 - (1 + \exp x)^{-1}$ is an order isomorphism of $^*\mathbb{R}$ onto the interval $]0, 1[\subset {}^*\mathbb{R}$. Each countable sequence in $^*\mathbb{R}$ is bounded; in $\mathfrak{R}$ this is not true.

(6) If $\{ b_\nu \mid \nu \in \mathbb{N} \}$ is a basis of $^*\mathbb{R}$, then, by 24.11, there is an element $b \in {}^*\mathbb{R}$ such that $b_\nu < b$ for all ν, moreover, $\mathbb{N} \cdot b < c$ for some $c \in {}^*\mathbb{R}$. Each element

in $^*\mathbb{R}$ has a representation $x = \sum_{\nu=1}^{n} r_\nu b_\nu$ with $r_\nu \in \mathbb{R}$. Choose $m \in \mathbb{N}$ with $r_\nu < m$ for $\nu = 1, 2, \ldots, n$. Then $x < n \cdot m \cdot b < c$, and c is not in the vector space spanned by the b_ν.

(7) Note that $v(x) = v(1) = 0 \Leftrightarrow x\mathfrak{R} = \mathfrak{R} \Leftrightarrow x \in \mathfrak{R}^\times$. Thus $v(x) > 0 \Leftrightarrow x \in \mathfrak{n}$, and $V_0 = \mathfrak{n} \in \tau$ because 0 is in the interior of $\mathfrak{n}$. If $\gamma = v(c)$, then $V_\gamma = c\mathfrak{n}$ is also open. On the other hand, each interval $]-c, c[$ contains V_γ.

25 Continuity and convergence

(1) Suppose that s converges to t and that $\{\nu \in \mathbb{N} \mid s(\nu) = 0\} \in \mathbf{\Psi}$. Then $t = 0$ and $^*s(m) \approx 0$ for each strictly increasing sequence m. This means that $\{\nu \mid s(\nu) = 1\}$ is finite or that s is finally constant.

(2) Put $x = (\nu)_\nu$ and $y = (\nu + \nu^{-1})_\nu$. Then $\overline{x} \approx \overline{y}$, but $\overline{y}^2 - \overline{x}^2 \approx 2$.

(3) It suffices to show that f is constant on some neighbourhood of each point $a \in \mathbb{R}$. Assume that there are elements $s_\nu \in (a - \nu^{-1}, a + \nu^{-1})$ with $f(s_\nu) \neq f(a)$. Then $\overline{s} = (s_\nu)_\nu/\mathbf{\Psi} \in a + \mathfrak{n}$ and $^*f(\overline{s}) \neq f(a)$.

26 Topology of the real numbers in non-standard terms

(1) The first part is easy, and the second part is a consequence of $(A \cap B) + \mathfrak{n} \subseteq$ $^*A \cap {}^*B = {}^*(A \cap B)$. Dually, $^*(A \cup B) = {}^*A \cup {}^*B$ by 21.2c, and the last claim follows with 26.2.

(2) We have $\overline{\mathbb{Q}} = \mathbb{R}$ by 26.2, since $\mathbb{R} \subseteq {}^*\mathbb{Q} + \mathfrak{n}$.

(3) The intersection of two dense open sets A, B is dense: if U is open and $U \neq \emptyset$, then $(A \cap B) \cap U = A \cap (B \cap U) \neq \emptyset$.

If the C_κ form a descending sequence of dense open sets, then $D = \bigcap_\kappa C_\kappa$ is dense: by 26.2 and translation invariance it suffices to show that $^*D \cap \mathfrak{n} \neq \emptyset$. Let $c_\kappa = (c_{\kappa\nu})/\mathbf{\Psi} \in {}^*C_\kappa \cap \mathfrak{n}$. Then $c_{\kappa\nu} \in C_\iota$ for $\iota \leq \kappa$ and $|c_{\kappa\nu}| < \kappa^{-1}$ for almost all ν. The element $c_{\kappa\kappa}$ may be chosen arbitrarily in C_κ, say such that $|c_{\kappa\kappa}| < \kappa^{-1}$. Obviously, $d = (c_{\nu\nu})_\nu/\mathbf{\Psi} \in {}^*D \cap \mathfrak{n}$.

Write $\mathbb{Q} = \{r_\nu \mid \nu \in \mathbb{N}\}$ and put $U = \bigcup_\nu]r_\nu - \varepsilon 2^{-\nu}, r_\nu + \varepsilon 2^{-\nu}[$. Then U has Lebesgue measure $< 4\varepsilon$.

(4) A closed interval C is compact. There are $x_\nu \in C$ such that the $f(x_\nu)$ converge to $t = \sup f(C)$, and $\overline{x} = (x_\nu)_\nu/\mathbf{\Psi} \in {}^*C$. By 26.5, there is an element $c \in C$ with $c \approx \overline{x}$. Continuity implies $f(c) \approx {}^*f(\overline{x}) \approx t$, and $f(c) = t$.

(5) By Exercise 4, there is some $c \in C$ such that $f(x) \leq f(c)$ for all $x \in C$. If $\overline{x} \in {}^*C$ and $\overline{x} = (x_\nu)_\nu/\mathbf{\Psi}$, then $^*f(\overline{x}) = (f(x_\nu))_\nu/\mathbf{\Psi} \leq f(c)$; see 21.6.

27 Differentiation

(1) There is some real $r > 0$ such that $f(a) \leq f(x)$ for $x \in [a - r, a + r]$. Consequently, $^*f(x \pm h) \geq f(a)$ for all $h \in \mathfrak{n}$, and condition $(*)$ implies that $f'(a) = 0$.

(2) One has $h^{-1}\big(g(a + h) - g(a)\big) \approx -h^{-1}\big(f(a + h) - f(a)\big)g(a)^2$.

31 The additive group of the rational numbers

(1) Every subgroup $A \cong \mathbb{Q}$ of $\mathbb{Q}$ is divisible. By 31.10 we may assume that $1 \in A$. Then A contains all fractions $1/n$ with $n \in \mathbb{N}$, hence $A = \mathbb{Q}$.

(2) See Section 1, Exercise 7.

(3) If p and q are distinct primes, then the map $(x \mapsto px)$ induces an automorphism of $\langle p^{\mathbb{Z}} \rangle = \{ ap^n \mid a, n \in \mathbb{Z} \}$, but it maps $\langle q^{\mathbb{Z}} \rangle$ onto a proper subgroup. Hence $\langle p^{\mathbb{Z}} \rangle \not\cong \langle q^{\mathbb{Z}} \rangle$.

(4) Let X be any set of primes and consider the subgroup A_X of $\mathbb{Q}^+$ generated by all sets $q^{\mathbb{Z}}$ with $q \in X$. If p is a prime, then the map $(a \mapsto pa)$ is surjective if, and only if, $p \in X$. Hence the groups A_X are mutually non-isomorphic.

32 The multiplication of the rational numbers

(1) The subgroup of $\mathbb{Q}^\times$ generated by -1 together with any infinite set of primes is isomorphic to $\mathbb{Q}^\times$.

(2) Exactly the infinite cyclic ones and $\{1\}$, because $\mathbb{Q}^+$ is locally cyclic.

(3) Consider the subgroups generated by an arbitrary set $X \subset \mathbb{P}$ such that $\mathbb{P} \smallsetminus X$ is infinite.

(4) By 32.1, $\mathbb{Q}_{\mathrm{pos}}^\times \cong \mathbb{Z}^{(\mathbb{P})}$. For $q \in \mathbb{P}$, define $(\delta_{pq})_p \in \mathbb{Z}^{(\mathbb{P})}$ by $\delta_{pp} = 1$ and $\delta_{pq} = 0$ for $p \neq q$. These elements generate the additive group $\mathbb{Z}^{(\mathbb{P})}$. An endomorphism $\varphi \in \mathrm{End}\,\mathbb{Z}^{(\mathbb{P})}$ is represented by the matrix $(z_{pq})_{p,q} \in \mathbb{Z}^{(\mathbb{P}) \times \mathbb{P}}$ whose q-th column is the image $\varphi((\delta_{pq})_p)$. Addition and composition of endomorphisms correspond to addition and multiplication of matrices in $\mathbb{Z}^{(\mathbb{P}) \times \mathbb{P}}$. The additive group of $\mathbb{Z}^{(\mathbb{P}) \times \mathbb{P}}$ clearly is isomorphic to $(\mathbb{Z}^{(\mathbb{P})})^{\mathbb{P}} \cong (\mathbb{Q}_{\mathrm{pos}}^\times)^{\mathbb{P}}$, in agreement with 32.4.

(5) There is at least one prime $q \equiv -1 \bmod 3$ which divides $(2 \cdot 3 \cdots p) - 1$ and at least one prime $r \equiv -1 \bmod 4$ which divides $(2 \cdot 3 \cdots p) + 1$.

(6) Let m be the product of all primes $\leq p$. Note that $(m-1)(m^2 + m + 1) = m^3 - 1$ and $\gcd(m - 1, m^2 + m + 1) = \gcd(m - 1, 2m + 1) = \gcd(m - 1, 3) = 1$. If the prime r divides $m^2 + m + 1$, then $r > p$ and $m^3 \equiv 1 \not\equiv m \bmod r$. Hence 3 is the order of m in $\mathbb{F}_r^\times$, whence $3 \mid r - 1$.

(7) Each prime $q \equiv -1 \bmod 4$ in $\mathbb{Z}$ is also prime in $\mathbb{J}$. In fact, if n is an odd natural number, then $n^2 \equiv 1 \bmod 4$, hence $c \in \mathbb{J}$ implies $c\bar{c} \not\equiv -1 \bmod 4$. Therefore $c\bar{c} \neq q$ and any factor c of q in $\mathbb{J}$ has norm 1 or norm q^2.

33 Ordering and topology of the rational numbers

(1) The additive group $S := \{ m \cdot 2^n \mid m, n \in \mathbb{Z} \}$ is strongly dense in itself and countable, hence $(S, <) \cong (\mathbb{Q}, <)$ by 3.4.

(2) Each interval $]n, n + 1[$ is isomorphic to the chain $(\mathbb{Q}, <)$. Denote by $\tau : \mathbb{Q} \to \mathbb{Q}$ the translation $x \mapsto x + 1$. There are $2^{\aleph_0}$ automorphisms of the chain $(\mathbb{Q}, <)$ which fix each integer n and induce on $]n, n + 1[$ a map which is equivalent to τ or to τ^{-1}.

(3) With the lexicographic ordering, the chain $\mathbb{Q}^n$ is strongly dense in itself and countable, hence isomorphic to the chain $(\mathbb{Q}, <)$ by 3.4.

(4) There is an involution $\eta \in \mathsf{H}$ which inverts the order on the open interval $]{-\sqrt{2}}, \sqrt{2}[$ (of length < 3) and induces the identity outside of this interval. If τ is the translation $x \mapsto x + 1$, then $\eta\tau^3\eta$ does not preserve the ordering; hence Γ is not normal in H.

Let H_I consist of all elements of H which permute the set I of rational intervals $]n + \sqrt{2}, n + 1 + \sqrt{2}[$ with $n \in \mathbb{N}$. The group H_I induces on I the full symmetric group $\mathrm{Sym}\,I$ (compare the proofs of 33.16 and 33.17), and

$\Gamma_I := \Gamma \cap H_I$ acts trivially on I. Hence $|H : \Gamma| \geq |H_I : \Gamma_I| \geq \operatorname{card} \operatorname{Sym} I = 2^{\aleph_0}$ by 61.16, and 33.15 yields $|H : \Gamma| = 2^{\aleph_0}$.

(5) A complete metric space X is a Baire space, i.e., any intersection of countably many dense open subsets is dense. However, the intersection of the complements of all singletons in X is empty.

(6) The chain $(\mathbb{Q}, <)$ is strongly dense in itself, hence each open interval is open in the Sorgenfrey topology σ. An interval $]a, b]$ is not open in the ordinary topology τ. Therefore τ is a proper subset of σ. Obviously, σ has a countable basis. An interval $]a, b]$ is also σ-closed. Consequently, $(\mathbb{Q}, \sigma)$ is a regular topological space, hence metrizable.

34 The rational numbers as a field

(1) The two fields $F_1 := \mathbb{Q}(\sqrt{-2})$ and $F_2 := \mathbb{Q}(\sqrt{-7})$ satisfy $F_1^+ \cong F_2^+$, because $[F_\nu : \mathbb{Q}] = 2$. We will show that $F_\nu^\times \cong \mathbb{Q}^\times$.

Note that F_ν is the field of fractions of the ring R_ν of algebraic integers in F_ν. We have $R_1 = \mathbb{Z} + \mathbb{Z}\sqrt{-2}$ and $R_2 = \{x + y\sqrt{-7} \mid 2x, 2y \in \mathbb{Z} \wedge x - y \in \mathbb{Z}\}$. The elements ± 1 are the only units in R_ν and R_ν is a Euclidean ring with respect to the norm N (for $a, b \in R_2$ there is some $q \in R_2$ such that $\mathrm{N}(ab^{-1} - q) \leq 7/4^2 + 1/4 < 1$). Hence both rings have unique prime decomposition. Moreover, there are infinitely many primes in R_ν: if $p \in \mathbb{P}$ and p is not a prime in R_ν, then $p = uv$ with $\mathrm{N}u = \mathrm{N}v = p$, and u, v are primes in the ring R_ν. Obviously, -2 is not a square in F_2. Therefore $F_1 \ncong F_2$.

See also *Amer. Math. Monthly* **93** 1986 p. 744 Problem 6489. A more general result due to Skolem is given in FUCHS 1973 Theorem 127.2.

(2) Assume that $x^2 + y^2 = 3$ with $x, y \in \mathbb{Q}$. Then there exist relatively prime integers a, b, c such that $a^2 + b^2 = 3c^2$. Because $m^2 \not\equiv -1 \bmod 3$ for all $m \in \mathbb{N}$, it follows that $a, b \equiv 0 \bmod 3$ and then also $c \equiv 0 \bmod 3$, which is a contradiction.

(3) It suffices to show that the projection of $C \cap \mathbb{Q}^2$ onto the x-axis is dense in $[-1, 1]$. For $u, v \in \mathbb{N}$, we have $\left((u^2 - v^2)/(u^2 + v^2), 2uv/(u^2 + v^2)\right) \in C$. Hence our claim says that the numbers $u^2/(u^2 + v^2)$ are dense in the interval $[0, 1]$ or, equivalently, that the numbers v^2/u^2 are dense in $[0, \infty[$. The latter is obviously true.

(4) The group Δ consists of all rational matrices $\begin{pmatrix} a & -b \\ b & a \end{pmatrix}$ with $a^2 + b^2 = 1$; it is sharply transitive on $C \cap \mathbb{Q}^2$.

(5) Let n be as in Corollary 34.9, and put $T := \{a^2 + b^2 \mid a, b \in \mathbb{N}\}$.

If v_2 is odd or if n is divisible by some prime number $p \equiv 1 \bmod 4$, then $n \in \{2m^2, km^2, 2km^2\}$, where $k > 1$ is a product of distinct primes $p \equiv 1 \bmod 4$. By 34.9 we have $k = a^2 + b^2$ with $a, b \in \mathbb{N}$, as k is not a square; thus $k \in T$. Moreover, $2 \in T$ and $2k = (a + b)^2 + (a - b)^2 \in T$, as $a \neq b$. We infer that $n \in T$.

For the converse, let $n \in T$. Aiming for a contradiction, we assume that $v_2 \equiv 0 \bmod 2$ and $v_p = 0$ for $p \equiv 1 \bmod 4$. Then n is a square that is not divisible by any prime $p \equiv 1 \bmod 4$; see 34.9. By repeated application of 34.6, we infer from $n \in T$ that $4^e \in T$ with $e \geq 0$. Let $4^e = a^2 + b^2$ with $a, b \in \mathbb{N}$; then $e > 0$ and $a^2 + b^2 \equiv 0 \bmod 4$. Hence a and b are even. We conclude that $4^{e-1} \in T$, hence $4^0 = 1 \in T$, which is absurd.

(6) There are two essentially different representations: $30 = 5^2 + 2^2 + 1^2 + 0^2$ and $30 = 4^2 + 3^2 + 2^2 + 1^2$. Counting permutations and signs, the total number is $8 \cdot 3 \cdot 24$.

(7) Let $a/b = s + ti \in \mathbb{Q}(i)$ and choose $z = x + yi \in \mathbb{J}$ such that $|x-s|, |y-t| \leq \frac{1}{2}$. Then $a - bz = b(a/b - z)$ and $N(a - bz) \leq \frac{1}{2} Nb$.

36 Addition and topologies of the rational numbers

(1) According to 31.10, each group automorphism of $\mathbb{Q}^+$ is given by multiplication with a rational number and therefore continuous. Let $0 \neq s \in S$. If the rational subspace $s\mathbb{Q}$ is a proper subgroup of S, then $S = s\mathbb{Q} \oplus H$ for some non-trivial rational subspace $H < S$ (here we use the axiom of choice). We obtain an automorphism α of S with $\alpha|_H = \mathrm{id}$ and $\alpha|_{s\mathbb{Q}} \neq \mathrm{id}$. Since H is dense in $\mathbb{R}$ and in S, such a map α is not continuous.

(2) Each neighbourhood of 0 generates the full group $\mathbb{Q}^+$, hence $\mathbb{Q}$ has no proper open subgroup. A cyclic subgroup is closed. If a subgroup A is not cyclic, then it is dense (1.4) and hence $A = \mathbb{Q}$ if A is closed.

37 Multiplication and topology of the rational numbers

(1) A typical neighbourhood in $\bigoplus_p p^{\mathbb{Z}}$ is given by specifying a finite number of entries, while the remaining ones are arbitrary; hence μ is not continuous. Write $r = \prod_p p^{v_p(r)}$. Continuity of μ^{-1} means that each map v_p is continuous on $\mathbb{Q}_{\mathrm{pos}}^{\times}$; obviously, this is not the case.

(2) Via logarithms, $\mathbb{Q}_{\mathrm{pos}}^{\times}$ is isomorphic to a dense subgroup S of $\mathbb{R}^+$; see 2.2 and 1.4. Each character $\alpha : S \to \mathbb{T}$ is uniformly continuous and hence has a unique extension to a character $\overline{\alpha} : \mathbb{R} \to \mathbb{T}$; compare 43.23. This proves that $S^* \cong \mathbb{R}^* \cong \mathbb{R}^+$.

51 The field of p-adic numbers

(1) The partial sums $\sum_{n=0}^{k} p^n = (1 - p^{k+1})/(1-p)$ converge for $k \to \infty$ to $1/(1-p)$, as p^{k+1} converges to 0.

(2) One has $\sum_{n=0}^{\infty} (p-1)p^n = (p-1)/(1-p) = -1$ in $\mathbb{Q}_p$ by Exercise 1.

In $\mathbb{Q}_3$, we want to solve $(1 + 3^2) \sum_n c_n 3^n = c_0 + c_1 \cdot 3 + \sum_{n \geq 2}(c_n + c_{n-2})3^n = 1$ with $c_n \in \{0, 1, 2\}$. Reading this equation modulo $3^k \mathbb{Z}_3$ and applying induction over k, we obtain $c_0 = 1$, $c_1 = 0$, $c_2 + c_0 = 3$, $c_n + c_{n-2} + 1 = 3$ for $n \geq 3$, hence $c_n = 2$ for $n \equiv 2, 3 \bmod 4$ and $c_n = 0$ for $0 < n \equiv 0, 1 \bmod 4$. In passing, this yields the expansion $1 = 1 + 3 \cdot 3^2 + 2 \cdot 3^3 \cdot \sum_{j \geq 0} 3^j$; compare Exercise 1.

Writing $\sum_{n \geq 0} c_n p^n = c_0, c_1 c_2 c_3 \ldots$ and using Exercise 1, quotients in $\mathbb{Q}_p$ can also be determined by the familiar 'long division' procedure.

(3) Each series $\sum_{n \geq k} z_n p^n$ with $k \in \mathbb{Z}$ and $z_n \in \mathbb{Z}$ converges in $\mathbb{Q}_p$, in view of $|z_n p^n|_p \leq p^{-n}$; compare Exercise 5. Hence by substituting p for x, we obtain an epimorphism $\mathbb{Z}((x)) \to \mathbb{Q}_p$ of rings. This epimorphism maps $\mathbb{Z}[[x]]$ onto $\mathbb{Z}_p$; see 51.8 and 51.9.

The kernel of this epimorphism contains $x - p$, hence also the ideal generated by $x - p$, and it remains to show that every $f = \sum_{n \geq k} a_n x^n \in \mathbb{Z}((x))$ with $f(p) = 0$ is divisible by $x - p$ in $\mathbb{Z}((x))$.

For this purpose we define integers b_n with $a_n = b_{n-1} - pb_n$ for $n \geq k$ as follows. Put $b_{k-1} = 0$. To define b_n inductively, we infer from $f(p) = 0$ that $p^{n+1}\mathbb{Z}_p$ contains $\sum_{j \leq n} a_j p^j = a_n p^n + \sum_{j < n}(b_{j-1} - pb_j)p^j = (a_n - b_{n-1})p^n$. Hence p divides $a_n - b_{n-1}$, and $b_n := (a_n - b_{n-1})/p$ is an integer. Then $f = \sum_{n \geq k}(b_{n-1} - pb_n)x^n = (x - p)\sum_{n \geq k} b_n x^n$.

Compare FALTIN *et al.* 1975.

(4) Let k_n be the order of a in the group of units of the finite ring $\mathbb{Z}/p^n\mathbb{Z}$. Then $|a^{k_n} - 1|_p \leq p^{-n}$, hence $\lim_{n \to \infty} a^{k_n} = 1$ in $\mathbb{Q}_p$.

(5) The ultrametric inequality implies that the partial sums of $\sum_n a_n$ form a Cauchy sequence whenever the sequence $(a_n)_n$ converges to 0.

Rearranging does not affect the property that $(a_n)_n$ converges to 0, hence the rearranged series $\sum_n a_{\pi(n)}$ converges for every permutation π of $\mathbb{N}$. For $n \in \mathbb{N}$ and $n' := \max \pi^{-1}(\{1, 2, \ldots, n\})$ we obtain $|\sum_{j \leq n} a_j - \sum_{j \leq n'} a_{\pi(j)}|_p \leq \sup\{|a_j|_p \mid j > n\}$, which converges to 0 for $n \to \infty$. Hence the rearranged series has the same value.

(6) If $\lim_n |a_{n+1} - a_n|_p = 0$, then the sequence $a_k = a_1 + \sum_{n=1}^{k-1}(a_{n+1} - a_n)$ converges by Exercise 5.

(7) Since $\mathbb{Z}_p \smallsetminus p\mathbb{Z}_p$ is closed in $\mathbb{Q}_p$ (see 51.6), it suffices to prove that $\mathbb{P} \smallsetminus \{p\}$ is dense in $\mathbb{Z}_p \smallsetminus p\mathbb{Z}_p$. Because $\mathbb{Z}$ is dense in $\mathbb{Z}_p$ by 51.9, every neighbourhood of an element of $\mathbb{Z}_p \smallsetminus p\mathbb{Z}_p$ contains a subset $a + p^n\mathbb{Z}_p$ with $a \in \mathbb{Z} \smallsetminus p\mathbb{Z}$ and $n \in \mathbb{N}$. The set $a + p^n\mathbb{Z}$ contains infinitely many prime numbers according to a famous result of Dirichlet; for a proof see SHAPIRO 1950, SERRE 1973 VI §4, BOREVICH–SHAFAREVICH 1966 Chapter 5 §3, LANG 1970 VIII §4 or NEUKIRCH 1992 VII.5.14.

(8) This map f (which can be found in WITT 1975) is bijective for combinatorial reasons. f is continuous, because the first n coordinates of $f(x)$ depend only on the first n coordinates of x. Since both products are compact Hausdorff spaces, f is a homeomorphism.

(9) This holds precisely if $a \in 1 + p\mathbb{Z}_p$. Indeed, if $a \in 1 + p\mathbb{Z}$, then $|a^n - a^m|_p = |a^{n-m} - 1|_p = |a \pm 1|_p \cdot |n - m|_p$ by 37.5. As $\mathbb{Z}$ is dense in $\mathbb{Z}_p$ (51.9), we have the same equation for $a \in 1 + p\mathbb{Z}_p$, hence f is continuous.

Conversely, if f is continuous, then $f(1 + p^n) = a \cdot a^{p^n}$ converges to $f(1) = a$ for $n \to \infty$, hence $\lim_{n \to \infty} a^{p^n} = 1$. Thus $|a|_p = 1$. Moreover the map $x \mapsto x^p$ acts as the identity on the quotient $\mathbb{Z}_p/p\mathbb{Z}_p \cong \mathbb{F}_p$, hence 1 belongs to the compact set $a + p\mathbb{Z}_p$.

See also 37.6(ii).

(10) The set $\mathbb{N}$ is p-adically dense in $\mathbb{Z}_p$ by 51.9 and f is continuous, hence $f(\mathbb{Z}) \subseteq f(\mathbb{Z}_p) \subseteq \mathbb{Z}_p$ for each prime p. Thus $f(\mathbb{Z}) \subseteq \bigcap_p \mathbb{Z}_p \cap \mathbb{Q} = \mathbb{Z}$.

(11) We claim that each power series $f_k = \sum_{n \geq 0} n^k x^n$ represents on $p\mathbb{Z}_p$ a rational function of x with rational coefficients (this implies that $\sum_{n \geq 0} n^k p^n = f_k(p) \in \mathbb{Q}$). We proceed by induction on k and observe that f_0 represents $1/(1 - x)$. The power series $\sum_{n \geq 1} n^{k+1} x^{n-1}$ represents the derivative f'_k on $p\mathbb{Z}_p$ (this is justified in ROBERT 2000 5.2.4). The derivative f'_k is a rational function with rational coefficients (by induction). Hence also $f_{k+1} = x f'_k$ represents a rational function with rational coefficients.

52 The additive group of p-adic numbers

(1) By 52.1, $\mathbb{R}$ and $\mathbb{Q}_p$ are vector spaces over $\mathbb{Q}$ of the same infinite dimension, and the quotient modulo a one-dimensional subspace has the same dimension.

(2) Let U be a subgroup of index n in $\mathbb{Z}_p^+$. Then $n\mathbb{Z}_p \subseteq U$, hence 52.6 implies that n is a power of p. Now use the proof of 52.3(i).

(3) By 52.2 we have to determine the sequences $c = (c_q)_q \in \bigtimes_{q \in \mathbb{P} \smallsetminus \{p\}} C_{q^\infty}$ which have finite order. This happens precisely if only finitely many components c_q are non-trivial, and this means that $c \in \bigoplus_{q \in \mathbb{P} \smallsetminus \{p\}} C_{q^\infty}$.

(4) The kernel of χ is $\mathbb{Z}_p$, and the image is isomorphic to the Prüfer group C_{p^∞}. Moreover, χ is continuous, as $\mathbb{Q}_p/\mathbb{Z}_p \cong C_{p^\infty}$ carries the discrete topology.

(5) The map χ_a is a character of F^+, since $x \mapsto ax$ is a continuous endomorphism of F. Hence $\chi : F \to F^* : a \mapsto \chi_a$ is a monomorphism, which is continuous by the definition of the compact-open topology on F^*.

In fact, χ is an embedding: let $(a_\nu)_\nu$ be a net in F such that χ_{a_ν} converges to χ_a; we have to show that $(a_\nu)_\nu$ converges to a. The net $(a_\nu)_\nu$ has an accumulation point c in the one-point compactification $F \cup \{\infty\}$ of F. If $c = \infty$, then $(a_\nu^{-1})_\nu$ accumulates at 0; see GRUNDHÖFER–SALZMANN 1990 XI.2.11; alternatively, F is of type V (compare WARNER 1989 Theorem 19.7(3)), thus one can use Exercise 3 of Section 57. Hence $(a_\nu^{-1}b)_\nu$ accumulates at 0, and $\chi_1(b) = \chi_{a_\nu}(a_\nu^{-1}b)$ accumulates at $\chi_a(0) = 0$ for every $b \in F$, a contradiction to $\chi_1 \neq 0$. Therefore $c \in F$, and then χ_{a_ν} accumulates at χ_c, whence $c = a$ by the injectivity of χ. This shows that $(a_\nu)_\nu$ converges to a. (Another argument uses the absolute value of F constructed in 58.5, as follows. It suffices to consider the case $a = 1$. A neighbourhood Ω of χ_1 consists of all $\varphi \in F^*$ mapping a ball $B = \{x \in F \mid |x| \leq r\}$ into an open subset $J \subseteq \mathbb{R}/\mathbb{Z}$ with $\chi_1(F) \not\subseteq J$. Choose c such that $\chi_1(c) \notin J$. If $\chi_a \in \Omega$, then $c \notin aB$ and $|a| < r^{-1}|c|$. Hence $\{a \in F \mid \chi_a \in \Omega\}$ is contained in some compact ball and the claim follows. See also the proof of 8.31(a) and HEWITT–ROSS 1963 25.1.)

Consequently, $\chi(F)$ is closed in F^*; see Section 62, Exercise 2. If $\chi(F) \neq F^*$, then by duality (63.27), there is some $x \in F^\times$ such that $\chi_a(x) = 0$ for all $a \in F$, but then $\chi_1(F) = 0$ and χ_1 would be trivial.

(6) The quotient S_p is compact, since it is the image of the compact space $[0,1] \times \mathbb{Z}_p$. It is a Hausdorff space by 62.8, because $\{(z,z) \mid z \in \mathbb{Z}\}$ is discrete and closed in $\mathbb{R} \times \mathbb{Z}_p$. The sets $\mathbb{R} \times \{0\}$ and $\mathbb{R} \times \mathbb{Z}$ have the same image in S_p, hence this image is connected. Moreover $\mathbb{R} \times \mathbb{Z}$ is dense in $\mathbb{R} \times \mathbb{Z}_p$ by 51.8, hence S_p is connected.

An element $(t,x) \in \mathbb{R} \times \mathbb{Z}_p$ is mapped to a torsion element in S_p if, and only if, $n \cdot t = n \cdot x \in \mathbb{Z}$ for some $n \in \mathbb{N}$. By 52.2, the torsion subgroup of S_p is isomorphic to $\bigoplus_{q \in \mathbb{P} \smallsetminus \{p\}} C_{q^\infty}$.

(7) We have $\mathrm{Aut}\, \mathbb{F}_{p^{q^n}} \cong C_{q^n}$; compare Exercise 6 of Section 64. Each element of $C_{q^n} = \mathbb{Z}/q^n\mathbb{Z}$ has the form $\sum_{j<n} c_j q^j + q^n\mathbb{Z}$ with uniquely determined numbers $c_j \in \{0,1,\ldots,q-1\}$. Any automorphism σ of $\mathbb{F}_{p,q}$ induces an automorphism σ_n on $\mathbb{F}_{p^{q^n}}$, and σ is uniquely determined by the sequence $(\sigma_n)_n$. Thus σ corresponds to $\sum_j c_j q^j \in \mathbb{Z}_q^+$, and this correspondence is a group isomorphism.

53   The multiplicative group of p-adic numbers

(1) The first part is 53.1(iii). If $n \geq 2$, then U_n is an open subgroup of U_1 and of U_2; see 53.1(i). By 53.2, $U_1 \cong \mathbb{Z}_p^+$ for $p \neq 2$ and $U_2 \cong \mathbb{Z}_2^+$ for $p = 2$. Result 52.3(i) implies that every open subgroup of $\mathbb{Z}_p^+$ is isomorphic to $\mathbb{Z}_p^+$.

(2) We have $v_p(n!) = \max\{e \in \mathbb{N}_0 \mid p^e \text{ divides } n!\} = \sum_{j \geq 1}[n/p^j]$, where $[r] := \max\{z \in \mathbb{Z} \mid z \leq r\}$. By Exercise 5 of Section 51, the series for $\exp(x)$ converges precisely if $|x^n/n!|_p = |x|_p^n \cdot p^{v_p(n!)}$ converges to 0 for $n \to \infty$.

The estimate $v_p(n!) \leq n\sum_{j \geq 1} p^{-j} = n/(p-1)$ gives $|x^n/n!|_p \leq |x|_p^n \cdot p^{n/(p-1)}$, hence $\exp(x)$ converges for $p \neq 2$ and $|x|_p \leq p^{-1}$, and for $p = 2$ and $|x|_2 \leq 2^{-2}$.

The estimates $v_p(n!) \geq [n/p] > (n/p) - 1$ and $|x^n/n!|_p \geq |x|_p^n \cdot p^{(n/p)-1}$ show that $\exp(x)$ does not converge for $|x|_p \geq 1$. If $p = 2$ and $|x|_2 = 2^{-1}$, then $\exp(x)$ does not converge either, since then $|x^n/n!|_2 = 2^{-n+v_2(n!)}$, and one has $-2^k + v_2((2^k)!) = -2^k + \sum_{j \geq 1}[2^{k-j}] = -1$ for all $k \in \mathbb{N}$.

(3) If $x \in \mathbb{N}$, then $\binom{x}{n} \in \mathbb{N}_0$. We infer that $\binom{x}{n} \in \mathbb{Z}_p$ for $x \in \mathbb{Z}_p$ from the density of $\mathbb{N}$ in $\mathbb{Z}_p$ (51.9) and the continuity of polynomials. This implies that the series $\sum_{n \geq 0} \binom{x}{n} p^n$ converges in $\mathbb{Z}_p$ for every $x \in \mathbb{Z}_p$. The mapping $\beta : \mathbb{Z}_p \to U_1 \subseteq \mathbb{Z}_p$ is continuous, because $\beta^{-1}(a + p^n\mathbb{Z}_p) = \{x \in \mathbb{Z}_p \mid \sum_{k=0}^{n-1} \binom{x}{k} p^k \in a + p^n\mathbb{Z}_p\}$ is the preimage of $a + p^n\mathbb{Z}_p$ under a polynomial. Furthermore, β satisfies $\beta(x + y) = \beta(x)\beta(y)$ for all $x, y \in \mathbb{Z}_p$, because this is true for $x, y \in \mathbb{N}$ by the binomial theorem, and then in general by the density of $\mathbb{N}$ in $\mathbb{Z}_p$ and the continuity of β.

(4) This can be proved by induction on n. Alternatively, one can use ideas from the proof of 37.5.

(5) If $p \neq 2$ and $x \in \mathbb{Z}_p$, then $1 + px^2 \in 1 + p\mathbb{Z}_p$, which consists of squares by 53.2 and 52.6. For the converse inclusion, consider $x \in \mathbb{Q}_p \smallsetminus \mathbb{Z}_p$. Then $|px^2|_p = p^{-1}|x|_p^2 \geq p^{-1+2} > 1$, hence $|1 + px^2|_p = |px^2|_p$ is a power of p with odd exponent, and we infer that $1 + px^2$ is not a square.

If $x \in \mathbb{Z}_2$, then $1 + 2x^3 \in 1 + 2\mathbb{Z}_2$; this group is 3-divisible by 53.2 and 52.6. For the converse let $x \in \mathbb{Q}_2 \smallsetminus \mathbb{Z}_2$. Then $|2x^3|_2 = 2^{-1}|x|_2^3 > 1$, hence $|1 + 2x^3|_2 = |2x^3|_2 = 2^n$ with n not divisible by 3. Thus $1 + 2x^3$ is not a cube.

These descriptions of $\mathbb{Z}_p$ show that $\mathbb{Z}_p$ is invariant under every ring endomorphism of $\mathbb{Q}_p$, and this invariance is the crucial step for proving 53.5.

(6) Use the direct decomposition 53.3 of $\mathbb{Q}_p^\times$, the isomorphism 53.2 and the divisibility property 52.6.

54   Squares of p-adic numbers and quadratic forms

(1) By 53.3 or 54.1, this happens if, and only if, $p \equiv 1 \bmod 4$.

(2) For $p \neq 2$ this is the special case $n = 2$ of Exercise 6 for Section 53; or use 54.1. For $p = 2$ we infer from 54.1 that $\mathbb{Q}_2^\square = \langle 4 \rangle \times (1 + 8\mathbb{Z}_2)$ consists of all 2-adic numbers $2^r(c_0 + 2c_1 + 4c_2 + 8c_3 + \cdots)$ such that r is even and $c_0 + 2c_1 + 4c_2 \equiv 1 \bmod 8$.

(3) One has to characterize when d is a square in $\mathbb{Q}_p$. By 54.1, this holds for $p \neq 2$ precisely if d is a non-zero square modulo p, and for $p = 2$ precisely if $d \equiv 1 \bmod 8$.

(4) Suppose that $x^2 + y^2 + z^2 + 1 = 0$ with $x, y, z \in \mathbb{Q}_2$. We may assume that $|x|_2 = \max\{|x|_2, |y|_2, |z|_2\}$. Then division by x^2 gives a relation of the same type with $x, y, z \in \mathbb{Z}_2$. Since $\mathbb{Z}_2/8\mathbb{Z}_2 \cong \mathbb{Z}/8\mathbb{Z}$ (see 51.7), we obtain the same relation in the finite ring $\mathbb{Z}/8\mathbb{Z}$. This is a contradiction, since $0, 1, 4$ are the only squares of $\mathbb{Z}/8\mathbb{Z}$.

The quadratic form $x^2 + y^2 + z^2$ takes the seven values 1, $1 + 1 = 2$, $1+1+1 = 3$, $4+1+1 = 6$, $9+4+1 = 14 = (-2)(-7)$, $16+4+1 = 21 = (-3)(-7)$ and $25 + 16 + 1 = 42 = (-6)(-7)$. Since $-7 = 1 - 8$ is a square in $\mathbb{Q}_2$, this shows that $x^2 + y^2 + z^2$ represents all square classes of $\mathbb{Q}_2$ except the square class of -1; see 54.1.

(5) By the Chinese remainder theorem we have $\mathbb{Z}/m\mathbb{Z} \cong \bigoplus_i \mathbb{Z}/p_i^{e_i}\mathbb{Z}$ as rings, with prime numbers p_i. Hence it suffices to consider the special case where $m = p^e$ is a power of a prime p. By 54.7 the quadratic form f is isotropic over $\mathbb{Q}_p$, and by homogeneity there exists a solution $x \in \mathbb{Z}_p^n \setminus (p\mathbb{Z}_p)^n$ of $f(x) = 0$. Since $\mathbb{Z}_p/p^e\mathbb{Z}_p \cong \mathbb{Z}/m\mathbb{Z}$ we obtain a non-trivial solution of $f(x) = 0$ in $(\mathbb{Z}/m\mathbb{Z})^n$ by reducing each coordinate of x modulo $p^e\mathbb{Z}_p$.

55 Absolute values

(1) The condition $\varphi_1(a) < 1$ is equivalent to $\lim_n a^n = 0$ with respect to the topology defined by φ_1, hence equivalent to $\varphi_2(a) < 1$ if the two absolute values define the same topology.

Assume that $\varphi_1(a) < 1 \Leftrightarrow \varphi_2(a) < 1$. For $x, y \in F^\times$, $m, n \in \mathbb{Z}$ and $a := x^m y^n$ we obtain $m \log \varphi_1(x) + n \log \varphi_1(y) < 0 \Leftrightarrow m \log \varphi_2(x) + n \log \varphi_2(y) < 0$, hence $\log \varphi_1(x)/\log \varphi_2(x) = \log \varphi_1(y)/\log \varphi_2(y)$ is a constant $s > 0$.

(2) By Exercise 1 we find $a, a' \in F$ with $\varphi_1(a) < 1 \leq \varphi_2(a)$ and $\varphi_2(a') < 1 \leq \varphi_1(a')$. Then $b := a/a'$ satisfies $\varphi_1(b) < 1 < \varphi_2(b)$. The elements $1 - (1+b^n)^{-1} = 1 - b^{-n}(1+b^{-n})^{-1}$ converge to 0 in F_1 and to 1 in F_2. Hence the closure C of the diagonal contains the pair $(0, 1)$. Replacing b by b^{-1} shows that $(1, 0) \in C$. Since C is an F-subspace of F^2, this yields $C = F^2$.

(3) Let $a \in \widehat{F}$. Since F is dense in $\widehat{F}$, also $F - a$ is dense in $\widehat{F}$, hence we find $b \in F$ with $\widehat{\varphi}(b-a) < \widehat{\varphi}(a)$. Then $\widehat{\varphi}(a) = \max\{\widehat{\varphi}(b-a), \widehat{\varphi}(a)\} = \widehat{\varphi}(b-a+a) = \varphi(b)$; see 55.3.

(4) For the subadditivity of φ^s it suffices to show that $a \leq b + c$ implies $a^s \leq b^s + c^s$, where $0 < a, b, c \in \mathbb{R}$. We may assume that $a = 1$; then the conclusion is true if $b \geq 1$ or $c \geq 1$. For $b, c \leq 1$ we have $1 \leq b + c \leq b^s + c^s$.

56 Valuations

(1) If F is algebraic over a finite field, then $F^\times$ consists of roots of unity, hence every homomorphism of $F^\times$ into an ordered group is trivial. If F is not an algebraic extension of a finite field, then F contains a subfield F_0 isomorphic to $\mathbb{Q}$ or $\mathbb{F}_p(x)$. By 56.3 we find valuations on F_0, which extend to F by 56.9. In order to obtain an absolute value on F, use Zorn's Lemma to enlarge any valuation ring of F to a maximal subring R of F; then R is a valuation ring, and by 56.8 the corresponding value group is Archimedean, which leads to an absolute value by 56.2.

(2) As in the proof of 56.7, one uses the fact that $k(x)$ is a principal ideal domain; for details see COHN 2003a 9.1 p. 312, RIBENBOIM 1999 3.1.K p. 89, ENGLER–PRESTEL 2005 Theorem 2.1.4b p. 30 or BOURBAKI 1972 VI.1.4 p. 380.

(3) If v is principal and π is a prime element, then every element of $F^\times$ has the form $a\pi^n$ with $n \in \mathbb{Z}$ and $v(a) = 0$, i.e. $a \in R^\times$. Hence $F^\times = \pi^{\mathbb{Z}} \times R^\times$, and the non-trivial ideals of R are the principal ideals $\pi^n R$ with $n \geq 0$.

Conversely, if R is a principal ideal domain, then its maximal ideal is of the form πR. One shows that $\bigcap_n \pi^n R = \{0\}$ and that $v(\pi)$ generates the value group; see COHN 2003a 9.1.3, WARNER 1989 Theorem 21.3 or BOURBAKI 1972 VI.3.6 Proposition 9 p. 392 for details.

If F is complete and F_v is finite, then one finds in R a root of unity ζ of order card $F_v^\times$ as in 53.1(ii). As in 53.1(iii) we obtain $R^\times = \langle \zeta \rangle \times (1 + M)$.

(4) Choose $e \in E$ with $v(e) > 0$. Then $e^n = \sum_{i=0}^{n-1} a_i e^i$ for some $n \in \mathbb{N}$ and suitable elements $a_i \in F$, $a_0 \neq 0$. The assumption that v is trivial on F leads to $v(\sum_i a_i e^i) = \min\{v(a_i e^i) \mid a_i \neq 0\} = v(a_0) = 0$, a contradiction to $v(e^n) = nv(e) \neq 0$.

(5) In order to prove that $w(fg) \leq w(f) + w(g)$ for polynomials $f = \sum a_i x^i$ and $g = \sum b_i x^i$, consider $i_0 = \min\{i \mid v(a_i) + i\delta = w(f)\}$ and $j_0 = \min\{j \mid v(b_j) + j\delta = w(g)\}$. For details see BOURBAKI 1972 VI.10.1 Lemma 1 p. 434, or ENGLER–PRESTEL 2005 Theorem 2.2.1.

(6) If R is a valuation ring of F, then $F = R \cup (R \smallsetminus \{0\})^{-1}$, hence F is the field of fractions of R. Conversely, let F be the field of fractions of a maximal subring R. Assume that $x \in F \smallsetminus R$ with $x^{-1} \notin R$. By maximality, the subring $R[x]$ generated by $R \cup \{x\}$ is F, and similarly $R[x^{-1}] = F$. Hence $1 \in R[x]$ and $1 \in R[x^{-1}]$, and this leads to a contradiction, as in the proof of 56.9.

(7) One example is the field $F = k((\Gamma))$ of Hahn power series; see 64.25. A smaller example is the field of fractions of the group ring $k\Gamma \subseteq k((\Gamma))$; compare RIBENBOIM 1999 13.1.C p. 368.

(8) Pure transcendency is ruled out by the fact that $\mathbb{Q}_p$ has trivial automorphism group; see 53.5 and 64.19. If $\mathbb{Q}_p|F$ is a finite extension, then $\mathbb{Q}_p = F(a)$ for some a by 64.12. The splitting field E of the minimal polynomial of a over F is a Galois extension of F of finite degree; see 64.10. We may assume that $E \subseteq \mathbb{Q}_p^\natural$; then $\mathbb{Q}_p = F(a) \subseteq E$. Result 56.15 gives a (unique) extension w of the p-adic valuation of $\mathbb{Q}_p$ to a principal valuation of E, and 56.13 shows that all field automorphisms of E are isometries and therefore continuous with respect to w. Hence $\mathrm{Gal}_F E$ fixes each element of $\mathbb{Q}$ and of its topological closure $\mathbb{Q}_p$, whence $\mathbb{Q}_p \subseteq F$. (See also RIBENBOIM 1999 6.2.K p. 167)

(9) The quaternion skew field $\mathbb{H}$ contains elements i, j with $i^2 = j^2 = -1$ and $ij = -ji$. Let R be a valuation ring of $\mathbb{H}$. Then R contains all elements of $\mathbb{H}^\times$ of order 4 (like i), hence all $a \in \mathbb{H}$ with $a + \bar{a} = 0$ and $a\bar{a} = 1$. These elements form a 2-sphere in $A := \{a \in \mathbb{H} \mid a + \bar{a} = 0\} \cong \mathbb{R}^3$, hence they generate A additively. We conclude that the ring R contains A and $iA \neq A$, hence $R \supseteq A + iA = \mathbb{H}$, a contradiction.

If v is an extension of v_p to the rational quaternions, then i, j are units of the valuation ring R of v. Since p belongs to the maximal ideal M of R, the image of the ring $\mathbb{Z} + \mathbb{Z}i + \mathbb{Z}j + \mathbb{Z}ij$ in the residue skew field R/M is a finite subring of R/M, hence a finite skew field. By Wedderburn's theorem (compare COHN 2003a 7.8.6 or JACOBSON 1985 7.7), that image is commutative. Thus the difference of 1 and -1 belongs to M, hence $2 \in M \cap \mathbb{Z} = p\mathbb{Z}$ and $p = 2$.

It remains to show that the 2-adic valuation of $\mathbb{Q}$ extends to the rational quaternions. In fact, an extension w of the 2-adic valuation v_2 of $\mathbb{Q}_2$ to the 2-

adic quaternions $\mathbb{H}_2 := \mathbb{Q}_2 + \mathbb{Q}_2 i + \mathbb{Q}_2 j + \mathbb{Q}_2 ij$ is given by $w(x) = v_2(x\bar{x})/2$, as we show now. The algebra $\mathbb{H}_2$ is a skew field, because its norm $x\bar{x} = \sum_{i=1}^4 x_i^2$ is anisotropic by Exercise 4 of Section 54. The mapping $w : \mathbb{H}_2^\times \to \mathbb{Q}^+$ is a group homomorphism. For every $a \in \mathbb{H}_2$, the restriction of w to the field $\mathbb{Q}_2(a)$ is a valuation of $\mathbb{Q}_2(a)$ by 56.15; in particular, $w(1+a) \geq \min\{w(a), 0\}$. This implies that w is a valuation (compare the proof of 56.5).

57 Topologies of valuation type

(1) One has field topologies by 13.2. If $U = \{a \in F \mid v(a) > \gamma\}$, then $(F \smallsetminus U)^{-1} = \{a \in F \mid v(a) \geq -\gamma\}$ is bounded; similar arguments apply to absolute values and ordered fields.

(2) As F is a field, each set aU with $a \in F^\times$ is a neighbourhood of 0. For any neighbourhood V of 0, there exists $a \in F^\times$ with $aU \subseteq V$ by 57.2(iii).

(3) The Hausdorff property holds for τ_∞ precisely if the points 0 and ∞ are separated by τ_∞; this is equivalent to the existence of a neighbourhood of 0 which is bounded and closed. By 13.4, τ is regular, hence each neighbourhood contains a closed neighbourhood.

Let $U \in \tau$ be an open neighbourhood of 0. The preimage $\{\infty\} \cup (U \smallsetminus \{0\})^{-1}$ under inversion belongs to τ_∞ precisely if $(F \smallsetminus U)^{-1}$ is contained in some bounded (closed) subset of F. This means that the closed set $(F \smallsetminus U)^{-1}$ is bounded; hence inversion is continuous at ∞ precisely if τ is of type V.

(4) Let τ_p be the p-adic topology of $\mathbb{Q}$, and let $B_p = \{a \in \mathbb{Q} \mid |a|_p < 1\}$. Then τ_X is the supremum of the topologies τ_p with $p \in X$; this means that $\bigcup\{\tau_p \mid p \in X\}$ is a subbasis of τ_X. Hence each τ_X is a field topology, and τ_p is of type V by Exercise 1. If X is finite, then $\bigcap_{p \in X} B_p$ belongs to τ_X and is bounded (by 57.2(iii)), hence τ_X is locally bounded.

Now let X be infinite. The sets $U_n := \bigcap_{p \in X, p \leq n} p^{n-1} B_p$ with $n \in \mathbb{N}$ form a neighbourhood base at 0 of τ_X. We show that none of the sets U_n is bounded (hence τ_X is not locally bounded), using 57.2(iii). Choose $q \in X$ with $q > n$. If $a \in \mathbb{Q}^\times$, then $x = |a|_q \prod_{p \in X, p \leq n} p^n \in U_n$ and $|ax|_q = 1$, which shows that $aU_n \not\subseteq B_q$.

Let X contain distinct primes p, q. The balls B_p and B_q belong to τ_X and contain 0, and $(\mathbb{Q} \smallsetminus B_p)^{-1} = \{x \in \mathbb{Q} \mid |x|_p \leq 1\}$. If $a \in \mathbb{Q}^\times$, then $|a|_q$ has p-adic absolute value 1, and $a|a|_q \in a(\mathbb{Q} \smallsetminus B_p)^{-1}$ has q-adic absolute value 1, hence $a(\mathbb{Q} \smallsetminus B_p)^{-1} \not\subseteq B_q$. This shows that τ_X is not of type V; see 57.2(iii).

Note that $\tau_{\{2,3\}}$ is the topology of the diagonal D in 44.12.

(5) If F is of type V and if W is a neighbourhood of 0, then we find a neighbourhood U of 0 with $U(F \smallsetminus W)^{-1} \subseteq W$. Now $xy \in U$ and $y \in F \smallsetminus W$ imply that $x = xyy^{-1} \in U(F \smallsetminus W)^{-1} \subseteq W$.

Conversely, assume that for every neighbourhood W of 0 there exists a neighbourhood U of 0 with $xy \in U \Rightarrow x \in W \lor y \in W$. Then $U(F \smallsetminus W)^{-1} \subseteq W$, because $y \in F \smallsetminus W$ and $x = (xy^{-1})y \in U$ imply $xy^{-1} \in W$. Hence $(F \smallsetminus W)^{-1}$ is bounded.

(6) Let F be a topological field of type V. We verify the condition in 44.8 for concentrated filterbases $\mathfrak{C}$ on F which do not converge to 0. There exist neighbourhoods W, W' of 0 with $W \notin \mathfrak{C}$ and $W' + W' \subseteq W$, and an element $B \in \mathfrak{C}$ with $B - B \subseteq W'$. If $x \in B \cap W'$, then $B = B - x + x \subseteq B - B + W' \subseteq W$, a contradiction to $W \notin \mathfrak{C}$; thus $B \cap W' = \emptyset$. Since F is of type V, the sets

$B^{-1} \subseteq (F \smallsetminus W')^{-1}$ are bounded, hence $B^{-1} \cdot B^{-1}$ is bounded by 57.2(ii). Hence for every neighbourhood U of 0 there exists a neighbourhood U' of 0 with $U' \cdot B^{-1} \cdot B^{-1} \subseteq U$, and then an element $C \in \mathfrak{C}$ with $C \subseteq B$ and $C - C \subseteq U'$. This yields $C^{-1} - C^{-1} \subseteq (C - C) \cdot C^{-1} \cdot C^{-1} \subseteq U' \cdot B^{-1} \cdot B^{-1} \subseteq U$; hence $\mathfrak{C}^{-1}$ is concentrated.

We remark that the completion $\widehat{F}$ is again of type V; compare WARNER 1989 Theorem 19.12.

(7) One checks that R_a is a subring of F; for example, if $x, y \in R_a$ and $n \in \mathbb{N}$, then $nv(xy) = nv(x) + nv(y) \geq 2n \min\{v(x), v(y)\} = \min\{2nv(x), 2nv(y)\} > -v(a)$. Moreover, R_a contains the valuation ring R of v, and $a^{-1} \notin R_a$. Hence R_a is a valuation ring of F, and R and R_a induce the same valuation topology on F by 56.10.

It remains to show that R_a is a maximal subring of F; then 56.8 implies that the value group determined by R_a is Archimedean. Let $b \in F \smallsetminus R_a$. Then $mv(b) + v(a) \leq 0$ for some $m \in \mathbb{N}$, hence $v(b^{-m}a^{-1}) \geq 0$ and $b^{-m}a^{-1} \in R$. We infer that $a^{-n} = (b^m(b^{-m}a^{-1}))^n$ belongs to the ring $R[b]$ generated by R and b for every $n \in \mathbb{N}$, hence $R[b] \supseteq \bigcup_{n \in \mathbb{N}} a^{-n}R = F$ and therefore $R_a[b] = F$.

58 Local fields and locally compact fields

(1) The polynomial $f(x+1) = \sum_{i=1}^{p} \binom{p}{i} x^{i-1}$ is an Eisenstein polynomial (see SCHÖNEMANN 1846). Reducing the coefficients modulo p gives the polynomial $\bar{f} = x^{p-1} \in \mathbb{F}_p[x]$. Hence each non-trivial factor of f in $\mathbb{Z}_p[x]$ has its constant term in $p\mathbb{Z}_p$. Since $f(0+1) = p$, this implies that f is irreducible in $\mathbb{Z}_p[x]$. The irreducibility in $\mathbb{Q}_p[x]$ is then a consequence of the Gauss Lemma; compare COHN 2003a 7.7.2, JACOBSON 1985 2.16 Lemma 2 or LANG 1993 IV §2.

(2) Let α be the generator of the Galois group of $F | \mathbb{Q}_p$. By 56.15 the p-adic valuation v_p of $\mathbb{Q}_p$ has a unique extension w to F which is given by $w(x) = v_p(xx^\alpha)/2$ for all $x \in F$. Thus we have $w(F) = v_p(\mathbb{Q}_p) = \mathbb{Z}$ precisely if $v_p(xx^\alpha)$ is even for all $x \in F$. Writing $x = a + b\sqrt{c}$, we obtain the condition that $v_p(a^2 - b^2 c)$ is even for $a, b \in \mathbb{Q}_p$. This is true for $b = 0$, and for $b \neq 0$ we have $v_p(a^2 - b^2 c) = v_p((a/b)^2 - c) + 2v_p(b)$.

If $p = 2$ and $c = 3$, then $v_2(1 - 3) = 1$, hence $\mathbb{Q}_2(\sqrt{3}) | \mathbb{Q}_2$ is ramified.

Let $p = 2$ and $c = 5$. We claim that $\mathbb{Q}_2(\sqrt{5}) | \mathbb{Q}_2$ is unramified. Otherwise $v_2(a^2 - 5)$ would be odd for some $a \in \mathbb{Q}_2$. If $v_2(a) \neq 0$, then $v_2(a^2 - 5) = 2 \min\{v_2(a), 0\}$ is even (see 56.1). Hence $v_2(a) = 0$, which means that $a \in 1 + 2\mathbb{Z}_2$. Therefore $a^2 - 5 \in 1 - 5 + 8\mathbb{Z}_2$ and $v_2(a^2 - 5) = 2$, a contradiction.

Alternatively, we observe that $-3 \cdot 5 = -15 \in 1 + 8\mathbb{Z}_2$ is a square in $\mathbb{Q}_2$ by 54.1, hence $\mathbb{Q}_2(\sqrt{5}) = \mathbb{Q}_2(\sqrt{-3}) = \mathbb{Q}_2((-1 + \sqrt{-3})/2)$. Since $(-1 + \sqrt{-3})/2$ is a root of unity of order $3 = 2^2 - 1$, the field $\mathbb{Q}_2(\sqrt{5})$ is the unique unramified quadratic extension of $\mathbb{Q}_2$; see 58.2.

(3) Each polynomial $x^2 + t^n x + t$ is separable over $F = \mathbb{F}_2((t))$, with roots $a_n, b_n = ta_n^{-1}$ in a fixed algebraic closure of F. By exchanging a_n and b_n we can achieve that a_n belongs to the natural compact valuation ring of $F(a_n) = F(a_n, b_n)$; see 56.15. If there were only finitely many extension types $F(a_n) | F$, then for some $k \in \mathbb{N}$ the field $F(a_k)$ would contain a_n for infinitely many $n \in \mathbb{N}$. Then some subsequence of $(a_n)_n$ has a limit $a \in F(a_k)$. Since $a_n^2 + t^n a_n + t = 0$ for $n \in \mathbb{N}$ and $\lim_{n \to \infty} t^n = 0$, we infer that $a^2 + t = 0$. Thus $F(a)$ is an inseparable quadratic extension of F contained in $F(a_k)$. Hence $F(a) = F(a_k)$, a contradiction, as $F(a_k) | F$ is separable.

(4) From 54.1 it follows that $17 = 1 + 8 \cdot 2$ is a square in $\mathbb{Q}_2$ (and in $\mathbb{F}_2$), and $2 = 6^2(1 - 17 \cdot 2 \cdot 6^{-2})$ is a square in $\mathbb{Q}_{17}$ (and in $\mathbb{F}_{17}$). If $p \neq 2, 17$, then at least one of the numbers $2, 17, 34 = 2 \cdot 17$ is a square in $\mathbb{F}_p$ (since the squares of $\mathbb{F}_p^\times$ form a subgroup of index 2; see Exercise 1 of Section 64) and hence also in $\mathbb{Q}_p$ (by Exercise 2 of Section 54).

(5) The mapping $x + yb \mapsto \begin{pmatrix} x & cy^\gamma \\ y & x^\gamma \end{pmatrix}$, where $x, y \in E$, is a Z-algebra isomorphism as required. Direct computation shows that the matrices of the shape $A = \begin{pmatrix} x & cy^\gamma \\ y & x^\gamma \end{pmatrix}$ form a Z-subalgebra of $E^{2 \times 2}$. Since $\det A = xx^\gamma - cyy^\gamma$, the non-zero matrices of this shape are invertible precisely if $c \neq xx^\gamma$ for all $x \in E$. The Cayley–Hamilton theorem implies that $A^{-1} \det A = (x + x^\gamma)I - A$, hence the inverses have the same shape, as $x + x^\gamma, \det A \in Z$.

(6) In the notation of 58.10, one has $Z_n = \mathbb{F}_{q^n}$ and $a^{\gamma^r} = a^{q^r}$ for $a \in \mathbb{F}_{q^n}$; compare 58.2. Substituting t for b in the definition of $[Z_n | Z; \gamma^r, x]$ gives an isomorphism as required.

61 Ordinals and cardinals

(1) Consider the set of elements with the smallest first coordinate.

(2) Countable subsets of $\mathbb{R}$ can be described by maps $\mathbb{N} \to \mathbb{R}$, and $\operatorname{card} \mathbb{R}^{\mathbb{N}} = \aleph^{\aleph_0} = \aleph$ by 61.13b.

(3) One may assume that $A_\iota \subset B_\iota$ and $B_\iota \cap B_\kappa = \emptyset$ for $\iota \neq \kappa$. Consider any map $\varphi : \bigcup_\iota A_\iota \to \mathop{\mathsf{X}}_\iota B_\iota : s \mapsto \varphi_s$ and choose $\psi(\iota) \in B_\iota \smallsetminus \{\varphi_s(\iota) \mid s \in A_\iota\}$. Then $\psi \neq \varphi_s$ for each s, and φ is not surjective.

(4) We have $\aleph_\omega = \sum_{\nu \in \omega} \aleph_\nu < \prod_{\nu \in \omega} \aleph_\nu \leq \aleph_\omega^{\aleph_0}$. With $\aleph_\omega = 2^{\aleph_0}$ this leads to a contradiction.

62 Topological groups

(1) If inversion is uniformly continuous, then for each neighbourhood U of 1 there is a symmetric neighbourhood S such that $yx^{-1} \in S$ implies $x^{-1}y \in U$. Hence $S \subseteq xUx^{-1}$, and this true for each x. Therefore, $V := \bigcap_x xUx^{-1}$ is an *invariant* neighbourhood of 1.

If V is invariant and $V^2 \subseteq U$, then $VaVb \subseteq Uab$ for all a and b, and this is equivalent with the assertion.

(2) Consider a symmetric (compact) neighbourhood V of 1 such that $H \cap V^2$ is compact. If $x \in \overline{H}$, then $H \cap xV \neq \emptyset$ and hence $\emptyset \neq Hx \cap V \subseteq \overline{H} \cap V \subseteq \overline{H \cap V^2} \subseteq H$.

(3) If U is a compact open neighbourhood of 1, then by the usual compactness arguments, there is a neighbourhood V of 1 such that $UV \subseteq U$. This implies that V generates an open (and hence compact) subgroup of U.

63 Locally compact abelian groups and Pontryagin duality

(1) Since $B \to C$ is surjective, the adjoint $C^* \to B^*$ is injective. In fact, $C^* \cong A^\perp \leq B^*$. Hence there is an exact sequence $0 \to A^\perp \to B^* \to X \to 0$. The first argument shows that $X^* \cong A^{\perp\perp} = A$, and reflexivity implies that $X \cong A^*$.

(2) As a rational vector space, D has a basis $\mathfrak{b}$ of cardinality $\aleph$. The values of a character of D can be chosen arbitrarily on $\mathfrak{b}$. Hence card $D^* = \aleph^\aleph = 2^\aleph > \aleph$. On the other hand, $\mathbb{R}^* \cong \mathbb{R}$ has cardinality $\aleph$.

(3) The dual A^* is discrete by 63.5. Hence A^* is torsion free if, and only if, each compact subgroup of A^* is trivial. The claim follows now from 63.30–63.32.

64 Fields

(1) Let d be a divisor of $|G| = n$. The polynomial $x^d - 1$ has at most d roots. If G contains an element of order d, then the set $\{\, g \in G \mid g^d = 1 \,\}$ has size d, hence the number of elements in G of order d is given by the Euler function $\varphi(d)$. Thus $\psi(d) := \mathrm{card}\{\, g \in G \mid g \text{ has order } d \,\} \in \{0, \varphi(d)\}$ for all divisors d of n. Now $n = \sum_d \psi(d) \leq \sum_d \varphi(d)$, and consideration of a cyclic group shows that the last sum has the value n. We conclude that $\psi = \varphi$, in particular $\psi(n) = \varphi(n) \neq 0$, whence G is cyclic. (This proof is due to Gauß.)

(2) The group Γ acts on $E[t]$ by acting on the coefficients of polynomials, fixing precisely the polynomials in $F[t]$. Therefore $f = \prod_{b \in B}(x - b)$ has coefficients in F. Clearly $f(a) = 0$. If $g \in F[t]$ satisfies $g(a) = 0$, then $x - a$ divides g in $E[t]$. Applying the elements of Γ we see that $x - b$ divides g for each $b \in B$, whence f divides g. This shows that f is the minimal polynomial of a over F.

(3) By Eisenstein's criterion (compare SCHÖNEMANN 1846), f is irreducible in $\mathbb{Q}[x]$. For $n = -3, 0, 1, 2$, the signs of $f(n)$ alternate, hence f has at least three real roots. The sum of the squares of the five roots x_k of f is 0, because $\sum_k x_k = 0 = \sum_{h \neq k} x_h x_k$ and therefore $\sum_k x_k^2 = 0$. Hence f has two roots $z, \bar{z} \in \mathbb{C}$ which are not real. Complex conjugation gives an element τ in the Galois group $\Gamma = \mathrm{Gal}_{\mathbb{Q}} E$ which acts as the transposition $(z, \bar{z})$ on the set of roots of f. By Exercise 2, Γ is transitive on the five roots. Hence Γ contains an element σ of order 5, which acts as a 5-cycle on the roots. Together, σ and τ generate the symmetric group of degree 5.

(4) By 61.16, the field F has at most $2^{\mathrm{card}\,F}$ automorphisms. On the other hand, the groups $\mathrm{Aut}\,\mathbb{Q}^\natural$ and $\mathrm{Aut}\,\mathbb{F}_p^\natural$ have cardinality $2^{\aleph_0}$; see 52.10 and 64.15. Any field is an algebraic extension of some purely transcendental field $\mathbb{Q}(T)$ or $\mathbb{F}_p(T)$; see 64.20. If F is countable, then the extension result 64.15 implies that F has at least $2^{\aleph_0}$ field automorphisms. If F is uncountable, then card $F = $ card T; see 64.20. There are $2^{\mathrm{card}\,T}$ permutations of T by 61.16, and by 64.15 each of these permutations extends to a field automorphism of F.

(5) Assume that $(t - a)^{-1}$ is an F-linear combination of rational functions $(t - b_i)^{-1}$ with $b_i \in F \smallsetminus \{a\}$. Then $(t - a)^{-1} = f / \prod_i (t - b_i)$ with $f \in F[t]$, hence $t - a$ divides $\prod_i (t - b_i)$ in $F[t]$, which is a contradiction to the fact that $F[t]$ is a unique factorization domain.

(6) The field of fixed elements of the Frobenius automorphism φ is the prime field $\mathbb{F}_p$, hence $\mathrm{Aut}\,\mathbb{F}_q = \mathrm{Gal}_{\mathbb{F}_p}\mathbb{F}_q$ has order n; compare 64.18. By Exercise 1, φ has order n, hence $\mathrm{Aut}\,\mathbb{F}_q = \langle\varphi\rangle$.

References

Numbers at the end of an entry indicate the pages where the entry is quoted.

U. ABEL 1980. On the group of piecewise linear monotone bijections of an arc. *Math. Z.* **171**, 155–161. 64

U. ABEL 1982. A homeomorphism of **Q** with **Q** as an orbit. *Elem. Math.* **37**, 108–109. 205

U. ABEL and J. MISFELD 1990. Congruences in decompositions of the real and rational numbers. *Rend. Mat. Appl.* (7) **10**, 279–285. 64

N. A'CAMPO 2003. A natural construction for the real numbers. Preprint arXiv math GN/0301015. 2

J. ACZÉL 1966. *Lectures on Functional Equations and their Applications.* New York: Academic Press. 6

J. F. ADAMS 1969. *Lectures on Lie Groups.* New York: Benjamin. 67, 90

P. ALEXANDROFF 1924. Über die Metrisation der im Kleinen kompakten topologischen Räume. *Math. Ann.* **92**, 294–301. 48

N. L. ALLING 1962. On exponentially closed fields. *Proc. Amer. Math. Soc.* **13**, 706–711. 117, 315

N. L. ALLING 1987. *Foundations of Analysis over Surreal Number Fields.* Amsterdam: North-Holland. 1

N. L. ALLING and S. KUHLMANN 1994. On η_α-groups and fields. *Order* **11**, 85–92. 78, 129

C. ALVAREZ 1999. On the history of Souslin's problem. *Arch. Hist. Exact Sci.* **54**, 181–242. 27

R. D. ANDERSON 1958. The algebraic simplicity of certain groups of homeomorphisms. *Amer. J. Math.* **80**, 955–963. 54, 206

G. E. ANDREWS, S. B. EKHAD, and D. ZEILBERGER 1993. A short proof of Jacobi's formula for the number of representations of an integer as a sum of four squares. *Amer. Math. Monthly* **100**, 274–276. 213

R. D. ARTHAN 2004. The Eudoxus real numbers. Preprint arXiv math HO/0405454. 2

H. BACHMANN 1967. *Transfinite Zahlen.* Berlin: Springer. Second edition. 337, 338

R. BAER 1970a. Dichte, Archimedizität und Starrheit geordneter Körper. *Math. Ann.* **188**, 165–205. 119

R. Baer 1970b. Die Automorphismengruppe eines algebraisch abgeschlossenen Körpers der Charakteristik 0. *Math. Z.* **117**, 7–17. 150

R. Baer and H. Hasse 1932. Zusammenhang und Dimension topologischer Körperräume. *J. Reine Angew. Math.* **167**, 40–45. 146

J. C. Baez 2002. The octonions.
Bull. Amer. Math. Soc. (N.S.) **39**, 145–205. xii

A. Baker 1975. *Transcendental Number Theory.*
Cambridge: Cambridge University Press. (Reprinted 1990) 201, 221

A. Baker 1984. *A Concise Introduction to the Theory of Numbers.*
Cambridge: Cambridge University Press. 201

J. A. Baker, C. T. Ng, J. Lawrence, and F. Zorzitto 1978. Sequence topologies on the real line. *Amer. Math. Monthly* **85**, 667–668. 69

B. Banaschewski 1957. Über die Vervollständigung geordneter Gruppen. *Math. Nachr.* **16**, 51–71. 264

B. Banaschewski 1998. On proving the existence of complete ordered fields. *Amer. Math. Monthly* **105**, 548–551. 248

H. Bauer 1968. *Wahrscheinlichkeitstheorie und Grundzüge der Masstheorie.*
Berlin: de Gruyter. (Translation: *Probability Theory and Elements of Measure Theory.* New York: Academic Press 1981) 202

J. E. Baumgartner and K. Prikry 1977. Singular cardinals and the generalized continuum hypothesis. *Amer. Math. Monthly* **84**, 108–113. 339

A. F. Beardon 2001. The geometry of Pringsheim's continued fractions. *Geom. Dedicata* **84**, 125–134. 31

E. Becker 1974. Euklidische Körper und euklidische Hüllen von Körpern. *J. Reine Angew. Math.* **268/269**, 41–52. 125

M. Benito and J. J. Escribano 2002. An easy proof of Hurwitz's theorem. *Amer. Math. Monthly* **109**, 916–918. 30

C. D. Bennett 1997. Explicit free subgroups of Aut($\mathbf{R}, \leq$). *Proc. Amer. Math. Soc.* **125**, 1305–1308. 64

E. Beth and A. Tarski 1956. Equilaterality as the only primitive notion of Euclidean geometry. *Indag. Math.* **18** (*Nederl. Akad. Wetensch. Proc. Ser. A* **59**), 462–467. 153

M. Bhargava 2000. On the Conway–Schneeberger fifteen theorem.
In: *Quadratic Forms and their Applications* (Dublin, 1999), vol. 272 of Contemporary Mathematics, pages 27–37. Providence, RI: American Mathematical Society. 213

G. Birkhoff 1948. *Lattice Theory.* Colloquium Publication 25. New York: American Mathematical Society. Second revised edition. 25, 26, 27, 164

G. D. Birkhoff and H. S. Vandiver 1904. On the integral divisors of $a^n - b^n$. *Ann. Math.* (*2*) **5**, 243–252. 231

A. Blass and J. M. Kister 1986. Free subgroups of the homeomorphism group of the reals. *Topology Appl.* **24**, 243–252. 64

T. S. Blyth 2005. *Lattices and Ordered Algebraic Structures.*
London: Springer. 77, 246

J. Bochnak, M. Coste, and M.-F. Roy 1998. *Real Algebraic Geometry.*
Berlin: Springer. 122, 133, 134

Z. I. Borevich and I. R. Shafarevich 1966. *Number Theory.* New York: Academic Press. (Third Russian edition: Moscow: Nauka 1985. German translation: Basel: Birkhäuser 1966) 284, 297, 323, 325, 363, 374

N. Bourbaki 1966. *General Topology.* Paris: Hermann.
(Reprints: Berlin: Springer 1989, 1998) 67, 164, 340, 341, 342

N. BOURBAKI 1972. *Commutative Algebra.* Paris: Hermann. (Second edition: Berlin: Springer 1989)   146, 304, 306, 326, 330, 377, 378

N. BOURBAKI 1990. *Algebra II.* *Chapters 4–7.* Berlin: Springer. (Reprinted 2003)   290, 350, 354, 355

G. E. BREDON 1963. A new treatment of the Haar integral. *Michigan Math. J.* **10**, 365–373.   326

H. BRENNER 1992. Ein überabzählbares, über $\mathbf{Q}$ linear unabhängiges System reeller Zahlen. *Math. Semesterber.* **39**, 89–93.   7

N. G. DE BRUIJN 1976. Defining reals without the use of rationals. *Indag. Math.* **38** (*Nederl. Akad. Wetensch. Proc. Ser. A* **79**), 100–108.   1

E. B. BURGER and T. STRUPPECK 1996. Does $\sum_{n=0}^{\infty} 1/n!$ really converge? Infinite Series and p-adic Analysis. *Amer. Math. Monthly* **103**, 565–577.   284

G. CANTOR 1895. Beiträge zur Begründung der transfiniten Mengenlehre. Erster Artikel. *Math. Ann.* **46**, 481–512. (Zweiter Artikel: *Math. Ann.* **49** 1897, 207–246)   25

J. W. S. CASSELS 1957. *An Introduction to Diophantine Approximation.* Cambridge: Cambridge University Press.   30

J. W. S. CASSELS 1978. *Rational Quadratic Forms.* London: Academic Press.   x, 297, 299, 325

J. W. S. CASSELS 1986. *Local Fields.* Cambridge: Cambridge University Press.   285, 289, 304, 315, 323, 325

C. C. CHANG and H. J. KEISLER 1990. *Model Theory.* Amsterdam: North-Holland. Third edition.   156, 170

M. D. CHOI and T. Y. LAM 1977. Extremal positive semidefinite forms. *Math. Ann.* **231**, 1–18.   134

C. O. CHRISTENSON and W. L. VOXMAN 1977. *Aspects of Topology.* New York: Dekker. (Revised second edition: Moscow, ID: BCS Associates 1998)   40, 42, 43, 50, 51, 53, 55, 283

J. C. CH'ÜAN and L. LIU 1981. Group topologies on the real line. *J. Math. Anal. Appl.* **81**, 391–398.   69, 89

K. CIESIELSKI 1997. *Set Theory for the Working Mathematician.* Cambridge: Cambridge University Press.   337

L. W. COHEN and G. EHRLICH 1963. *The Structure of the Real Number System.* Princeton, NJ: Van Nostrand.   xii

P. J. COHEN 1966. *Set Theory and the Continuum Hypothesis.* New York: Benjamin.   339

S. D. COHEN and A. M. W. GLASS 1997. Free groups from fields. *J. London Math. Soc.* (*2*) **55**, 309–319.   64

D. L. COHN 1993. *Measure Theory.* Boston, MA: Birkhäuser.   104

P. M. COHN 1995. *Skew Fields. Theory of General Division Rings.* Cambridge: Cambridge University Press.   330, 332

P. M. COHN 2003a. *Basic Algebra.* London: Springer.   16, 20, 131, 133, 145, 285, 289, 293, 306, 315, 323, 333, 350, 352, 353, 354, 355, 362, 367, 377, 378, 380

P. M. COHN 2003b. *Further Algebra and Applications.* London: Springer.   332, 333, 368

W. W. COMFORT and S. NEGREPONTIS 1974. *The Theory of Ultrafilters.* New York: Springer.   155

A. CONNES 1998. A new proof of Morley's theorem. In: *Les relations entre les mathématiques et la physique théorique*, pages 43–46. Bures-sur-Yvette: Institut des Hautes Études Scientifiques. 151

M. CONTESSA, J. L. MOTT, and W. NICHOLS 1999. Multiplicative groups of fields. In: *Advances in Commutative Ring Theory* (Fez, 1997), vol. 205 of Lecture Notes in Pure and Applied Mathematics, pages 197–216. New York: Dekker. 17, 19, 72

J. H. CONWAY 1976. *On Numbers and Games.* London: Academic Press. (Second edition: Natick, MA: A K Peters 2001) 1

J. H. CONWAY 1997. *The Sensual (Quadratic) Form.* Washington, DC: Mathematical Association of America. 212, 279, 297

J. H. CONWAY 2000. Universal quadratic forms and the fifteen theorem. In: *Quadratic Forms and their Applications* (Dublin, 1999), vol. 272 of Contemporary Mathematics, pages 23–26. Providence, RI: American Mathematical Society. 213

J. H. CONWAY and D. A. SMITH 2003. *On Quaternions and Octonions: their Geometry, Arithmetic, and Symmetry.* Natick, MA: A K Peters. xii, 211

L. CORWIN 1976. Uniqueness of topology for the p-adic integers. *Proc. Amer. Math. Soc.* **55**, 432–434. 290

H. DALES and W. WOODIN 1996. *Super-real Fields. Totally Ordered Fields with Additional Structure.* Oxford: Oxford University Press. 1, 166, 169, 246, 359

G. DARBOUX 1880. Sur le théorème fondamental de la géométrie projective. *Math. Ann.* **17**, 55–61. 74

R. DEDEKIND 1872. *Stetigkeit und irrationale Zahlen.* Braunschweig: Vieweg. (Sixth edition 1960. Translation: *Essays on the Theory of Numbers.* New York: Dover Publications 1963) xii

O. DEISER 2005. Der Multiplikationssatz der Mengenlehre. *Jahresber. Deutsch. Math.-Verein.* **107**, 88–109. 338

B. DESCHAMPS 2001. À propos d'un théorème de Frobenius. *Ann. Math. Blaise Pascal* **8**, 61–66. 129

K. J. DEVLIN and H. JOHNSBRÅTEN 1974. *The Souslin Problem*, vol. 405 of Lecture Notes in Mathematics. Berlin: Springer. 27

J. DIEUDONNÉ 1944. Sur la complétion des groupes topologiques. *C. R. Acad. Sci. Paris* **218**, 774–776. 263

J. DIEUDONNÉ 1945. Sur les corps topologiques connexes. *C. R. Acad. Sci. Paris* **221**, 396–398. 146

D. DIKRANJAN 1979. Topological characterization of p-adic numbers and an application to minimal Galois extensions. *Ann. Univ. Sofia Fac. Math. Méc.* **73**, 103–110. 287

J. D. DIXON and B. MORTIMER 1996. *Permutation Groups.* New York: Springer. 72

E. K. VAN DOUWEN and W. F. PFEFFER 1979. Some properties of the Sorgenfrey line and related spaces. *Pacific J. Math.* **81**, 371–377. 69

E. K. VAN DOUWEN and H. H. WICKE 1977. A real, weird topology on the reals. *Houston J. Math.* **3**, 141–152. 69

M. DROSTE and R. GÖBEL 2002. On the homeomorphism groups of Cantor's discontinuum and the spaces of rational and irrational numbers. *Bull. London Math. Soc.* **34**, 474–478. 54

M. DUGAS and R. GÖBEL 1987. All infinite groups are Galois groups over any field. *Trans. Amer. Math. Soc.* **304**, 355–384. 75

J. DUGUNDJI 1966. *Topology.* Boston, MA: Allyn and Bacon.
    32, 57, 58, 90, 164, 200, 202, 335, 336, 337, 338, 342, 343, 345

A. DURAND 1975. Un système de nombres algébriquement indépendents.
    C. R. Acad. Sci. Paris Sér. A **280**, 309–311.   7

H. EBBINGHAUS, H. HERMES, F. HIRZEBRUCH, M. KOECHER, K. MAINZER,
    J. NEUKIRCH, A. PRESTEL, and R. REMMERT 1991. *Numbers.*
    New York: Springer. (Translation of: *Zahlen.* Berlin: Springer 1988)
    xii, 128, 138, 144, 151, 211, 216, 278, 304, 315, 330

P. EHRLICH (ed.) 1994. *Real Numbers, Generalizations of the Reals, and
    Theories of Continua,* vol. 242 of Synthese Library. Dordrecht: Kluwer.
    xii, 1

P. EHRLICH 2001. Number systems with simplicity hierarchies: a general-
    ization of Conway's theory of surreal numbers. *J. Symbolic Logic* **66**,
    1231–1258. Corrigendum: *J. Symbolic Logic* **70** 2005, 1022.   1, 76, 124

R. ELLIS 1957. Locally compact transformation groups.
    Duke Math. J. **24**, 119–125.   340

C. ELSHOLTZ 2003. Kombinatorische Beweise des Zweiquadratesatzes und
    Verallgemeinerungen. *Math. Semesterber.* **50**, 77–93.   209

C. ELSNER 2000. Über eine effektive Konstruktion großer Mengen algebraisch
    unabhängiger Zahlen. *Math. Semesterber.* **47**, 243–256.   7

R. ENGELKING 1969. On closed images of the space of irrationals.
    Proc. Amer. Math. Soc. **21**, 583–586.   111

R. ENGELKING 1978. *Dimension Theory.*
    Amsterdam: North-Holland.   197

A. J. ENGLER and A. PRESTEL 2005. *Valued Fields.*
    Berlin: Springer.   184, 279, 304, 306, 359, 377, 378

D. B. A. EPSTEIN 1970. The simplicity of certain groups of homeomorphisms.
    Compositio Math. **22**, 165–173.   64

P. ERDÖS 1940. The dimension of the rational points in Hilbert space.
    Ann. Math. (*2*) **41**, 734–736.   197

P. ERDÖS, L. GILLMAN, and M. HENRIKSEN 1955. An isomorphism theorem
    for real-closed fields. *Ann. Math.* (*2*) **61**, 542–554.   167

D. M. EVANS and D. LASCAR 1997. The automorphism group of the field of
    complex numbers is complete. In: *Model Theory of Groups and Auto-
    morphism Groups* (Blaubeuren, 1995), vol. 244 of London Mathematical
    Society Lecture Notes, pages 115–125. Cambridge: Cambridge University
    Press.   150

F. FALTIN, N. METROPOLIS, B. ROSS, and G.-C. ROTA 1975. The real num-
    bers as a wreath product. *Adv. Math.* **16**, 278–304.   1, 374

A. FEDELI and A. LE DONNE 2001. The Sorgenfrey line has a locally pathwise
    connected connectification. *Proc. Amer. Math. Soc.* **129**, 311–314.   69

S. FEFERMAN 1964. *The Number Systems. Foundations of Algebra and Anal-
    ysis.* Reading, MA: Addison-Wesley.   xii

U. FELGNER 1976. Das Problem von Souslin für geordnete algebraische Struk-
    turen. In: *Set Theory and Hierarchy Theory* (Bierutowice, 1975), vol. 537
    of Lecture Notes in Mathematics, pages 83–107. Berlin: Springer.   27

W. FELSCHER 1978/79. *Naive Mengen und abstrakte Zahlen I.*
    Naive Mengen und abstrakte Zahlen II: Algebraische und reelle Zahlen.
    Naive Mengen und abstrakte Zahlen III: Transfinite Methoden.
    Mannheim: Bibliographisches Institut.   xii

B. FINE and G. ROSENBERGER 1997. *The Fundamental Theorem of Algebra*.
New York: Springer. 128, 144

N. J. FINE and G. E. SCHWEIGERT 1955. On the group of homeomorphisms
of an arc. *Ann. Math.* (*2*) **62**, 237–253. 58, 63, 64

G. FLEGG 1983. *Numbers. Their History and Meaning*.
New York: Schocken Books. xii, 1

L. FUCHS 1970. *Infinite Abelian Groups*, vol. I.
New York: Academic Press. 184, 186

L. FUCHS 1973. *Infinite Abelian Groups*, vol. II.
New York: Academic Press. 96, 207, 372

H. GEIGES 2001. Beweis des Satzes von Morley nach A. Connes.
Elem. Math. **56**, 137–142. 151

M. GERSTENHABER and C. T. YANG 1960. Division rings containing a real
closed field. *Duke Math. J.* **27**, 461–465. 129

L. GILLMAN 2002. Two classical surprises concerning the axiom of choice and
the continuum hypothesis. *Amer. Math. Monthly* **109**, 544–553. 339

L. GILLMAN and M. JERISON 1960. *Rings of Continuous Functions*.
Princeton, NJ: Van Nostrand. (Reprint: New York: Springer 1976) 118

A. M. W. GLASS 1981. *Ordered Permutation Groups*.
Cambridge: Cambridge University Press. 27, 205

A. M. W. GLASS and P. RIBENBOIM 1994. Automorphisms of the ordered
multiplicative group of positive rational numbers. *Proc. Amer. Math.
Soc.* **122**, 15–18. 221

W. GLATTHAAR 1971. Unterebenen und Kollineationen der komplexen Ebene.
Zulassungsarbeit, University of Tübingen. 152

A. M. GLEASON and R. S. PALAIS 1957. On a class of transformation groups.
Amer. J. Math. **79**, 631–648. 66

K. GÖDEL 1940. *The Consistency of the Axiom of Choice and of the Generalized Continuum Hypothesis*, vol. 3 of Annals of Mathematics Studies.
Princeton, NJ: Princeton University Press. 339

R. GOLDBLATT 1998. *Lectures on the Hyperreals*.
New York: Springer. 154

S. W. GOLOMB 1959. A connected topology for the integers.
Amer. Math. Monthly **66**, 663–665. 45

H. GONSHOR 1986. *An Introduction to the Theory of Surreal Numbers*. Cambridge: Cambridge University Press. 1

R. A. GORDON 1994. *The Integrals of Lebesgue, Denjoy, Perron, and Hen-
stock*. Providence, RI: American Mathematical Society. 202

C. GOURION 1992. À propos du groupe des automorphismes de $(\mathbf{Q}; \leq)$.
C. R. Acad. Sci. Paris Sér. I Math. **315**, 1329–1331. 204

F. Q. GOUVÊA 1997. *p-Adic Numbers: an Introduction*. Berlin: Springer.
Second edition. 278, 283, 284

M. J. GREENBERG 1967. *Lectures on Algebraic Topology*.
New York: Benjamin. 86, 87

M. J. GREENBERG 1969. *Lectures on Forms in Many Variables*.
New York: Benjamin. 279

G. R. GREENFIELD 1983. Sums of three and four integer squares.
Rocky Mountain J. Math. **13**, 169–175. 212

E. GROSSWALD 1985. *Representations of Integers as Sums of Squares*.
New York: Springer. 212, 213, 215

T. GRUNDHÖFER 2005. Describing the real numbers in terms of integers. *Arch. Math.* (*Basel*) **85**, 79–81. 2

T. GRUNDHÖFER and H. SALZMANN 1990. Locally compact double loops and ternary fields. In: *Quasigroups and Loops: Theory and Applications* (eds O. Chein, H. O. Pflugfelder and J. D. H. Smith), Chapter XI, pages 313–355. Berlin: Heldermann. 139, 375

V. S. GUBA 1986. A finitely generated complete group (Russian). *Izv. Akad. Nauk SSSR Ser. Mat.* **50**, 883–924. Translated in *Math. USSR Izv.* **29**, 233–277. 5

Y. GUREVICH and W. C. HOLLAND 1981. Recognizing the real line. *Trans. Amer. Math. Soc.* **265**, 527–534. 27

J. A. GUTHRIE, H. E. STONE, and M. L. WAGE 1978. Maximal connected expansions of the reals. *Proc. Amer. Math. Soc.* **69**, 159–165. 69

L. HAHN 1994. *Complex Numbers and Geometry.* Washington, DC: Mathematical Association of America. 151

H. HALBERSTAM 1974. Transcendental numbers. *Math. Gaz.* **58**, 276–284. 221

P. R. HALMOS 1944. Comment on the real line. *Bull. Amer. Math. Soc.* **50**, 877–878. 99

P. R. HALMOS 1950. *Measure Theory.* New York: Van Nostrand. (Reprint: New York: Springer 1974) 104, 202, 326

F. HALTER-KOCH 1982. Darstellung natürlicher Zahlen als Summe von Quadraten. *Acta Arith.* **42**, 11–20. 212

G. HAMEL 1905. Eine Basis aller Zahlen und die unstetigen Lösungen der Funktionalgleichung: $f(x + y) = f(x) + f(y)$. *Math. Ann.* **60**, 459–462. 6

G. H. HARDY and E. M. WRIGHT 1971. *An Introduction to the Theory of Numbers.* Oxford: Oxford University Press. Fourth edition. (Fifth edition 1979) 28, 159, 192, 201, 213

W. S. HATCHER and C. LAFLAMME 1983. On the order structure of the hyperreal line. *Z. Math. Logik Grundlag. Math.* **29**, 197–202. 169

F. HAUSDORFF 1914. *Grundzüge der Mengenlehre.* Leipzig: Veit & Comp. (Reprint: New York: Chelsea 1949) 25, 166, 167

F. HAUSDORFF 1935. *Mengenlehre.* Berlin: de Gruyter. Third edition. (Translation: *Set Theory.* New York: Chelsea 1957) 109, 110

U. HECKMANNS 1991. Beispiel eines topologischen Körpers, dessen Vervollständigung ein Integritätsring, aber kein Körper ist. *Arch. Math.* (*Basel*) **57**, 144–148. 268

I. N. HERSTEIN 1975. *Topics in Algebra.* Lexington, MA: Xerox. Second edition. 213

E. HEWITT and K. A. ROSS 1963. *Abstract Harmonic Analysis*, vol. I. Springer: Berlin. 90, 99, 283, 284, 326, 340, 341, 342, 344, 345, 349, 375

D. HILBERT 1903. *Grundlagen der Geometrie.* Leipzig: Teubner. Second edition. (14th edition 1999. Translation: *Foundations of Geometry.* Chicago: Open Court 1988) 125

D. HILBERT and S. COHN-VOSSEN 1932. *Anschauliche Geometrie.* Berlin: Springer. (Reprinted 1996. Translation: *Geometry and the Imagination.* New York: Chelsea 1952) 215

M. D. HIRSCHHORN 1987. A simple proof of Jacobi's four-square theorem. *Proc. Amer. Math. Soc.* **101**, 436–438. 213

G. HJORTH and M. MOLBERG 2006. Free continuous actions on zero-dimensional spaces. *Topology Appl.* **153**, 1116–1131.   54

E. HLAWKA, J. SCHOISSENGEIER, and R. TASCHNER 1991. *Geometric and Analytic Number Theory.* Berlin: Springer.   68

W. HODGES 1993. *Model Theory.*
Cambridge: Cambridge University Press.   362

J. E. HOFMANN 1958. Zur elementaren Dreiecksgeometrie in der komplexen Ebene. *Enseignement Math.* (*2*) **4**, 178–211.   151

K. H. HOFMANN and S. A. MORRIS 1998. *The Structure of Compact Groups.*
Berlin: de Gruyter.   66, 90, 99, 343, 344, 345, 346, 347, 348

K. H. HOFMANN and C. TERP 1994. Compact subgroups of Lie groups and locally compact groups. *Proc. Amer. Math. Soc.* **120**, 623–634.   343

O. HÖLDER 1901. Die Axiome der Quantität und die Lehre vom Maß.
Ber. Verh. Kgl. sächs. Ges. Wiss. Leipzig, math.-phys. Kl. **53**, 1–64.
English translation (in two parts): J. Mitchell and C. Ernst, *J. Math. Psychology* **40** 1996, 235–252 and **41** 1997, 345–356.   77

W. C. HOLLAND 1992. Partial orders of the group of automorphisms of the real line. In: *Proceedings of the International Conference on Algebra, Part 1* (Novosibirsk, 1989), vol. 131 of Contemporary Mathematics, pages 197–207. Providence, RI: American Mathematical Society.   59

M. HOLZ, K. STEFFENS, and E. WEITZ 1999. *Introduction to Cardinal Arithmetic.* Basel: Birkhäuser.   337

A. E. HURD and P. A. LOEB 1985. *An Introduction to Nonstandard Real Analysis.* Orlando, FL: Academic Press.   154

W. HUREWICZ and H. WALLMAN 1948. *Dimension Theory.* Princeton, NJ: Princeton University Press. Second revised edition.   200

A. HURWITZ 1898. Über die Komposition der quadratischen Formen von beliebig vielen Variablen. *Nachr. Ges. Wiss. Göttingen*, 309–316.   216

T. HUSAIN 1966. *Introduction to Topological Groups.*
Philadelphia, PA: Saunders.   342

K. IWASAWA 1949. On some types of topological groups.
Ann. Math. (*2*) **50**, 507–558.   343

K. JACOBS 1978. *Measure and Integral.*
New York: Academic Press.   110

N. JACOBSON 1985. *Basic Algebra. I.* New York: Freeman. Second edition.
    16, 127, 130, 138, 145, 190, 330, 350, 352, 353, 378, 380

N. JACOBSON 1989. *Basic Algebra. II.* New York: Freeman. Second edition.
    122, 133, 134, 167, 184, 303, 304, 306, 315, 322, 323, 326, 331, 332, 333, 350, 353, 355, 366, 367, 368

F. B. JONES 1939. Concerning certain linear abstract spaces and simple continuous curves. *Bull. Amer. Math. Soc.* **45**, 623–628.   42

R. R. KALLMAN 1976. A uniqueness result for topological groups.
Proc. Amer. Math. Soc. **54**, 439–440.   290

R. R. KALLMAN and F. W. SIMMONS 1985. A theorem on planar continua and an application to automorphisms of the field of complex numbers. *Topology Appl.* **20**, 251–255.   147

I. KAPUANO 1946. Sur les corps de nombres à une dimension distincts du corps réel. *Rev. Fac. Sci. Univ. Istanbul* (*A*) **11**, 30–39.   146

H. J. KEISLER 1994. The hyperreal line.
In: EHRLICH 1994, pages 207–237.   154

B. VON KERÉKJÁRTÓ 1931. Geometrische Theorie der zweigliedrigen kontinuierlichen Gruppen. *Abh. Math. Sem. Hamburg* **8**, 107–114.   141

H. KESTELMAN 1951. Automorphisms of the field of complex numbers.
Proc. London Math. Soc. (2) **53**, 1–12.   148

A. YA. KHINCHIN 1964. *Continued Fractions.* Chicago, IL: University of Chicago Press.   28, 30

J. O. KILTINEN 1973. On the number of field topologies on an infinite field.
Proc. Amer. Math. Soc. **40**, 30–36.   138, 166

A. M. KIRCH 1969. A countable, connected, locally connected Hausdorff space.
Amer. Math. Monthly **76**, 169–171.   45

G. KLAAS, C. R. LEEDHAM-GREEN, and W. PLESKEN 1997. *Linear Pro-p-Groups of Finite Width*, vol. 1674 of Lecture Notes in Mathematics. Berlin: Springer.   324

H. KNESER 1960. Eine kontinuumsmächtige, algebraisch unabhängige Menge reeller Zahlen. *Bull. Soc. Math. Belg.* **12**, 23–27.   7

H. KNESER and M. KNESER 1960. Reell-analytische Strukturen der Alexandroff-Geraden und der Alexandroff-Halbgeraden.
Arch. Math. (*Basel*) **11**, 104–106.   48

N. KOBLITZ 1977. *p-Adic Numbers, p-adic Analysis, and Zeta-Functions.*
New York: Springer. (Second edition 1984)   284

S. KOPPELBERG 1989. *Handbook of Boolean Algebras, Vol. 1* (ed. J. D. Monk). Amsterdam: North-Holland.   362

H.-J. KOWALSKY 1958. Kennzeichnung von Bogen.
Fund. Math. **46**, 103–107.   36, 42

L. KRAMER 2000. Splitting off the real line and plane.
Results Math. **37**, 119.   142

F.-V. KUHLMANN, S. KUHLMANN, and S. SHELAH 1997. Exponentiation in power series fields. *Proc. Amer. Math. Soc.* **125**, 3177–3183.   117

S. KUHLMANN 2000. *Ordered Exponential Fields.* Providence, RI: American Mathematical Society.   117

L. KUIPERS and H. NIEDERREITER 1974. *Uniform Distribution of Sequences.*
New York: Wiley.   68

M. LACZKOVICH 1998. Analytic subgroups of the reals. *Proc. Amer. Math. Soc.* **126**, 1783–1790.   7, 112

T. Y. LAM 2005. *Introduction to Quadratic Forms over Fields.* Providence, RI: American Mathematical Society.
      122, 125, 130, 133, 134, 216, 297, 298, 299, 325

E. LANDAU 1930. *Grundlagen der Analysis.* Leipzig: Akademische Verlagsgesellschaft. (Reprints: New York: Chelsea 1951, Darmstadt: Wissenschaftliche Buchgesellschaft 1970, and Lemgo: Heldermann 2004.
Translation: *Foundations of Analysis.* New York: Chelsea 1966)   xii

S. LANG 1966. *Introduction to Transcendental Numbers.* Reading, MA: Addison-Wesley.   221

S. LANG 1970. *Algebraic Number Theory.* Reading, MA: Addison-Wesley.
(Second edition: Berlin: Springer 1994)   279, 289, 323, 324, 374

S. LANG 1971. Transcendental numbers and diophantine approximations.
Bull. Amer. Math. Soc. **77**, 635–677.   201, 221

S. LANG 1993. *Algebra.* Reading, MA: Addison Wesley. Third edition.
(Revised third edition: Berlin: Springer 2002)   16, 20, 122, 131, 133, 288, 289, 292, 303, 304, 350, 352, 353, 354, 368, 380

D. LASCAR 1992. Les automorphismes d'un ensemble fortement minimal.
J. Symbolic Logic **57**, 238–251. 150

D. LASCAR 1997. The group of automorphisms of the field of complex numbers
leaving fixed the algebraic numbers is simple. In: *Model Theory of Groups
and Automorphism Groups* (Blaubeuren, 1995), vol. 244 of London Mathematical Society Lecture Notes, pages 110–114. Cambridge: Cambridge
University Press. 150

H. LÄUCHLI 1962. Auswahlaxiom in der Algebra.
Comment. Math. Helvet. **37**, 1–18. 134, 354

P. LAX 1973. The differentiability of Pólya's function.
Adv. Math. **10**, 456–464. 362

C. R. LEEDHAM-GREEN and S. McKAY 2002. *The Structure of Groups of
Prime Power Order.* Oxford: Oxford University Press. 324

J. B. LEICHT 1966. Zur Charakterisierung reell abgeschlossener Körper.
Monatsh. Math. **70**, 452–453. 130

H. W. LENSTRA, JR. and P. STEVENHAGEN 1989. Über das Fortsetzen von
Bewertungen in vollständigen Körpern.
Arch. Math. (Basel) **53**, 547–552. 315

A. LÉVY 1979. *Basic Set Theory.*
Berlin: Springer. 337

T. LINDSTRØM 1988. An invitation to nonstandard analysis.
In: *Nonstandard Analysis and its Applications* (Hull, 1986), pages 1–105.
Cambridge: Cambridge University Press. 154

E. LIVENSON 1937. An example of a non-closed connected subgroup of the
two-dimensional vector space. *Ann. Math. (2)* **38**, 920–922. 93

H. LOMBARDI and M.-F. ROY 1991. Elementary constructive theory of ordered fields. In: *Effective Methods in Algebraic Geometry* (Castiglioncello,
1990), vol. 94 of Progress in Mathematics, pages 249–262. Boston, MA:
Birkhäuser. 133

L. H. LOOMIS 1945. Abstract congruence and the uniqueness of Haar measure.
Ann. Math. (2) **46**, 348–355. 326

M. LÓPEZ PELLICER 1994. Las construcciones de los números reales.
In: *History of Mathematics in the XIXth Century, Part 2* (Madrid, 1993),
pages 11–33. Madrid: Real Academia de Ciencias Exactas, Físicas y Nat-
urales. xii, 1

R. LÖWEN 1985. Lacunary free actions on the real line.
Topology Appl. **20**, 135–141. 60

H. LÜNEBURG 1973. *Einführung in die Algebra.*
Berlin: Springer. 284

D. J. LUTZER 1980. Ordered topological spaces. In: G. M. REED (ed.), *Surveys
in General Topology*, pages 247–295. New York: Academic Press. 164

G. MALLE and B. MATZAT 1999. *Inverse Galois Theory.*
Berlin: Springer. 145, 150

A. MARKOV 1945. On free topological groups (Russian).
Izvestia Akad. Nauk SSSR **9**, 3–64. 341

D. A. MARTIN 1976. Hilbert's first problem: the continuum hypothesis.
In: F. E. BROWDER (ed.), *Mathematical Developments arising from
Hilbert Problems (Proc. Sympos. Pure Math.* **28**, 1974), pages 81–92.
Providence, RI: American Mathematical Society. 339

G. E. MARTIN 1998. *Geometric Constructions.*
New York: Springer. 124, 125

B. MAZUR 1993. On the passage from local to global in number theory.
Bull. Amer. Math. Soc. (N.S.) **29**, 14–50. 325

G. H. MEISTERS and J. D. MONK 1973. Construction of the reals via ultra-powers. *Rocky Mountain J. Math.* **3**, 141–158. 160

A. H. MEKLER 1986. Groups embeddable in the autohomeomorphisms of **Q**.
J. London Math. Soc. (2) **33**, 49–58. 205

J. VAN MILL 1982a. Homogeneous subsets of the real line.
Compositio Math. **46**, 3–13. 57

J. VAN MILL 1982b. Homogeneous subsets of the real line which do not admit the structure of a topological group. *Indag. Math.* **44** (*Nederl. Akad. Wetensch. Proc. Ser. A* **85**), 37–43. 57

J. VAN MILL 1992. Sierpiński's technique and subsets of **R**.
Topology Appl. **44**, 241–261. 26

A. W. MILLER 1979. On the length of Borel hierarchies.
Ann. Math. Logic **16**, 233–267. 109

A. W. MILLER 1984. Special subsets of the real line. In: *Handbook of Set-theoretic Topology*, pages 201–233. Amsterdam: North-Holland. 51

A. W. MILLER 1993. Special sets of reals. In: *Set Theory of the Reals* (Ramat Gan, 1991), vol. 6 of *Israel Mathematical Conference Proceedings*, pages 415–431. Ramat Gan: Bar-Ilan University. 51

D. MONTGOMERY 1948. Connected one-dimensional groups.
Ann. Math. (2) **49**, 110–117. 91

D. MONTGOMERY and L. ZIPPIN 1955. *Topological Transformation Groups.*
New York: Interscience. 88, 141

R. L. MOORE 1920. Concerning simple continuous curves.
Trans. Amer. Math. Soc. **21**, 333–347. 42

P. MORANDI 1996. *Field and Galois Theory.*
New York: Springer. 145, 350, 353, 354, 355

S. A. MORRIS 1977. *Pontryagin Duality and the Structure of Locally Compact Abelian Groups.* Cambridge: Cambridge University Press. 344, 345

S. A. MORRIS 1986. A characterization of the topological group of real numbers. *Bull. Austral. Math. Soc.* **34**, 473–475. 90

S. A. MORRIS and S. OATES-WILLIAMS 1987. A characterization of the topological group of *p*-adic integers. *Bull. London Math. Soc.* **19**, 57–59. 287

S. A. MORRIS, S. OATES-WILLIAMS, and H. B. THOMPSON 1990. Locally compact groups with every closed subgroup of finite index.
Bull. London Math. Soc. **22**, 359–361. 287

A. F. MUTYLIN 1966. Imbedding of discrete fields into connected ones (Russian). *Dokl. Akad. Nauk SSSR* **168**, 1005–1008. Translated in *Soviet Math. Dokl.* **7**, 772–775. 138

A. F. MUTYLIN 1968. Connected complete locally bounded fields. Complete not locally bounded fields (Russian). *Mat. Sbornik (N.S.)* **76 (118)**, 454–472. Translated in *Math. USSR Sbornik* **5**, 433–449. 138, 139, 321

K. NAGAMI 1970. *Dimension Theory.*
New York: Academic Press. 91, 92

M. DI NASSO and M. FORTI 2002. On the ordering of the nonstandard real line. In: *Logic and Algebra*, vol. 302 of Contemporary Mathematics, pages 259–273. Providence, RI: American Mathematical Society. 164

J. NEUKIRCH 1992. *Algebraische Zahlentheorie.* Berlin: Springer. (Translation: *Algebraic Number Theory.* Berlin: Springer 1999)
284, 289, 290, 304, 322, 374

B. H. NEUMANN 1949a. On ordered division rings.
Trans. Amer. Math. Soc. **66**, 202–252. 359

B. H. NEUMANN 1949b. On ordered groups.
Amer. J. Math. **71**, 1–18. 76

J. VON NEUMANN 1928. Ein System algebraisch unabhängiger Zahlen.
Math. Ann. **99**, 134–141. 7

P. M. NEUMANN 1985. Automorphisms of the rational world.
J. London Math. Soc. (*2*) **32**, 439–448. 200, 205

I. NIVEN 1956. *Irrational Numbers.*
New York: Mathematical Association of America. 28

I. NIVEN 1961. *Numbers: Rational and Irrational.*
New York: Random House. 201

P. J. NYIKOS 1992. Various smoothings of the long line and their tangent
bundles. *Adv. Math.* **93**, 129–213. 48

J. E. NYMANN 1993. The sum of the Cantor set with itself.
Enseign. Math. (*2*) **39**, 177–178. 106

J. C. OXTOBY 1971. *Measure and Category.*
New York: Springer. 108

R. S. PALAIS 1968. The classification of real division algebras.
Amer. Math. Monthly **75**, 366–368. 138, 331

V. PAMBUCCIAN 1990. On the Pythagorean hull of $\mathbb{Q}$.
Extracta Math. **5**, 29–31. 125

S. PAULI and X.-F. ROBLOT 2001. On the computation of all extensions of a
p-adic field of a given degree. *Math. Comput.* **70**, 1641–1659. 324

A. R. PEARS 1975. *Dimension Theory of General Spaces.*
Cambridge: Cambridge University Press. 197

J. PETRO 1987. Real division algebras of dimension > 1 contain **C**.
Amer. Math. Monthly **94**, 445–449. 138

A. PFISTER 1995. *Quadratic Forms with Applications to Algebraic Geometry
and Topology.* Cambridge: Cambridge University Press. 122, 134, 216

G. PÓLYA 1913. Über eine *Peano*sche Kurve.
Bull. Acad. Sci. Cracovie Ser. A, 305–313. 362

L. S. PONTRYAGIN 1932. Über stetige algebraische Körper.
Ann. Math. (*2*) **33**, 163–174. 138

L. S. PONTRYAGIN 1986. *Topological Groups.* Selected works vol. 2 (ed. R. V.
Gamkrelidze). New York: Gordon and Breach. Third edition. (German
translation: Leipzig: Teubner 1957/1958) 90, 139, 340, 344, 345, 346

K. POTTHOFF 1981. *Einführung in die Modelltheorie und ihre Anwendungen.*
Darmstadt: Wissenschaftliche Buchgesellschaft. 170

V. POWERS 1996. Hilbert's 17th problem and the champagne problem.
Amer. Math. Monthly **103**, 879–887. 134

K. PRACHAR and H. SAGAN 1996. On the differentiability of the coordinate
functions of Pólya's space-filling curve.
Monatsh. Math. **121**, 125–138. 362

A. PRESTEL 1984. *Lectures on Formally Real Fields,* vol. 1093 of Lecture
Notes in Mathematics. Berlin: Springer. 122, 133, 134

A. PRESTEL and C. N. DELZELL 2001. *Positive Polynomials. From Hilbert's
17th Problem to Real Algebra.* Berlin: Springer. 122, 133, 134

S. PRIESS-CRAMPE 1983. *Angeordnete Strukturen: Gruppen, Körper, projektive Ebenen.* Berlin: Springer. 76, 77, 166, 167, 246, 359

S. Priess-Crampe and P. Ribenboim 2000. A general Hensel's lemma.
J. Algebra **232**, 269–281. 279

W. Rautenberg 1987. Über den Cantor-Bernsteinschen Äquivalenzsatz.
Math. Semesterber. **34**, 71–88. 337

I. Reiner 1975. *Maximal Orders*. London: Academic Press. (Corrected
reprint: Oxford: Oxford University Press 2003) 323, 324, 331, 332, 333

G. Ren 1992. A note on $(\mathbf{Q} \times \mathbf{Q}) \cup (\mathbf{I} \times \mathbf{I})$.
Questions Answers Gen. Topology **10**, 157–158. 70

P. Ribenboim 1985. Equivalent forms of Hensel's lemma.
Expo. Math. **3**, 3–24. 279

P. Ribenboim 1992. Fields: algebraically closed and others.
Manuscripta Math. **75**, 115–150. 124, 125, 133, 359

P. Ribenboim 1993. Some examples of lattice-orders in real closed fields.
Arch. Math. (Basel) **61**, 59–63. 124, 133

P. Ribenboim 1999. *The Theory of Classical Valuations*. New York: Springer.
289, 290, 304, 306, 313, 315, 323, 324, 357, 358, 377, 378

A. M. Robert 2000. *A Course in p-adic Analysis*.
New York: Springer. 278, 284, 285, 323, 324, 326, 374

L. Robertson and B. M. Schreiber 1968. The additive structure of integer
groups and p-adic number fields.
Proc. Amer. Math. Soc. **19**, 1453–1456. 287

A. M. Rockett and P. Szüsz 1992. *Continued Fractions*. River Edge, NJ:
World Scientific. 28, 30

W. Roelcke and S. Dierolf 1981. *Uniform Structures on Topological Groups
and their Quotients*. New York: McGraw-Hill. 341

C. A. Rogers and T. E. Jayne 1980. *Analytic Sets*.
London: Academic Press. 110, 112

K. F. Roth 1955. Rational approximations to algebraic numbers.
Mathematika **2**, 1–20. Corrigendum page 168. 30

K. F. Roth 1960. Rational approximations to algebraic numbers. In: *Proceedings of the International Congress of Mathematicians* (Edinburgh, 1958),
pages 203–210. Cambridge: Cambridge University Press. 30

G. Rousseau 1987. On a construction for the representation of a positive
integer as the sum of four squares. *Enseign. Math.* (2) **33**, 301–306. 213

J. E. Rubin 1967. *Set Theory for the Mathematician*.
San Francisco, CA: Holden-Day. 335, 337

H. Rubin and J. E. Rubin 1985. *Equivalents of the Axiom of Choice II*.
Amsterdam: North-Holland. 336

M. E. Rudin 1969. Souslin's conjecture.
Amer. Math. Monthly **76**, 1113–1119. 27

H. Sagan 1994. *Space-filling Curves*.
New York: Springer. 362

S. Saks 1937. *Theory of the Integral*. Warszawa–New York: Stechert.
(Second revised edition: New York: Dover Publications 1964) 112

H. Salzmann 1958. Kompakte zweidimensionale projektive Ebenen.
Arch. Math. (Basel) **9**, 447–454. 59

H. Salzmann 1969. Homomorphismen komplexer Ternärkörper.
Math. Z. **112**, 23–25. 147

H. Salzmann 1971. *Zahlbereiche. I: Die reellen Zahlen*. University of Tübingen. Mimeographed lecture notes prepared by R. Löwen. xii, 39

H. SALZMANN 1973. *Zahlbereiche. II: Die rationalen Zahlen. III: Die komplexen Zahlen.* University of Tübingen. Mimeographed lecture notes prepared by H. Hähl.   xii, 221

H. SALZMANN, D. BETTEN, T. GRUNDHÖFER, H. HÄHL, R. LÖWEN, and M. STROPPEL 1995. *Compact Projective Planes.*
Berlin: de Gruyter.   xii, 72, 91, 92, 141, 211

T. SANDER 1991. Existence and uniqueness of the real closure of an ordered field without Zorn's lemma. *J. Pure Appl. Algebra* **73**, 165–180.   133

W. SCHARLAU 1985. *Quadratic and Hermitian Forms.*
Springer, Berlin.   122, 125, 133, 134, 216, 297, 299, 325, 331, 332, 333

B. SCHNOR 1992. Involutions in the group of automorphisms of an algebraically closed field. *J. Algebra* **152**, 520–524.   150

T. SCHÖNEMANN 1846. Von denjenigen Moduln, welche Potenzen von Primzahlen sind. *J. Reine Angew. Math.* **32**, 93–105.   360, 380, 382

E. SCHÖNHARDT 1963. Die Sätze von Pappus, Pascal und Desargues als Identitäten in komplexen Zahlen. *Tensor (N.S.)* **13**, 223–231.   151

H. SCHWERDTFEGER 1979. *Geometry of Complex Numbers.* New York: Dover Publications. Corrected reprint of the 1962 edition.   151

J.-P. SERRE 1973. *A Course in Arithmetic.* New York: Springer.
(Corrected second printing 1978)   284, 293, 297, 325, 374

J.-P. SERRE 1979. *Local Fields.* New York: Springer.
(Corrected second printing 1995)   315, 323, 324, 331, 332, 333, 358

D. B. SHAKHMATOV 1983. Imbeddings in topological fields and the construction of a field whose space is not normal (Russian).
Comment. Math. Univ. Carolin. **24**, 525–540.   136

D. B. SHAKHMATOV 1987. The structure of topological fields, and cardinal invariants (Russian). *Trudy Moskov. Mat. Obshch.* **50**, 249–259, 262. Translated in *Trans. Moscow Math. Soc.* 1988, 251–261.   136

H. N. SHAPIRO 1950. On primes in arithmetic progressions.
Ann. Math. (2) **52**, 217–230, 231–243.   374

S. SHELAH 1983. Models with second order properties. IV. A general method and eliminating diamonds. *Ann. Pure Appl. Logic* **25**, 183–212.   125

N. SHELL 1990. *Topological Fields and Near Valuations.*
New York: Dekker.   135, 136, 139, 146, 321, 359

W. SIERPIŃSKI 1920. Sur une propriété topologique des ensembles dénombrables denses en soi. *Fund. Math.* **1**, 11–16.   197

R. SIKORSKI 1964. *Boolean Algebras.*
Berlin: Springer. Second edition.   54

C. SMALL 1982. A simple proof of the four-squares theorem.
Amer. Math. Monthly **89**, 59–61.   213

C. SMALL 1986. Sums of three squares and levels of quadratic number fields.
Amer. Math. Monthly **93**, 276–279.   212

H. J. SMITH 1855. De compositione numerorum primorum formae $4\lambda + 1$ ex duobus quadratis. *J. reine angew. Math.* **50**, 91–92.   209

R. M. SMULLYAN and M. FITTING 1996. *Set Theory and the Continuum Problem.* Oxford: Oxford University Press.   335, 338, 339

S. SOLECKI and S. M. SRIVASTAVA 1997. Automatic continuity of group operations. *Topology Appl.* **77**, 65–75.   340

R. H. SORGENFREY 1947. On the topological product of paracompact spaces.
Bull. Amer. Math. Soc. **53**, 631–632.   69

T. Soundararajan 1969. The topological group of the p-adic integers.
Publ. Math. Debrecen **16**, 75–78. 290

T. Soundararajan 1991. On the cardinality of the group of automorphisms of algebraically closed extension fields.
Acta Math. Vietnam. **16**, 147–153. 150

M. Souslin 1920. Problème 3.
Fund. Math. **1**, 223. 27

E. Specker 1954. Verallgemeinerte Kontinuumshypothese und Auswahlaxiom. *Arch. Math. (Basel)* **5**, 332–337. 339

L. A. Steen and J. A. Seebach, Jr. 1978. *Counterexamples in Topology*. New York: Springer. Second edition. (Reprint: New York: Dover Publications 1995) 45

E. Steinitz 1910. Algebraische Theorie der Körper.
J. reine angew. Math. **137**, 167–309. (Re-edited by R. Baer and H. Hasse, Berlin: de Gruyter 1930 and New York: Chelsea 1950) 355

J. Stillwell 2002. The continuum problem.
Amer. Math. Monthly **109**, 286–297. 339

M. Stroppel 2006. *Locally Compact Groups*.
Zürich: EMS Publishing House.   66, 90, 249, 340, 344, 345, 346, 349

T. Szele 1949. Die Abelschen Gruppen ohne eigentliche Endomorphismen.
Acta Univ. Szeged. Sect. Sci. Math. **13**, 54–56. 96

S. Tennenbaum 1968. Souslin's problem.
Proc. Nat. Acad. Sci. U.S.A. **59**, 60–63. 27

J. Tits 1974. *Buildings of Spherical Type and Finite BN-Pairs*, vol. 386 of Lecture Notes in Mathematics. Berlin: Springer. 152

J. K. Truss 1997. Conjugate homeomorphisms of the rational world.
Forum Math. **9**, 211–227. 205

P. Ullrich 1998. The genesis of Hensel's p-adic numbers. In: *Charlemagne and his Heritage. 1200 Years of Civilization and Science in Europe, vol. 2* (Aachen, 1995), pages 163–178. Turnhout: Brepols. 278

H. Völklein 1996. *Groups as Galois Groups*.
Cambridge: Cambridge University Press.   145, 150, 358

S. Wagon 1990. The Euclidean algorithm strikes again.
Amer. Math. Monthly **97**, 125–129. 209

M. Waldschmidt 1992. *Linear Independence of Logarithms of Algebraic Numbers*, vol. 116 of IMSc Reports. Madras: Institute of Mathematical Sciences. 7

J. A. Ward 1936. The topological characterization of an open linear interval.
Proc. London Math. Soc. **41**, 191–198. 42

J. P. Ward 1997. *Quaternions and Cayley Numbers*.
Kluwer, Dordrecht. xii

S. Warner 1989. *Topological Fields*. Amsterdam: North-Holland.
135, 139, 146, 249, 263, 268, 276, 277, 278, 293, 303, 304, 306, 313, 315, 321, 330, 375, 378, 380

W. C. Waterhouse 1985. When one equation solves them all.
Amer. Math. Monthly **92**, 270–273. 130

A. G. Waterman and G. M. Bergman 1966. Connected fields of arbitrary characteristic. *J. Math. Kyoto Univ.* **5**, 177–184. 138

A. Weil 1967. *Basic Number Theory*. New York: Springer. (Second edition 1973)   166, 289, 293, 314, 323, 324, 326, 330, 332, 333

W. WIĘSŁAW 1988. *Topological Fields*. New York: Dekker. (Earlier version: Acta Univ. Wratislav. 1982)   138, 139, 146, 321, 329

E. WITT 1975. Homöomorphie einiger Tychonoffprodukte. In: *Topology and its Applications* (St. John's, Newfoundland, 1973), vol. 12 of Lecture Notes in Pure and Applied Mathematics, pages 199–200. New York: Dekker.   283, 374

W. H. WOODIN 2001. The continuum hypothesis I, II. *Notices Amer. Math. Soc.* **48**, 567–576, 681–690.   339

T.-S. WU 1962. Continuity in topological groups.
Proc. Amer. Math. Soc. **13**, 452–453.   340

S. W. YOUNG 1994. The representation of homeomorphisms on the interval as finite compositions of involutions.
Proc. Amer. Math. Soc. **121**, 605–610.   61

D. ZAGIER 1990. A one-sentence proof that every prime $p \equiv 1 \pmod 4$ is a sum of two squares. *Amer. Math. Monthly* **97**, 144.   209

D. ZAGIER 1997. Newman's short proof of the prime number theorem.
Amer. Math. Monthly **104**, 705–708.   159, 192

H. ZASSENHAUS 1970. A real root calculus. In: *Computational Problems in Abstract Algebra* (Oxford, 1967), pages 383–392. Oxford: Pergamon.   133

Index